Strongly Elliptic Systems and Boundary Integral Equations

Partial differential equations provide mathematical models of many important problems in the physical sciences and engineering. This book treats one class of such equations, concentrating on methods involving the use of surface potentials. It provides the first detailed exposition of the mathematical theory of boundary integral equations of the first kind on non-smooth domains. Included are chapters on three specific examples: the Laplace equation, the Helmholtz equation and the equations of linear elasticity.

The book is designed to provide an ideal preparation for studying the modern research literature on boundary element methods.

Dr. McLean received his PhD from the Australian National University, and is currently a Senior Lecturer in Applied Mathematics at the University of New South Wales.

Strongly Elliptic Systems and Boundary Integral Equations

WILLIAM McLEAN

University of New South Wales

CAMBRIDGE UNIVERSITY PRESS
Cambridge, New York, Melbourne, Madrid, Cape Town, Singapore,
São Paulo, Delhi, Dubai, Tokyo, Mexico City

Cambridge University Press
32 Avenue of the Americas, New York, NY 10013-2473, USA

www.cambridge.org
Information on this title: www.cambridge.org/9780521663755

© Cambridge University Press 2000

First published 2000

A catalog record for this publication is available from the British Library

Library of Congress Cataloging in Publication data
McLean, William Charles Hector, 1960–
Strongly elliptic systems and boundary integral equations /
William McLean.
p.   cm.
Includes index.
ISBN 0-521-66332-6 (hc). – ISBN 0-521-66375-X (pbk.)
1. Differential equations, Elliptic.   2. Boundary element methods.
I. Title.
QA377.M3227   2000
515'.353 – dc21   99-30938
                                                                    CIP

ISBN 978-0-521-66332-8 Hardback
ISBN 978-0-521-66375-5 Paperback

To Meg

Contents

Preface

The study of integral equations in connection with elliptic boundary value problems has a long history, during which a variety of approaches has emerged. Rather than attempt a broad survey, I have chosen to pursue in detail just one approach, in which both the differential and integral formulations of a given boundary value problem are viewed abstractly as linear equations involving a bounded operator from a Hilbert space into its dual. The decisive property of this operator is that its associated sesquilinear form is positive and bounded below, apart perhaps from a compact perturbation.

In the classical Fredholm method, the solvability of the Dirichlet and Neumann problems is proved by reformulating them as integral equations of the second kind. Here, we effectively reverse this strategy, deriving key properties of the boundary integral equations from previously established results for the associated partial differential equations. Moreover, our approach leads to Fredholm integral equations of the *first* kind. The theory of such first-kind integral equations can be traced back to Gauss (see Chapter 1), and developed into the form presented here during the 1970s, in the work of Nedelec and Planchard [74], [76]; Le Roux [56], [57], [58]; and Hsiao and Wendland [42]. Those authors were all studying Galerkin boundary element methods, and although this book does not deal at all with numerical techniques, it is written very much from the perspective of a numerical analyst.

A major difficulty in a work such as this is the large amount of background material needed to present the main topics. Aware that readers differ in their prior knowledge, I have tried to adopt a middle path between, on the one hand, writing a textbook on functional analysis, distributions and function spaces, and on the other hand just stating, without proof or exposition, a litany of definitions and theorems. The result is that more than one-third of the text is made up of what might be considered technical preliminaries. My hope is that the book will be suitable for someone interested in finite or boundary element methods who wants a deeper understanding of the relevant non-numerical theory. I have aimed to keep the exposition as simple, concise and self-contained as possible, while at the same time avoiding assumptions that would be unrealistic for applications. Thus, I felt it essential to allow non-smooth domains, to consider systems and not just scalar equations, and to treat mixed boundary conditions.

Here is an outline of the contents.

Chapter 1 has two purposes. Firstly, it attempts to sketch the early history of the ideas from which the theory of this book developed. Secondly, it serves to introduce those ideas in an informal way, and to acquaint the reader with some of the notation used later.

The second chapter presents topics from linear functional analysis that are immediately relevant to what follows. I assume that the reader is already familiar with elementary facts about the topology of normed spaces, and of a few fundamental, deeper results such as the open mapping theorem and the Hahn–Banach theorem.

Chapter 3 develops the theory of Sobolev spaces on Lipschitz domains. After a quick treatment of distributions and Fourier transforms, we study in detail fractional- and negative-order spaces based on L_2. These spaces play an essential role in nearly all of the subsequent theory.

In Chapter 4, we begin our investigations of elliptic systems. A key tool is the first Green identity, used to arrive at the abstract (weak) formulation of a boundary value problem mentioned above. The centrepiece of the chapter is the Fredholm alternative for the mixed Dirichlet and Neumann problem on a bounded Lipschitz domain. We go on to prove some standard results on regularity of solutions, including the transmission property. The final section of the chapter proves some difficult estimates of Nečas [72] that relate the H^1-norm of the trace of a solution to the L_2-norm of its conormal derivative. These estimates are used later when showing that, even for general Lipschitz domains, the basic mapping properties of the surface potentials and boundary integral operators hold in a range of Sobolev spaces.

Chapter 5 is something of a technical digression on homogeneous distributions. As well as dealing with standard material such as the calculation of Fourier transforms, we include results from the thesis of Kieser [48], including the change-of-variables formula for finite-part integrals.

Chapters 6 and 7 form the heart of the book. Here, we study potentials and boundary integral operators associated with a strongly elliptic system of partial differential equations. Our overall approach is essentially that of Costabel [14], allowing us to handle Lipschitz domains. The first part of Chapter 6 deals with parametrices and fundamental solutions, and uses the results of Chapter 5. We then prove the third Green identity, and establish the main properties of the single- and double-layer potentials, including the familiar jump relations. Chapter 7 derives the boundary integral equations for the Dirichlet, Neumann and mixed problems, treating interior as well as exterior problems. The Fredholm alternative for the various boundary integral equations is established by showing positive-definiteness up to a compact perturbation, a property that is intimately related to the strong ellipticity of the associated partial differential operator.

Chapters 8–10 treat three of the simplest and most important examples of elliptic operators. For these specific cases, we can refine the general theory in certain respects. Chapter 8 deals with the Laplace equation, and includes a few classical topics such as spherical harmonics and capacity. Chapter 9 deals with the Helmholtz (or reduced wave) equation, and Chapter 10 gives a brief treatment of the linearised equilibrium equations for a homogeneous and isotropic elastic medium.

The book concludes with three appendices. The first of these proves Calderón's extension theorem for Sobolev spaces on Lipschitz domains, including the fractional-order case. The second gives a rapid but self-contained treatment of interpolation spaces and establishes the interpolation properties of Sobolev spaces on Lipschitz domains. The third proves a few facts about spherical harmonics.

At the end of each chapter and appendix is a set of exercises. These are of various types. Some are simple technical lemmas or routine calculations used at one or more points in the main text. Others present explicit solutions or examples, intended to help give a better feeling for the general theory. A few extend results in the text, or introduce related topics.

Some mention of what I have *not* covered also seems in order.

Many books treat Fredholm integral equations of the second kind. Well-known older texts include Kellogg [45] and Günter [35], and we also mention Smirnov [95] and Mikhlin [65, Chapter 18]. Problems on non-smooth domains are treated by Král [49] and Burago and Maz'ya [6], using methods from geometric measure theory, and by Verchota [102] and Kenig [46], [47] using harmonic analysis techniques. Works oriented towards numerical analysis include Kress [50], Hackbusch [36] and Atkinson [3]. Boundary value problems can also be reformulated as Cauchy singular integral equations, as in the pioneering work by Muskhelishvili [71]; for a modern approach, see Gohberg and Krupnik [28] or Mikhlin and Prößdorf [66].

Even for boundary integral equations of the first kind, the material presented in this book is by no means exhaustive. For instance, Costabel and Wendland [15] have generalised the approach used here to higher-order strongly elliptic equations. One can also study boundary integral equations as special cases of pseudodifferential equations; see, e.g., Chazarain and Piriou [10]. We make contact with the theory of pseudodifferential operators on several occasions, but do not attempt a systematic account of this topic. Other significant matters not treated include the L_p theory for $p \neq 2$, various alternative boundary conditions, especially non-linear ones, and a detailed study of the dominant singularities in a solution at corner points or edges of the domain.

During the period I have worked on this book, the Australian Research Council has provided support for a number of related research projects. I thank David Elliott for reading an early draft of the complete manuscript and making a number of helpful suggestions. I also thank Werner Ricker and Jan Brandts for the care with which they read through later versions of some of the chapters. Alan McIntosh and Marius Mitrea helped me negotiate relevant parts of the harmonic analysis literature. Visits to Mark Ainsworth at Leicester University, U.K., to Youngmok Jeon at Ajou University, Korea, and to the Mittag–Leffler Institute, Stockholm, provided valuable opportunities to work without the usual distractions, and made it possible for me to complete the book sooner than would otherwise have been the case. Needless to say, I am also indebted to many other people, who helped by suggesting references, discussing technical questions, and passing on their knowledge through seminars.

Sydney,
December 1998

1

Introduction

The theory of elliptic partial differential equations has its origins in the eighteenth century, and the present chapter outlines a few of the most important historical developments up to the beginning of the twentieth century. We concentrate on those topics that will play an important role in the main part of the book, and change the notation of the original authors, wherever necessary, to achieve consistency with what comes later. Such a brief account cannot pretend to be a balanced historical survey, but this chapter should at least serve to introduce the main ideas of the book in a readable manner.

To limit subsequent interruptions, we fix some notational conventions at the outset. Let Ω denote a bounded, open subset of $\mathbb{R}^n$ (where $n = 2$ or 3 in this chapter), and assume that the boundary $\Gamma = \partial\Omega$ is sufficiently regular for the outward unit normal ν and the element of surface area $d\sigma$ to make sense. Given a function u defined on Ω, we denote the normal derivative by $\partial_\nu u$ or $\partial u / \partial \nu$. Sometimes we shall work with both the interior and the exterior domains (see Figure 1)

$$\Omega^- = \Omega \quad \text{and} \quad \Omega^+ = \mathbb{R}^n \backslash (\Omega^- \cup \Gamma),$$

in which case, if the function u is defined on $\Omega^\pm$, we write

$$\gamma^\pm u(x) = \lim_{y \to x, y \in \Omega^\pm} u(y) \text{ and}$$

$$\partial_\nu^\pm u(x) = \lim_{y \to x, y \in \Omega^\pm} \nu(x) \cdot \operatorname{grad} u(y) \qquad \text{for } x \in \Gamma,$$

whenever these limits exist. The Euclidean norm of $x \in \mathbb{R}^n$ is denoted by $|x|$.

The prototype of an elliptic partial differential equation is $\Delta u = 0$, where Δ denotes the Laplace operator (or Laplacian), defined, in n dimensions, by

$$\Delta u(x) = \sum_{j=1}^{n} \frac{\partial^2 u}{\partial x_j^2}. \tag{1.1}$$

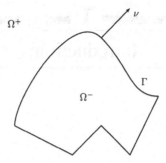

Figure 1. Interior and exterior domains Ω^- and Ω^+ with boundary Γ.

When $\Delta u = 0$ on Ω, we say that the function u is *harmonic* on Ω. In two dimensions, there is a close connection between the Laplace equation and complex-analytic functions. Indeed, $u + iv$ is differentiable as a function of the complex variable $x_1 + ix_2$ if and only if u and v satisfy the Cauchy–Riemann equations,

$$\frac{\partial u}{\partial x_1} = \frac{\partial v}{\partial x_2} \quad \text{and} \quad \frac{\partial u}{\partial x_2} = -\frac{\partial v}{\partial x_1}, \tag{1.2}$$

in which case $\Delta u = 0 = \Delta v$ and we say that u and v are *conjugate harmonic functions*.

The pair of equations (1.2) appeared in Jean-le-Rond d'Alembert's *Essai d'une Nouvelle Théorie de la Résistance des Fluides*, published in 1752. At around the same time, Leonhard Euler derived the equations of motion for an irrotational fluid in three dimensions. He showed that the fluid velocity has the form grad u, and that for a steady flow the velocity potential satisfies $\Delta u = 0$. This work of d'Alembert and Euler is discussed by Truesdell [100]; see also Dauben [18, p. 311].

In 1774, Joseph-Louis Lagrange won the *Prix de l'Academie Royale des Sciences* for a paper [51] on the motion of the moon; see also [30, pp. 478–479, 1049]. This paper drew attention to two functions that later came to be known as the *fundamental solution*,

$$G(x, y) = \frac{1}{4\pi |x - y|} \quad \text{for } x, y \in \mathbb{R}^3 \text{ and } x \neq y, \tag{1.3}$$

and the *Newtonian potential*,

$$u(x) = \int_{\mathbb{R}^3} G(x, y) f(y) \, dy. \tag{1.4}$$

Up to an appropriate constant of proportionality, $G(x, y)$ is the gravitational potential at x due to a unit point mass at y, and thus u is the gravitational potential due to a continuous mass distribution with density f. The Coulomb force law in electrostatics has the same inverse-square form as Newton's law of gravitational attraction. Thus, u also describes the electrostatic potential due to a charge distribution with density f; mathematically, the only change is that f may be negative.

In a paper of 1782 entitled *Théorie des attractions des sphéroïdes et de la figure des planètes*, Pierre Simon de Laplace observed that the Newtonian potential (1.4) satisfies $\Delta u = 0$ outside the support of f, writing Δu in spherical polar coordinates. Later, in a paper of 1787 on the rings of Saturn, he gave the same result in Cartesian and cylindrical coordinates. Birkhoff and Merzbach [7, pp. 335–338] give English translations of relevant excerpts from these two works.

By transforming to polar coordinates centred at x, i.e., by using the substitution $y = x + \rho\omega$ where $\rho = |y - x|$, it is easy to see that the Newtonian potential (1.4) makes sense even if x lies within the support of f, because $dy = \rho^2\, d\rho\, d\omega$. However, the second partial derivatives of G are $O(\rho^{-3})$, and this singularity is too strong to allow a direct calculation of Δu by simply differentiating under the integral sign. In fact, it turns out that

$$-\Delta u = f$$

everywhere on $\mathbb{R}^3$, an equation derived by Siméon-Denis Poisson [7, pp. 342–346] in 1813; see Exercise 1.1 for the special case when f is radially symmetric.

Poisson made other important contributions to potential theory. A paper [18, p. 360] of 1812 dealt with the distribution of electric charge on a conductor Ω. In equilibrium, mutual repulsion causes all of the charge to reside on the surface Γ of the conducting body, and Γ is an equipotential surface. The electrical potential at $x \in \mathbb{R}^3$ due to a charge distribution with surface density ψ on Γ is given by the integral

$$\mathrm{SL}\,\psi(x) = \int_\Gamma G(x, y)\psi(y)\, do_y, \tag{1.5}$$

so $\mathrm{SL}\,\psi$ is constant on Γ if ψ is the equilibrium distribution. The function $\mathrm{SL}\,\psi$ is known as the *single-layer potential* with density ψ, and satisfies the Laplace equation on the complement of Γ, i.e., on $\Omega^+ \cup \Omega^-$. Although $\mathrm{SL}\,\psi$ is continuous everywhere, Poisson found that its normal derivative has a jump discontinuity:

$$\partial_\nu^+ \mathrm{SL}\,\psi - \partial_\nu^- \mathrm{SL}\,\psi = -\psi \quad \text{on } \Gamma. \tag{1.6}$$

Exercise 1.2 proves an easy special case of this result.

A further stimulus to the study of the Laplace equation was Jean-Baptiste-Joseph Fourier's theory of heat diffusion. In 1807, he published a short note containing the heat equation,

$$\frac{\partial u}{\partial t} - a \Delta u = 0,$$

where $u = u(x, t)$ is the temperature at position x and time t, and $a > 0$ is the thermal conductivity (here assumed constant). For a body Ω in thermal equilibrium, $\partial u / \partial t = 0$, so if one knows the temperature distribution g on the bounding surface Γ, then one can determine the temperature distribution u in the interior by solving the boundary value problem

$$\begin{aligned} \Delta u &= 0 \quad \text{on } \Omega, \\ u &= g \quad \text{on } \Gamma. \end{aligned} \tag{1.7}$$

This problem later became known as the *Dirichlet problem*, and for particular, simple choices of Ω, Fourier constructed solutions using separation of variables; see [7, pp. 132–138]. His book, *Théorie analytique de la chaleur*, was published in 1822.

In 1828, George Green published *An Essay on the Application of Mathematical Analysis to the Theories of Electricity and Magnetism* [31], [33, pp. 1–115]; an extract appears in [7, pp. 347–358]. In his introduction, Green discusses previous work by other authors including Poisson, and writes that

although many of the artifices employed in the works before mentioned are remarkable for their elegance, it is easy to see they are adapted only to particular objects, and that some general method, capable of being employed in every case, is still wanting.

Green's "general method" was based on his two integral identities:

$$\int_\Omega \operatorname{grad} w \cdot \operatorname{grad} u \, dx = \int_\Gamma w \frac{\partial u}{\partial \nu} \, d\sigma - \int_\Omega w \Delta u \, dx \tag{1.8}$$

and

$$\int_\Omega (w \Delta u - u \Delta w) \, dx = \int_\Gamma \left(w \frac{\partial u}{\partial \nu} - u \frac{\partial w}{\partial \nu} \right) d\sigma, \tag{1.9}$$

where u and w are arbitrary, sufficiently regular functions. Using (1.9) with

$w(y) = G(x, y)$, he obtained a third identity,

$$u(x) = -\int_\Omega G(x, y)\Delta u(y)\, dy - \int_\Gamma u(y)\frac{\partial}{\partial \nu_y}G(x, y)\, d\sigma_y$$

$$+ \int_\Gamma G(x, y)\frac{\partial u}{\partial \nu}(y)\, d\sigma_y \quad \text{for } x \in \Omega. \tag{1.10}$$

Actually, Green derived a more general result, showing that (1.10) is valid when $G(x, y)$ is replaced by a function of the form

$$\mathrm{Gr}(x, y) = G(x, y) + V(x, y),$$

where V is any smooth function satisfying $\Delta_y V(x, y) = 0$ for $x, y \in \Omega$. In other words, $\mathrm{Gr}(x, y)$ has the same singular behaviour as $G(x, y)$ when $y = x$, and satisfies $\Delta_y \mathrm{Gr}(x, y) = 0$ for $y \neq x$. Green gave a heuristic argument for the existence of a unique such Gr satisfying $\mathrm{Gr}(x, y) = 0$ for all $y \in \Gamma$: physically, $\mathrm{Gr}(x, y)$ represents the electrostatic potential at y due to a point charge at x when Γ is an earthed conductor. This particular Gr became known as the *Green's function* for the domain Ω, and yields an integral representation formula for the solution of the Dirichlet problem (1.7),

$$u(x) = -\int_\Gamma g(y)\frac{\partial}{\partial \nu_y}\mathrm{Gr}(x, y)\, d\sigma_y \quad \text{for } x \in \Omega. \tag{1.11}$$

In practice, finding an explicit formula for Gr is possible only for very simple domains. For instance, if Ω is the open ball with radius $r > 0$ centred at the origin, then

$$\mathrm{Gr}(x, y) = \frac{1}{4\pi|x - y|} - \frac{1}{4\pi}\frac{r}{|x||x^\sharp - y|},$$

where $x^\sharp = (r/|x|)^2 x$ is the image of x under a reflection in the sphere Γ. In this case, the integral (1.11) is given by

$$u(x) = \frac{1}{4\pi r}\int_{|y|=r} g(y)\frac{r^2 - |x|^2}{|x - y|^3}\, d\sigma_y \quad \text{for } |x| < r,$$

a formula obtained by Poisson [18, p. 360] in 1813 by a different method. Green also used (1.10) to derive a kind of converse to the jump relation (1.6), by showing that if a function u satisfies the Laplace equation on $\Omega^+ \cup \Omega^-$, is continuous everywhere and decays appropriately at infinity, then $u = \mathrm{SL}\,\psi$, where $\psi = -(\partial_\nu^+ u - \partial_\nu^- u)$.

Green's *Essay* did not begin to become widely known until 1845, when William Thomson (Lord Kelvin) introduced it to Joseph Liouville in Paris [99, pp. 113–121]. Eventually, Thomson had the work published in three parts during 1850–1854 in Crelle's *Journal für die reine und angewandte Mathematik* [31].

Meanwhile, Poisson and others continued to apply the method of separation of variables to a variety of physical problems. A key step in many such calculations is to solve a two-point boundary value problem with a parameter $\lambda > 0$,

$$-\frac{d}{dx}\left(a\frac{du}{dx}\right) + (b - \lambda w)u = 0 \quad \text{for } 0 < x < 1,$$

$$\frac{du}{dx} - m_0 u = 0 \quad \text{at } x = 0, \qquad (1.12)$$

$$\frac{du}{dx} - m_1 u = 0 \quad \text{at } x = 1,$$

where a, b and w are known real-valued functions of x such that $a > 0$ and $w > 0$, and where m_0 and m_1 are known constants (possibly ∞, in which case the boundary condition is to be interpreted as $u = 0$). The main features of the problem (1.12) can be seen in the simplest example: $a = w = 1$ and $b = 0$. The general solution of the differential equation is then a linear combination of $\sin(\sqrt{\lambda}x)$ and $\cos(\sqrt{\lambda}x)$, and the boundary conditions imply that the solution is identically zero unless the parameter λ satisfies a certain transcendental equation having a sequence of positive solutions $\lambda_1 \le \lambda_2 \le \lambda_3 \le \cdots$ with $\lambda_j \to \infty$. In the general case, the number λ_j was subsequently called an *eigenvalue* for the problem, and any corresponding, non-trivial solution $u = \phi_j$ of the differential equation was called an *eigenfunction*. For the special case $a = w = 1$, Poisson showed in 1826 that eigenfunctions corresponding to distinct eigenvalues are orthogonal, i.e.,

$$\int_0^1 \phi_j(x)\phi_k(x)w(x)\,dx = 0 \quad \text{if } \lambda_j \ne \lambda_k,$$

and that all eigenvalues are real; see [62, p. 433]. A much deeper analysis was given by Charles François Sturm in 1836, who established many important properties of the eigenfunctions, as well as proving the existence of infinitely many eigenvalues. Building on Sturm's work, Liouville showed in two papers from 1836 and 1837 that an arbitrary function f could be expanded in a generalised

Fourier series,

$$f(x) = \sum_{j=1}^{\infty} c_j \phi_j(x), \qquad \text{where } c_j = \frac{\int_0^1 \phi_j(x) f(x) w(x)\, dx}{\int_0^1 \phi_j(x)^2 w(x)\, dx},$$

thereby justifying many applications of the method of separation of variables. Excerpts from the papers of Sturm and Liouville are reproduced in [7, pp. 258–281]; see also [62, Chapter X].

Carl Friedrich Gauss wrote a long paper [26] on potential theory in 1839; see also [7, pp. 358–361] and Lützen [62, pp. 583–586]. He re-derived many of Poisson's results, including (1.6), using more rigorous arguments, and was apparently unaware of Green's work. Gauss sought to find, for an arbitrary conductor Ω, the equilibrium charge distribution with total charge M, i.e., in mathematical terms, he sought to find a function ψ whose single-layer potential SL ψ is constant on Γ, subject to the constraint that $\int_\Gamma \psi\, d\sigma = M$. Introducing an arbitrary function g, he considered the quadratic functional

$$J_g(\phi) = \int_\Gamma (S\phi - 2g)\phi\, d\sigma,$$

where $S\phi = \gamma^+ \mathrm{SL}\, \phi = \gamma^- \mathrm{SL}\, \phi$ denotes the boundary values of the single-layer potential, or, explicitly,

$$S\phi(x) = \int_\Gamma G(x, y)\phi(y)\, d\sigma_y \quad \text{for } x \in \Gamma.$$

In the case $g = 0$, the quantity $J_g(\phi)$ has a physical meaning: it is proportional to the self-energy of the charge distribution ϕ; see Kellogg [45, pp. 79–80].

One easily sees that $J_g(\phi)$ is bounded below for all ϕ in the class V_M of functions satisfying $\int_\Gamma \phi\, d\sigma = M$ and $\phi \geq 0$ on Γ. Also, $J_g(\phi + \delta\phi) = J_g(\phi) + \delta J_g + O(\delta\phi^2)$, where the first variation of J_g is given by

$$\delta J_g = \delta J_g(\phi, \delta\phi) = 2 \int_\Gamma (S\phi - g)\, \delta\phi\, d\sigma.$$

Suppose that the minimum value of J_g over the class V_M is achieved when $\phi = \psi$. It follows that $\delta J_g(\psi, \delta\phi) = 0$ for all $\delta\phi$ satisfying $\int_\Gamma \delta\phi\, d\sigma = 0$ and $\psi + \delta\phi \geq 0$ on Γ, and therefore $S\psi - g$ is constant on any component of Γ where $\psi > 0$. Gauss showed that if $g = 0$, then $\psi > 0$ everywhere on Γ, and thus deduced the existence of an equilibrium potential from the existence of a minimiser for J_0. He also showed that this minimiser is unique, and gave an argument for the existence of a solution ψ to the boundary integral equation

$$S\psi = g \quad \text{on } \Gamma. \tag{1.13}$$

The single-layer potential of this ψ is the solution of the Dirichlet problem for the Laplace equation, i.e., $u = \text{SL}\,\psi$ satisfies (1.7).

In a series of papers from 1845 to 1846, Liouville studied the single-layer potential when Γ is an ellipsoid, solving the integral equation (1.13) by adapting his earlier work on the eigenvalue problem (1.12). Let w be the equilibrium density for Γ, normalised so that $Sw = 1$. Liouville showed that if Γ is an ellipsoid, then

$$S(w\psi_j) = \mu_j\psi_j \quad \text{for } j = 1, 2, 3, \ldots,$$

where the ψ_j are *Lamé functions*, and the μ_j are certain constants satisfying

$$\mu_1 \geq \mu_2 \geq \mu_3 \geq \cdots > 0 \quad \text{with } \mu_j \to 0 \text{ as } j \to \infty.$$

He established the orthogonality property

$$\int_\Gamma \psi_j(x)\psi_k(x)w(x)\,d\sigma_x = 0 \quad \text{for } j \neq k,$$

and concluded that the solution of (1.13) is

$$\psi(x) = w(x)\sum_{j=1}^{\infty} c_j\psi_j(x), \quad \text{where } c_j = \frac{\int_\Gamma \psi_j(x)g(x)w(x)\,d\sigma_x}{\int_\Gamma \psi_j(x)^2 w(x)\,d\sigma_x}.$$

In his unpublished notebooks (described in [62, Chapter XV]) Liouville went a considerable distance towards generalising these results to the case of an arbitrary surface Γ, inventing in the process the Rayleigh–Ritz procedure for finding the eigenvalues and eigenfunctions, 20 years before Rayleigh [97] and 60 years before Ritz [98].

During the 1840s, Thomson and Peter Gustav Lejeune Dirichlet separately advanced another type of existence argument [7, pp. 379–387] that became widely known on account of its use by Riemann in his theory of complex analytic functions. Riemann introduced the term *Dirichlet's principle* for this method of establishing the existence of a solution to the Dirichlet problem, although a related variational argument had earlier been used by Green [32]. If one considers the functional

$$J(v) = \int_\Omega |\text{grad } v|^2 \, dx$$

for v in a class of sufficiently regular functions V_g satisfying $v = g$ on Γ, then it seems obvious, because $J(v) \geq 0$ for all $v \in V_g$, that there exists a $u \in V_g$ satisfying

$$J(u) \leq J(v) \quad \text{for all } v \in V_g. \tag{1.14}$$

Given any w such that $w = 0$ on Γ, and any constant h, the function $v = u + hw$ belongs to V_g, and, assuming the validity of the first Green identity (1.8), simple manipulations yield

$$J(v) = J(u) - 2h \int_\Omega w \Delta u \, dx + h^2 J(w).$$

Here, the constant h is arbitrary, so the minimum condition (1.14) implies that

$$\int_\Omega w \Delta u \, dx = 0 \quad \text{whenever } w = 0 \text{ on } \Gamma.$$

By choosing w to take the same sign as Δu throughout Ω, we conclude that u is a solution of the Dirichlet problem for the Laplace equation. Conversely, each solution of the Dirichlet problem minimises the integral. Dirichlet also established the uniqueness of the minimiser u. In fact, if both u_1 and u_2 minimise J in the class of functions V_g, then the difference $w = u_1 - u_2$ vanishes on Γ, and, arguing as above with $h = 1$, we find that $J(u_1) = J(u_2) + J(w)$. Thus, $J(w) = 0$, so w is constant, and hence identically zero, implying that $u_1 = u_2$ on Ω.

In 1869, H. Weber [104] employed the quadratic functional $J(v)$ in a Rayleigh–Ritz procedure to show the existence of eigenfunctions and eigenvalues for the Laplacian on a general bounded domain. He minimised $J(v)$ subject to *two* constraints: $v = 0$ on Γ, and $\int_\Omega v(x)^2 \, dx = 1$. If we suppose that a minimum is achieved when $v = u_1$, then by arguing as above we see that

$$\int_\Omega w \Delta u_1 \, dx = 0 \quad \text{whenever } w = 0 \text{ on } \Gamma \text{ and } \int_\Omega w(x) u_1(x) \, dx = 0.$$

Here, the extra restriction on w arises from the second of the constraints in the minimisation problem. Weber showed that $-\Delta u_1 = \lambda_1 u_1$ on Ω, where $\lambda_1 = J(u_1)$. In fact, for an arbitrary v satisfying $v = 0$ on Γ, if we put

$a = \int_\Omega v u_1 \, dx$ and $w = v - a u_1$, then $w = 0$ on Γ, and $\int_\Omega w u_1 \, dx = 0$, so by the first Green identity,

$$\int_\Omega v(-\Delta u_1) \, dx = -\int_\Omega (w + a u_1) \Delta u_1 \, dx$$

$$= a\left(J(u_1) - \int_\Gamma u_1 \frac{\partial u_1}{\partial v} \, d\sigma \right) = \lambda_1 \int_\Omega v u_1 \, dx,$$

remembering that $u_1 = 0$ on Γ. Next, Weber minimised $J(v)$ subject to three constraints: the two previous ones and in addition $\int_\Omega v u_1 \, dx = 0$. The minimiser u_2 is the next eigenfunction, satisfying $-\Delta u_2 = \lambda_2 u_2$ on Ω, where $\lambda_2 = J(u_2) \geq \lambda_1$. Continuing in this fashion, he obtained sequences of (orthonormal) eigenfunctions u_j and corresponding eigenvalues λ_j, with $0 < \lambda_1 \leq \lambda_2 \leq \lambda_3 \leq \cdots$.

Although simple and beautiful, Dirichlet's principle (in its naïve form) is based on a false assumption, namely, that a minimiser $u \in V_g$ must exist because $J(v) \geq 0$ for all $v \in V_g$. This error was pointed out by Karl Theodore Wilhelm Weierstraß [7, pp. 390–391] in 1870, and the same objection applies to the variational arguments of Gauss, Liouville and Weber. During the period from 1870 to 1890, alternative existence proofs for the Dirichlet problem were devised by Hermann Amandus Schwarz, Carl Gottfried Neumann and Jules Henri Poincaré; see Gårding [25] and Kellogg [45, pp. 277–286]. We shall briefly describe the first of these proofs, Neumann's *Methode des arithmetischen Mittels*, after first introducing some important properties of the *double-layer potential*,

$$\mathrm{DL}\, \psi(x) = \int_\Gamma \psi(y) \frac{\partial}{\partial v_y} G(x, y) \, d\sigma_y \quad \text{for } x \notin \Gamma.$$

A surface potential of this type appears in the third Green identity (1.10), with $\psi = u|_\Gamma$; note the similarity with the general Poisson integral formula (1.11).

The double layer potential has a very simple form when the density is constant on Γ. In fact

$$\mathrm{DL}\, 1(x) = \begin{cases} -1 & \text{for } x \in \Omega^-, \\ 0 & \text{for } x \in \Omega^+, \end{cases} \tag{1.15}$$

as one sees by taking $u = 1$ in (1.10) if $x \in \Omega^-$, and by applying the divergence theorem if $x \in \Omega^+$. Obviously, $\mathrm{DL}\, \psi$ is harmonic on $\Omega^\pm$, but the example $\psi = 1$ shows that the double-layer potential can have a jump discontinuity, and it turns out that in general

$$\gamma^+ \operatorname{DL} \psi - \gamma^- \operatorname{DL} \psi = \psi \quad \text{on } \Gamma;$$

cf. (1.6). Thus, if we let

$$T\psi = \gamma^+ \operatorname{DL} \psi + \gamma^- \operatorname{DL} \psi, \tag{1.16}$$

then

$$\gamma^\pm \operatorname{DL} \psi = \tfrac{1}{2}(\pm\psi + T\psi) \quad \text{on } \Gamma. \tag{1.17}$$

The operator T may be written explicitly as

$$T\psi(x) = -\psi(x) + 2\int_\Gamma [\psi(y) - \psi(x)]\frac{\partial}{\partial \nu_y}G(x, y)\, d\sigma_y \quad \text{for } x \in \Gamma,$$

and we see in particular that $T1 = -1$, in agreement with (1.15).

Neumann's existence proof built on earlier work by A. Beer [4], who, in 1856, sought a solution to the Dirichlet problem (1.7) in the form of a double-layer potential $u = \operatorname{DL}\psi$. Beer worked in two dimensions, and so used the fundamental solution

$$G(x, y) = \frac{1}{2\pi}\log\frac{1}{|x - y|} \quad \text{for } x, y \in \mathbb{R}^2 \text{ and } x \neq y.$$

In view of (1.17), the boundary condition $\gamma^- u = g$ on Γ leads to the integral equation

$$-\psi + T\psi = 2g \quad \text{on } \Gamma. \tag{1.18}$$

The form of this equation suggests application of the method of successive approximations, a technique introduced by Liouville in 1830 to construct the solution to a two-point boundary value problem; see [62, p. 447]. Beer defined a sequence $\psi_0, \psi_1, \psi_2, \ldots$ by

$$\psi_0 = -2g \quad \text{and} \quad \psi_j = T\psi_{j-1} - 2g \quad \text{for } j \geq 1,$$

which, *if it converged uniformly*, would yield the desired solution $\psi = \lim_{j\to\infty}\psi_j$. However, Beer did not attempt to prove convergence; see Hellinger and Toeplitz [38, pp. 1345–1349].

The kernel appearing in the double layer potential has the form

$$\frac{\partial}{\partial \nu_y}G(x, y) = \frac{1}{\Upsilon_n}\frac{\nu_y \cdot (x - y)}{|x - y|^n},$$

where $\Upsilon_2 = 2\pi$ is the length of the unit circle, and $\Upsilon_3 = 4\pi$ is the area of the unit sphere. For his proof, Neumann [77] assumed that Ω^- is *convex*. In this case, $\nu_y \cdot (x - y) \leq 0$ for all $x, y \in \Gamma$, so

$$\min_{\Gamma} \psi \leq -(T\psi)(x) \leq \max_{\Gamma} \psi \quad \text{for } x \in \Gamma,$$

and it can be shown that (provided the convex domain Ω^- is not the intersection of two cones) for every continuous g there exists a constant a_g such that

$$\max_{x \in \Gamma} \left| (T^m g)(x) - (-1)^m a_g \right| \leq C r^m, \quad \text{with } 0 < r < 1,$$

where the constants C and r depend only on Γ. We define a density function

$$\psi = \sum_{j=0}^{\infty} (T^{2j} g + T^{2j+1} g),$$

noting that the series converges uniformly on Γ because

$$|T^{2j} g + T^{2j+1} g| \leq |T^{2j} g - (-1)^{2j} a_g| + |T^{2j+1} g - (-1)^{2j+1} a_g|$$
$$\leq C(r^{2j} + r^{2j+1}).$$

Also, the identity

$$g + T \sum_{j=0}^{m} (T^{2j} g + T^{2j+1} g) = T^{2m+2} g + \sum_{j=0}^{m} (T^{2j} g + T^{2j+1} g)$$

implies that $g + T\psi = a_g + \psi$, so by (1.17) we have $\gamma^- \operatorname{DL} \psi = \frac{1}{2}(a_g - g)$. Therefore, the desired solution of the Dirichlet problem (1.7) is the function $u = a_g - 2\operatorname{DL}\psi$.

In a paper of 1888 dealing with the Laplace equation, P. du Bois-Reymond [20] expressed the view that a general theory of integral equations would be of great value, but confessed his inability to see even the outline of such a theory. (This paper, incidentally, contains the first use of the term "integral equation", or rather *Integralgleichung*.) The various results known at that time all seemed to rely on special properties of the particular equation under investigation. Only during the final decade of the nineteenth century did a way forward begin to emerge. In 1894, Le Roux [55] successfully analysed an integral equation of the form

$$\int_a^x K(x, y)u(y)\,dy = f(x) \quad \text{for } a \leq x \leq b,$$

with a sufficiently smooth but otherwise quite *general* kernel K, and a right-hand side satisfying $f(a) = 0$. He constructed a solution by first differentiating with respect to x, and then applying the method of successive approximations. Two years later, Volterra [103, Volume 2, pp. 216–262] independently considered the same problem, using the same approach, and remarked in passing that the integral equation could be looked upon as the continuous limit of an $n \times n$ linear algebraic system as $n \to \infty$.

Volterra's remark was taken up by Ivar Fredholm [23] in a short paper of 1900, which was subsequently expanded into a longer work [24] in 1903. Fredholm considered an integral equation of the form

$$u(x) + \lambda \int_0^1 K(x, y)u(y)\, dy = f(x) \quad \text{for } 0 \le x \le 1, \tag{1.19}$$

with a general continuous kernel K and a complex parameter λ. As motivation, he mentions a problem discussed a few years earlier in an influential paper of Poincaré [82], namely, for a given function f on Γ to find a double-layer potential $u = \mathrm{DL}\,\psi$ satisfying

$$\gamma^- u - \gamma^+ u = \lambda(\gamma^- u + \gamma^+ u) + 2f \quad \text{on } \Gamma.$$

In view of (1.16) and (1.17), this problem amounts to finding a density function ψ satisfying

$$-\psi - \lambda T\psi = 2f \quad \text{on } \Gamma. \tag{1.20}$$

The special case $\lambda = -1$ and $f = g$ is just Beer's equation (1.18) arising from the interior Dirichlet problem, and similarly $\lambda = +1$ and $f = -g$ gives the analogous equation arising from the *exterior* Dirichlet problem. Poincaré had shown that both equations are solvable for a wide class of smooth but *not necessarily convex* domains.

Fredholm began his analysis of (1.19) by introducing a function $D(\lambda)$ defined by the series

$$D(\lambda) = 1 + \lambda \int_0^1 K(y, y)\, dy$$

$$+ \frac{\lambda^2}{2!} \int_0^1 \int_0^1 \begin{vmatrix} K(y_1, y_1) & K(y_1, y_2) \\ K(y_2, y_1) & K(y_2, y_2) \end{vmatrix} dy_1\, dy_2 + \cdots, \tag{1.21}$$

which he called the *determinant* of the integral equation. In fact, if we put $x_j = j/n$ for $1 \le j \le n$, and replace the integral in (1.19) by the obvious

Riemann sum, then we obtain the discrete system

$$u(x_j) + \frac{\lambda}{n} \sum_{k=1}^{n} K(x_j, x_k) u(x_k) = f(x_j) \quad \text{for } 1 \le j \le n,$$

whose determinant can be written as

$$1 + \frac{\lambda}{n} \sum_{k=1}^{n} K(x_k, x_k) + \frac{\lambda^2}{2! n^2} \sum_{k_1=1}^{n} \sum_{k_2=1}^{n} \begin{vmatrix} K(x_{k_1}, x_{k_1}) & K(x_{k_1}, x_{k_2}) \\ K(x_{k_2}, x_{k_1}) & K(x_{k_2}, x_{k_2}) \end{vmatrix} + \cdots$$

$$+ \frac{\lambda^n}{n! n^n} \sum_{k_1=1}^{n} \cdots \sum_{k_n=1}^{n} \begin{vmatrix} K(x_{k_1}, x_{k_1}) & \cdots & K(x_{k_1}, x_{k_n}) \\ \vdots & & \vdots \\ K(x_{k_n}, x_{k_1}) & \cdots & K(x_{k_n}, x_{k_n}) \end{vmatrix}.$$

Formally at least, in the limit as $n \to \infty$ the determinant of the discrete system tends to $D(\lambda)$. (This heuristic derivation does not appear in Fredholm's papers, but see [38, p. 1356] and [19, p. 99].) Fredholm proved that the series (1.21) converges uniformly for λ in any compact subset of the complex plane, and so defines an entire function. By generalising Cramer's rule for finite linear systems, Fredholm showed that if $D(\lambda) \ne 0$, then (1.19) has a unique continuous solution u for each continuous f. He applied this result to the boundary integral equation (1.20), and so proved the existence of a solution to the Dirichlet problem on any bounded C^3 domain in the plane.

Fredholm also gave a complete account of the case when $D(\lambda) = 0$, by considering the *transposed* integral equation

$$v(x) + \lambda \int_0^1 K(y, x) v(y) \, dy = g(x) \quad \text{for } 0 \le x \le 1, \tag{1.22}$$

which has the same determinant as the original equation (1.19). He proved that if $D(\lambda)$ has a zero of multiplicity m at $\lambda = \lambda_0$, then for this value of the parameter the two *homogeneous* equations, i.e., (1.19) and (1.22) with f and g identically zero, each have m linearly independent solutions. In this case, the *in*homogeneous equation (1.19) has a (non-unique) solution u if and only if $\int_0^1 f(x) v(x) \, dx = 0$ for every solution v of the transposed homogenenous equation. The above dichotomy in the behaviour of the two integral equations, corresponding to the cases $D(\lambda) \ne 0$ and $D(\lambda) = 0$, is today known as the *Fredholm alternative*.

The simplicity and generality of Fredholm's theory made an immediate and lasting impression, not least on David Hilbert, who, during the period 1904–1906, made important contributions that later appeared in his influential book [39] on integral equations. Hilbert was especially interested in the case when the kernel is *symmetric*, i.e., when K is real-valued and satisfies $K(y, x) = K(x, y)$ for all x and y. The zeros of the determinant are then purely real, and form a nondecreasing sequence $\lambda_1, \lambda_2, \ldots$, counting multiplicities. For each j there is a non-trivial solution ψ_j of the homogeneous equation with $\lambda = \lambda_j$, and the sequence $\psi_1, \psi_2, \ldots$ can be chosen in such a way that the functions are orthonormal:

$$\int_0^1 \psi_j(x)\psi_k(x)\,dx = \delta_{jk} = \begin{cases} 1 & \text{if } j = k, \\ 0 & \text{if } j \neq k. \end{cases}$$

Of course, ψ_j is an eigenfunction of the integral operator with kernel K, and the corresponding eigenvalue is $1/\lambda_j$. Hilbert proved the identity

$$\int_0^1 \int_0^1 K(x, y)u(x)v(y)\,dx\,dy = \sum_{j \geq 1} \frac{1}{\lambda_j} \int_0^1 \psi_j(x)u(x)\,dx$$
$$\times \int_0^1 \psi_j(y)v(y)\,dy,$$

which is the continuous analogue of the reduction to principal axes of the quadratic form associated with a real symmetric matrix. He also studied the convergence of eigenfunction expansions.

Our story has now arrived at a natural stopping point. The period of classical analysis is about to be overtaken by the geometric spirit of functional analysis. By 1917, F. Reisz [87] had effectively subsumed Fredholm's results in the general theory of compact linear operators, a topic we shall take up in the next chapter.

Exercises

1.1 Show that if f is a radially symmetric function, say $f(y) = F(\tau)$ where $\tau = |y|$, then the Newtonian potential (1.4) is radially symmetric, and is given by $u(x) = U(\rho)$, where $\rho = |x|$ and

$$U(\rho) = \frac{1}{\rho} \int_0^\rho F(\tau)\tau^2\,d\tau + \int_\rho^\infty F(\tau)\tau\,d\tau.$$

Hence verify Poisson's equation:

$$-\Delta u(x) = -\frac{1}{\rho^2}\frac{d}{d\rho}\left(\rho^2\frac{dU}{d\rho}\right) = F(\rho) = f(x).$$

1.2 Let $\Gamma = \{y \in \mathbb{R}^3 : |y| = a\}$ denote the sphere of radius $a > 0$ centred at the origin. Show that if the density ψ is constant on Γ, then the single-layer potential (1.5) is radially symmetric, i.e., a function of $\rho = |x|$. Show in particular that

$$\text{SL } 1(x) = \begin{cases} a & \text{if } \rho < a, \\ a^2/\rho & \text{if } \rho > a, \end{cases}$$

and verify that the jump relation (1.6) holds in this case.

1.3 Fix $x, y \in \Omega$ with $x \neq y$, and for any sufficiently small $\epsilon > 0$ let Ω_ϵ denote the region obtained from Ω by excising the balls with radius ϵ centred at x and y. By applying the second Green identity (1.9) to the functions $\text{Gr}(x, \cdot)$ and $\text{Gr}(y, \cdot)$ over Ω_ϵ, and then sending $\epsilon \downarrow 0$, show that $\text{Gr}(x, y) = \text{Gr}(y, x)$.

2

Abstract Linear Equations

Later in this book, we shall reduce elliptic boundary value problems, and also their equivalent boundary integral formulations, to operator equations of the form $Au = f$, with A a bounded linear operator from a Hilbert space into its dual. The ellipticity of the partial differential equation will imply that A is the sum of a positive-definite operator and a compact operator (the latter possibly zero). Our aim now is to study such operators abstractly, using techniques from functional analysis. We begin by considering some topics that can be understood more clearly in a less restricted setting. In fact, we shall develop the concept of a Fredholm operator acting between two Banach spaces, even though it would suffice for our later applications to consider only operators with index zero acting between Hilbert spaces. At the end of the chapter is a short treatment of spectral theory, covering just the simplest cases, namely, self-adjoint operators that are compact or have a compact inverse.

We shall use c and C to denote small and large generic constants, whose values may change even within a single chain of estimates, but with c always bounded away from zero, and C always bounded away from infinity. If $\| \cdot \|$ and $\| \cdot \|'$ are norms on a vector space X, then we write

$$\|u\| \sim \|u\|' \quad \text{for all } u \in X,$$

to indicate equivalence of the norms, i.e., $c\|u\|_X \le \|u\|' \le C\|u\|_X$ for all $u \in X$. The reader should also note that our sesquilinear forms, and in particular our inner products, are conjugate-linear in the *first* argument, and linear in the second.

A familiarity with basic concepts and results from general topology and linear functional analysis is assumed, but some effort will be made to refresh the reader's memory. The theorems that we cite without proof can be found in virtually any textbook on functional analysis; Yosida [106] and Simmons [94] will serve as our standard references.

The Kernel and Image

Suppose that X and Y are complex vector spaces, and let $A : X \to Y$ be a linear map. The *kernel* (or null space) of A is the subspace of X defined by

$$\ker A = \{u \in X : Au = 0\},$$

and the *image* of A is the subspace of Y defined by

$$\operatorname{im} A = \{f \in Y : \text{there exists } u \in X \text{ such that } f = Au\}.$$

Given $f \in Y$, we can seek a solution $u \in X$ to the linear equation

$$Au = f.$$

It follows at once from the definitions above that a solution exists if and only if $f \in \operatorname{im} A$, in which case u is unique modulo $\ker A$, i.e., any two solutions differ by an element of the kernel. Thus, the inverse A^{-1} exists if and only if $\ker A = \{0\}$ and $\operatorname{im} A = Y$.

Recall that if W is a subspace of X, then the elements of the quotient space X/W are the cosets $u + W = \{u + w : w \in W\}$, and the vector space operations in X/W are given by

$$\lambda(u + W) = (\lambda u) + W \quad \text{and} \quad (u_1 + W) + (u_2 + W) = (u_1 + u_2) + W,$$

for $\lambda \in \mathbb{C}$ and $u, u_1, u_2 \in X$. The dimension of X/W is called the *codimension* of W in X. Every linear map $A : X \to Y$ induces an isomorphism

$$A_/ : X/\ker A \to \operatorname{im} A$$

defined by

$$A_/(u + \ker A) = Au \quad \text{for } u \in X.$$

Assume now that X and Y are *normed* spaces. A linear map $A : X \to Y$ is continuous if and only if it is bounded, i.e., if and only if $\|Au\|_Y \leq C\|u\|_X$ for all $u \in X$. The vector space $\mathcal{L}(X, Y)$ consisting of all such bounded linear maps is itself a normed space, with

$$\|A\|_{\mathcal{L}(X,Y)} = \sup_{0 \neq u \in X} \frac{\|Au\|_Y}{\|u\|_X}.$$

Recall that if Y is a Banach space (i.e., if Y is complete), then so is $\mathcal{L}(X, Y)$.

If W is a *closed* subspace of X, then we can make X/W into a normed space by defining

$$\|u + W\|_{X/W} = \inf_{w \in W} \|u + w\|_X.$$

Furthermore, X/W is a Banach space when X is a Banach space. If $A \in \mathcal{L}(X, Y)$, then ker A is closed, and the induced isomorphism is bounded, i.e., $A_/ \in \mathcal{L}(X/\ker A, \text{im } A)$. When is the inverse $(A_/)^{-1}$ bounded? We can answer this question with the help of the open mapping theorem.

Theorem 2.1 Suppose that X and Y are Banach spaces, and let $A \in \mathcal{L}(X, Y)$. If im $A = Y$, then A is an open mapping, i.e., A maps each open subset of X to an open subset of Y.

The proof of this result uses a Baire category argument, and can be found in [106, p. 75] or [94, p. 236]. Since a function $Y \to X$ is continuous if and only if the pre-image of every open set in X is open in Y, and since a closed subspace of a Banach space is again a Banach space, the next corollary follows at once.

Corollary 2.2 Suppose that X and Y are Banach spaces. If $A \in \mathcal{L}(X, Y)$, then the following conditions are equivalent:

(i) The subspace im A is closed in Y.
(ii) The induced map $A_/ : X/\ker A \to \text{im } A$ has a bounded inverse.
(iii) There is a constant C such that

$$\|u + \ker A\|_{X/\ker A} \le C\|Au\|_Y \quad \text{for } u \in X.$$

In particular, there exists a bounded inverse $A^{-1} \in \mathcal{L}(Y, X)$ if and only if im $A = Y$ and ker $A = \{0\}$.

When A is as in Corollary 2.2, our problem $Au = f$ is essentially well posed: for each $f \in \text{im } A$, the solutions form a coset $u + \ker A$ that depends continuously on f. Of course, in applications, one typically starts with a concrete integral or differential operator A, and then seeks complete spaces X and Y such that $A : X \to Y$ is bounded and im A is closed. Satisfying both conditions simultaneously requires a tight fit between operator and spaces. If $\|\cdot\|_Y$ is too strong (i.e., too big) relative to $\|\cdot\|_X$, then the space Y will be too small to serve as the codomain, and typically one obtains an unbounded operator by restricting

A to some dense subspace of *X*. However, if $\| \cdot \|_Y$ is too weak (i.e., too small), then the space *Y* will be too large, and im *A* will fail to be closed.

If *V* and *W* are subspaces of *X*, and if each $u \in X$ can be written as $u = v + w$ for a unique $v \in V$ and a unique $w \in W$, then *X* is said to be the (internal) *direct sum* of *V* and *W*, and we indicate this fact by writing $X = V \oplus W$. A *projection* is a bounded linear operator $P : X \to X$ having the property that $P^2 = P$. If *P* is a projection, then $(I - P)^2 = I - P$ and $P(I - P) = (I - P)P = 0$, where *I* is the identity operator, i.e., $Iu = u$. Thus, $I - P$ is also a projection, and by writing $u = Pu + (I - P)u$, we obtain a direct sum decomposition $X = V \oplus W$, where *V* and *W* are the closed subspaces

$$V = \operatorname{im} P = \ker(I - P) \quad \text{and} \quad W = \operatorname{im}(I - P) = \ker P.$$

This state of affairs is described by saying that *P* is the projection of *X* onto *V*, parallel to *W*. Exercise 2.2 shows that there exists a projection onto every finite-dimensional subspace, and onto every closed subspace with finite codimension.

Duality

If *X* is a normed space, then we denote its *dual space* by X^*. Thus, $X^* = \mathcal{L}(X, \mathbb{C})$ is the space of bounded linear functionals $g : X \to \mathbb{C}$. We shall write

$$\langle g, u \rangle = g(u)$$

for the value of the functional $g \in X^*$ at the vector $u \in X$. By the definition of the norm in $\mathcal{L}(X, \mathbb{C})$,

$$|\langle g, u \rangle| \leq \|g\|_{X^*} \|u\|_X \quad \text{and} \quad \|g\|_{X^*} = \sup_{0 \neq u \in X} \frac{|\langle g, u \rangle|}{\|u\|_X}. \tag{2.1}$$

The dual space X^* is a Banach space even if *X* is not complete.

A key tool in the study of duality is the Hahn–Banach theorem, one version of which is as follows.

Theorem 2.3 If W is a subspace of a normed space X, then every functional in W^* can be extended to a functional in X^* having the same norm.

For a proof, see [106, p. 106] or [94, p. 228]. Zorn's lemma or another equivalent of the axiom of choice is needed unless one introduces some extra assumption(s), such as that *X* is separable.

We now establish a few simple consequences of Theorem 2.3 that will be used later. If $u \in X$ and $W \subseteq X$, then the distance between *u* and *W* is defined by

$$\text{dist}(u, W) = \inf_{w \in W} \|u - w\|_X;$$

notice that if W is a closed subspace, then $\text{dist}(u, W) = \|u + W\|_{X/W}$.

Theorem 2.4 *Let W be a subspace of a normed linear space X, and let $u \in X$. If $\text{dist}(u, W) > 0$, then there exists a functional $g \in X^*$ such that*

$$\langle g, u \rangle = \text{dist}(u, W), \quad \|g\|_{X^*} = 1 \quad \text{and} \quad \langle g, w \rangle = 0 \qquad \text{for } w \in W.$$

Proof. Put $d = \text{dist}(u, W)$, and assume that $d > 0$. It follows that $u \notin W$, and we may form the direct sum $W_1 = W \oplus \text{span}\{u\}$. Define $g \in W_1^*$ by

$$\langle g, w + \lambda u \rangle = \lambda d \qquad \text{for } w \in W \text{ and } \lambda \in \mathbb{C},$$

and observe that

$$|\langle g, w + \lambda u \rangle| \leq |\lambda| \|u - (-w/\lambda)\|_X = \|w + \lambda u\|_X \qquad \text{for } w \in W \text{ and } \lambda \neq 0,$$

so $\|g\|_{W_1^*} \leq 1$. Moreover, given $\epsilon > 0$, there is a $w \in W$ such that $d \leq \|u - w\|_X < d + \epsilon$, and thus $d = |(-1)d| = |\langle g, w - u \rangle| \leq \|g\|_{W_1^*} \|w - u\|_X < \|g\|_{W_1^*}(d + \epsilon)$, implying that $\|g\|_{W_1^*} > d/(d + \epsilon)$. Hence, $\|g\|_{W_1^*} = 1$, and we can complete the proof by applying Theorem 2.3. $\qquad \square$

Corollary 2.5 *Let W be a closed subspace of X, and let $u \in X$. If $u \notin W$, then there exists a functional $g \in X^*$ such that*

$$\langle g, u \rangle > 0, \quad \|g\|_{X^*} = 1 \quad \text{and} \quad \langle g, w \rangle = 0 \qquad \text{for } w \in W.$$

Proof. If W is closed and $u \notin W$, then $\text{dist}(u, W) > 0$. $\qquad \square$

Corollary 2.6 *If $0 \neq u \in X$, then there exists a functional $g \in X^*$ such that*

$$\langle g, u \rangle = \|u\|_X \quad \text{and} \quad \|g\|_{X^*} = 1.$$

Proof. Take $W = \{0\}$, so that $\text{dist}(u, W) = \|u\|_X$. $\qquad \square$

Corollary 2.7 *The dual space X^* separates the points in X, i.e., for all $u, v \in X$, if $u \neq v$, then there exists a functional $g \in X^*$ such that $\langle g, u \rangle \neq \langle g, v \rangle$.*

Corollary 2.8 *Let $u \in X$. If $\langle g, u \rangle = 0$ for all $g \in X^*$, then $u = 0$.*

Consider $X^{**} = (X^*)^*$, the second dual of X, and define $\iota : X \to X^{**}$ by

$$\langle \iota u, g \rangle = \langle g, u \rangle \qquad \text{for } u \in X \text{ and } g \in X^*.$$

From (2.1) and Corollary 2.6, we see that $\|\iota u\|_{X^{**}} = \|u\|_X$, so ι is an isometric isomorphism from X onto a subspace $\iota(X)$ in X^{**}. This fact allows us to identify X with $\iota(X)$, and write

$$X \subseteq X^{**}.$$

Obviously, X is closed in the complete space X^{**} if and only if X is itself complete. If $X = X^{**}$, then X is said to be *reflexive*; thus, every reflexive space is complete.

For any linear map $A : X \to Y$, the *transpose* $A^t : Y^* \to X^*$ is the linear map defined by

$$\langle A^t v, u \rangle = \langle v, Au \rangle \qquad \text{for all } v \in Y^* \text{ and } u \in X.$$

With the help of Corollary 2.6, we can show the following.

Lemma 2.9 The transpose A^t is bounded if and only if A is bounded. In fact,

$$\|A^t\|_{\mathcal{L}(Y^*, X^*)} = \|A\|_{\mathcal{L}(X,Y)}.$$

Proof. If A is bounded, then the definition of A^t gives

$$|\langle A^t v, u \rangle| = |\langle v, Au \rangle| \leq \|v\|_{Y^*} \|A\|_{\mathcal{L}(X,Y)} \|u\|_X \qquad \text{for } u \in X \text{ and } v \in Y^*,$$

so $\|A^t v\|_{X^*} \leq \|A\|_{\mathcal{L}(X,Y)} \|v\|_{Y^*}$ and hence $\|A^t\|_{\mathcal{L}(Y^*, X^*)} \leq \|A\|_{\mathcal{L}(X,Y)}$. Conversely, suppose that A^t is bounded, and let $u \in X$. If $Au \neq 0$, then by Corollary 2.6 there is a $v \in Y^*$ such that

$$\langle v, Au \rangle = \|Au\|_X \quad \text{and} \quad \|v\|_{Y^*} = 1.$$

Using the definition of A^t once again, we have

$$\|Au\|_X = |\langle A^t v, u \rangle| \leq \|A^t v\|_{X^*} \|u\|_X$$
$$\leq \|A^t\|_{\mathcal{L}(Y^*, X^*)} \|v\|_{Y^*} \|u\|_X = \|A^t\|_{\mathcal{L}(Y^*, X^*)} \|u\|_X,$$

and since the inequality $\|Au\|_X \leq \|A^t\|_{\mathcal{L}(Y^*, X^*)} \|u\|_X$ is trivial if $Au = 0$, we deduce that $\|A\|_{\mathcal{L}(X,Y)} \leq \|A^t\|_{\mathcal{L}(Y^*, X^*)}$. $\qquad\square$

In studying solutions of the equation $Au = f$, it is helpful to consider at the same time the transposed equation

$$A^t v = g$$

for a given $g \in X^*$ and an unknown $v \in Y^*$. To describe the relationship between the two equations, we use the following terminology. For any subset $W \subseteq X$, the *annihilator* W^a is the closed subspace of X^* defined by

$$W^a = \{g \in X^* : \langle g, u \rangle = 0 \text{ for all } u \in W\}.$$

Dually, for $V \subseteq X^*$ the annihilator $^a V$ is the closed subspace of X defined by

$$^a V = \{u \in X : \langle g, u \rangle = 0 \text{ for all } g \in V\}.$$

Lemma 2.10 The kernels and images of A and A^t satisfy

$$\ker A = {}^a(\operatorname{im} A^t) \quad \text{and} \quad \ker A^t = (\operatorname{im} A)^a.$$

Proof. Applying the various definitions gives

$$
\begin{aligned}
{}^a(\operatorname{im} A^t) &= \{u \in X : \langle g, u \rangle = 0 \text{ for all } g \in \operatorname{im} A^t\} \\
&= \{u \in X : \langle A^t v, u \rangle = 0 \text{ for all } v \in Y^*\} \\
&= \{u \in X : \langle v, Au \rangle = 0 \text{ for all } v \in Y^*\} \\
&= \{u \in X : Au = 0\} = \ker A,
\end{aligned}
$$

and a similar argument shows that $(\operatorname{im} A)^a = \ker A^t$. $\qquad\square$

One sees directly from the definition of the annihilator that

$$W \subseteq {}^a(W^a) \quad \text{for any subset } W \subseteq X,$$

and likewise

$$V \subseteq ({}^a V)^a \quad \text{for any subset } V \subseteq X^*,$$

so by Lemma 2.10,

$$\operatorname{im} A \subseteq {}^a(\ker A^t) \quad \text{and} \quad \operatorname{im} A^t \subseteq (\ker A)^a.$$

The question now arises as to when the reverse inclusions also hold. The next two lemmas will help to provide us with the answer.

Lemma 2.11 Let X be a normed space. A subset $W \subseteq X$ satisfies $W = {}^a(W^a)$ if and only if W is a closed subspace of X.

Proof. The condition is obviously necessary. To prove sufficiency, assume that W is a closed subspace of X, and let $u \in X$. If $u \notin W$, then we can find $g \in X^*$ as in Corollary 2.5. Since $g \in W^a$ and $\langle g, u \rangle \neq 0$, we see that $u \notin {}^a(W^a)$. Thus, ${}^a(W^a) \subseteq W$, and the result follows. $\qquad\square$

Dual to Lemma 2.11 is the result that a subset $V \subseteq X^*$ satisfies $V = ({}^aV)^a$ if and only if V is a weak* closed subspace of X^*; see Schechter [91, p. 192]. However, we shall not use this fact.

Lemma 2.12 Suppose that W is a subspace of a normed space X.

(i) There exists an isometric isomorphism $J_1 : X^*/W^a \to W^*$ given by

$$\langle J_1(g + W^a), w \rangle = \langle g, w \rangle \qquad \text{for } g \in X^* \text{ and } w \in W.$$

(ii) If W is closed, then there exists an isometric isomorphism $J_2 : W^a \to (X/W)^*$ given by

$$\langle J_2 g, u + W \rangle = \langle g, u \rangle \qquad \text{for } g \in W^a \text{ and } u \in X.$$

Proof. Let $g \in X^*$. For any $h \in W^a$ and $w \in W$, we have $\langle g, w \rangle = \langle g + h, w \rangle$, so J_1 is well defined and $|\langle J_1(g + W^a), w \rangle| \leq \|g + h\|_{X^*} \|w\|_X$, implying that

$$|\langle J_1(g + W^a), w \rangle| \leq \inf_{0 \neq h \in W^a} \|g + h\|_{X^*} \|w\|_X = \|g + W^a\|_{X^*/W^a} \|w\|_W.$$

Hence, $\|J_1(g + W^a)\|_{W^*} \leq \|g + W^a\|_{X^*/W^a}$.

Next, if $J_1(g + W^a) = 0$, then $g \in W^a$, and so $\ker J_1 = \{0 + W^a\}$, i.e., J_1 is one–one. To see that J_1 is onto, let $g \in W^*$. By Theorem 2.3, there is a functional $g_1 \in X^*$ such that $g_1 = g$ on W, and $\|g_1\|_{X^*} = \|g\|_{W^*}$. Hence, $g = J_1(g_1 + W^a) \in \operatorname{im} J_1$, and moreover

$$\|J_1^{-1} g\|_{X^*/W^a} = \|g_1 + W^a\|_{X^*/W^a} \leq \|g_1\|_{X^*} = \|g\|_{W^*},$$

completing the proof of (i).

Turning to part (ii), we assume that W is closed and let $g \in W^a$. For any $u \in X$ and $w \in W$, $\langle g, u \rangle = \langle g, u + w \rangle$, so J_2 is well defined, and $|\langle J_2 g, u + W \rangle| \leq \|g\|_{X^*} \|u + w\|_X$, implying that

$$|\langle J_2 g, u + W \rangle| \leq \|g\|_{X^*} \inf_{w \in W} \|u + w\|_X = \|g\|_{X^*} \|u + W\|_{X/W}.$$

Hence, $\|J_2 g\|_{(X/W)^*} \leq \|g\|_{X^*}$. Obviously, $\ker J_2 = \{0\}$, so J_2 is one–one. To see that J_2 is onto, let $h \in (X/W)^*$. We define a linear functional $g : X \to \mathbb{C}$ by $\langle g, u \rangle = \langle h, u + W \rangle$, and observe that

$$|\langle g, u \rangle| \leq \|h\|_{(X/W)^*} \|u + W\|_{X/W} \leq \|h\|_{(X/W)^*} \|u\|_X .$$

Hence, g is bounded with $\|g\|_{X^*} \leq \|h\|_{(X/W)^*}$. Also, if $u \in W$, then $u + W = 0 + W$ so $\langle g, u \rangle = 0$, showing that $g \in W^a$ and $h = J_2 g \in \operatorname{im} J_2$. Finally,

$$\left\| J_2^{-1} h \right\|_{W^a} = \|g\|_{X^*} \leq \|h\|_{(X/W)^*} ,$$

so J_2 is an isometric isomorphism. □

We are now ready to prove the main result for this section. Here, the key point is that if the image of A is closed, then a necessary and sufficient condition for the equation $Au = f$ to be solvable is that the given right-hand side f be annihilated by every solution v of the homogeneous transposed equation $A^t v = 0$.

Theorem 2.13 Suppose that X and Y are Banach spaces. For $A \in \mathcal{L}(X, Y)$, the following conditions are equivalent:

(i) $\operatorname{im} A$ is closed in Y.
(ii) $\operatorname{im} A^t$ is closed in X^*.
(iii) $\operatorname{im} A = {}^a(\ker A^t)$.
(iv) $\operatorname{im} A^t = (\ker A)^a$.

When these conditions hold, there are isometric isomorphisms

$$(Y/\operatorname{im} A)^* \simeq \ker A^t \quad and \quad X^*/\operatorname{im} A^t \simeq (\ker A)^* .$$

Proof. We have seen that A gives rise to a bounded linear operator,

$$A_/ : X/\ker A \to \operatorname{im} A .$$

Likewise, A^t gives rise to a bounded linear operator

$$(A^t)_/ : Y^*/\ker A^t \to \operatorname{im} A^t ,$$

related to the transpose of $A_/$,

$$(A_/)^t : (\operatorname{im} A)^* \to (X/\ker A)^* ,$$

in the following way. Since $\ker A^t = (\operatorname{im} A)^a$, and since $\ker A$ is a closed subspace of X, Lemma 2.12 yields isometric isomorphisms

$$J_1 : Y^*/\ker A^t \to (\operatorname{im} A)^* \quad \text{and} \quad J_2 : (\ker A)^a \to (X/\ker A)^*,$$

so if $u \in X$ and $v \in Y^*$, then

$$
\begin{aligned}
\langle (A_/)^t J_1 (v + \ker A^t), u + \ker A \rangle &= \langle J_1(v + \ker A^t), A_/(u + \ker A) \rangle \\
&= \langle J_1(v + \ker A^t), Au \rangle = \langle v, Au \rangle \\
&= \langle A^t v, u \rangle = \langle (A^t)_/(v + \ker A^t), u \rangle \\
&= \langle J_2(A^t)_/(v + \ker A^t), u + \ker A \rangle,
\end{aligned}
$$

where, in the final step, we used the fact that $\operatorname{im}(A^t)_/ = \operatorname{im} A^t \subseteq (\ker A)^a$. Thus,

$$(A_/)^t J_1 = J_2(A^t)_/ : Y^*/\ker A^t \to (X/\ker A)^*, \qquad (2.2)$$

and we have the chain of equivalences

$$
\begin{array}{llll}
\operatorname{im} A \text{ is closed} & \Longleftrightarrow & A_/ \text{ has a bounded inverse} & \text{(Corollary 2.2)} \\
& \Longleftrightarrow & (A_/)^t \text{ has a bounded inverse} & \text{(Exercise 2.3)} \\
& \Longleftrightarrow & (A^t)_/ \text{ has a bounded inverse} & \text{(2.2)} \\
& \Longleftrightarrow & \operatorname{im} A^t \text{ is closed} & \text{(Corollary 2.2)},
\end{array}
$$

showing that (i) is equivalent to (ii).

Next, Lemmas 2.10 and 2.11 show that (i) implies (iii). Since every annihilator is closed, (iii) implies (i), and (iv) implies (ii). Hence, the equivalence of the four conditions will follow if we prove that (i) implies (iv).

Thus, assume that $\operatorname{im} A$ is a closed subspace of Y. To prove (iv), it suffices to show the inclusion $(\ker A)^a \subseteq \operatorname{im} A^t$, so let $g \in (\ker A)^a$. We define a bounded linear functional v on $\operatorname{im} A$ by

$$\langle v, f \rangle = \langle J_2 g, A_/^{-1} f \rangle \quad \text{for } f \in \operatorname{im} A,$$

and then use the Hahn–Banach theorem to extend v to a bounded linear functional $\tilde{v}$ on all of Y. In this way,

$$\langle g, u \rangle = \langle J_2 g, u + \ker A \rangle = \langle J_2 g, A_/^{-1} Au \rangle = \langle v, Au \rangle = \langle \tilde{v}, Au \rangle = \langle A^t \tilde{v}, u \rangle$$

for all $u \in X$, so $g = A^t \tilde{v} \in \operatorname{im} A^t$, and thus $(\ker A)^a \subseteq \operatorname{im} A^t$, as required.

Finally, if (i)–(iv) hold, then Lemma 2.12 yields isometric isomorphisms

$$X^*/\operatorname{im} A^t = X^*/(\ker A)^a \to (\ker A)^* \quad \text{and}$$
$$\ker A^t = (\operatorname{im} A)^a \to (Y/\operatorname{im} A)^*.$$

$\square$

In many applications, it is natural to work with a space Z that is only *isomorphic* to X^* in the following way. Suppose there is a bilinear form $\langle\!\langle \cdot, \cdot \rangle\!\rangle$: $Z \times X \to \mathbb{C}$ that is bounded in the obvious sense, i.e.,

$$|\langle\!\langle g, u \rangle\!\rangle| \le C \|g\|_Z \|u\|_X \quad \text{for } g \in Z \text{ and } u \in X,$$

allowing us to define a bounded linear operator $\iota : Z \to X^*$ by

$$\langle \iota g, u \rangle = \langle\!\langle g, u \rangle\!\rangle.$$

If ι has a bounded inverse, i.e., if ι is an isomorphism of Banach spaces, then we call Z a *realisation* of the dual space X^*. In this case,

$$\|\iota g\|_{X^*} \sim \|g\|_Z \sim \sup_{0 \ne u \in X} \frac{|\langle\!\langle g, u \rangle\!\rangle|}{\|u\|_X},$$

and we routinely identify Z with X^* by suppressing the distinction between g and ιg, and between $\langle \cdot, \cdot \rangle$ and $\langle\!\langle \cdot, \cdot \rangle\!\rangle$.

Compactness

Let us recall some facts about compact subsets of metric spaces. Suppose that a set X is equipped with a metric $|\cdot, \cdot|_X$. An *open cover* of a subset $W \subseteq X$ is a family of open subsets of X whose union contains W. We say that W is *compact* if every open cover of W has a finite subcover. Every compact set is closed. If $\overline{W}$, the closure of W, is compact, then W is said to be *relatively compact*. For $\epsilon > 0$, an *ϵ-net* for W is a finite subset $\{w_1, \dots, w_n\} \subseteq W$ with the property that for each $w \in W$ there exists an index $i = i(w) \in \{1, \dots, n\}$ such that $|w, w_i|_X < \epsilon$. If W has an ϵ-net for *every* $\epsilon > 0$, then W is said to be *totally bounded*. Every totally bounded set is bounded.

Theorem 2.14 In any metric space X, the following three statements are equivalent:

(i) The subset W is relatively compact.
(ii) Every sequence in W has a subsequence that converges in X.
(iii) The subset W is totally bounded.

In particular, any relatively compact set is bounded.

For a proof, see [106, p. 13] and [94, pp. 120–125].

Suppose now that the whole metric space X is compact. The set $C(X)$ of all continuous functions $f : X \to \mathbb{C}$ is a Banach space with norm

$$\|f\|_{C(X)} = \max_{x \in X} |f(x)|.$$

A subset $F \subseteq C(X)$ is said to be *equicontinuous* if for every $\epsilon > 0$ there exists a $\delta > 0$ such that, for all $x, y \in X$ and for every $f \in F$,

$$|x, y|_X < \delta \quad \text{implies} \quad |f(x) - f(y)| < \epsilon.$$

The importance of this property stems from the following theorem of Arzela and Ascoli.

Theorem 2.15 *Let X be a compact metric space. A subset of $C(X)$ is relatively compact if and only if it is bounded and equicontinuous.*

For a proof, see [106, p. 85] or [94, p. 126]. There is an analogous characterisation of the relatively compact subsets of $L_p(\mathbb{R}^n)$; see [106, p. 275].

Theorem 2.16 *For $1 \le p < \infty$, a subset W is relatively compact in $L_p(\mathbb{R}^n)$ if and only if the following three conditions are satisfied:*

   (i) *W is bounded, i.e., $\|f\|_{L_p(\mathbb{R}^n)} \le C$ for $f \in W$.*
  (ii) *W is p-mean equicontinuous, i.e., $\|f(\cdot + h) - f\|_{L_p(\mathbb{R}^n)} \to 0$ as $h \to 0$, uniformly for $f \in W$.*
 (iii) *$\|f\|_{L_p(\mathbb{R}^n \setminus B_\rho)} \to 0$ as $\rho \to \infty$, uniformly for $f \in W$, where $B_\rho = \{x \in \mathbb{R}^n : |x| < \rho\}$.*

Suppose now that X and Y are normed spaces. A linear operator from X into Y is said to be *compact* (or *completely continuous*) if it maps every bounded subset of X to a relatively compact subset of Y. Every compact operator is bounded. Also, any linear operator with a finite-dimensional image is compact, because in a finite-dimensional normed space every bounded set is totally bounded. It follows from Theorem 2.14 that a linear map $K : X \to Y$ is compact if and only if every bounded sequence u_j in X has a subsequence $u_{j'}$ such that $Ku_{j'}$ converges in Y.

In the light of the Arzela–Ascoli theorem, if $K : C[0, 1] \to C[0, 1]$ is compact, then we expect Ku to be smoother than u, and so it is not surprising that many integral operators are compact. Similarly, we shall see later that, in the case of a partial differential operator acting between appropriate Sobolev spaces, the lower-order terms give rise to only a compact perturbation of the principal part.

Given our earlier study of duality, it is natural to ask about the compactness of the transpose.

Theorem 2.17 Consider a linear map $K : X \to Y$ and its transpose $K^t : Y^* \to X^*$.

(i) If K is compact, then K^t is compact.
(ii) If K^t is compact, and if Y is complete, then K is compact.

Proof. Assume that K is compact, and take a bounded sequence w_j in Y^*. To prove (i), it suffices to show that a subsequence of $K^t w_j$ converges in X^*. We denote the closed unit ball in X by $U = \{u \in X : \|u\|_X \le 1\}$, and let v_j denote the restriction of the functional w_j to the compact set $\overline{K(U)}$. For $f = Ku$ and $u \in U$, we have

$$|v_j(f)| = |\langle w_j, Ku \rangle| \le \|w_j\|_{Y^*} \|K\|_{\mathcal{L}(X,Y)} \|u\|_X \le C,$$

because the w_j are bounded in Y^*, and $\|u\|_X \le 1$. Thus, $|v_j(f)| \le C$ for $f \in \overline{K(U)}$, and

$$|v_j(f_1) - v_j(f_2)| = |\langle w_j, f_1 - f_2 \rangle| \le C\|f_1 - f_2\|_Y \quad \text{for } f_1, f_2 \in \overline{K(U)},$$

so the v_j are bounded and equicontinuous. Applying the Arzela–Ascoli theorem, we deduce that a subsequence $v_{j'}$ converges uniformly on $\overline{K(U)}$. Given any nonzero $u \in X$, put $\tilde{u} = \|u\|_X^{-1} u \in U$, and observe that

$$\left|\langle K^t w_{j'} - K^t w_{k'}, u \rangle\right| = \|u\|_X \left|\langle w_{j'} - w_{k'}, K\tilde{u} \rangle\right| = \|u\|_X |v_{j'}(K\tilde{u}) - v_{k'}(K\tilde{u})|,$$

so

$$\|K^t w_{j'} - K^t w_{k'}\|_{X^*} \le \max_{f \in \overline{K(U)}} |v_{j'}(f) - v_{k'}(f)|.$$

Therefore, the subsequence $K^t w_{j'}$ is Cauchy in the complete space X^*.

To prove (ii), we assume now that K^* is compact, and let $\iota_X : X \to X^{**}$ and $\iota_Y : Y \to Y^{**}$ be the natural imbeddings. For all $u \in X$ and $g \in Y^*$,

$$\langle \iota_Y Ku, g \rangle = \langle g, Ku \rangle = \langle K^* g, u \rangle = \langle \iota_X u, K^* g \rangle = \langle K^{**} \iota_X u, g \rangle,$$

which shows that $\iota_Y K = K^{**} \iota_X$. Suppose that u_j is a bounded sequence in X. The sequence $\iota_X u_j$ is bounded in X^{**}, and $K^{**} : X^{**} \to Y^{**}$ is compact by part (i), so there is a subsequence $u_{j'}$ such that $K^{**} \iota_X u_{j'}$ converges in Y^{**}. Thus, $Ku_{j'} = \iota_Y^{-1} K^{**} \iota_X u_{j'}$ is Cauchy, and hence convergent, provided Y is complete. $\qquad\square$

We now turn our attention to operators of the form $I + K : X \to X$, where I is the identity operator and K is compact. The results obtained for this special case will be used in the next section to deduce important properties of a much wider class of operators. As we saw in Chapter 1, Fredholm developed the first general theory of equations of the form $u + Ku = f$, albeit with K a concrete integral operator. His method of Fredholm determinants used only the techniques of classical analysis, but is briefly described by Riesz and Sz.-Nagy in their well-known textbook on functional analysis [88, pp. 172–176]. This book also sets out an abstract theory due to its first author [87], and which we now follow. Until the end of this section, we write for brevity $\| \cdot \| = \| \cdot \|_X$; no other norms occur.

In the next lemma, one thinks of u as being nearly orthogonal to W, even though the norm might not arise from an inner product.

Lemma 2.18 *If W is a closed subspace of a normed space X, and if $W \neq X$, then for each $\epsilon \in (0, 1)$ there exists $u \in X$ such that $\|u\| = 1$ and $\mathrm{dist}(u, W) \geq 1 - \epsilon$.*

Proof. Choose any $v \in X \setminus W$, put $d = \mathrm{dist}(v, W)$, and note that $d > 0$ because W is closed. Given $\epsilon \in (0, 1)$, choose $w_\epsilon \in W$ such that $d \leq \|v - w_\epsilon\| \leq d/(1 - \epsilon)$, and put $u = \|v - w_\epsilon\|^{-1}(v - w_\epsilon)$. Obviously $\|u\| = 1$, and for all $w \in W$,

$$\|v - w_\epsilon\|(u - w) = v - w_\epsilon - \|v - w_\epsilon\|w = v - (\text{an element of } W),$$

so $\|v - w_\epsilon\|\|u - w\| \geq \mathrm{dist}(v, W) = d$, and hence $\|u - w\| \geq d/\|v - w_\epsilon\| \geq 1 - \epsilon$. $\qquad\square$

For emphasis, the symbol $\subsetneqq$ is used below to denote *strict* inclusion, i.e., for any sets V and W, we write $V \subsetneqq W$ if and only if $V \subseteq W$ and $V \neq W$.

Theorem 2.19 *Let X be a normed space, assume that $K : X \to X$ is compact, and define $A : X \to X$ by*

$$A = I + K.$$

(i) *For each $n \geq 0$, the subspace $V_n = \ker A^n$ is finite-dimensional.*
(ii) *For each $n \geq 0$, the subspace $W_n = \mathrm{im}\, A^n$ is closed.*
(iii) *There is a finite number r such that*

$$\{0\} = V_0 \subsetneqq V_1 \subsetneqq \cdots \subsetneqq V_r = V_{r+1} = \cdots$$

and

$$X = W_0 \supsetneq W_1 \supsetneq \cdots \supsetneq W_r = W_{r+1} = \cdots.$$

(iv) $X = V_r \oplus W_r$.

Proof. Suppose for a contradiction that V_1 is not finite-dimensional. Using Lemma 2.18, we can recursively construct a sequence u_j in V_1 such that

$$\|u_j\| = 1 \quad \text{and} \quad \|u_j - u_k\| \geq \tfrac{1}{2} \quad \text{for } j \neq k.$$

Since K is compact, there is a subsequence $u_{j'}$ and an element $\phi \in X$ such that $K u_{j'} \to \phi$. But $u_j + K u_j = A u_j = 0$ so $u_{j'} = -K u_{j'} \to -\phi$, which is impossible because $u_{j'}$ is not a Cauchy sequence. Hence, V_1 must be finite-dimensional. Part (i) follows at once, because

$$A^n = I + L \quad \text{where } L = \sum_{m=1}^{n} \binom{n}{m} K^m, \tag{2.3}$$

and L is compact by Exercise 2.6. (The case $n = 0$ is trivial because $A^0 = I$.)

To prove that W_1 is closed, suppose that $f_j = A u_j \to f$. Let $d_j = \operatorname{dist}(u_j, \ker A)$, and choose $v_j \in \ker A$ such that

$$d_j \leq \|u_j - v_j\| \leq (1 + j^{-1}) d_j. \tag{2.4}$$

If the d_j are bounded, then so is the sequence $w_j = u_j - v_j$, and there is a subsequence $w_{j'}$ such that $K w_{j'} \to \phi$. Since $A w_j = A u_j = f_j$, it follows that $w_{j'} = f_{j'} - K w_{j'} \to f - \phi$, and thus $f = \lim A w_{j'} = A(f - \phi) \in W_1$. If the d_j are not bounded, then by passing to a subsequence we can assume that $d_j \to \infty$ and $d_j > 0$. Define $w_j = \|u_j - v_j\|^{-1}(u_j - v_j)$ so that $\|w_j\| = 1$ and $K w_{j'} \to \phi$. Since $\|A w_j\| = \|u_j - v_j\|^{-1}\|A(u_j - v_j)\| \leq d_j^{-1}\|f_j\|$, we have $A w_j \to 0$ and $w_{j'} = A w_{j'} - K w_{j'} \to -\phi$. Moreover, $A\phi = -\lim A w_{j'} = 0$, so $\phi \in \ker A$. Consider $\psi_j = w_j + \phi$. We have

$$\|u_j - v_j\|\psi_j = u_j - v_j + \|u_j - v_j\|\phi = u_j - \text{(an element of } \ker A),$$

so the definition of d_j gives $\|u_j - v_j\|\|\psi_j\| \geq d_j$, and hence $\|\psi_j\| \geq d_j / \|u_j - v_j\| \geq 1/(1 + j^{-1})$ by (2.4). But $\psi_{j'} \to 0$, and we conclude from this contradiction that the d_j must be bounded. Hence, $f \in W_1$, so W_1 is closed, and part (ii) follows by (2.3).

One easily verifies that $V_n \subseteq V_{n+1}$, and that if $V_r = V_{r+1}$, then $V_{n+1} = V_n$ for all $n \geq r$. Suppose for a contradiction that no such r exists, i.e., assume

$V_n \subsetneq V_{n+1}$ for all n. By Lemma 2.18, we can choose $u_n \in V_{n+1}$ such that $\|u_n\| = 1$ and dist$(u_n, V_n) > \frac{1}{2}$. If $n > m$, then

$$K u_m - K u_n = u_n - (A u_n + u_m - A u_m) = u_n - \text{(an element of } V_n),$$

so $\|K u_m - K u_n\| \geq \text{dist}(u_n, V_n) > \frac{1}{2}$ and hence no subsequence of $K u_n$ converges. This contradiction implies that $V_n = V_{n+1}$ for some n, and we define $r = \min\{n : V_n = V_{n+1}\}$.

Next, one easily verifies that $W_n \supseteq W_{n+1}$, and that if $W_{r'} = W_{r'+1}$ for some r', then $W_n = W_{n+1}$ for all $n \geq r'$. Suppose for a contradiction that no such r' exists, i.e., assume $W_n \supsetneq W_{n+1}$ for all n. By Lemma 2.18, we can choose $u_n \in W_n$ such that $\|u_n\| = 1$ and dist$(u_n, W_{n+1}) > \frac{1}{2}$. If $n > m$, then

$$K u_n - K u_m = u_m - (A u_m + u_n - A u_n) = u_m - \text{(an element of } W_{m+1}),$$

so $\|K u_n - K u_m\| \geq \text{dist}(u_m, W_{m+1}) > \frac{1}{2}$, giving the desired contradiction. We may therefore define $r' = \min\{n : W_n = W_{n+1}\}$.

At this point, a simple lemma is needed: for each $k \geq 0$ and $f \in W_{r'}$, the equation $A^k u = f$ has a unique solution u in $W_{r'}$. Indeed, u exists because $W_{r'+k} = W_{r'}$; to prove uniqueness, suppose for a contradiction that the homogeneous equation $A^k u = 0$ has a non-trivial solution $u = u_1 \in W_{r'}$. For $n \geq 1$, we choose recursively $u_n \in W_{r'}$ such that $A u_n = u_{n-1}$. Since $A^n u_n = A u_1 = 0$ but $A^{n-1} u_n = u_1 \neq 0$, we see that $u_n \in V_n$ but $u_n \notin V_{n-1}$, a contradiction if $n > r$.

To complete the proof of part (iii), we now show that $r = r'$. If $f \in V_{r'+1}$, then the homogeneous equation $A u = 0$ has a solution $u = A^{r'} f \in W_{r'}$, and by the argument above, $u = 0$, so $f \in V_{r'}$. Therefore, $V_{r'+1} \subseteq V_{r'}$, and hence $r' \geq r$. In particular, if $r' = 0$ then $r = r' = 0$. Suppose now that $r' \geq 1$, and choose $f = A^{r'-1} v \in W_{r'-1}$ such that $f \notin W_{r'}$. The equation $A^{r'} u = A^{r'} v$ has a solution u in $W_{r'}$, and we have $A^{r'}(v - u) = 0$, but $A^{r'-1}(v - u) = f - A^{r'-1} u \neq 0$ because $A^{r'-1} u \in W_{2r'-1} = W_{r'}$. Thus, $v - u \in V_{r'}$ but $v - u \notin V_{r'-1}$, showing that $V_{r'} \neq V_{r'-1}$ and hence $r' \leq r$.

Finally, we turn to part (iv). The homogeneous equation $A^r u = 0$ has only the trivial solution in W_r, so $V_r \cap W_r = \{0\}$. Given $f \in X$, let $u \in W_r$ be the solution of the equation $A^{2r} u = A^r f$, and put $v = A^r u \in W_r$. Since $A^r(f - v) = A^r f - A^{2r} u = 0$, we have $f = (f - v) + v \in V_r + W_r$. □

Fredholm Operators

Throughout this section, we shall assume that X and Y are Banach spaces. A bounded linear operator $A : X \rightarrow Y$ is said to be *Fredholm* if

1. the subspace im A is closed in Y;
2. the subspaces ker A and $Y/\text{im } A$ are finite-dimensional.

In this case, the *index* of A is the integer defined by

$$\text{index } A = \dim \ker A - \dim(Y/\text{im } A).$$

As a consequence of Theorem 2.13, A is Fredholm if and only if A^t is Fredholm, in which case

$$\text{index } A = \dim \ker A - \dim \ker A^t = -\text{index } A^t.$$

F. Noether [80] introduced the term "index" in the above sense for a concrete class of singular integral operators.

In finite dimensions, the index depends only on the spaces, and not on the operator.

Theorem 2.20 *If $A : \mathbb{C}^n \to \mathbb{C}^m$ is a linear map, then A is Fredholm, and* index $A = n - m$.

Proof. By performing elementary row and column operations, we can find bases for $\mathbb{C}^n$ and $\mathbb{C}^m$ relative to which the action of A is given by an $m \times n$ diagonal matrix $[a_{jk}]$ with entries

$$a_{jk} = \begin{cases} 1 & \text{if } 1 \le j = k \le r, \\ 0 & \text{otherwise,} \end{cases}$$

where $r = \dim \text{im } A = \dim \text{im } A^t$ is the rank of A. The number of zero columns in the matrix $[a_{jk}]$ is $\dim \ker A = n - r$, so

$$\text{index } A = \dim \ker A - \dim(\mathbb{C}^m/\text{im } A) = (n - r) - (m - r) = n - m.$$

$\square$

The preceding result might lead one to expect the following.

Theorem 2.21 *If $A : X \to Y$ and $B : Y \to Z$ are Fredholm, then so is $BA : X \to Z$, and*

$$\text{index}(BA) = \text{index } A + \text{index } B.$$

Proof. The operator BA factors naturally into a product of Fredholm operators, each of which is either one–one or onto, as follows:

$$u \mapsto u + \ker A \mapsto Au \mapsto Au + \ker B \mapsto BAu,$$
$$X \to X/\ker A \to Y \to Y/\ker B \to Z.$$

Therefore, it suffices to consider two special cases:

(i) $\operatorname{im} A = Y$ and $\ker B = \{0\}$ (onto followed by one–one).
(ii) $\ker A = \{0\}$ and $\operatorname{im} B = Z$ (one–one followed by onto).

In case (i), $\ker(BA) = \ker A$ and $\operatorname{im}(BA) = \operatorname{im} B$, so the result is obvious. Thus, assume case (ii).

Since $Y/\operatorname{im} A$ and $\ker B$ are finite-dimensional, there is a finite-dimensional subspace $Y_A \subseteq Y$ and a closed subspace $Y_B \subseteq Y$ such that

$$Y = Y_A \oplus \operatorname{im} A = Y_B \oplus \ker B.$$

We define

$$Y_1 = \operatorname{im} A \cap Y_B, \quad Y_2 = \operatorname{im} A \cap \ker B, \quad Y_3 = Y_A \cap Y_B, \quad Y_4 = Y_A \cap \ker B,$$

so that Y_1 is closed; Y_2, Y_3 and Y_4 are finite-dimensional; and

$$\operatorname{im} A = Y_1 \oplus Y_2, \quad \ker B = Y_2 \oplus Y_4, \quad Y = Y_1 \oplus Y_2 \oplus Y_3 \oplus Y_4.$$

Next, we define

$$X_1 = A^{-1}(Y_1), \quad X_2 = A^{-1}(Y_2), \quad Z_1 = B(Y_1), \quad Z_2 = B(Y_3).$$

By restriction, the operators A and B define Banach space isomorphisms

$$X_1 \simeq Y_1 \simeq Z_1, \quad X_2 \simeq Y_2, \quad Y_3 \simeq Z_2,$$

so X_1 and Z_1 are closed; X_2 and Z_2 are finite-dimensional; and

$$X = X_1 \oplus X_2, \quad \ker(BA) = X_2, \quad Z = Z_1 \oplus Z_2, \quad \operatorname{im}(BA) = Z_1.$$

We see at once that BA is Fredholm, with

$$\begin{aligned}
\operatorname{index}(BA) &= \dim \ker(BA) - \dim[Z/\operatorname{im}(BA)] = \dim X_2 - \dim Z_2 \\
&= \dim Y_2 - \dim Y_3 = \dim(Y_2 \oplus Y_4) - \dim(Y_3 \oplus Y_4) \\
&= \dim \ker B - \dim(Y/\operatorname{im} A),
\end{aligned}$$

and, remembering that $\dim(Z/\operatorname{im} B) = 0 = \dim \ker A$, the formula for the index of BA follows. □

The next result encapsulates, in the present abstract setting, the main conclusions of Fredholm's original theory of second-kind integral equations.

Theorem 2.22 *If $A = I + K$, where $K : X \to X$ is compact, then $A : X \to X$ is Fredholm and* index $A = 0$.

Proof. Theorem 2.19 shows that ker A is finite-dimensional and that im A is closed. Moreover, since $K^t : X^* \to X^*$ is compact by Theorem 2.17, we may apply Theorem 2.19 to $A^t = I + K^t$, and deduce that ker A^t (and hence also $X/\operatorname{im} A$) is finite-dimensional. Thus, A is Fredholm. To compute the index, let r be as in Theorem 2.19. If $r = 0$, then ker $A = \{0\}$ and im $A = X$, so index $A = 0$. If $r \geq 1$, then since X is the direct sum of ker A^r and im A^r, we have ker $A^r \simeq X/\operatorname{im} A^r$ and hence index $A^r = 0$. Theorem 2.21 gives index $A^r = r$ index A, so again index $A = 0$. □

Suppose $A \in \mathcal{L}(X, Y)$ and $B \in \mathcal{L}(Y, X)$. If $BA = I + K_X$ where $K_X : X \to X$ is compact, then B is called a *left regulariser* for A. Likewise, if $AB = I + K_Y$ where $K_Y : Y \to Y$ is compact, then B is called a *right regulariser* for A. If B is both a left and a right regulariser for A, then we say that B is a *two-sided regulariser* for A. Notice that by Theorem 2.17, B is a left (respectively, right) regulariser for A if and only if B^t is a right (respectively, left) regulariser for A^t. The next lemma leads to a characterization of Fredholm operators.

Lemma 2.23 *Let $A \in \mathcal{L}(X, Y)$.*

(i) *If A has a left regulariser, then* ker A *is finite-dimensional, and* im A *is closed.*

(ii) *If A has a right regulariser, then* im A *is closed, and* $Y/\operatorname{im} A$ *is finite-dimensional.*

Proof. Suppose that B is a left regulariser for A. Since ker $A \subseteq \ker(BA)$, and since $BA = I + K_X$ has a finite-dimensional kernel by Theorem 2.19, we see that ker A is finite-dimensional. Now suppose for a contradiction that im A is not closed. By Corollary 2.2, there is a sequence $u_j \in X$ such that

$$\|u_j + \ker A\|_{X/\ker A} = 1 \quad \text{and} \quad \|Au_j\|_Y \to 0.$$

By Exercise 2.4, we can assume that the u_j are bounded, and hence obtain a subsequence $u_{j'}$ such that $K_X u_{j'}$ converges. Since $Au_{j'} \to 0$, it follows that

$BAu_{j'} \to 0$ and so $u_{j'} = BAu_{j'} - K_X u_{j'}$ converges; say $u_{j'} \to u$. On the one hand, $Au = \lim Au_{j'} = 0$, so $u \in \ker A$, but on the other hand $u_{j'} + \ker A \to u + \ker A$, so $\|u + \ker A\|_{X/\ker A} = 1$, a contradiction. Thus, (i) holds, and then (ii) follows by duality using Theorem 2.13. □

Theorem 2.24 For $A \in \mathcal{L}(X, Y)$, the following are equivalent:

 (i) A is Fredholm.
 (ii) A has a left regulariser and a right regulariser.
(iii) A has a two-sided regulariser.

Proof. Since (iii) trivially implies (ii), and since Lemma 2.23 shows that (ii) implies (i), it suffices to prove that (i) implies (iii). Thus, assume A is Fredholm. Since $\ker A$ and $Y/\operatorname{im} A$ are finite-dimensional, there is a closed subspace $X_A \subseteq X$ and a finite-dimensional subspace $Y_A \subseteq Y$ such that $X = X_A \oplus \ker A$ and $Y = Y_A \oplus \operatorname{im} A$. Let P denote the projection of X onto $\ker A$ parallel to X_A, and let Q denote the projection of Y onto Y_A parallel to $\operatorname{im} A$. The operators P and Q are compact because their images are finite-dimensional. We define a Banach space isomorphism $A_1 : X_A \to \operatorname{im} A$ by $A_1 u = Au$ for $u \in X_A$, and then define $B = A_1^{-1}(I - Q) \in \mathcal{L}(Y, X)$. Since

$$BAu = A_1^{-1}(I-Q)Au = A_1^{-1}Au = A_1^{-1}A_1(I-P)u = (I-P)u \quad \text{for } u \in X,$$

and

$$ABf = AA_1^{-1}(I - Q)f = A_1 A_1^{-1}(I - Q)f = (I - Q)f \quad \text{for } f \in Y,$$

the operator B is a two-sided regulariser for A. □

Corollary 2.25 If B is a two-sided regulariser for A, then $\operatorname{index} B = -\operatorname{index} A$.

Proof. Since A is a two-sided regulariser for B, it follows that B is Fredholm, and then Theorems 2.21 and 2.22 imply that $\operatorname{index} B + \operatorname{index} A = 0$. □

As an important application of regularisation, we now prove that the index is stable under compact perturbations, and thereby generalise Theorem 2.22.

Theorem 2.26 If $A : X \to Y$ is Fredholm, and if $K : X \to Y$ is compact, then their sum $A + K : X \to Y$ is Fredholm, and $\operatorname{index}(A + K) = \operatorname{index} A$.

Proof. By Theorem 2.24, A has a two-sided regulariser $B \in \mathcal{L}(Y, X)$, i.e., $BA = I + L_X$ and $AB = I + L_Y$ with $L_X : X \to X$ and $L_Y : Y \to Y$ compact.

Since

$$B(A + K) = I + (L_X + BK) \quad \text{and} \quad (A + K)B = I + (L_Y + KB),$$

and since $L_X + BK : X \to X$ and $L_Y + KB : Y \to Y$ are compact by Exercise 2.6, we see that B is also a two-sided regulariser for $A + K$. Hence, by Theorem 2.24 and Corollary 2.25, the operator $A + K$ is Fredholm and has the same index as A. □

Suppose that X is equipped with a *conjugation*, i.e., with a continuous, unary operation $u \mapsto \bar{u}$ satisfying

$$\overline{u + v} = \bar{u} + \bar{v}, \quad \overline{\lambda u} = \bar{\lambda}\bar{u} \quad \text{and} \quad \overline{(\bar{u})} = u,$$

for all $u, v \in X$ and $\lambda \in \mathbb{C}$. The third condition implies that the map $u \mapsto \bar{u}$ has a continuous inverse, and we have

$$\|\bar{u}\|_X \sim \|u\|_X.$$

In practice, when X is a function space, $\bar{u}$ is just the usual pointwise complex conjugate, i.e., $\bar{u}(x) = \overline{u(x)}$ for all x. In general, a conjugation on X induces a conjugation on X^*, defined by $\langle \bar{g}, u \rangle = \overline{\langle g, \bar{u} \rangle}$, and this expression defines a bounded *sesqui*linear form on $X^* \times X$, which we denote by

$$(g, u) = \langle \bar{g}, u \rangle.$$

If X is reflexive, and if we identify X with X^{**} in the usual way, then $(\cdot, \cdot)$ is naturally defined on $X \times X^*$, and satisfies

$$(u, g) = \overline{(g, u)}.$$

Now consider a linear map $A : X \to Y$. If X and Y are each equipped with a conjugation, then we define the *adjoint* $A^* : Y^* \to X^*$ by

$$(A^* v, u) = (v, Au) \quad \text{for } v \in Y^* \text{ and } u \in X.$$

Throughout this and the preceding two sections, we could have worked with $(\cdot, \cdot)$ and A^* in place of $\langle \cdot, \cdot \rangle$ and A^t. For instance, we can state the celebrated *Fredholm alternative* as follows.

Theorem 2.27 Assume that $A : X \to Y$ is Fredholm with index $A = 0$. There are two, mutually exclusive possibilities:

(i) *The homogeneous equation $Au = 0$ has only the trivial solution $u = 0$. In this case,*

    (a) *for each $f \in Y$, the inhomogeneous equation $Au = f$ has a unique solution $u \in X$;*

    (b) *for each $g \in X^*$, the adjoint equation $A^*v = g$ has a unique solution $v \in Y^*$.*

(ii) *The homogeneous equation $Au = 0$ has exactly p linearly independent solutions $u_1, \ldots, u_p$ for some finite $p \geq 1$. In this case,*

    (a) *the homogeneous adjoint equation $A^*v = 0$ has exactly p linearly independent solutions $v_1, \ldots, v_p$;*

    (b) *the inhomogeneous equation $Au = f$ is solvable if and only if the right-hand side f satisfies $(v_j, f) = 0$ for $j = 1, \ldots, p$;*

    (c) *the inhomogeneous adjoint equation $A^*v = g$ is solvable if and only if the right-hand side g satisfies $(g, u_j) = 0$ for $j = 1, \ldots, p$.*

Proof. The result follows at once from Theorem 2.13 and the definition of the index. □

Hilbert Spaces

Let H be a vector space equipped with an inner product $(\cdot, \cdot)_H$, and denote the induced norm by

$$\|u\|_H = \sqrt{(u, u)_H} \quad \text{for } u \in H;$$

see Exercise 2.8. Recall that H is said to be a *Hilbert space* if it is complete with respect to $\| \cdot \|_H$. Remember also our convention that inner products are conjugate-linear in the *first* argument.

For a general normed space X, given $u \in X$ and $W \subseteq X$, we call $w \in W$ a *best approximation* to u from W if $\text{dist}(u, W) = \|u - w\|_X$. Also, recall that the set W is said to be *convex* if $\lambda u + \mu v \in W$ whenever $u, v \in W$ and $\lambda, \mu \in [0, 1]$ with $\lambda + \mu = 1$. The following property of Hilbert spaces turns out to have important consequences.

Lemma 2.28 Suppose that W is a non-empty, closed, convex subset of a Hilbert space H. For each $u \in H$, there exists a unique best approximation to u from W.

Proof. Let $d = \text{dist}(u, W)$, and choose a sequence $v_n \in W$ such that $\|u - v_n\|_H \to d$. By the parallelogram law (Exercise 2.8),

$$\|(v_m - u) + (u - v_n)\|_H^2 + \|(v_m - u) - (u - v_n)\|_H^2$$
$$= 2\|v_m - u\|_H^2 + 2\|u - v_n\|_H^2,$$

and this equation simplifies to give

$$\|v_m - v_n\|_H^2 = 2\|v_m - u\|_H^2 + 2\|u - v_n\|_H^2 - 4\left\|\tfrac{1}{2}(v_m + v_n) - u\right\|_H^2.$$

Since W is convex, we have $\tfrac{1}{2}(v_m + v_n) \in W$, so

$$\left\|\tfrac{1}{2}(v_m + v_n) - u\right\|_H \geq \mathrm{dist}(u, W) = d,$$

and hence

$$\|v_m - v_n\|_H^2 \leq 2\bigl(\|v_m - u\|_H^2 - d^2\bigr) + 2\bigl(\|v_n - u\|_H^2 - d^2\bigr).$$

Therefore, v_n is a Cauchy sequence, and since W is closed, we deduce that $v_n \to w$ for some $w \in W$, with $\|u - w\|_H = \lim_{n\to\infty} \|u - v_n\|_H = d$, as required.

To show uniqueness, suppose that w_1, $w_2 \in W$ satisfy

$$\|u - w_1\|_H = d = \|u - w_2\|_H.$$

Using the parallelogram law as above, with v_m and v_n replaced by w_1 and w_2, we find that

$$\|w_1 - w_2\|_H^2 \leq 2\bigl(\|w_1 - u\|_H^2 - d^2\bigr) + 2\bigl(\|w_2 - u\|_H^2 - d^2\bigr) = 0,$$

so $w_1 = w_2$. □

We now focus on the special case when W is a subspace of H. Write $u \perp v$ if u and v are orthogonal, i.e., if $(u, v)_H = 0$. More generally, for a subset $W \subseteq H$, write $u \perp W$ if $(u, v)_H = 0$ for all $v \in W$.

Lemma 2.29 *Let W be a subspace of an inner product space H, and let $u \in H$. A vector $w \in W$ is a best approximation to u from W if and only if $u - w \perp W$.*

Proof. Suppose that $\mathrm{dist}(u, W) = \|u - w\|_H$, and let $v \in W$. In showing that $(u - w, v)_H = 0$, we can assume $\|v\|_H = 1$. Since W is a subspace, $w + v \in W$, and so, with $d = \mathrm{dist}(u, W)$,

$$d^2 \leq \|u - (w + v)\|_H^2 = \|u - w\|_H^2 - 2\,\mathrm{Re}(u - w, v)_H + 1$$
$$= d^2 - [\mathrm{Re}(u - w, v)_H]^2 + [1 - \mathrm{Re}(u - w, v)_H]^2$$
$$\leq d^2 - [\mathrm{Re}(u - w, v)_H]^2,$$

giving $\text{Re}(u - w, v)_H = 0$. Replacing v with iv, we have

$$0 = \text{Re}(u - w, iv)_H = \text{Re}[i(u - w, v)_H] = -\text{Im}(u - w, v)_H,$$

showing that $(u - w, v)_H = 0$, as required.

To prove the converse, suppose that $u - w \perp W$. For every $v \in W$, we have $w - v \in W$ so $(u - w, w - v)_H = 0$, giving

$$\begin{aligned}
\|u - v\|_H^2 &= \|(u - w) + (w - v)\|_H^2 \\
&= \|u - w\|_H^2 + 2\,\text{Re}(u - w, w - v)_H + \|w - v\|_H^2 \\
&= \|u - w\|_H^2 + \|w - v\|_H^2 \geq \|u - w\|_H^2.
\end{aligned}$$

Hence, $d^2 \geq \|u - w\|_H^2$, implying that $\text{dist}(u, W) = \|u - w\|_H$.   □

If W is a *closed* subspace of a Hilbert space H, then we can define $P : H \to H$ by $Pu = w$, where $w \in W$ is the best approximation to u from W. With the help of Lemma 2.29, we see that P is linear, and that

$$P^2 = P, \quad \|P\|_{\mathcal{L}(H,H)} = 1, \quad \text{im } P = W, \quad \ker P = W^\perp,$$

where the closed subspace $W^\perp = \{u \in H : u \perp W\}$ is called the *orthogonal complement* of W in H. Thus,

$$H = W \oplus W^\perp$$

The operator P is called the *orthogonal projection* of H onto W.

Given $u \in H$, the inner product determines a bounded linear functional $\iota_1 u \in H^*$, given by

$$\langle \iota_1 u, v \rangle = (u, v)_H \quad \text{for } v \in H.$$

In fact, from the Cauchy–Schwarz inequality (Exercise 2.8),

$$|(u, v)_H| \leq \|u\|_H \|v\|_H, \tag{2.5}$$

we see that $\|\iota_1 u\|_H = \|u\|_H$, so $\iota_1 : H \to H^*$ is a conjugate-linear isometry. The next result, known as the Riesz representation theorem, shows that ι_1 is onto, and hence invertible.

Theorem 2.30 *Let H be a Hilbert space. For each $f \in H^*$ there exists a unique $u \in H$ such that*

$$\langle f, v \rangle = (u, v)_H \quad \text{for all } v \in H.$$

Furthermore, $\|f\|_{H^*} = \|u\|_H$.

Proof. Consider the closed subspace $W = \ker f$. If $W = H$, then $f = 0$ and so $u = 0$. Thus, we suppose that $W \neq H$, and choose $u_0 \in W^\perp$ with $u_0 \neq 0$. Since $u_0 \notin W$, we have $\langle f, u_0 \rangle \neq 0$. Put $u_1 = \langle f, u_0 \rangle^{-1} u_0$ so that $\langle f, u_1 \rangle = 1$, and observe that for every $v \in H$,

$$\langle f, v - \langle f, v \rangle u_1 \rangle = \langle f, v \rangle - \langle f, v \rangle \langle f, u_1 \rangle = 0.$$

Therefore, $u_1 \in W^\perp$ and $v - \langle f, v \rangle u_1 \in W$, implying

$$0 = \left(u_1, v - \langle f, v \rangle u_1 \right)_H = (u_1, v) - \langle f, v \rangle \| u_1 \|_H^2,$$

so $u = \| u_1 \|_H^{-2} u_1$ has the required property. Uniqueness is immediate from Exercise 2.9. $\qquad \square$

The Riesz representation theorem shows that we can make H^* into a Hilbert space by defining

$$(f, g)_{H^*} = \left(\iota_1^{-1} g, \iota_1^{-1} f \right)_H,$$

and then obtain a conjugate-linear isometry $\iota_2 : H^* \to H^{**}$ given by

$$\langle \iota_2 f, g \rangle = (f, g)_{H^*}.$$

The composite map $\iota_2 \iota_1 : H \to H^{**}$ is *linear* and invertible, with

$$\langle \iota_2 \iota_1 u, f \rangle = (\iota_1 u, f)_{H^*} = \left(\iota_1^{-1} f, u \right)_H = \langle f, u \rangle.$$

Thus, $\iota_2 \iota_1$ coincides with the natural imbedding of H in H^{**}, and we see that every Hilbert space is reflexive. When H is equipped with a conjugation, the norm in the dual space can be written as

$$\| f \|_{H^*} = \sup_{0 \neq u \in H} \frac{|(f, u)|}{\| u \|_H} \qquad \text{for } f \in H^* \text{ and } u \in H.$$

It follows that we can then define a *linear* isometry $\iota : H \to H^*$ by

$$\langle \iota u, v \rangle = (\bar{u}, v)_H \quad \text{for } u, v \in H,$$

or equivalently, $(\iota u, v) = (u, v)_H$. Using ι, we can identify H with H^*, observing that in this way the inner product $(\cdot, \cdot)_H$ coincides with the sesquilinear form $(\cdot, \cdot)$.

We conclude this section with one other important fact about Hilbert spaces. (See [106, p. 126] or [94, p. 233] for generalisations to Banach spaces.) Given any sequence u_j in H, if

$$(u_j, v)_H \to (u, v)_H \quad \text{for each } v \in H,$$

then we say that the u_j *converge weakly* to u in H, and write $u_j \rightharpoonup u$ in H. Exercise 2.9 shows a weak limit is unique, when it exists.

Theorem 2.31 Let H be a Hilbert space. Every bounded sequence in H has a weakly convergent subsequence.

Proof. Let u_j be a bounded sequence in H. To begin with, we assume that H is separable, i.e., we assume that there exists a countable dense subset $S = \{v_1, v_2, \ldots\}$. The Cauchy–Schwarz inequality shows that the sequence of scalars $(u_j, v_1)_H$ is bounded, so there is a subsequence u_j^1 of u_j such that $(u_j^1, v_1)_H$ converges. Likewise, there is a subsequence u_j^2 of u_j^1 such that $(u_j^2, v_2)_H$ converges. Proceeding in this fashion, we obtain successive subsequences $u_j^3, u_j^4, \ldots$ such that $(u_j^k, v_l)_H$ has a finite limit as $j \to \infty$, for each fixed k and l satisfying $1 \le l \le k$. Hence, the diagonal subsequence u_j^j has the property that $\lim_{j\to\infty}(u_j^j, v)_H$ exists for each $v \in S$. Since S is dense in H, and since the u_j^j are bounded, Exercise 2.11 shows that $u_j^j \rightharpoonup u$.

Now drop the assumption that H is separable. We let H_0 denote the closure of $\mathrm{span}\{u_j\}$, and let P denote the orthogonal projection of H onto H_0. Since H_0 is separable, there exists $u \in H_0$ such that $(u_j, v)_H \to (u, v)_H$ for each $v \in H_0$. Hence, if $v \in H$, then $(u_j, v)_H = (u_j, Pv)_H \to (u, Pv)_H = (u, v)_H$, or in other words $u_j \rightharpoonup u$ in H. $\qquad\square$

Coercivity

Consider a Hilbert space V and a bounded, sesquilinear form $\Phi : V \times V \to \mathbb{C}$. Thus, $\Phi(u, v)$ is conjugate-linear in u, is linear in v, and satisfies

$$|\Phi(u, v)| \le C \|u\|_V \|v\|_V \quad \text{for } u, v \in V.$$

We can therefore define a bounded, *linear* operator $A : V \to V^*$ by

$$(Au, v) = \Phi(u, v) \quad \text{for } u, v \in V.$$

Conversely, each bounded linear operator from V into V^* determines a unique bounded sesquilinear form on V. In this context, we *do not* identify V with its dual V^*, but we *do* identify V with its second dual V^{**}, so the adjoint

$A^* : V \rightarrow V^*$ is given by

$$(A^*u, v) = (u, Av) = \overline{(Av, u)} \quad \text{for } u \in V \text{ and } v \in V.$$

We also define $\Phi^* : V \times V \rightarrow \mathbb{C}$, the adjoint of Φ, by

$$\Phi^*(u, v) = \overline{\Phi(v, u)},$$

so that Φ^* corresponds to A^*, i.e., $(A^*u, v) = \Phi^*(u, v)$.

Given $f \in V^*$, one can seek a solution $u \in V$ to the equation $Au = f$, or equivalently, to the variational problem

$$\Phi(u, v) = (f, v) \quad \text{for all } v \in V. \tag{2.6}$$

We say that Φ and A are *positive and bounded below* on V if

$$\text{Re } \Phi(u, u) \geq c\|u\|_V^2 \quad \text{for } u \in V.$$

Notice that $\text{Re } \Phi^*(u, u) = \text{Re } \Phi(u, u)$, so Φ and A are positive and bounded below on V if and only if the same is true of Φ^* and A^*. In this case, we can apply the celebrated lemma of Lax and Milgram [54].

Lemma 2.32 *If the bounded linear operator $A : V \rightarrow V^*$ is positive and bounded below, then it has a bounded inverse $A^{-1} : V^* \rightarrow V$.*

Proof. Since $c\|u\|_V^2 \leq \text{Re } \Phi(u, u) \leq |\Phi(u, u)| = |(Au, u)| \leq C\|Au\|_{V^*}\|u\|_V$, we have an estimate of the form

$$\|u\|_V \leq C\|Au\|_{V^*} \quad \text{for } u \in V.$$

Hence, $\ker A = \{0\}$, and Corollary 2.2 shows that $\text{im } A$ is closed. Likewise, $\ker A^* = \{0\}$, so in fact $\text{im } A = V^*$ by Theorem 2.13. $\square$

Combining Theorem 2.26 and Lemma 2.32, we immediately obtain the main result for this section.

Theorem 2.33 *If $A = A_0 + K$, where $A_0 : V \rightarrow V^*$ is positive and bounded below, and $K : V \rightarrow V^*$ is compact, then $A : V \rightarrow V^*$ is Fredholm with zero index, and hence the Fredholm alternative holds for the equation $Au = f$.*

If $A^* = A$, or equivalently, if $\Phi^* = \Phi$, then we say that A is *self-adjoint*, and that Φ is *Hermitian*. In this case, (Au, u) is real for all $u \in V$, and if

$$(Au, u) > 0 \quad \text{for all } u \in V \setminus \{0\},$$

then we say that A is *strictly positive-definite*. Such an operator A determines a new inner product and norm on V,

$$(u, v)_A = \Phi(u, v) \quad \text{and} \quad \|u\|_A = \sqrt{(u, u)_A}.$$

In many applications, $(u, u)_A$ can be interpreted as a measure of the energy of the system whose state is represented by u, so $(\cdot, \cdot)_A$ is often called the *energy inner product* associated with the operator A. The boundedness of A immediately implies that the energy norm is weaker than the norm in V, i.e., $\|u\|_A \leq C\|u\|_V$, and it is easy to see that the two norms are equivalent if and only if A is positive and bounded below on V. Thus, one can view the Riesz representation theorem (Theorem 2.30) as a special case of Lemma 2.32.

Now consider *two* Hilbert spaces, V and H, such that V is a dense subspace of H with

$$\|u\|_H \leq C\|u\|_V \quad \text{for all } u \in V.$$

Assume that H is equipped with a conjugation inducing, by restriction, a conjugation on V. In particular, $\|\bar{u}\|_V \sim \|u\|_V$ as well as $\|\bar{u}\|_H \sim \|u\|_H$. We identify H with its dual H^*, but do not identify V with V^*. It will help to keep in mind the standard example where $H = L_2(\Omega)$, and V is a closed subspace of the Sobolev space $H^1(\Omega)$.

With the assumptions above, the inclusion map $V \to H$ is bounded and one–one with dense image. Consequently, the same is true for the transposed map $H = H^* \to V^*$, as one sees from Lemma 2.10 and Exercise 2.5. Using the transposed map to identify H with a dense subspace of V^*, we write

$$V \subseteq H \subseteq V^*,$$

and say that H acts as a *pivot space* for V. Note that the original meaning of (u, v), as the inner product of the vectors u and v in H, is consistent with its second meaning, as $\langle \bar{u}, v \rangle$ for a functional $u \in V^*$ and a vector $v \in V$, i.e., the two interpretations agree if $u \in V$.

We say that Φ and A are *coercive* on V (with respect to the pivot space H) if

$$\text{Re } \Phi(u, u) \geq c\|u\|_V^2 - C\|u\|_H^2 \quad \text{for } u \in V. \tag{2.7}$$

The next result then follows at once, as a special case of Theorem 2.33.

Theorem 2.34 Let H act as a pivot space for V. If Φ is coercive on V, and if the inclusion map $V \to H$ is compact, then $A : V \to V^*$ is a Fredholm operator with zero index.

Elementary Spectral Theory

Suppose that X is a dense subspace of a normed space Y, and that $A : X \to Y$ is a (possibly unbounded) linear operator. The *resolvent set* of A consists of those $\lambda \in \mathbb{C}$ for which the operator $\lambda I - A$ has a bounded inverse on Y, or, more precisely, for which

1. $\lambda I - A : X \to \mathrm{im}(\lambda I - A)$ is one–one and onto;
2. $\mathrm{im}(\lambda I - A)$ is dense in Y;
3. $\|(\lambda I - A)^{-1} f\|_Y \leq C \|f\|_Y$ for $f \in \mathrm{im}(\lambda I - A)$.

In this case, $(\lambda I - A)^{-1}$ has a unique extension to an operator in $\mathcal{L}(Y, Y)$. The complement of the resolvent set, i.e., the set of those λ for which a bounded inverse does *not* exist, is called the *spectrum*. We denote the spectrum of A by $\mathrm{spec}(A)$.

If λ is an *eigenvalue* of A, i.e., if $A\phi = \lambda\phi$ for some non-zero $\phi \in X$, then $\lambda \in \mathrm{spec}(A)$ because $\ker(\lambda I - A)$ contains a non-zero element, namely, the eigenvector ϕ. In fact,

$$\ker(\lambda I - A) = \{\phi \in X : A\phi = \lambda\phi\}$$

is precisely the eigenspace corresponding to the eigenvalue λ. In general, it can happen that some elements of the spectrum are not eigenvalues of A.

Now suppose that H is a pivot space for V, take $X = V$ and $Y = V^*$, and assume that $A : V \to V^*$ is self-adjoint. The eigenvalues of A must be purely real, and any two of its eigenvectors with distinct eigenvalues must be orthogonal, as the following elementary arguments show. If $A\phi = \lambda\phi$ and $\phi \neq 0$, then

$$\lambda\|\phi\|^2 = (\phi, \lambda\phi) = (\phi, A\phi) = (A\phi, \phi) = (\lambda\phi, \phi) = \bar{\lambda}\|\phi\|^2,$$

so $\lambda = \bar{\lambda}$, i.e., λ is real. Also, if $A\phi_1 = \lambda_1\phi_1$ and $A\phi_2 = \lambda_2\phi_2$, with $\lambda_1 \neq \lambda_2$, then

$$(\lambda_1 - \lambda_2)(\phi_1, \phi_2) = (\lambda_1\phi_1, \phi_2) - (\phi_1, \lambda_2\phi_2) = (A\phi_1, \phi_2) - (\phi_1, A\phi_2) = 0,$$

so $(\phi_1, \phi_2) = 0$.

Next, we further restrict our attention to the case when $V = H = V^*$, and consider a *compact* self-adjoint operator $K : H \to H$. The spectral theory for such operators follows in a remarkably simple manner from the next lemma. Here, and in the proof of the next theorem, $\|\cdot\|$ always denotes either the norm in H or the norm in $\mathcal{L}(H, H)$; no other norms occur.

Lemma 2.35 If $K : H \to H$ is compact and self-adjoint, then

(i) at least one of the numbers $\|K\|$ and $-\|K\|$ is an eigenvalue of K,
(ii) every eigenvalue of K lies in the closed interval $[-\|K\|, \|K\|]$.

Proof. We can assume that $\|K\| > 0$, because otherwise the result is trivial. Choose a sequence u_j in H such that

$$\|Ku_j\| \to \|K\| \quad \text{and} \quad \|u_j\| = 1.$$

On the one hand, the self-adjointness of K implies that

$$(K^2 u_j, u_j) = (Ku_j, Ku_j) = \|Ku_j\|^2,$$

so

$$\begin{aligned}\left\|K^2 u_j - \|K\|^2 u_j\right\|^2 &= \|K^2 u_j\|^2 - 2\|K\|^2 \|Ku_j\|^2 + \|K\|^4 \\ &\le \|K\|^2 \left(\|K\|^2 - \|Ku_j\|^2\right),\end{aligned}$$

and therefore

$$K^2 u_j - \|K\|^2 u_j \to 0.$$

On the other hand, the compactness of K implies that, after passing to a subsequence $u_{j'}$, there exists $u \in H$ such that

$$K^2 u_{j'} \to \|K\|^2 u.$$

Hence, $\|K\|^2 u_{j'} \to \|K\|^2 u$, so $u_{j'} \to u$, and we have

$$K^2 u = \|K\|^2 u \quad \text{and} \quad \|u\| = 1,$$

which means that u is an eigenvector of K^2 with eigenvalue $\|K\|^2$. Finally, since

$$\left(K + \|K\|\right)v = 0, \qquad \text{where } v = \left(K - \|K\|\right)u,$$

we see that either $v = 0$, in which case $\|K\|$ is an eigenvalue of K with eigenvector u, or else $v \ne 0$, in which case $-\|K\|$ is an eigenvalue of K with eigenvector v. Thus, (i) holds, and (ii) follows at once from Exercise 2.15. □

Recall that $\overline{W}$ denotes the closure of a subset W.

Theorem 2.36 If $K : H \to H$ is a compact, self-adjoint linear operator on a Hilbert space H, then there exist (possibly finite) sequences of vectors ψ_1, $\psi_2, \psi_3, \ldots$ in H, and of real numbers $\mu_1, \mu_2, \mu_3, \ldots$, having the following properties:

(i) *Each ψ_j is an eigenvector of K with eigenvalue μ_j.*
(ii) *The eigenvectors $\psi_1, \psi_2, \psi_3, \ldots$ are orthonormal.*
(iii) *The eigenvalues satisfy $|\mu_1| \geq |\mu_2| \geq |\mu_3| \geq \cdots > 0$.*
(iv) *If the sequences are infinite, then $\mu_j \to 0$ as $j \to \infty$.*
(v) *For every $u \in H$,*

$$Ku = \sum_{j \geq 1} \mu_j (\psi_j, u)\psi_j,$$

with convergence in H when the sum is infinite.
(vi) *We have the orthogonal direct sum decomposition*

$$H = \overline{W} \oplus \ker K, \qquad where \ \ W = \text{span}\{\psi_1, \psi_2, \ldots\}.$$

In particular, if $\ker K = \{0\}$ then W is dense in H.

Proof. Assume $\|K\| > 0$, because otherwise the result is trivial. Note also that if λ is any non-zero eigenvalue of K, then the corresponding eigenspace $\ker(\lambda I - K) = \ker(I - \lambda^{-1}K)$ is finite-dimensional by Theorem 2.19. Thus, we have orthonormal bases $\psi_1, \ldots, \psi_l$ and $\psi_{l+1}, \ldots, \psi_{m_1}$ for the eigenspaces associated with the eigenvalues $\|K\|$ and $-\|K\|$, respectively. (It may happen that $l = 0$ or m_1, but $m_1 \geq 1$.) Put

$$\mu_1 = \cdots = \mu_l = \|K\| \quad \text{and} \quad \mu_{l+1} = \cdots = \mu_{m_1} = -\|K\|,$$

so that

$$K\psi_j = \mu_j \psi_j, \quad (\psi_j, \psi_k) = \delta_{jk} \text{ and } |\mu_j| = \|K\| \qquad \text{for } j, k \in \{1, \ldots, m_1\}.$$

Let $V_1 = \text{span}\{\psi_1, \ldots, \psi_{m_1}\}$, and observe that if $u \in V_1^\perp$ then $Ku \in V_1^\perp$ because

$$(Ku, \psi_j) = (u, K\psi_j) = \mu_j(u, \psi_j) = 0 \quad \text{for } 1 \leq j \leq m_1.$$

Hence, the restriction $K_1 = K|_{V_1^\perp}$ is a compact, self-adjoint linear operator on $V_1^\perp$, and we claim that

$$\|K_1\| < \|K\|.$$

Indeed, it is clear that $\|K_1\| \leq \|K\|$; suppose for a contradiction that $\|K_1\| = \|K\|$. By Lemma 2.35, K_1 has an eigenvalue μ with $|\mu| = \|K\|$, so there exists a non-zero vector $\psi \in V_1^\perp$ with $K_1\psi = \mu\psi$. However, $K\psi = K_1\psi = \mu\psi$, so $\psi \in V_1$, which is impossible because $V_1 \cap V_1^\perp = \{0\}$.

If $\|K_1\| > 0$, then we repeat the construction above to obtain ψ_j and μ_j for $m_1 + 1 \leq j \leq m_2$ satisfying

$$K\psi_j = \mu_j\psi_j \quad\text{and}\quad (\psi_j, \psi_k) = \delta_{jk} \qquad \text{for } j, k \in \{1, \ldots, m_2\},$$

with $|\mu_j| = \|K_1\|$ for $m_1 + 1 \leq j \leq m_2$. Putting $V_2 = \text{span}\{\psi_1, \ldots, \psi_{m_2}\}$, the restriction $K_2 = K|_{V_2^\perp}$ is a compact, self-adjoint operator on $V_2^\perp$, and

$$\|K_2\| < \|K_1\|.$$

There are two possibilities: either we can continue indefinitely, constructing an infinite sequence of operators $K_1, K_2, K_3, \ldots$, or else the process halts after (say) r steps because $\|K_r\| = 0$. In either case, it is clear that (i), (ii) and (iii) hold.

To prove (iv), suppose that the sequences are infinite, and observe that

$$\|K\psi_j/\mu_j - K\psi_k/\mu_k\| = \|\psi_j - \psi_k\| = \sqrt{2} \quad \text{if } j \neq k,$$

so the sequence $(K\psi_j/\mu_j)_{j=1}^\infty$ has no convergent subsequence. Since K is compact, it follows that the sequence $(\psi_j/\mu_j)_{j=1}^\infty$ is unbounded, and so $1/\mu_j = \|\psi_j/\mu_j\| \to \infty$.

Next, for any $u \in H$, if K_l exists then

$$u = \sum_{j=1}^{m_l}(\psi_j, u)\psi_j + v_l \quad \text{with } v_l \in V_l^\perp,$$

where $V_l = \text{span}\{\psi_1, \ldots, \psi_{m_l}\}$ as above, and so

$$Ku = \sum_{j=1}^{m_l}\mu_j(\psi_j, u)\psi_j + K_l v_l.$$

If $\|K_r\| = 0$ for some r, then we are done with the proof of (v). Otherwise,

$$\|K_l v_l\| \leq \|K_l\|\|v_l\| \leq \|K_l\|\|u\| = |\mu_{m_l}|\|u\|,$$

so $K_l v_l \to 0$ as $l \to \infty$, and again (v) holds.

Finally, consider (vi). If $u \in \ker K$, then $\mu_j(\psi_j, u) = (K\psi_j, u) = (\psi_j, Ku) = 0$ and so $(\psi_j, u) = 0$ for all j, implying $(w, u) = 0$ for all $w \in W$, and hence also for all $w \in \overline{W}$, i.e., $u \in \overline{W}^{\perp}$. Conversely, if $u \in \overline{W}^{\perp}$, then $(\psi_j, u) = 0$ for all j, so $Ku = 0$ by (v), i.e., $u \in \ker K$. Thus, $\ker K = \overline{W}^{\perp}$. ☐

Theorem 2.36 has its historical origins in the Hilbert–Schmidt theory of integral operators with symmetric kernels. Our next result applies to self-adjoint partial differential operators; see Theorem 4.12. In part (ii), the term "complete orthonormal system" means that the ϕ_j are orthonormal and span a dense subspace of H; consequently,

$$u = \sum_{j=1}^{\infty} (\phi_j, u)\phi_j \quad \text{for } u \in H,$$

with the sum converging in H.

Theorem 2.37 Let H act as a pivot space for V, and assume that H is infinite-dimensional and that the inclusion map $V \to H$ is compact. If the bounded linear operator $A : V \to V^*$ is self-adjoint and coercive, then there exist a sequence of vectors $\phi_1, \phi_2, \phi_3, \ldots$ in V and a sequence of real numbers $\lambda_1, \lambda_2, \lambda_3, \ldots$, having the following properties:

 (i) *For each $j \geq 1$, ϕ_j is an eigenvector of A with eigenvalue λ_j.*
 (ii) *The eigenvectors $\phi_1, \phi_2, \phi_3, \ldots$ form a complete orthonormal system in H.*
 (iii) *The eigenvalues satisfy $\lambda_1 \leq \lambda_2 \leq \lambda_3 \leq \cdots$ with $\lambda_j \to \infty$ as $j \to \infty$.*
 (iv) *For each $u \in V$,*

$$Au = \sum_{j=1}^{\infty} \lambda_j (\phi_j, u)\phi_j \quad \text{(convergence in } V^*\text{).}$$

Proof. Let the constant C be as in the coercivity bound (2.7), so that the bounded, self-adjoint linear operator $A + C : V \to V^*$ is positive and bounded below. By Lemma 2.32, the inverse $(A + C)^{-1} : V^* \to V$ exists and is bounded, and the restriction $K = (A + C)^{-1}|_V$ is compact from V into V, because the inclusion $V \subseteq V^*$ is compact. The operator K is also self-adjoint with respect to the energy inner product for $A + C$, because

$$(Ku, v)_{A+C} = \big((A + C)Ku, v\big) = (u, v) = \big(u, (A + C)Kv\big)$$
$$= (u, Kv)_{A+C} \quad \text{for } u, v \in V.$$

We apply Theorem 2.36, noting that $\psi_j = (A + C)K\psi_j = \mu_j(A + C)\psi_j$ and, since the ψ_j are orthonormal with respect to the energy inner product for $A+C$,

$$(\psi_j, \psi_k) = \big((A + C)K\psi_j, \psi_k\big) = (\mu_j \psi_j, \psi_k)_{A+C} = \mu_j \delta_{jk}.$$

In particular, $\mu_j = \|\psi_j\|_H^2 > 0$, and we define

$$\phi_j = \frac{1}{\sqrt{\mu_j}}\psi_j \quad \text{and} \quad \lambda_j = \frac{1}{\mu_j} - C,$$

so that $(\phi_j, \phi_k) = \delta_{jk}$ and $A\phi_j = (A + C)\phi_j - C\phi_j = \lambda_j\phi_j$. The eigenfunctions ψ_j, and hence also the ϕ_j, span a dense subspace of V because $\ker K = \{0\}$. Parts (i)–(iii) now follow because V is dense in H, and $\mu_j \downarrow 0$ as $j \to \infty$. Moreover, if $u \in V$, then

$$u = \sum_{j=1}^{\infty}(\psi_j, u)_{A+C}\psi_j = \sum_{j=1}^{\infty}(\phi_j, u)\phi_j, \tag{2.8}$$

with the sum converging in V, and since $A : V \to V^*$ is bounded, we obtain the expansion of Au in part (iv). $\qquad\qquad\square$

Corollary 2.38 If, in addition to the assumptions of Theorem 2.37, the operator A is strictly positive-definite, i.e.,

$$(Au, u) > 0 \quad \text{for all } u \in V \setminus \{0\},$$

then

(i) the eigenvalues are all strictly positive: $0 < \lambda_1 \leq \lambda_2 \leq \lambda_3 \leq \cdots$;

(ii) the operator A is positive and bounded below on V, and the energy norm for A is given by

$$\|u\|_A = \sqrt{(Au, u)} = \left(\sum_{j=1}^{\infty}\lambda_j|(\phi_j, u)|^2\right)^{1/2} \quad \text{for } u \in V;$$

(iii) the inverse operator $A^{-1} : V^* \to V$ exists and is bounded, self-adjoint and positive and bounded below, with energy norm

$$\|f\|_{A^{-1}} = \sqrt{(A^{-1}f, f)} = \left(\sum_{j=1}^{\infty}\lambda_j^{-1}|(\phi_j, f)|^2\right)^{1/2} \quad \text{for } f \in V^*.$$

Proof. The positive-definiteness of A implies that $\lambda_j = (A\phi_j, \phi_j) > 0$ for all j. Since zero is not an eigenvalue of A, we have $\ker A = \{0\}$, and hence by Theorem 2.34 the inverse $A^{-1} : V^* \to V$ exists and is bounded. By Exercise 2.3, A^{-1} is also self-adjoint, and it is easy to see that A^{-1} is strictly positive-definite. Thus, the two energy norms make sense and satisfy $\|u\|_A \leq C\|u\|_V$

and $\|f\|_{A^{-1}} \le C\|f\|_{V^*}$. Furthermore, since the sum (2.8) converges in V,

$$(f, u) = \sum_{j=1}^{\infty} (f, \phi_j)(\phi_j, u) \qquad \text{for } f \in V^* \text{ and } u \in V.$$

Taking $f = Au$ gives the formulae for $\|u\|_A$ and $\|f\|_{A^{-1}}$, because $(f, \phi_j) = (Au, \phi_j) = (u, A\phi_j) = \lambda_j(u, \phi_j)$ and hence $(\phi_j, u) = \lambda_j^{-1}(\phi_j, f)$. Finally, by the Cauchy–Schwarz inequality,

$$|(f, u)| \le \left| \sum_{j=1}^{\infty} \lambda_j^{-1/2}(f, \phi_j)\lambda_j^{1/2}(\phi_j, u) \right| \le \|f\|_{A^{-1}}\|u\|_A \le \begin{cases} C\|f\|_{A^{-1}}\|u\|_V, \\ C\|f\|_{V^*}\|u\|_A, \end{cases}$$

which shows that $\|f\|_{V^*} \le C\|f\|_{A^{-1}}$ and $\|u\|_V = \|u\|_{V^{**}} \le C\|u\|_A$, so A and A^{-1} are positive and bounded below. $\qquad\square$

Our final result can be viewed as a special case of the Fredholm alternative.

Corollary 2.39 Let $\lambda \in \mathbb{C}$ and $f \in V^*$, and consider the equation

$$(A - \lambda)u = f,$$

where the assumptions of Theorem 2.37 hold.

(i) If $\lambda \notin \{\lambda_1, \lambda_2, \lambda_3, \ldots\}$, then the operator $A - \lambda : V \to V^*$ has a bounded inverse, and the unique solution $u \in V$ is given by

$$u = (A - \lambda)^{-1}f = \sum_{j=1}^{\infty} \frac{(\phi_j, f)}{\lambda_j - \lambda}\phi_j.$$

(ii) If $\lambda \in \{\lambda_1, \lambda_2, \lambda_3, \ldots\}$, then a necessary and sufficient condition for the existence of a solution $u \in V$ is that

$$(\phi_j, f) = 0 \quad \text{for all } j \text{ with } \lambda_j = \lambda.$$

In this case, the general solution is

$$u = \sum_{\lambda_j = \lambda} a_j\phi_j + \sum_{\lambda_j \ne \lambda} \frac{(\phi_j, f)}{\lambda_j - \lambda}\phi_j,$$

where the a_j are arbitrary constants.

The infinite sums in parts (i) and (ii) converge in V.

Proof. It is instructive to avoid making direct use of Theorem 2.27. Without loss of generality, we can assume that $\lambda_j > 0$ for all j, as in Corollary 2.38.

Define a linear operator B_λ by

$$B_\lambda f = \sum_{\lambda_j \neq \lambda} \frac{(\phi_j, f)}{\lambda_j - \lambda} \phi_j,$$

and observe that $B_\lambda : V^* \to V$ is bounded because, by Corollary 2.38,

$$\|B_\lambda f\|_V^2 \sim \sum_{j=1}^{\infty} \lambda_j |(\phi_j, B_\lambda f)|^2 = \sum_{\lambda_j \neq \lambda} \lambda_j \left| \frac{(\phi_j, f)}{\lambda_j - \lambda} \right|^2$$

$$\leq \sup_{\lambda_j \neq \lambda} \left(\frac{\lambda_j}{\lambda_j - \lambda} \right)^2 \sum_{\lambda_j \neq \lambda} \lambda_j^{-1} |(\phi_j, f)|^2 \leq C \|f\|_{V^*}^2.$$

Since the sums

$$(A - \lambda)B_\lambda f = B_\lambda (A - \lambda)f = \sum_{\lambda_j \neq \lambda} (\phi_j, f)\phi_j \quad \text{and} \quad f = \sum_{j=1}^{\infty} (\phi_j, f)\phi_j$$

converge in V^*, we see that if λ is not an eigenvalue of A, then $B_\lambda = (A - \lambda)^{-1}$.

Assume now that λ *is* an eigenvalue. If a solution $u \in V$ exists, then the right-hand side f satisfies

$$(\phi_j, f) = \big(\phi_j, (A - \lambda)u\big) = \big((A - \lambda)\phi_j, u\big)$$

$$= (\lambda_j - \lambda)(\phi_j, u) = 0 \quad \text{whenever } \lambda_j = \lambda.$$

Conversely, if $(\phi_j, f) = 0$ whenever $\lambda_j = \lambda$, then $(A - \lambda)B_\lambda f = f$ so $B_\lambda f$ is a solution, and to complete the proof it suffices to show that

$$\ker(A - \lambda) = \text{span}\{\phi_j : \lambda_j = \lambda\}. \tag{2.9}$$

In fact, $(A - \lambda)u = 0$ if and only if $\big(\phi_j, (A - \lambda)u\big) = 0$ for all j, and since

$$\big(\phi_j, (A - \lambda)u\big) = \big((A - \lambda)\phi_j, u\big) = (\lambda_j - \lambda)(\phi_j, u),$$

we conclude that $u \in \ker(A - \lambda)$ if and only if $u \perp \phi_j$ whenever $\lambda_j \neq \lambda$. Hence, (2.9) holds because of (2.8). $\qquad\qquad\qquad\qquad\qquad\qquad \square$

Exercises

2.1 Show that for a normed space X to be complete it is necessary and sufficient that, for every sequence u_j in X, if the numerical sum $\sum_{j=1}^{\infty} \|u_j\|_X$

converges in $\mathbb{R}$, then the vector sum $\sum_{j=1}^{\infty} u_j$ converges in the norm of X, i.e., there exists a $v \in X$ such that $\|v - \sum_{j=1}^{m} u_j\|_X \to 0$ as $m \to \infty$.

2.2 Let X be a normed space.

(i) Show that the vectors $u_1, \ldots, u_n \in X$ are linearly independent if and only if there exist functionals $g_1, \ldots, g_n \in X^*$ such that

$$\langle g_j, u_k \rangle = \delta_{jk} = \begin{cases} 1 & \text{if } j = k, \\ 0 & \text{if } j \neq k. \end{cases}$$

(ii) Show that with u_j and g_j as in (i), we can define a projection $P : X \to X$ by

$$Pu = \sum_{j=1}^{n} \langle g_j, u \rangle u_j \quad \text{for } u \in X.$$

(iii) Let $V = \text{span}\{u_1, \ldots u_n\}$ and $W = \bigcap_{j=1}^{n} \ker g_j$. Show that P is the projection of X onto V, parallel to W.

(iv) Suppose that W is a closed subspace of X, and that the cosets $u_1 + W, \ldots, u_n + W$ form a basis for the quotient space X/W. We may therefore define functionals $g_1, \ldots, g_n$ in $(X/W)^*$ by $\langle g_j, u_k + W \rangle = \delta_{jk}$. Show that any set of representatives $\{u_1, \ldots, u_n\}$ must be linearly independent, and deduce that

$$Pu = \sum_{j=1}^{n} \langle g_j, u + W \rangle u_j$$

is the projection of X onto $V = \text{span}\{u_1, \ldots, u_n\}$, parallel to W.

2.3 Let $A \in \mathcal{L}(X, Y)$, where X and Y are normed spaces. Show that A has a bounded inverse if and only if A^t has a bounded inverse, in which case $(A^{-1})^t = (A^t)^{-1}$.

2.4 Let W be a closed subspace of a normed space X, and let $\epsilon \in (0, 1)$. Show that each coset $u + W \in X/W$ has a representative $v \in u + W$ satisfying

$$(1 - \epsilon)\|v\|_X \leq \|u + W\|_{X/W} \leq \|v\|_X.$$

2.5 Let X be a normed space.

(i) Show that a subset $W \subseteq X$ is dense if and only if $W^a = \{0\}$.

(ii) Suppose now that V is a subset of the dual space X^*.

(a) Show that if V is dense, then $^a V = \{0\}$.

(b) Show that if $^a V = \{0\}$, and if X is reflexive, then V is dense.

2.6 Let X, Y and Z be Banach spaces.

    (i) Show that if $K : X \to Y$ and $L : X \to Y$ are compact, then so are $K + L : X \to Y$ and $\mu K : X \to Y$, for any scalar μ.

    (ii) Show that if $A : X \to Y$ is bounded, and if $K : Y \to Z$ is compact, then $KA : X \to Z$ is compact.

    (iii) Show that if $K : X \to Y$ is compact, and if $A : Y \to Z$ is bounded, then $AK : X \to Z$ is compact.

2.7 Assume that X and Y are Banach spaces.

    (i) Show that if $\|A\|_{\mathcal{L}(X,X)} < 1$, then $I - A : X \to X$ has a bounded inverse, and

$$\|(I - A)^{-1}\|_{\mathcal{L}(X,X)} \le \frac{1}{1 - \|A\|_{\mathcal{L}(X,X)}}.$$

    (ii) Show that if $A \in \mathcal{L}(X, Y)$ has a bounded inverse $A^{-1} \in \mathcal{L}(Y, X)$, and if $E \in \mathcal{L}(X, Y)$ satisfies $\|E\|_{\mathcal{L}(X,Y)} < 1/\|A^{-1}\|_{\mathcal{L}(Y,X)}$, then $A + E$ has a bounded inverse, and

$$\|(A + E)^{-1}\|_{\mathcal{L}(Y,X)} \le \frac{\|A^{-1}\|_{\mathcal{L}(Y,X)}}{1 - \|A^{-1}\|_{\mathcal{L}(Y,X)}\|E\|_{\mathcal{L}(X,Y)}}.$$

    (iii) Suppose that $B \in \mathcal{L}(Y, X)$ is a two-sided regulariser for a Fredholm operator $A \in \mathcal{L}(X, Y)$. Show that if $\|E\|_{\mathcal{L}(X,Y)} < 1/\|B\|_{\mathcal{L}(Y,X)}$, then $A + E : X \to Y$ is Fredholm, and index$(A + E) = $ index A.

    (iv) Let $t \mapsto A_t$ be a continuous function from $[0, 1]$ into $\mathcal{L}(X, Y)$. Show that if A_t is Fredholm for each $t \in [0, 1]$, then index A_t is constant for $t \in [0, 1]$, and in particular index $A_1 = $ index A_0.

2.8 Let H be an inner product space, and let u, $v \in H$.

    (i) Prove the Cauchy–Schwarz inequality (2.5). [Hint: first reduce to the case when $\|u\|_H = \|v\|_H = 1$.]

    (ii) Deduce the triangle inequality $\|u + v\|_H \le \|u\|_H + \|v\|_H$.

    (iii) Prove the parallelogram law: $\|u + v\|_H^2 + \|u - v\|_H^2 = 2\|u\|_H^2 + 2\|v\|_H^2$.

2.9 Let H be an inner product space and let $u \in H$. Show that if $(u, v)_H = 0$ for all $v \in H$, then $u = 0$.

2.10 Let H be a Hilbert space.

    (i) Show that if W is a closed subspace of H, and if P is the orthogonal projection of H onto W, then P is self-adjoint, i.e., $(Pu, v)_H = (u, Pv)_H$ for all u, $v \in H$.

    (ii) Show that if $P : H \to H$ is a self-adjoint projection, and if $W = $ im P, then P is the orthogonal projection onto W.

2.11 Let u_j be a bounded sequence in a Hilbert space H.

(i) Prove that if the sequence of scalars $(u_j, v)_H$ converges for each v in a dense subset of H, then there exists a unique $u \in H$ such that $u_j \rightharpoonup u$ in H, and moreover $\|u\|_H \le \limsup_{j \to \infty} \|u_j\|_H$.

(ii) Prove that $u_j \to u$ in H if and only if $u_j \rightharpoonup u$ in H and $\|u_j\|_H \to \|u\|$.

2.12 Let H_1 and H_2 be Hilbert spaces.

(i) Show that if $A : H_1 \to H_2$ is bounded, and if $u_j \rightharpoonup u$ in H_1, then $Au_j \rightharpoonup Au$ in H_2.

(ii) Show that if $K : H_1 \to H_2$ is compact, and if $u_j \rightharpoonup u$ in H_1, then $Au_j \to Au$ in H_2.

(iii) Prove (conversely) that if a bounded linear operator $K : H_1 \to H_2$ has the property that $u_j \rightharpoonup 0$ in H_1 implies $Ku_j \to 0$ in H_2, then K is compact.

2.13 Suppose that V is a *real* Hilbert space, and that $A : V \to V^*$ is a bounded, self-adjoint linear operator. Assume further that A is strictly positive-definite. Let $f \in V^*$, and define the quadratic functional $J_f : V \to \mathbb{R}$ by

$$J_f(v) = \tfrac{1}{2}(Av, v) - (f, v).$$

(i) Show that if $Au = f$, then $J_f(v) = J_f(u) + \tfrac{1}{2}\|v - u\|_A^2$.

(ii) Show that $J_f(v + \delta v) = J_f(v) + (Av - f, \delta v) + \tfrac{1}{2}\|\delta v\|_A^2$ for all v, $\delta v \in V$.

(iii) Deduce that $u \in V$ satisfies $Au = f$ if and only if $J_f(u) = \min_{v \in V} J_f(v)$.

2.14 Suppose that Φ is a bounded sesquilinear form on a *Banach* space X. Show that if Φ is positive and bounded below, then X is in fact a Hilbert space. [Hint: consider $(u, v) = \Phi(u, v) + \overline{\Phi(v, u)}$.]

2.15 Let $A : X \to X$ be a bounded linear operator on a Banach space X. Show that if $\mu \in \mathbb{C}$ satisfies $|\mu| > \|A\|_{\mathcal{L}(X,X)}$, then the operator $\mu I - A : X \to X$ has a bounded inverse.

2.16 Let $K : H \to H$ be a compact, self-adjoint linear operator on a Hilbert space H, and let $\mu_1, \mu_2, \ldots$ and $\psi_1, \psi_2, \ldots$ be the eigenvalues and eigenvectors of K, as given by Theorem 2.36. Suppose also that K is strictly positive-definite:

$$(Ku, u) > 0 \quad \text{for } u \in H \setminus \{0\},$$

and define V to be the Hilbert space obtained by completing H in the energy norm

$$\|u\|_V = \sqrt{(Ku, u)} = \left(\sum_{j=1}^{n} \mu_j |(\psi_j, u)|^2 \right)^{1/2}$$

(i) Show that the inclusion map $H \to V$ is compact with dense image. Deduce that $V^* \subseteq H \subseteq V$, i.e., H is a pivot space for V^*.

(ii) Show that K has a unique extension to a bounded linear operator $K : V \to V^*$, and that this extension has a bounded inverse.

(iii) Obviously, $\psi_j \in V$ for all j, and $\mu_1 > \mu_2 > \cdots > 0$. Show that

$$Ku = \sum_{j=1}^{\infty} \mu_j (\psi_j, u) \psi_j \quad \text{for } u \in V,$$

with the sum converging in V^*.

2.17 Let H act as a pivot space for V, and let $A : V \to V^*$ be a self-adjoint, Fredholm operator with index zero. Define $V_0 = V \cap (\ker A)^{\perp}$, where $\perp$ means the orthogonal complement in H.

(i) Show that $V = V_0 \oplus \ker A$ and $V^* = \operatorname{im} A \oplus \ker A$, where the direct sums are orthogonal with respect to $(\cdot, \cdot)$.

(ii) Let $A_0 = A|_{V_0}$, and show that $A_0 : V_0 \to V_0^* \simeq \operatorname{im} A$ is invertible.

(iii) Show that if, in addition, A is coercive and $(Au, u) > 0$ for all $u \in V_0 \setminus \{0\}$, then

$$(Au, u) \geq c\|u\|_V^2 \quad \text{for all } u \in V_0.$$

and thus A_0 is positive and bounded below on V_0.

3

Sobolev Spaces

In the context of variational methods, one naturally seeks a solution to a linear second-order elliptic boundary value problem in a space of functions that are square-integrable and have square-integrable first partial derivatives. Physically, the functions in such a *Sobolev space* typically represent the system states for which the total energy is finite. Before commencing our study of elliptic problems, we shall therefore treat Sobolev spaces as a separate topic. Lützen [61] discusses the historical developments that led to the modern ideas of weak or distributional derivatives, on which the theory of Sobolev spaces is built; see also Dieudonné [19, pp. 248–252].

The first four sections of this chapter cover relevant parts of the theory of distributions and Fourier transforms. The reader will need to understand this material at a practical level before proceeding any further. Next, we define the Sobolev space $W_p^s(\Omega)$ based on $L_p(\Omega)$, for $1 \leq p \leq \infty$, but soon focus almost exclusively on the case $p = 2$. The space $H^s(\mathbb{R}^n)$, which coincides with $W^s(\mathbb{R}^n) = W_2^s(\mathbb{R}^n)$, is then defined via the Fourier transform in the usual way, and after that we introduce $H^s(\Omega)$ and $\widetilde{H}^s(\Omega)$ for a general open set $\Omega \subseteq \mathbb{R}^n$. We go on to develop the standard density, imbedding, duality and trace theorems, mostly assuming that Ω is a Lipschitz domain. The classical reference for this theory is Lions and Magenes [59]; however, our approach is more along the lines of Chazarain and Piriou [10, Chapter 2]. For results on Sobolev spaces $W_p^s(\Omega)$ with $p \neq 2$, we refer to Adams [2], Bergh and Löfström [5] and Grisvard [34]. Only one result in this chapter is not completely standard: a theorem of Costabel [14] stating that, for a Lipschitz domain Ω with boundary Γ, the trace operator is bounded from $H^s(\Omega)$ to $H^{s-1/2}(\Gamma)$ for $\frac{1}{2} < s < \frac{3}{2}$ (not just for $\frac{1}{2} < s \leq 1$). This fact will be used when we consider the mapping properties of boundary integral operators.

Convolution

In this section only, Ω may be any (Lebesgue) measurable subset of $\mathbb{R}^n$ ($n \geq 1$) with strictly positive measure; throughout the rest of the book, Ω is always assumed to be open. The Banach space $L_p(\Omega)$ is defined in the usual way, with

$$\|u\|_{L_p(\Omega)} = \left(\int_\Omega |u(x)|^p \, dx \right)^{1/p} \quad \text{for } 1 \leq p < \infty,$$

and with $\|u\|_{L_\infty(\Omega)}$ equal to the essential supremum of u over Ω. We define

$$\langle u, v \rangle_\Omega = \int_\Omega u(x) v(x) \, dx$$

whenever the integrand belongs to $L_1(\Omega)$, and usually just write $\langle u, v \rangle$ if $\Omega = \mathbb{R}^n$. Let p^* denote the conjugate exponent to p, i.e.,

$$\frac{1}{p} + \frac{1}{p^*} = 1, \tag{3.1}$$

and recall Hölder's inequality:

$$|\langle u, v \rangle_\Omega| \leq \|u\|_{L_{p^*}(\Omega)} \|v\|_{L_p(\Omega)} \quad \text{for } 1 \leq p \leq \infty.$$

By Exercise 3.1,

$$\|u\|_{L_{p^*}(\Omega)} = \sup_{0 \neq v \in L_p(\Omega)} \frac{|\langle u, v \rangle_\Omega|}{\|v\|_{L_p(\Omega)}} \quad \text{for } u \in L_{p^*}(\Omega) \text{ and } 1 \leq p \leq \infty,$$

$$\tag{3.2}$$

so $L_{p^*}(\Omega)$ is isometrically imbedded in the dual of $L_p(\Omega)$. In fact,

$$L_{p^*}(\Omega) = [L_p(\Omega)]^* \quad \text{for } 1 \leq p < \infty \text{ (but not } p = \infty); \tag{3.3}$$

see Yosida [106, p. 115].

If the (measurable) functions u and v are defined on the whole of $\mathbb{R}^n$, then we define their *convolution* $u * v$ by

$$(u * v)(x) = \int_{\mathbb{R}^n} u(x - y) v(y) \, dy$$

whenever the integrand belongs to $L_1(\mathbb{R}^n)$. Convolution turns out to be a very effective tool for approximating non-smooth functions by smooth ones, a technique that we shall use repeatedly. The substitution $y = x - z$ shows that $(u * v)(x)$ exists if and only if $(v * u)(x)$ exists, in which case they are the same. Thus, convolution is commutative:

$$v * u = u * v.$$

The following theorem gives simple criteria for the existence of the convolution.

Theorem 3.1 Let $1 \leq p \leq \infty$, $1 \leq q \leq \infty$ *and* $1 \leq r \leq \infty$, *and suppose that*

$$\frac{1}{p} + \frac{1}{q} = 1 + \frac{1}{r}. \tag{3.4}$$

If $u \in L_p(\mathbb{R}^n)$ *and* $v \in L_q(\mathbb{R}^n)$, *then* $u * v$ *exists almost everywhere in* $\mathbb{R}^n$ *and belongs to* $L_r(\mathbb{R}^n)$, *with*

$$\|u * v\|_{L_r(\mathbb{R}^n)} \leq \|u\|_{L_p(\mathbb{R}^n)} \|v\|_{L_q(\mathbb{R}^n)}. \tag{3.5}$$

Proof. First note that if $p = q = r = 1$, then

$$\int_{\mathbb{R}^n} \int_{\mathbb{R}^n} |u(x - y)v(y)| \, dy \, dx = \int_{\mathbb{R}^n} |v(y)| \left(\int_{\mathbb{R}^n} |u(x - y)| \, dx \right) dy$$

$$= \int_{\mathbb{R}^n} |v(y)| \|u\|_{L_1(\mathbb{R}^n)} \, dy = \|u\|_{L_1(\mathbb{R}^n)} \|v\|_{L_1(\mathbb{R}^n)},$$

so $u * v$ exists almost everywhere and belongs to $L_1(\mathbb{R}^n)$, and (3.5) holds.

Now let $u \in L_p(\mathbb{R}^n)$, $v \in L_q(\mathbb{R}^n)$ and $\phi \in L_{r^*}(\mathbb{R}^n)$, for p, q and r satisfying (3.4), with $r > 1$. Assume first that u, v and ϕ have compact supports, and so belong to $L_1(\mathbb{R}^n)$. Since $1/p + 1/q + 1/r^* = 2$, if we put $\psi(x) = \phi(-x)$, then by Exercises 3.2 and 3.3,

$$|\langle u * v, \phi \rangle| = |(u * v * \psi)(0)| \leq \|u\|_{L_p(\mathbb{R}^n)} \|v\|_{L_q(\mathbb{R}^n)} \|\psi\|_{L_{r^*}(\mathbb{R}^n)}$$

$$= \|u\|_{L_p(\mathbb{R}^n)} \|v\|_{L_q(\mathbb{R}^n)} \|\phi\|_{L_{r^*}(\mathbb{R}^n)}.$$

Now drop the assumption that u, v and ϕ have compact support, define

$$\chi_j(x) = \begin{cases} 1 & \text{if } |x| \leq j, \\ 0 & \text{if } |x| > j, \end{cases}$$

and put $u_j = \chi_j u$, $v_j = \chi_j v$ and $\phi_j = \chi_j \phi$. Since

$$\langle |u_j| * |v_j|, |\phi_j| \rangle \leq \|u_j\|_{L_p(\mathbb{R}^n)} \|v_j\|_{L_q(\mathbb{R}^n)} \|\phi_j\|_{L_{r^*}(\mathbb{R}^n)}$$

$$\leq \|u\|_{L_p(\mathbb{R}^n)} \|v\|_{L_q(\mathbb{R}^n)} \|\phi\|_{L_{r^*}(\mathbb{R}^n)}$$

for all j, the monotone convergence theorem implies that

$$|\langle u * v, \phi \rangle| \leq \langle |u| * |v|, |\phi| \rangle \leq \|u\|_{L_p(\mathbb{R}^n)} \|v\|_{L_q(\mathbb{R}^n)} \|\phi\|_{L_{r^*}(\mathbb{R}^n)},$$

giving the desired bound for $\|u * v\|_{L_r(\mathbb{R}^n)}$. $\qquad \square$

Taking $r = \infty$ in (3.4), we see that $u * v$ is bounded if $u \in L_p(\mathbb{R}^n)$ and $v \in L_{p^*}(\mathbb{R}^n)$. To prove a stronger result, we introduce the *p-mean modulus of continuity*,

$$\omega_p(t, u) = \sup_{|h| \leq t} \left(\int_{\mathbb{R}^n} |u(x + h) - u(x)|^p \, dx \right)^{1/p} \qquad \text{for } t > 0 \text{ and}$$

$$1 \leq p < \infty. \quad (3.6)$$

It can be shown that, for $1 \leq p < \infty$, if $u \in L_p(\mathbb{R}^n)$, then $\omega_p(t, u) \to 0$ as $t \downarrow 0$.

Theorem 3.2 *Let* $1 \leq p < \infty$. *If* $u \in L_p(\mathbb{R}^n)$ *and* $v \in L_{p^*}(\mathbb{R}^n)$, *then* $u * v$ *is uniformly continuous on* $\mathbb{R}^n$. *If, in addition,* $p \neq 1$, *then* $(u * v)(x) \to 0$ *as* $|x| \to \infty$.

Proof. By Hölder's inequality,

$$|(u * v)(x) - (u * v)(y)| = \left| \int_{\mathbb{R}^n} [u(x - z) - u(y - z)]v(z) \, dz \right|$$

$$\leq \omega_p(|x - y|, u) \|v\|_{L_{p^*}(\mathbb{R}^n)},$$

so $u * v$ is uniformly continuous on $\mathbb{R}^n$. Assume now that $p \neq 1$, and let $\epsilon > 0$. Since p and p^* are both finite, there exists $R > 0$ such that

$$\int_{|y| > R} |u(y)|^p \, dy < \epsilon^p \quad \text{and} \quad \int_{|y| > R} |v(y)|^{p^*} \, dy < \epsilon^{p^*},$$

and by Hölder's inequality,

$$|(u * v)(x)| \leq \left| \int_{|y| \leq R} u(x - y)v(y) \, dy \right| + \left| \int_{|y| > R} u(x - y)v(y) \, dy \right|$$

$$\leq \left(\int_{|y| \leq R} |u(x - y)|^p \, dy \right)^{1/p} \|v\|_{L_{p^*}(\mathbb{R}^n)}$$

$$+ \|u\|_{L_p(\mathbb{R}^n)} \left(\int_{|y| > R} |v(y)|^{p^*} \, dy \right)^{1/p^*}$$

$$\leq \left(\int_{|y - x| \leq R} |u(y)|^p \, dy \right)^{1/p} \|v\|_{L_{p^*}(\mathbb{R}^n)} + \|u\|_{L_p(\mathbb{R}^n)} \epsilon.$$

If $|x| > 2R$ and $|y - x| \leq R$, then $|y| \geq |x| - |x - y| > 2R - R = R$. Thus,

$$|(u * v)(x)| \leq \epsilon \|v\|_{L_{p^*}(\mathbb{R}^n)} + \|u\|_{L_p(\mathbb{R}^n)} \epsilon \quad \text{for } |x| > 2R,$$

and so $(u * v)(x) \to 0$ as $|x| \to \infty$. $\qquad \square$

Differentiation

Let Ω be a non-empty open subset of $\mathbb{R}^n$. If a function $u : \Omega \to \mathbb{C}$ is sufficiently smooth for them to exist, then we denote the partial derivatives of u by

$$\partial^\alpha u(x) = \left(\frac{\partial}{\partial x_1}\right)^{\alpha_1} \cdots \left(\frac{\partial}{\partial x_n}\right)^{\alpha_n} u(x),$$

where $\alpha = (\alpha_1, \ldots, \alpha_n)$ is a *multi-index*, i.e., an n-tuple of non-negative integers. The *order* of the partial derivative $\partial^\alpha u$ is the number

$$|\alpha| = \alpha_1 + \cdots + \alpha_n,$$

and we put

$$u^{(k)}(x; y) = \sum_{|\alpha|=k} \frac{k!}{\alpha!} \partial^\alpha u(x) y^\alpha, \tag{3.7}$$

where $\alpha! = \alpha_1! \cdots \alpha_n!$ and $y^\alpha = y_1^{\alpha_1} \cdots y_n^{\alpha_n}$. In this way, Taylor's formula may be written as

$$u(x + y) = \sum_{j=0}^{k} \frac{1}{j!} u^{(j)}(x; y) + \frac{1}{k!} \int_0^1 (1 - t)^k u^{(k+1)}(x + ty; y)\, dt; \tag{3.8}$$

see Exercise 3.5.

For any integer $r \geq 0$ we let

$$C^r(\Omega) = \{u : \partial^\alpha u \text{ exists and is continuous on } \Omega \text{ for } |\alpha| \leq r\},$$

and put

$$C^\infty(\Omega) = \bigcap_{r \geq 0} C^r(\Omega).$$

The support of u, denoted by $\operatorname{supp} u$, is the closure in Ω of the set $\{x \in \Omega : u(x) \neq 0\}$. If $K \Subset \Omega$, i.e., if K is a compact subset of Ω, then we put

$$C_K^r(\Omega) = \{u \in C^r(\Omega) : \operatorname{supp} u \subseteq K\} \quad \text{and} \quad C_K^\infty(\Omega) = \bigcap_{r \geq 0} C_K^r(\Omega).$$

Next, we define

$$C_{\text{comp}}^r(\Omega) = \left\{u : u \in C_K^r(\Omega) \text{ for some } K \Subset \Omega\right\}$$

and

$$C_{\text{comp}}^\infty(\Omega) = \left\{u : u \in C_K^\infty(\Omega) \text{ for some } K \Subset \Omega\right\}.$$

Exercise 3.6 gives some examples of compactly supported C^∞ functions.

We now show that differentiation commutes with convolution; see also Exercise 3.10.

Theorem 3.3 *Let $r \geq 0$ and $1 \leq p \leq \infty$. If $u \in C^r_{\text{comp}}(\mathbb{R}^n)$ and $v \in L_p(\mathbb{R}^n)$, then $u * v \in C^r(\mathbb{R}^n)$ and*

$$\partial^\alpha(u * v) = (\partial^\alpha u) * v$$

for $|\alpha| \leq r$.

Proof. The case $r = 0$ follows from Theorem 3.2. Suppose $r = 1$, and consider the difference quotient with respect to the lth variable,

$$\Delta_{l,h} u(x) = \frac{u(x + he_l) - u(x)}{h} \qquad \text{for } 1 \leq l \leq n \text{ and } h \in \mathbb{R},$$

where $e_1, \ldots, e_n$ are the standard basis vectors for $\mathbb{R}^n$. It is easy to see that $\Delta_{l,h}(u * v) = (\Delta_{l,h} u) * v$, so for every $x \in \mathbb{R}^n$,

$$|[\Delta_{l,h}(u * v)](x) - [(\partial_l u) * v](x)| = |[(\Delta_{l,h} u - \partial_l u) * v](x)|$$

$$\leq \|\Delta_{l,h} u - \partial_l u\|_{L_{p^*}(\mathbb{R}^n)} \|v\|_{L_p(\mathbb{R}^n)}.$$

Since u is C^1 and has compact support, we have $\Delta_{l,h} u \to \partial_l u$ in $L_{p^*}(\mathbb{R}^n)$, and thus $\partial_l(u * v) = (\partial_l u) * v$. A simple inductive argument proves the result for the general case $r \geq 1$. $\square$

The connection between convolution and approximation by smooth functions, mentioned earlier, comes about via the following construction. Let $\psi \in C^\infty_{\text{comp}}(\mathbb{R}^n)$ satisfy

$$\psi \geq 0 \text{ on } \mathbb{R}^n, \quad \psi(x) = 0 \text{ for } |x| > 1, \quad \int_{\mathbb{R}^n} \psi(x)\,dx = 1; \qquad (3.9)$$

such a function exists by Exercise 3.6. We define

$$\psi_\epsilon(x) = \epsilon^{-n}\psi(\epsilon^{-1}x) \qquad \text{for } x \in \mathbb{R}^n \text{ and } \epsilon > 0, \qquad (3.10)$$

so that, as one sees via the substitution $x = \epsilon z$,

$$\psi_\epsilon \geq 0 \text{ on } \mathbb{R}^n, \quad \psi_\epsilon(x) = 0 \text{ for } |x| > \epsilon, \quad \int_{\mathbb{R}^n} \psi_\epsilon(x)\,dx = 1.$$

These properties of ψ_ϵ mean that $(\psi_\epsilon * u)(x)$ is a kind of local average of u around x, and for this reason $\psi_\epsilon * u \approx u$ when ϵ is small.

Theorem 3.4 Let $1 \leq p < \infty$. If ψ_ϵ is as above, and if $u \in L_p(\mathbb{R}^n)$, then

$$\|\psi_\epsilon * u\|_{L_p(\mathbb{R}^n)} \leq \|u\|_{L_p(\mathbb{R}^n)} \quad \text{and} \quad \|\psi_\epsilon * u - u\|_{L_p(\mathbb{R}^n)} \leq \omega_p(\epsilon, u),$$

so $\psi_\epsilon * u \to u$ in $L_p(\mathbb{R}^n)$ as $\epsilon \downarrow 0$.

Proof. Since $\|\psi_\epsilon\|_{L_1(\mathbb{R}^n)} = 1$, we see from Theorem 3.1 that

$$\|\psi_\epsilon * u\|_{L_p(\mathbb{R}^n)} \leq \|\psi_\epsilon\|_{L_1(\mathbb{R}^n)}\|u\|_{L_p(\mathbb{R}^n)} = \|u\|_{L_p(\mathbb{R}^n)}.$$

Also,

$$\psi_\epsilon * u(x) - u(x) = u * \psi_\epsilon(x) - u(x) \int_{\mathbb{R}^n} \psi_\epsilon(y)\, dy$$

$$= \int_{|y| \leq \epsilon} [u(x - y) - u(x)]\psi_\epsilon(y)\, dy$$

and so, for any $\phi \in L_{p^*}(\mathbb{R}^n)$, Hölder's inequality implies that

$$|\langle \psi_\epsilon * u - u, \phi \rangle| = \left| \int_{|y| \leq \epsilon} \psi_\epsilon(y) \int_{\mathbb{R}^n} [u(x - y) - u(x)]\phi(x)\, dx\, dy \right|$$

$$\leq \int_{|y| \leq \epsilon} \psi_\epsilon(y)\omega_p(\epsilon, u)\|\phi\|_{L_{p^*}(\mathbb{R}^n)}\, dy = \omega_p(\epsilon, u)\|\phi\|_{L_{p^*}(\mathbb{R}^n)},$$

and the desired estimate for $\psi_\epsilon * u - u$ follows. $\square$

Corollary 3.5 If $1 \leq p < \infty$, then $C_{\text{comp}}^\infty(\Omega)$ is dense in $L_p(\Omega)$.

Proof. We choose any $K_j \Subset \Omega$ such that $K_1 \subseteq K_2 \subseteq \cdots$ and $\Omega = \bigcup_{j=1}^\infty K_j$; for example,

$$K_j = \{x \in \Omega : \text{dist}(x, \mathbb{R}^n \setminus \Omega) \geq 1/j \text{ and } |x| \leq j\}.$$

Let $u \in L_p(\Omega)$, define $u_j \in L_p(\mathbb{R}^n)$ by

$$u_j(x) = \begin{cases} u(x) & \text{if } x \in K_j, \\ 0 & \text{if } x \in \mathbb{R}^n \setminus K_j, \end{cases}$$

and consider the function $u_{j,\epsilon} = \psi_\epsilon * u_j$. Theorem 3.3 shows that $u_{j,\epsilon} \in C_{\text{comp}}^\infty(\mathbb{R}^n)$, Exercise 3.4 shows that

$$\text{supp}\, u_{j,\epsilon} \Subset \Omega \quad \text{if } \epsilon < \text{dist}(K_j, \mathbb{R}^n \setminus \Omega),$$

and by Theorem 3.4

$$\|u_{j,\epsilon} - u\|_{L_p(\Omega)} \leq \|u_{\epsilon,j} - u_j\|_{L_p(\mathbb{R}^n)} + \|u_j - u\|_{L_p(\Omega)}$$

$$\leq \omega_p(\epsilon, u_j) + \|u\|_{L_p(\Omega \setminus K_j)}.$$

Since $\|u\|_{L_p(\Omega \setminus K_j)} \to 0$ as $j \to \infty$, and since $\omega_p(\epsilon, u_j) \to 0$ as $\epsilon \to 0$ for each fixed j, the result follows. $\qquad\qquad\qquad\qquad\qquad\qquad\qquad\qquad$ $\square$

We can also use convolution to smooth out the characteristic function of a set.

Theorem 3.6 *Let F be a closed subset of $\mathbb{R}^n$. For each $\epsilon > 0$, there exists $\chi_\epsilon \in C^\infty(\mathbb{R}^n)$ satisfying*

$$\begin{aligned}
\chi_\epsilon(x) &= 1 & &\textit{if } x \in F, \\
0 \le \chi_\epsilon(x) \le 1 \textit{ and } |\partial^\alpha \chi_\epsilon(x)| &\le C\epsilon^{-|\alpha|} & &\textit{if } 0 < \mathrm{dist}(x, F) < \epsilon, \\
\chi_\epsilon(x) &= 0 & &\textit{if } \mathrm{dist}(x, F) \ge \epsilon.
\end{aligned}$$

Proof. Define $v_\epsilon \in L_\infty(\mathbb{R}^n)$ by

$$v_\epsilon(x) = \begin{cases} 1 & \text{if } \mathrm{dist}(x, F) < \epsilon, \\ 0 & \text{if } \mathrm{dist}(x, F) \ge \epsilon, \end{cases}$$

so that $v_\epsilon = 1$ on a neighbourhood of F, and choose $\psi \in C^\infty_{\mathrm{comp}}(\mathbb{R}^n)$ satisfying (3.9). With the help of Exercise 3.4, one sees that the function $\chi_\epsilon = \psi_{\epsilon/4} * v_{\epsilon/2}$ has the required properties. $\qquad\qquad\qquad\qquad$ $\square$

Schwartz Distributions

A (measurable) function $u : \Omega \to \mathbb{C}$ is said to be *locally integrable* if u is absolutely integrable on every compact subset of Ω. We denote the set of all such functions by $L_{1,\mathrm{loc}}(\Omega)$. The following observation is the starting point for the theory of distributions.

Theorem 3.7 *If $u, v \in L_{1,\mathrm{loc}}(\Omega)$ satisfy*

$$\int_\Omega u(x)\phi(x)\, dx = \int_\Omega v(x)\phi(x)\, dx \quad \textit{for all } \phi \in C^\infty_{\mathrm{comp}}(\Omega),$$

then $u = v$ almost everywhere on Ω.

Proof. Let $K \Subset \Omega$, and choose an open set Ω_1 such that $K \Subset \Omega_1 \subseteq \overline{\Omega}_1 \Subset \Omega$. We define $f \in L_1(\mathbb{R}^n)$ by

$$f(x) = \begin{cases} u(x) - v(x) & \text{if } x \in \Omega_1, \\ 0 & \text{if } x \in \mathbb{R}^n \setminus \Omega_1, \end{cases}$$

and let ψ_ϵ be as in Theorem 3.4. There is an $\epsilon_0 > 0$ such that if $x \in K$ and $0 < \epsilon < \epsilon_0$ then $\psi_\epsilon(x - \cdot) \in C^\infty_{\mathrm{comp}}(\Omega_1)$ and therefore $(\psi_\epsilon * f)(x) = \langle f, \psi_\epsilon(x - \cdot)\rangle_{\mathbb{R}^n} = \langle u - v, \psi_\epsilon(x - \cdot)\rangle_\Omega = 0$. Since $\psi_\epsilon * f \to f$ in $L_1(\mathbb{R}^n)$, it follows that $f = 0$ almost everywhere on K, i.e., $u = v$ almost everywhere on K. □

Theorem 3.7 shows that a locally integrable function u is uniquely determined by its associated linear functional $\phi \mapsto \langle u, \phi \rangle_\Omega$. We wish to introduce a larger class of linear functionals on $C^\infty_{\text{comp}}(\Omega)$. Following the notation introduced by Schwartz [92], put

$$\mathcal{E}(\Omega) = C^\infty(\Omega), \quad \mathcal{D}_K(\Omega) = C^\infty_K(\Omega), \quad \mathcal{D}(\Omega) = C^\infty_{\text{comp}}(\Omega),$$

for any $K \Subset \Omega$. Since we want our functionals to be continuous in an appropriate sense, we now define convergence of sequences in each of these function spaces. No deeper properties of the underlying locally convex topologies will be used; cf. [106].

Let $(\phi_j)_{j=1}^\infty$ be a sequence in $\mathcal{E}(\Omega)$. We write

$$\phi_j \to 0 \text{ in } \mathcal{E}(\Omega)$$

if, for each compact set K and for each multi-index α,

$$\partial^\alpha \phi_j \to 0 \quad \text{uniformly on } K.$$

When, for a *fixed* K, this condition holds for all α, and in addition $\operatorname{supp} \phi_j \subseteq K$ for all j, we write

$$\phi_j \to 0 \text{ in } \mathcal{D}_K(\Omega).$$

Finally, $\phi_j \to 0$ in $\mathcal{D}(\Omega)$ means that $\phi_j \to 0$ in $\mathcal{D}_K(\Omega)$ for some $K \Subset \Omega$. Convergence to a non-zero function is then defined in the obvious way. For instance, $\phi_j \to \phi$ in $\mathcal{D}_K(\Omega)$ means $\phi \in \mathcal{D}_K(\Omega)$, $\phi_j \in \mathcal{D}_K(\Omega)$ for all j, and $\phi_j - \phi \to 0$ in $\mathcal{D}_K(\Omega)$.

Consider an abstract linear functional $\ell : \mathcal{D}(\Omega) \to \mathbb{C}$. For the moment, we write $\ell(\phi)$ to denote the value of ℓ at $\phi \in \mathcal{D}(\Omega)$. If ℓ is sequentially continuous, i.e., if for every sequence ϕ_j in $\mathcal{D}(\Omega)$,

$$\phi_j \to 0 \text{ in } \mathcal{D}(\Omega) \quad \text{implies} \quad \ell(\phi_j) \to 0,$$

then ℓ is called a (Schwartz) *distribution* on Ω. In this context, the elements of $\mathcal{D}(\Omega)$ are referred to as *test functions* on Ω. The set of all distributions on Ω is denoted by $\mathcal{D}^*(\Omega)$.

Associated with each $u \in L_{1,\text{loc}}(\Omega)$ is the linear functional ιu defined by

$$(\iota u)(\phi) = \langle u, \phi \rangle_\Omega \quad \text{for } \phi \in \mathcal{D}(\Omega). \tag{3.11}$$

It is clear that $\iota u : \mathcal{D}(\Omega) \to \mathbb{C}$ is sequentially continuous, and hence a distribution on Ω. Furthermore, Theorem 3.7 shows that the linear map $\iota : L_{1,\mathrm{loc}}(\Omega) \to \mathcal{D}^*(\Omega)$ is one–one. Hence, we may identify u with ιu, and thereby make $L_{1,\mathrm{loc}}(\Omega)$ into a subspace of $\mathcal{D}^*(\Omega)$. Those distributions that are not locally integrable can then be viewed as *generalised functions*.

As an example, fix $x \in \mathbb{R}^n$, and let $\ell \in \mathcal{D}^*(\Omega)$ be the associated point evaluation functional defined by

$$\ell(\phi) = \phi(x) \quad \text{for } \phi \in \mathcal{D}(\mathbb{R}^n).$$

It is not possible to represent this ℓ by a locally integrable function. To see why, suppose for a contradiction that $\ell = \iota u$ for some $u \in L_{1,\mathrm{loc}}(\mathbb{R}^n)$. It follows from Theorem 3.7 that $u = 0$ almost everywhere on $\mathbb{R}^n \setminus \{x\}$, and since the set $\{x\}$ has measure zero, this means that $u = 0$ almost everywhere on $\mathbb{R}^n$. Hence, $\phi(x) = \langle u, \phi \rangle = 0$ for all $\phi \in \mathcal{D}(\mathbb{R}^n)$, a contradiction. Following the convention introduced by Dirac, we denote the point evaluation functional for x by δ_x, or in the case $x = 0$, just by δ.

In keeping with the philosophy that distributions are generalised functions, we henceforth write $\langle u, \phi \rangle_\Omega$ for the value of $u \in \mathcal{D}^*(\Omega)$ at $\phi \in \mathcal{D}(\Omega)$, whether or not u is locally integrable on Ω. If $\Omega = \mathbb{R}^n$, then we usually omit the subscript, and just write $\langle u, \phi \rangle$; for instance,

$$\langle \delta_x, \phi \rangle = \phi(x) \quad \text{for } \phi \in \mathcal{D}(\mathbb{R}^n).$$

Suppose that Ω_1 is an open subset of Ω, and for any $\phi \in \mathcal{D}(\Omega_1)$ let $\tilde{\phi} \in \mathcal{D}(\Omega)$ denote the extension of ϕ by zero. For any distribution $u \in \mathcal{D}^*(\Omega)$ the *restriction* $u|_{\Omega_1} \in \mathcal{D}^*(\Omega_1)$ is defined by

$$\langle u|_{\Omega_1}, \phi \rangle_{\Omega_1} = \langle u, \tilde{\phi} \rangle_\Omega \quad \text{for } \phi \in \mathcal{D}(\Omega_1).$$

We say that $u = 0$ on Ω_1 if $u|_{\Omega_1} = 0$, and define $\operatorname{supp} u$ to be the largest relatively closed subset of Ω such that $u = 0$ on $\Omega \setminus \operatorname{supp} u$. If $u \in L_{1,\mathrm{loc}}(\Omega)$, then the distributional support of u is the same as its *essential* support as a function, i.e., $\operatorname{supp} u$ is the largest closed set such that $u = 0$ *almost everywhere* on $\Omega \setminus \operatorname{supp} u$.

We define $\mathcal{E}^*(\Omega)$ in the obvious way, i.e., a linear functional $u : \mathcal{E}(\Omega) \to \mathbb{C}$ belongs to $\mathcal{E}^*(\Omega)$ if it is sequentially continuous: $\langle u, \phi_j \rangle_\Omega \to 0$ whenever $\phi_j \to 0$ in $\mathcal{E}(\Omega)$. Exercise 3.8 shows that the inclusion $\mathcal{D}(\Omega) \subseteq \mathcal{E}(\Omega)$ is continuous and dense, so $\mathcal{E}^*(\Omega) \subseteq \mathcal{D}^*(\Omega)$. In fact, we can characterise $\mathcal{E}^*(\Omega)$ as follows.

Theorem 3.8 The space $\mathcal{E}^*(\Omega)$ coincides with the space of distributions having compact support, i.e.,

$$\mathcal{E}^*(\Omega) = \{u \in \mathcal{D}^*(\Omega) : \operatorname{supp} u \Subset \Omega\}.$$

Proof. Suppose that $u \in \mathcal{D}^*(\Omega)$ and $\operatorname{supp} u \Subset \Omega$. By Theorem 3.6, there is a $\chi \in \mathcal{D}(\Omega)$ with $\chi = 1$ on a neighbourhood of $\operatorname{supp} u$. We define a linear functional $\tilde{u}$ on $\mathcal{E}(\Omega)$ by putting

$$\langle \tilde{u}, \phi \rangle_\Omega = \langle u, \chi\phi \rangle_\Omega \quad \text{for } \phi \in \mathcal{E}(\Omega),$$

and claim that

(i) $\phi_j \to \phi$ in $\mathcal{E}(\Omega)$ implies $\langle \tilde{u}, \phi_j \rangle_\Omega \to \langle \tilde{u}, \phi \rangle_\Omega$;
(ii) $\langle \tilde{u}, \phi \rangle_\Omega = \langle u, \phi \rangle_\Omega$ for $\phi \in \mathcal{D}(\Omega)$.

Together, these facts mean that $u = \tilde{u} \in \mathcal{E}^*(\Omega)$ when $\mathcal{E}^*(\Omega)$ is viewed as a subspace of $\mathcal{D}^*(\Omega)$. We remark that $\tilde{u}$ does not depend on the choice of χ.

To prove (i), assume $\phi_j \to \phi$ in $\mathcal{E}(\Omega)$. It follows at once that $\chi\phi_j \to \chi\phi$ in $\mathcal{D}(\Omega)$, so $\langle \tilde{u}, \phi_j \rangle_\Omega = \langle u, \chi\phi_j \rangle \to \langle u, \chi\phi \rangle_\Omega = \langle \tilde{u}, \phi \rangle_\Omega$.

To prove (ii), assume $\phi \in \mathcal{D}(\Omega)$. Since $(1 - \chi)\phi = 0$ on a neighbourhood of $\operatorname{supp} u$, we have $\langle u, (1 - \chi)\phi \rangle_\Omega = 0$ and therefore $\langle \tilde{u}, \phi \rangle_\Omega = \langle u, \phi \rangle_\Omega - \langle u, (1 - \chi)\phi \rangle_\Omega = \langle u, \phi \rangle_\Omega$.

Conversely, let $u \in \mathcal{E}^*(\Omega)$, and suppose for a contradiction that $\operatorname{supp} u$ is not compact. Choose an increasing sequence of compact sets $K_1 \subseteq K_2 \subseteq \cdots$ with $\Omega = \bigcup_{j=1}^\infty K_j$, then $u|_{\Omega \setminus K_j} \neq 0$ for all j. Thus, we can find $\phi_j \in \mathcal{D}(\Omega)$ such that $\langle u, \phi_j \rangle_\Omega = 1$ and $\operatorname{supp} \phi_j \Subset \Omega \setminus K_j$. It follows that $\phi_j \to 0$ in $\mathcal{E}(\Omega)$, so $\langle u, \phi_j \rangle_\Omega \to 0$, a contradiction. $\square$

The restriction map $u \mapsto u|_{\Omega_1}$ is just one of many linear operations that can be extended from functions to distributions. For instance, to define partial differentiation of distributions one formally integrates by parts:

$$\langle \partial^\alpha u, \phi \rangle_\Omega = (-1)^{|\alpha|} \langle u, \partial^\alpha \phi \rangle_\Omega \quad \text{for } u \in \mathcal{D}^*(\Omega) \text{ and } \phi \in \mathcal{D}(\Omega).$$

Here, the sequential continuity of $\partial^\alpha u$ follows at once from the fact that if $\phi_j \to 0$ in $\mathcal{D}(\Omega)$, then $\partial^\alpha \phi_j \to 0$ in $\mathcal{D}(\Omega)$. Also, we define the complex conjugate $\bar{u} \in \mathcal{D}^*(\Omega)$ of $u \in \mathcal{D}^*(\Omega)$ by

$$\langle \bar{u}, \phi \rangle = \overline{\langle u, \bar{\phi} \rangle} \quad \text{for } \phi \in \mathcal{D}(\Omega),$$

and we generalise the meaning of the inner product in $L_2(\Omega)$,

$$(u, v)_\Omega = \int_\Omega \overline{u(x)} v(x) \, dx,$$

by putting

$$(u, \phi)_\Omega = \langle \bar{u}, \phi \rangle_\Omega \qquad \text{for } u \in \mathcal{D}^*(\Omega) \text{ and } \phi \in \mathcal{D}(\Omega).$$

When $\Omega = \mathbb{R}^n$, we just write (u, ϕ). The convolution of a distribution with a test function is defined in the obvious way,

$$(u * \phi)(x) = \langle u, \phi(x - \cdot) \rangle \qquad \text{for } x \in \mathbb{R}^n, u \in \mathcal{D}^*(\mathbb{R}^n) \text{ and } \phi \in \mathcal{D}(\mathbb{R}^n).$$
(3.12)

Thus, in particular, $\delta * \phi = \phi$, or formally $\int_{\mathbb{R}^n} \delta(x - y)\phi(y) \, dy = \phi(x)$, so we can think of $\delta(x - y)$ as the continuous analogue of Kronecker's δ_{jk}. Finally, multiplication of a distribution $u \in \mathcal{D}^*(\Omega)$ by a smooth function $\psi \in C^\infty(\Omega)$ is defined by

$$\langle \psi u, \phi \rangle_\Omega = \langle u, \psi \phi \rangle_\Omega \quad \text{for } \phi \in \mathcal{D}(\Omega).$$

In each of the above examples, Theorem 3.7 guarantees that the generalised concept is consistent with the classical one. For instance, if the classical partial derivative $\partial_j u$ exists and is locally integrable on Ω, then $\partial_j(\iota u) = \iota(\partial_j u)$, where ι is the imbedding defined in (3.11).

Distributions are a powerful conceptual tool for the study of partial differential equations, and provide, in particular, a very effective system of notation. Fortunately, we shall require few technical results from the theory of distributions. However, the following fact will be used.

Theorem 3.9 *Suppose that $u \in \mathcal{D}^*(\Omega)$ and $x \in \Omega$. If $\operatorname{supp} u \subseteq \{x\}$, then for some $m \geq 0$,*

$$u = \sum_{|\alpha| \leq m} a_\alpha \partial^\alpha \delta_x \quad \text{on } \Omega,$$

where the coefficients are given by $a_\alpha = (-1)^{|\alpha|} \langle u, (\cdot - x)^\alpha \rangle / \alpha!$.

Proof. Denote the open ball with centre x and radius $\epsilon > 0$ by

$$B_\epsilon(x) = \{ y \in \mathbb{R}^n : |y - x| < \epsilon \},$$

and choose $\epsilon_0 > 0$ such that the closed ball $K = \overline{B_{\epsilon_0}(x)}$ is a subset of Ω. By

Exercise 3.9, there exists an integer $m \geq 0$ such that

$$|\langle u, \phi \rangle| \leq C \sum_{|\alpha| \leq m} \sup_K |\partial^\alpha \phi| \quad \text{for } \phi \in \mathcal{D}_K(\Omega). \tag{3.13}$$

Given an arbitrary $\phi \in \mathcal{E}(\Omega)$, we write the Taylor expansion of ϕ about x as $\phi = \phi_1 + \phi_2$, where

$$\phi_1(y) = \sum_{j=1}^m \frac{1}{j!} \phi^{(j)}(x; y - x) = \sum_{|\alpha| \leq m} \frac{1}{\alpha!} \partial^\alpha \phi(x)(y - x)^\alpha$$

and

$$\phi_2(y) = \frac{1}{m!} \int_0^1 (1 - t)^m \phi^{(m+1)}(x + t(y - x); y - x) \, dt.$$

Since $\langle u, \phi \rangle = \langle u, \phi_1 \rangle + \langle u, \phi_2 \rangle$ and

$$\langle u, \phi_1 \rangle = \sum_{|\alpha| \leq m} \frac{1}{\alpha!} \langle u, \partial^\alpha \phi(x)(\cdot - x)^\alpha \rangle = \sum_{|\alpha| \leq m} (-1)^{|\alpha|} a_\alpha \partial^\alpha \phi(x)$$

$$= \sum_{|\alpha| \leq m} a_\alpha \langle \partial^\alpha \delta_x, \phi \rangle,$$

it suffices to show that $\langle u, \phi_2 \rangle = 0$.

By Theorem 3.6, we can choose a function $\chi \in C^\infty(\mathbb{R}^n)$ such that $\chi(y) = 1$ for $|y| \leq \frac{1}{2}$, and $\chi(y) = 0$ for $|y| \geq 1$. Define $\chi_\epsilon(y) = \chi(\epsilon^{-1}(y - x))$ so that $\chi_\epsilon = 1$ on $B_{\epsilon/2}(x)$, and $\chi_\epsilon = 0$ outside $B_\epsilon(x)$. Hence, $(1 - \chi_\epsilon)\phi_2 = 0$ on a neighbourhood of $\text{supp}\, u = \{x\}$, so $\langle u, (1 - \chi_\epsilon)\phi_2 \rangle = 0$ and therefore $\langle u, \phi_2 \rangle = \langle u, [\chi_\epsilon + (1 - \chi_\epsilon)]\phi_2 \rangle = \langle u, \chi_\epsilon \phi_2 \rangle$. Since $|\partial^\alpha \chi_\epsilon| \leq C\epsilon^{-|\alpha|}$ and $|\partial^\alpha \phi_2| \leq C\epsilon^{m+1-|\alpha|}$ on $B_\epsilon(x)$, the estimate (3.13) implies that

$$|\langle u, \phi_2 \rangle| \leq C \sum_{|\alpha| \leq m} \sup_K |\partial^\alpha(\chi_\epsilon \phi_2)| \leq C \sum_{|\alpha| \leq m} \epsilon^{|\alpha|-m} \sup_{|y-x| \leq \epsilon} |\partial^\alpha \phi_2(y)| \leq C\epsilon$$

for $0 < \epsilon \leq \epsilon_0$. Thus, $\langle u, \phi_2 \rangle = 0$, as required. $\qquad \square$

Fourier Transforms

To motivate the definition of the Fourier transform, we begin with some heuristic remarks on multiple Fourier *series*. Given $L > 0$, we say that a function $u : \mathbb{R}^n \to \mathbb{C}$ is *L-periodic* if

$$u(x + kL) = u(x) \qquad \text{for } x \in \mathbb{R}^n \text{ and } k \in \mathbb{Z}^n,$$

or in other words, if u is L-periodic in each of its n variables. We can think of such a function as being defined on the additive quotient group $\mathbb{T}_L^n = \mathbb{R}^n/(L\mathbb{Z}^n)$, and introduce the L_2 inner product

$$(u, v)_{\mathbb{T}_L^n} = \int_{\mathbb{T}_L^n} \overline{u(x)} v(x)\, dx,$$

where integration over $\mathbb{T}_L^n$ just means integration over any translate of the cube $(0, L)^n$. Using separation of variables, we can easily find the normalised L-periodic eigenfunctions of the Laplacian (1.1): for $k \in \mathbb{Z}^n$, the function $\phi_k(x) = L^{-n/2} e^{i2\pi k \cdot x/L}$ satisfies

$$-\Delta \phi_k = \left(\frac{2\pi |k|}{L} \right)^2 \phi_k \quad \text{on } \mathbb{T}_L^n,$$

and $\|\phi_k\|_{L_2(\mathbb{T}_L^n)} = 1$. Since $-\Delta$ is self-adjoint, we expect from Theorem 2.37 that the ϕ_k will form a complete orthonormal system in $L_2(\mathbb{T}_L^n)$. Thus, for a general L-periodic function u, we should have

$$u(x) = \sum_{k \in \mathbb{Z}^n} (\phi_k, u)_{\mathbb{T}_L^n} \phi_k(x) = \frac{1}{L^n} \sum_{k \in \mathbb{Z}^n} \hat{u}_L(k/L) e^{i2\pi (k/L) \cdot x}, \tag{3.14}$$

where

$$\hat{u}_L(\xi) = \int_{\mathbb{T}_L^n} e^{-i2\pi \xi \cdot x} u(x)\, dx.$$

The analogous expansion for a non-periodic function can be viewed as arising in the limit as $L \to \infty$. If $u \in L_1(\mathbb{R}^n)$, then we define its *Fourier transform* $\hat{u} = \mathcal{F}u$ by

$$\hat{u}(\xi) = \mathcal{F}_{x \to \xi}\{u(x)\} = \int_{\mathbb{R}^n} e^{-i2\pi \xi \cdot x} u(x)\, dx \quad \text{for } \xi \in \mathbb{R}^n,$$

and expect from (3.14) that, under appropriate conditions, $u = \mathcal{F}^* \hat{u}$, where $\mathcal{F}^*$ is the adjoint of the integral operator $\mathcal{F}$, i.e.,

$$u(x) = \mathcal{F}^*_{\xi \to x}\{\hat{u}(\xi)\} = \int_{\mathbb{R}^n} \hat{u}(\xi) e^{i2\pi \xi \cdot x}\, d\xi, \tag{3.15}$$

for $x \in \mathbb{R}^n$. In fact, we can readily prove the following.

Theorem 3.10 *If both u and $\hat{u} = \mathcal{F}u$ belong to $L_1(\mathbb{R}^n)$, then the Fourier inversion formula (3.15) is valid at every point x where u is continuous.*

Proof. Let $\psi(x) = e^{-\pi|x|^2}$ and $\psi_\epsilon(x) = \epsilon^{-n}\psi(\epsilon^{-1}x)$. Exercise 3.11 shows that ψ is invariant under the Fourier transform: $\mathcal{F}\psi = \psi = \mathcal{F}^*\psi$. Therefore, by Exercise 3.12, $\hat{\psi}_\epsilon(\xi) = \hat{\psi}(\epsilon\xi)$ and $\mathcal{F}^*\hat{\psi}_\epsilon = \psi_\epsilon$. In other words, the inversion formula is valid for ψ_ϵ, implying that

$$\int_{\mathbb{R}^n} \hat{u}(\xi)\hat{\psi}_\epsilon(\xi)e^{i2\pi\xi\cdot x}\,d\xi = \int_{\mathbb{R}^n}\left(\int_{\mathbb{R}^n} e^{-i2\pi\xi\cdot y}u(y)\,dy\right)\hat{\psi}_\epsilon(\xi)e^{i2\pi\xi\cdot x}\,d\xi$$

$$= \int_{\mathbb{R}^n} u(y)\int_{\mathbb{R}^n} e^{i2\pi(x-y)\cdot\xi}\hat{\psi}_\epsilon(\xi)\,d\xi\,dy$$

$$= \int_{\mathbb{R}^n} u(y)\psi_\epsilon(x-y)\,dy.$$

Since $\hat{\psi}_\epsilon(\xi) = \psi(\epsilon\xi) \to 1$ as $\epsilon \downarrow 0$ for each $\xi \in \mathbb{R}^n$, we deduce from the dominated convergence theorem that

$$\int_{\mathbb{R}^n} \hat{u}(\xi)e^{i2\pi\xi\cdot x}\,d\xi = \lim_{\epsilon\downarrow 0}(\psi_\epsilon * u)(x). \tag{3.16}$$

Assuming u is continuous at x, let $\epsilon_0 > 0$ and choose $\delta_0 > 0$ such that $|u(x-y) - u(x)| < \epsilon_0$ for $|y| < \delta_0$. Observe that $\int_{\mathbb{R}^n} \psi_\epsilon(y)\,dy = \hat{\psi}_\epsilon(0) = 1$, so for $0 < \epsilon < \epsilon_0$,

$$|(\psi_\epsilon * u)(x) - u(x)| = \left|\int_{\mathbb{R}^n} [u(x-y) - u(x)]\psi_\epsilon(y)\,dy\right|$$

$$\leq \int_{|y|<\delta_0} |u(x-y) - u(x)|\psi_\epsilon(y)\,dy$$

$$+ \int_{|y|\geq\delta_0} |u(x-y) - u(x)|\psi_\epsilon(y)\,dy$$

$$\leq \epsilon_0 \int_{\mathbb{R}^n} \psi_\epsilon(y)\,dy$$

$$+ \left(\int_{\mathbb{R}^n} |u(x-y) - u(x)|\,dy\right)\sup_{|y|\geq\delta_0}\psi_\epsilon(y)$$

$$\leq \epsilon_0 + 2\|u\|_{L_1(\mathbb{R}^n)}\epsilon^{-n}e^{-\pi(\delta_0/\epsilon)^2},$$

and the result follows at once from (3.16). $\qquad\square$

Corollary 3.11 *If both u and $\hat{u}$ are continuous everywhere and belong to $L_1(\mathbb{R}^n)$, then $\mathcal{F}^*\mathcal{F}u = u = \mathcal{F}\mathcal{F}^*u$.*

We now consider the actions of $\mathcal{F}$ and $\mathcal{F}^*$ on the Schwartz space of rapidly decreasing, C^∞ functions,

$$\mathcal{S}(\mathbb{R}^n) = \{\phi \in C^\infty(\mathbb{R}^n) : \sup_{x \in \mathbb{R}^n} |x^\alpha \partial^\beta \phi(x)| < \infty$$

for all multi-indices α and $\beta\}$.

Sequential convergence in this space is defined by interpreting the statement

$$\phi_j \to 0 \quad \text{in } \mathcal{S}(\mathbb{R}^n)$$

to mean that, for all multi-indices α and β,

$$x^\alpha \partial_j^\beta \phi(x) \to 0 \quad \text{uniformly for } x \in \mathbb{R}^n.$$

Elementary calculations show that if $\phi \in \mathcal{S}(\mathbb{R}^n)$, then

$$\mathcal{F}_{x \to \xi}\{\partial^\alpha \phi(x)\} = (i2\pi\xi)^\alpha \hat{\phi}(\xi) \quad \text{and} \quad \mathcal{F}_{x \to \xi}\{(-i2\pi x)^\alpha \phi(x)\} = \partial^\alpha \hat{\phi}(\xi),$$

$$(3.17)$$

so the Fourier transform defines a (sequentially) continuous linear operator

$$\mathcal{F} : \mathcal{S}(\mathbb{R}^n) \to \mathcal{S}(\mathbb{R}^n).$$

Moreover, by Corollary 3.11, this operator has a continuous inverse, namely the adjoint $\mathcal{F}^* : \mathcal{S}(\mathbb{R}^n) \to \mathcal{S}(\mathbb{R}^n)$.

By Exercise 3.8, the inclusions $\mathcal{D}(\mathbb{R}^n) \subseteq \mathcal{S}(\mathbb{R}^n) \subseteq \mathcal{E}(\mathbb{R}^n)$ are continuous with dense image, so we have

$$\mathcal{E}^*(\mathbb{R}^n) \subseteq \mathcal{S}^*(\mathbb{R}^n) \subseteq \mathcal{D}^*(\mathbb{R}^n),$$

and the elements of $\mathcal{S}^*(\mathbb{R}^n)$, i.e., the continuous linear functionals on $\mathcal{S}(\mathbb{R}^n)$, are called *temperate distributions*. A sufficient condition for a function $u \in L_{1,\text{loc}}(\mathbb{R}^n)$ to be a temperate distribution is that it is *slowly growing*: $u(x) = O(|x|^r)$ as $|x| \to \infty$, for some r. The formulae

$$\langle \mathcal{F}u, \phi \rangle = \langle u, \mathcal{F}\phi \rangle \quad \text{and} \quad \langle \mathcal{F}^*u, \phi \rangle = \langle u, \mathcal{F}^*\phi \rangle$$

are obviously valid if both u and ϕ belong to $\mathcal{S}(\mathbb{R}^n)$, and serve to define extensions

$$\mathcal{F} : \mathcal{S}^*(\mathbb{R}^n) \to \mathcal{S}^*(\mathbb{R}^n) \quad \text{and} \quad \mathcal{F}^* : \mathcal{S}^*(\mathbb{R}^n) \to \mathcal{S}^*(\mathbb{R}^n).$$

We also have the following result, known as Plancherel's theorem.

Theorem 3.12 The Fourier transform and its adjoint determine bounded linear operators

$$\mathcal{F} : L_2(\mathbb{R}^n) \to L_2(\mathbb{R}^n) \quad and \quad \mathcal{F}^* : L_2(\mathbb{R}^n) \to L_2(\mathbb{R}^n),$$

with $\mathcal{F}^{-1} = \mathcal{F}^*$. Furthermore, these operators are unitary:

$$(\mathcal{F}u, \mathcal{F}v) = (u, v) = (\mathcal{F}^*u, \mathcal{F}^*v) \quad for\ u, v \in L_2(\mathbb{R}^n).$$

Proof. If $u, v \in \mathcal{S}(\mathbb{R}^n)$, then $(\mathcal{F}u, \mathcal{F}v) = (\mathcal{F}^*\mathcal{F}u, v)$, and $\mathcal{F}^*\mathcal{F}u = u$ by Corollary 3.11, because $\hat{u} \in \mathcal{S}(\mathbb{R}^n) \subseteq L_1(\mathbb{R}^n)$, so $(\mathcal{F}u, \mathcal{F}v) = (u, v)$. In particular, taking $v = u$, we see that $\|\mathcal{F}u\|_{L_2(\mathbb{R}^n)} = \|u\|_{L_2(\mathbb{R}^n)}$ for $u \in \mathcal{S}(\mathbb{R}^n)$. Corollary 3.5 implies that $\mathcal{S}(\mathbb{R}^n)$ is dense in $L_2(\mathbb{R}^n)$, so $\mathcal{F}$ has a unique extension from $\mathcal{S}(\mathbb{R}^n)$ to a unitary operator on $L_2(\mathbb{R}^n)$. Furthermore, this extension satisfies $\langle \mathcal{F}u, \phi \rangle = \langle u, \mathcal{F}\phi \rangle$ for all $\phi \in \mathcal{S}(\mathbb{R}^n)$, consistent with the definition of $\mathcal{F}u$ as a temperate distribution. In other words, the extension from $\mathcal{S}(\mathbb{R}^n)$ agrees with the restriction from $\mathcal{S}^*(\mathbb{R}^n)$. Similar arguments yield the same results for $\mathcal{F}^*$. □

Corollary 3.13 The Fourier transform preserves the L_2-norm: $\|\hat{u}\|_{L_2(\mathbb{R}^n)} = \|u\|_{L_2(\mathbb{R}^n)}$.

Another important fact about the Fourier transform is its effect on convolutions: if $u, v \in L_1(\mathbb{R}^n)$ then

$$\mathcal{F}_{\xi \to x}\{(u * v)(x)\} = \int_{\mathbb{R}^n} e^{-i2\pi\xi \cdot x} \int_{\mathbb{R}^n} u(x - y)v(y)\,dy\,dx$$

$$= \int_{\mathbb{R}^n} \left(\int_{\mathbb{R}^n} e^{-i2\pi(x-y)\cdot\xi} u(x - y)\,dx \right) e^{-i2\pi\xi \cdot y} v(y)\,dy$$

$$= \hat{u}(\xi)\hat{v}(\xi).$$

Sobolev Spaces – First Definition

Suppose $1 \le p \le \infty$, and let Ω be a non-empty open subset of $\mathbb{R}^n$. The *Sobolev space* $W_p^r(\Omega)$ of order r based on $L_p(\Omega)$ is defined by

$$W_p^r(\Omega) = \{u \in L_p(\Omega) : \partial^\alpha u \in L_p(\Omega) \text{ for } |\alpha| \le r\}.$$

Here, of course, $\partial^\alpha u$ is viewed as a distribution on Ω, so the condition $\partial^\alpha u \in L_p(\Omega)$ means that there exists a function $g_\alpha \in L_p(\Omega)$ such that $\langle u, \partial^\alpha \phi \rangle_\Omega = (-1)^{|\alpha|} \langle g_\alpha, \phi \rangle_\Omega$ for all $\phi \in \mathcal{D}(\Omega)$, or equivalently $\partial^\alpha u = \iota g_\alpha$ where $\iota : L_{1,\mathrm{loc}}(\Omega)$

$\rightarrow \mathcal{D}^*(\Omega)$ is the imbedding defined by (3.11). Such a function g_α is often described as a *weak partial derivative* of u.

The completeness of $L_p(\Omega)$ implies that $W_p^r(\Omega)$ becomes a Banach space on putting

$$\|u\|_{W_p^r(\Omega)} = \left(\sum_{|\alpha| \le r} \int_\Omega |\partial^\alpha u(x)|^p \, dx \right)^{1/p} .$$

To define Sobolev spaces of fractional order, we denote the Slobodeckiĭ seminorm by

$$|u|_{\mu,p,\Omega} = \left(\int_\Omega \int_\Omega \frac{|u(x) - u(y)|^p}{|x - y|^{n+p\mu}} \, dx \, dy \right)^{1/p} \quad \text{for } 0 < \mu < 1. \quad (3.18)$$

Notice that the integrand is the pth power of $|u(x) - u(y)|/|x - y|^{\mu+n/p}$, so for $p = \infty$ we get the usual Hölder seminorm. For $s = r + \mu$, we define

$$W_p^s(\Omega) = \left\{ u \in W_p^r(\Omega) : |\partial^\alpha u|_{\mu,p,\Omega} < \infty \text{ for } |\alpha| = r \right\},$$

and equip this space with the norm

$$\|u\|_{W_p^s(\Omega)} = \left(\|u\|^p_{W_p^r(\Omega)} + \sum_{|\alpha|=r} |\partial^\alpha u|^p_{\mu,p,\Omega} \right)^{1/p} .$$

For any integer $r \ge 1$, the negative-order space $W_p^{-r}(\Omega)$ is defined to be the space of distributions $u \in \mathcal{D}^*(\Omega)$ that admit a representation

$$u = \sum_{|\alpha| \le r} \partial^\alpha f_\alpha \quad \text{with } f_\alpha \in L_p(\Omega). \quad (3.19)$$

This space is equipped with the norm

$$\|u\|_{W_p^{-r}(\Omega)} = \inf \left(\sum_{|\alpha| \le r} \|f_\alpha\|^p_{L_p(\Omega)} \right)^{1/p} ,$$

where the infimum is taken over all representations of the form (3.19). Using Hölder's inequality, it is easy to verify that

$$|(u, v)_\Omega| \le \|u\|_{W_{p^*}^{-r}(\Omega)} \|v\|_{W_p^r(\Omega)} \quad \text{for } u \in W_{p^*}^{-r}(\Omega) \text{ and } v \in \mathcal{D}(\Omega),$$

where p^* is given by (3.1).

In this book, we will rarely use Sobolev spaces with $p \neq 2$, and so adopt the abbreviation $W^s(\Omega) = W_2^s(\Omega)$. For any integer $r \geq 0$, the norm in $W^r(\Omega)$ arises from the inner product

$$(u, v)_{W^r(\Omega)} = \sum_{|\alpha| \leq r} \int_\Omega \overline{\partial^\alpha u(x)} \partial^\alpha v(x)\, dx,$$

and likewise if $s = r + \mu$ then the norm in $W^s(\Omega)$ arises from the inner product

$(u, v)_{W^s(\Omega)}$

$$= (u, v)_{W^r(\Omega)} + \sum_{|\alpha|=r} \int_\Omega \int_\Omega \frac{[\overline{\partial^\alpha u(x)} - \overline{\partial^\alpha u(y)}][\partial^\alpha v(x) - \partial^\alpha v(y)]}{|x - y|^{n+2\mu}}\, dx\, dy.$$

Thus, $W^s(\Omega)$ is a Hilbert space for all real $s \geq 0$.

Sobolev Spaces – Second Definition

In this section, we introduce a second family of Sobolev spaces, which later will turn out to be equivalent to the one given in the preceding section.

For $s \in \mathbb{R}$, we define a continuous linear operator $\mathcal{J}^s : \mathcal{S}(\mathbb{R}^n) \to \mathcal{S}(\mathbb{R}^n)$, called the *Bessel potential* of order s, by

$$\mathcal{J}^s u(x) = \int_{\mathbb{R}^n} (1 + |\xi|^2)^{s/2} \hat{u}(\xi) e^{\mathrm{i}2\pi \xi \cdot x}\, d\xi \quad \text{for } x \in \mathbb{R}^n.$$

In this way,

$$\mathcal{F}_{x \to \xi}\{\mathcal{J}^s u(x)\} = (1 + |\xi|^2)^{s/2} \hat{u}(\xi), \tag{3.20}$$

so under Fourier transformation the action of $\mathcal{J}^s$ is to multiply $\hat{u}(\xi)$ by a function that is $O(|\xi|^s)$ for large ξ. We can therefore think of $\mathcal{J}^s$ as a kind of differential operator of order s; cf. (3.17). Notice that for all $s, t \in \mathbb{R}$,

$$\mathcal{J}^{s+t} = \mathcal{J}^s \mathcal{J}^t, \quad (\mathcal{J}^s)^{-1} = \mathcal{J}^{-s}, \quad \mathcal{J}^0 = \text{identity operator}.$$

It follows from (3.20) that

$$\langle \mathcal{J}^s u, v \rangle = \langle u, \mathcal{J}^s v \rangle \quad \text{and} \quad (\mathcal{J}^s u, v) = (u, \mathcal{J}^s v)$$

for all $u, v \in \mathcal{S}(\mathbb{R}^n)$, giving a natural extension of the Bessel potential to a linear operator $\mathcal{J}^s : \mathcal{S}^*(\mathbb{R}^n) \to \mathcal{S}^*(\mathbb{R}^n)$ on the space of temperate distributions.

For any $s \in \mathbb{R}$, we define $H^s(\mathbb{R}^n)$, the Sobolev space of order s on $\mathbb{R}^n$, by

$$H^s(\mathbb{R}^n) = \{u \in \mathcal{S}^*(\mathbb{R}^n) : \mathcal{J}^s u \in L_2(\mathbb{R}^n)\},$$

and equip this space with the inner product

$$(u, v)_{H^s(\mathbb{R}^n)} = (\mathcal{J}^s u, \mathcal{J}^s v)$$

and the induced norm

$$\|u\|_{H^s(\mathbb{R}^n)} = \sqrt{(u, u)_{H^s(\mathbb{R}^n)}} = \|\mathcal{J}^s u\|_{L_2(\mathbb{R}^n)}. \tag{3.21}$$

Notice that the Bessel potential

$$\mathcal{J}^s : H^s(\mathbb{R}^n) \to L_2(\mathbb{R}^n)$$

is a unitary isomorphism, and in particular, since $\mathcal{J}^0 u = u$,

$$H^0(\mathbb{R}^n) = L_2(\mathbb{R}^n).$$

Several facts about $H^s(\mathbb{R}^n)$ follow immediately from standard properties of $L_2(\mathbb{R}^n)$. For instance, $H^s(\mathbb{R}^n)$ is a separable Hilbert space, and $\mathcal{D}(\mathbb{R}^n)$ is dense in $H^s(\mathbb{R}^n)$ because $\mathcal{J}^s[\mathcal{S}(\mathbb{R}^n)] = \mathcal{S}(\mathbb{R}^n)$ is dense in $L_2(\mathbb{R}^n)$, and the inclusion $\mathcal{D}(\mathbb{R}^n) \subseteq \mathcal{S}(\mathbb{R}^n)$ is continuous with dense image. Also, one sees from (3.2) and (3.3), with $p = 2$, that $H^{-s}(\mathbb{R}^n)$ is an isometric realisation of the dual space of $H^s(\mathbb{R}^n)$, i.e.,

$$H^{-s}(\mathbb{R}^n) = [H^s(\mathbb{R}^n)]^* \quad \text{for } s \in \mathbb{R}, \tag{3.22}$$

and

$$\|u\|_{H^{-s}(\mathbb{R}^n)} = \sup_{0 \neq v \in H^s(\mathbb{R}^n)} \frac{|\langle u, v \rangle|}{\|v\|_{H^s(\mathbb{R}^n)}} = \sup_{0 \neq v \in H^s(\mathbb{R}^n)} \frac{|(u, v)|}{\|v\|_{H^s(\mathbb{R}^n)}}$$

for $u \in H^{-s}(\mathbb{R}^n)$. Plancherel's theorem (Theorem 3.12) and (3.20) imply that

$$\|u\|^2_{H^s(\mathbb{R}^n)} = \int_{\mathbb{R}^n} (1 + |\xi|^2)^s |\hat{u}(\xi)|^2 \, d\xi,$$

so if $s \leq t$ then $\|u\|_{H^s(\mathbb{R}^n)} \leq \|u\|_{H^t(\mathbb{R}^n)}$ and hence $H^t(\mathbb{R}^n) \subseteq H^s(\mathbb{R}^n)$. This inclusion is continuous with dense image.

For any closed set $F \subseteq \mathbb{R}^n$, we define the associated Sobolev space of order s by

$$H^s_F = \{u \in H^s(\mathbb{R}^n) : \operatorname{supp} u \subseteq F\},$$

whereas for any non-empty open set $\Omega \subseteq \mathbb{R}^n$ we define

$$H^s(\Omega) = \{u \in \mathcal{D}^*(\Omega) : u = U|_\Omega \text{ for some } U \in H^s(\mathbb{R}^n)\}.$$

We see at once that H_F^s is a closed subspace of $H^s(\mathbb{R}^n)$, and is therefore a Hilbert space when equipped with the restriction of the inner product of $H^s(\mathbb{R}^n)$. A Hilbert structure for $H^s(\Omega)$ is defined with the help of the orthogonal projection

$$P = P_{s,\Omega} : H^s(\mathbb{R}^n) \to H_{\mathbb{R}^n \setminus \Omega}^s,$$

which satisfies

$$PU|_\Omega = 0 \text{ and } (I - P)U|_\Omega = U|_\Omega \qquad \text{for all } U \in H^s(\mathbb{R}^n).$$

Noting that if $U|_\Omega = 0$ then $PU = U$ because $U \in H_{\mathbb{R}^n \setminus \Omega}^s$, we see that a well-defined inner product on $H^s(\Omega)$ arises by putting

$$(u, v)_{H^s(\Omega)} = \big((I - P)U, (I - P)V\big)_{H^s(\mathbb{R}^n)} \quad \text{if } u = U|_\Omega \text{ and } v = V|_\Omega,$$

for $U, V \in H^s(\mathbb{R}^n)$. Notice that the induced norm satisfies

$$\|u\|_{H^s(\Omega)} = \sqrt{(u, u)_{H^s(\Omega)}} = \min_{U|_\Omega = u, U \in H^s(\mathbb{R}^n)} \|U\|_{H^s(\mathbb{R}^n)}, \qquad (3.23)$$

because if $U|_\Omega = u$ then

$$\|U\|_{H^s(\mathbb{R}^n)}^2 = \|PU\|_{H^s(\mathbb{R}^n)}^2 + \|(I - P)U\|_{H^s(\mathbb{R}^n)}^2$$
$$\geq \|(I - P)U\|_{H^s(\mathbb{R}^n)}^2 = \|u\|_{H^s(\mathbb{R}^n)}^2.$$

The map $U \mapsto (I - P)U|_\Omega$ is a unitary isomorphism from the orthogonal complement of $H_{\mathbb{R}^n \setminus \Omega}^s$ onto $H^s(\Omega)$. Therefore, $H^s(\Omega)$ is a separable Hilbert space. Also, (3.23) shows that the restriction operator $U \mapsto U|_\Omega$ is continuous from $H^s(\mathbb{R}^n)$ to $H^s(\Omega)$, and thus the space

$$\mathcal{D}(\overline{\Omega}) = \{u : u = U|_\Omega \text{ for some } U \in \mathcal{D}(\mathbb{R}^n)\}$$

is dense in $H^s(\Omega)$ because $\mathcal{D}(\mathbb{R}^n)$ is dense in $H^s(\mathbb{R}^n)$.

We also define two other Sobolev spaces on Ω,

$$\widetilde{H}^s(\Omega) = \text{closure of } \mathcal{D}(\Omega) \text{ in } H^s(\mathbb{R}^n),$$
$$H_0^s(\Omega) = \text{closure of } \mathcal{D}(\Omega) \text{ in } H^s(\Omega),$$

which we make into Hilbert spaces in the obvious way, by restriction of the inner products in $H^s(\mathbb{R}^n)$ and in $H^s(\Omega)$, respectively. It is clear that

$$\widetilde{H}^s(\Omega) \subseteq H^s_{\overline{\Omega}} \quad \text{and} \quad \widetilde{H}^s(\Omega) \subseteq H^s_0(\Omega),$$

and later we shall establish the reverse inclusions subject to conditions on Ω and s. Note that an element of $H^s_{\overline{\Omega}}$ is a distribution on $\mathbb{R}^n$, but, provided the n-dimensional Lebesgue measure of the boundary of Ω is zero, the restriction operator $u \mapsto u|_\Omega$ defines an imbedding

$$H^s_{\overline{\Omega}} \subseteq L_2(\Omega) \quad \text{for } s \geq 0. \tag{3.24}$$

(If $u \in H^s_{\overline{\Omega}}$ and $u|_\Omega = 0$, then $\operatorname{supp} u \subseteq \partial\Omega = \overline{\Omega} \setminus \Omega$, implying $u = 0$ on $\mathbb{R}^n$.) In general, if $s < -\frac{1}{2}$, then $H^s_{\partial\Omega} \neq \{0\}$ no matter how smooth the boundary of Ω, so we cannot imbed $H^s_{\overline{\Omega}}$ in a space of distributions on Ω; see Lemma 3.39.

The necessity of introducing more than one kind of Sobolev space on Ω can be seen already from the next theorem, which extends our earlier observation that $H^{-s}(\mathbb{R}^n)$ is an isometric realisation of the dual space of $H^s(\mathbb{R}^n)$; see also Theorem 3.30.

Theorem 3.14 *Let Ω be a non-empty open subset of $\mathbb{R}^n$, and let $s \in \mathbb{R}$.*

(i) If $\widetilde{H}^{-s}(\Omega) = H^{-s}_{\overline{\Omega}}$, and if we define

$$(u, v)_\Omega = (u, V) \quad \text{for } u \in \widetilde{H}^{-s}(\Omega) \text{ and } v = V|_\Omega \text{ with } V \in H^s(\mathbb{R}^n),$$

then $\widetilde{H}^{-s}(\Omega)$ is an isometric realisation of $H^s(\Omega)^*$.

(ii) If $\widetilde{H}^s(\Omega) = H^s_{\overline{\Omega}}$, and if we define

$$(u, v)_\Omega = (U, v) \quad \text{for } u = U|_\Omega \text{ with } U \in H^{-s}(\mathbb{R}^n), \text{ and } v \in \widetilde{H}^s(\Omega),$$

then $H^{-s}(\Omega)$ is an isometric realisation of $\widetilde{H}^s(\Omega)^*$.

Proof. First note that

$$(u, V) = 0 \quad \text{if } u \in \mathcal{D}(\Omega), V \in H^s(\mathbb{R}^n) \text{ and } V|_\Omega = 0,$$

so $(u, v)_\Omega$ is well defined for $u \in \widetilde{H}^{-s}(\Omega)$ and $v \in H^s(\Omega)$. We claim that $(\iota u)(v) = (u, v)_\Omega$ defines an isometric isomorphism $\iota : \widetilde{H}^{-s}(\Omega) \to H^s(\Omega)^*$. Indeed,

$$|(\iota u)(v)| = |(u, V)| \leq \|u\|_{H^{-s}(\mathbb{R}^n)} \|V\|_{H^s(\mathbb{R}^n)} \quad \text{whenever } v = V|_\Omega,$$

so $|(\iota u)(v)| \leq \|u\|_{\widetilde{H}^{-s}(\Omega)} \|v\|_{H^s(\Omega)}$ and hence $\|\iota u\|_{H^s(\Omega)^*} \leq \|u\|_{\widetilde{H}^{-s}(\Omega)}$. Furthermore, given a functional $\ell \in H^s(\Omega)^*$, the map $V \mapsto \ell(V|_\Omega)$ is bounded on $H^s(\mathbb{R}^n)$ because the restriction map $V \mapsto V|_\Omega$ is bounded from $H^s(\mathbb{R}^n)$ onto $H^s(\Omega)$. We know already that $H^{-s}(\mathbb{R}^n)$ is an isometric realisation of $[H^s(\mathbb{R}^n)]^*$, so there exists $u \in H^{-s}(\mathbb{R}^n)$ satisfying

$$\ell(V|_\Omega) = (u, V) \quad \text{for all } V \in H^s(\mathbb{R}^n),$$

and

$$\|u\|_{H^{-s}(\mathbb{R}^n)} = \sup_{0 \neq V \in H^s(\mathbb{R}^n)} \frac{|\ell(V|_\Omega)|}{\|V\|_{H^s(\mathbb{R}^n)}}.$$

If $V \in \mathcal{D}(\mathbb{R}^n \setminus \Omega)$, then $V|_\Omega = 0$, so $(u, V) = 0$, showing that $\operatorname{supp} u \subseteq \overline{\Omega}$, i.e., $u \in H^{-s}_{\overline{\Omega}}$. Assume now that $\widetilde{H}^{-s}(\Omega) = H^{-s}_{\overline{\Omega}}$. In this case, $u \in \widetilde{H}^{-s}(\Omega)$, and we see from

$$\ell(v) = \ell(V|_\Omega) = (u, V) = (\iota u)(v) \quad \text{for } v = V|_\Omega \text{ and } V \in H^s(\mathbb{R}^n)$$

that $\ell = \iota u$, so ι is onto. Finally,

$$|\ell(V|_\Omega)| \leq \|\ell\|_{H^s(\Omega)^*} \|V|_\Omega\|_{H^s(\Omega)} \leq \|\iota u\|_{H^s(\Omega)^*} \|V\|_{H^s(\mathbb{R}^n)},$$

implying that $\|u\|_{\widetilde{H}^{-s}(\Omega)} = \|u\|_{H^{-s}(\mathbb{R}^n)} \leq \|\iota u\|_{H^s(\Omega)^*}$. Thus, ι is an isometry. Part (ii) then follows because every Hilbert space is reflexive. $\qquad\square$

Equivalence of the Norms

When $\Omega = \mathbb{R}^n$ and $s \geq 0$, a simple argument based on Plancherel's theorem shows the equivalence of the two kinds of Sobolev norms; see also Exercise 3.13. We use the abbreviation

$$|u|_{\mu,\Omega} = \left(\int_\Omega \int_\Omega \frac{|u(x) - u(y)|^2}{|x - y|^{n+2\mu}} \, dx \, dy \right)^{1/2}$$

for the Slobodeckiĭ seminorm (3.18) with $p = 2$, and write $|u|_\mu = |u|_{\mu,\mathbb{R}^n}$ in the usual way.

Lemma 3.15 If $0 < \mu < 1$, then

$$|u|_\mu^2 = a_\mu \int_{\mathbb{R}^n} |\xi|^{2\mu} |\hat{u}(\xi)|^2 \, d\xi,$$

where

$$a_\mu = \int_{t>0} t^{-2\mu-1} \int_{|\omega|=1} \left| e^{i2\pi\omega_1 t} - 1 \right|^2 d\omega \, dt.$$

Proof. Define the forward difference operator δ_h by

$$\delta_h u(x) = u(x+h) - u(x),$$

and consider the Fourier transform

$$\mathcal{F}\delta_h u(\xi) = (e^{i2\pi\xi \cdot h} - 1)\hat{u}(\xi). \tag{3.25}$$

By making the substitution $h = y - x$, applying Plancherel's theorem and then reversing the order of integration, one finds that

$$|u|_\mu^2 = \int_{\mathbb{R}^n} \frac{\|\delta_h u\|_{L_2(\mathbb{R}^n)}^2}{|h|^{2\mu+n}}\, dh = \int_{\mathbb{R}^n} |\hat{u}(\xi)|^2 \int_{\mathbb{R}^n} \frac{|e^{i2\pi\xi \cdot h} - 1|^2}{|h|^{2\mu+n}}\, dh\, d\xi.$$

We transform the inner integral to polar coordinates, letting $h = \rho\omega$, where $\rho = |h|$ and $\omega = h/|h|$. Since $dh = \rho^{n-1}\, d\rho\, d\omega$,

$$\int_{\mathbb{R}^n} \frac{|e^{i2\pi\xi \cdot h} - 1|^2}{|h|^{2\mu+n}}\, dh = \int_{\rho>0} \rho^{-2\mu-1} \int_{|\omega|=1} |e^{i2\pi\rho\xi \cdot \omega} - 1|^2\, d\omega\, d\rho = a_\mu |\xi|^{2\mu},$$

where we used the substitution $\rho = |\xi|^{-1}t$ and exploited radial symmetry. Note that $\int_{|\omega|=1} |e^{i2\pi\omega_1 t} - 1|^2\, d\omega$ is $O(t^2)$ as $t \downarrow 0$, and is $O(1)$ as $t \uparrow \infty$, so that a_μ is a finite, positive real number for $0 < \mu < 1$.   □

Theorem 3.16 *If $s \geq 0$, then $W^s(\mathbb{R}^n) = H^s(\mathbb{R}^n)$ with equivalent norms.*

Proof. Let r be a non-negative integer, and let $0 < \mu < 1$. In view of (3.17), Plancherel's theorem gives

$$\|u\|_{W^r(\mathbb{R}^n)}^2 = \sum_{|\alpha|\leq r} \|\partial^\alpha u\|_{L_2(\mathbb{R}^n)}^2 = \int_{\mathbb{R}^n} b_r(\xi)|\hat{u}(\xi)|^2\, d\xi$$

where

$$b_r(\xi) = \sum_{|\alpha|\leq r} (2\pi\xi)^{2\alpha} \sim (1 + |\xi|^2)^r,$$

proving the result if $s = r$. By Lemma 3.15, if $s = r + \mu$, then

$$\|u\|_{W^s(\mathbb{R}^n)}^2 = \|u\|_{W^r(\mathbb{R}^n)}^2 + \sum_{|\alpha|=r} |\partial^\alpha u|_\mu^2$$

$$= \int_{\mathbb{R}^n} b_r(\xi)|\hat{u}(\xi)|^2\, d\xi + \sum_{|\alpha|=r} \int_{\mathbb{R}^n} a_\mu |\xi|^{2\mu} |(2\pi\xi)^\alpha \hat{u}(\xi)|^2\, d\xi$$

$$\sim \int_{\mathbb{R}^n} (1 + |\xi|^2)^{r+\mu} |\hat{u}(\xi)|^2\, d\xi = \|u\|_{H^{r+\mu}(\mathbb{R}^n)}^2.$$

□

Corollary 3.17 For any non-empty open subset $\Omega \subseteq \mathbb{R}^n$, there is a continuous inclusion

$$H^s(\Omega) \subseteq W^s(\Omega) \quad for\ s \geq 0.$$

Proof. Given $u \in H^s(\Omega)$, we can find $U \in H^s(\mathbb{R}^n)$ such that $u = U|_\Omega$ and $\|u\|_{H^s(\Omega)} = \|U\|_{H^s(\mathbb{R}^n)}$. By the theorem, $U \in W^s(\mathbb{R}^n)$, so $u \in W^s(\Omega)$ and

$$\|u\|_{W^s(\Omega)} \leq \|U\|_{W^s(\mathbb{R}^n)} \sim \|U\|_{H^s(\mathbb{R}^n)} = \|u\|_{H^s(\Omega)}. \qquad \square$$

The next theorem shows that the reverse inclusion holds if there exists an *extension operator* for Ω. The existence of such operators is proved in Appendix A, under appropriate assumptions on Ω; see also Theorem 3.30.

Theorem 3.18 For any non-empty open set $\Omega \subseteq \mathbb{R}^n$ and any real $s \geq 0$, if there exists a continuous linear operator $E : W^s(\Omega) \to W^s(\mathbb{R}^n)$ such that $Eu|_\Omega = u$ for all $u \in W^s(\Omega)$, then

$$H^s(\Omega) = W^s(\Omega)$$

with equivalent norms.

Proof. If $u \in W^s(\Omega)$, then $u = U|_\Omega$ for $U = Eu \in W^s(\mathbb{R}^n) = H^s(\mathbb{R}^n)$, so $u \in H^s(\Omega)$ and

$$\|u\|_{H^s(\Omega)} \leq \|U\|_{H^s(\mathbb{R}^n)} \sim \|Eu\|_{W^s(\mathbb{R}^n)} \leq C\|u\|_{W^s(\Omega)},$$

giving a continuous inclusion $W^s(\Omega) \subseteq H^s(\Omega)$. $\qquad \square$

No extension operator is needed to establish the corresponding result when s is a negative integer.

Theorem 3.19 For any non-empty open set $\Omega \subseteq \mathbb{R}^n$, and for any integer $r \geq 0$,

$$H^{-r}(\Omega) = W^{-r}(\Omega)$$

with equivalent norms.

Proof. First consider the case $\Omega = \mathbb{R}^n$, and recall that $[H^r(\mathbb{R}^n)]^* = H^{-r}(\mathbb{R}^n)$. We define a Banach space isomorphism $J : W^r(\mathbb{R}^n) \to H^{-r}(\mathbb{R}^n)$ by

$$(Ju, v) = (u, v)_{W^r(\mathbb{R}^n)} \quad for\ u, v \in W^r(\mathbb{R}^n) = H^r(\mathbb{R}^n),$$

and introduce another inner product and norm for $H^{-r}(\mathbb{R}^n)$,

$$((u, v))_{-r} = (J^{-1}u, J^{-1}v)_{W^r} \quad \text{and} \quad \|u\|_{-r} = \sqrt{((u, u))_{-r}} = \|J^{-1}u\|_{W^r(\mathbb{R}^n)}.$$

Obviously, $\|u\|_{-r} \sim \|u\|_{H^{-r}(\mathbb{R}^n)}$.

If $u \in H^{-r}(\mathbb{R}^n)$ and $v \in H^r(\mathbb{R}^n)$, then

$$(u, v) = (J^{-1}u, v)_{W^r(\mathbb{R}^n)} = \sum_{|\alpha| \leq r} (\partial^\alpha J^{-1}u, \partial^\alpha v) = \sum_{|\alpha| \leq r} (\partial^\alpha f_\alpha, v),$$

where $f_\alpha = (-1)^{|\alpha|} \partial^\alpha J^{-1}u \in W^{r-|\alpha|}(\mathbb{R}^n) \subseteq L_2(\mathbb{R}^n)$. Thus, $u = \sum_{|\alpha| \leq r} \partial^\alpha f_\alpha \in W^{-r}(\mathbb{R}^n)$ with

$$\|u\|^2_{W^{-r}(\mathbb{R}^n)} \leq \sum_{|\alpha| \leq r} \|f_\alpha\|^2_{L_2(\mathbb{R}^n)} = \sum_{|\alpha| \leq r} \|\partial^\alpha J^{-1}u\|^2_{L_2(\mathbb{R}^n)}$$
$$= (J^{-1}u, J^{-1}u)_{W^r(\mathbb{R}^n)} = \|u\|^2_{-r}.$$

Conversely, if $u \in W^{-r}(\mathbb{R}^n)$ has a representation $u = \sum_{|\alpha| \leq r} \partial^\alpha f_\alpha$, then

$$|(u, v)| = \left| \sum_{|\alpha| \leq r} (\partial^\alpha f_\alpha, v) \right| = \left| \sum_{|\alpha| \leq r} (-1)^{|\alpha|}(f_\alpha, \partial^\alpha v) \right|$$
$$\leq C \left(\sum_{|\alpha| \leq r} \|f_\alpha\|^2_{L_2(\mathbb{R}^n)} \right)^{1/2} \|v\|_{W^r(\mathbb{R}^n)}$$

for all $v \in W^r(\mathbb{R}^n) = H^r(\mathbb{R}^n)$, so $u \in H^{-r}(\mathbb{R}^n)$ and $\|u\|_{H^{-r}(\mathbb{R}^n)} \leq C\|u\|_{W^{-r}(\mathbb{R}^n)}$.

Now consider the case when $\Omega \neq \mathbb{R}^n$, and let $u \in H^{-r}(\Omega)$. We choose $U \in H^{-r}(\mathbb{R}^n) = W^{-r}(\mathbb{R}^n)$ such that $u = U|_\Omega$ and $\|u\|_{H^{-r}(\Omega)} = \|U\|_{H^{-r}(\mathbb{R}^n)}$, and then choose $F_\alpha \in L_2(\mathbb{R}^n)$ such that

$$U = \sum_{|\alpha| \leq r} \partial^\alpha F_\alpha \quad \text{and} \quad \|U\|^2_{W^{-r}(\mathbb{R}^n)} \leq \sum_{|\alpha| \leq r} \|F_\alpha\|^2_{L_2(\mathbb{R}^n)} \leq 2\|U\|^2_{W^{-r}(\mathbb{R}^n)}.$$

Put $f_\alpha = F_\alpha|_\Omega \in L_2(\Omega)$, so that $u = \sum_{|\alpha| \leq r} \partial^\alpha f_\alpha$ on Ω. Obviously, $u \in W^{-r}(\Omega)$, and

$$\|u\|^2_{W^{-r}(\Omega)} \leq \sum_{|\alpha| \leq r} \|f_\alpha\|^2_{L_2(\Omega)} \leq \sum_{|\alpha| \leq r} \|F_\alpha\|^2_{L_2(\mathbb{R}^n)}$$
$$\leq 2\|U\|^2_{W^{-r}(\mathbb{R}^n)} \sim \|U\|^2_{H^{-r}(\mathbb{R}^n)} = \|u\|^2_{H^{-r}(\Omega)}.$$

Conversely, suppose that $u \in W^{-r}(\Omega)$ satisfies $u = \sum_{|\alpha| \leq r} \partial^\alpha f_\alpha$ on Ω with $f_\alpha \in L_2(\Omega)$. Extend f_α by zero outside Ω, and denote this extension by $F_\alpha \in L_2(\mathbb{R}^n)$.

Put $U = \sum_{|\alpha|\leq r} \partial^\alpha F_\alpha$, so that $u = U|_\Omega$ and $U \in W^{-r}(\mathbb{R}^n)$, with

$$\|u\|^2_{H^{-r}(\Omega)} \leq \|U\|^2_{H^{-r}(\mathbb{R}^n)} \sim \|U\|^2_{W^{-r}(\mathbb{R}^n)} \leq \sum_{|\alpha|\leq r} \|F_\alpha\|^2_{L_2(\mathbb{R}^n)} = \sum_{|\alpha|\leq r} \|f_\alpha\|^2_{L_2(\Omega)}.$$

Thus, $u \in H^{-r}(\Omega)$ and $\|u\|_{H^{-r}(\Omega)} \leq C\|u\|_{W^{-r}(\Omega)}$. $\qquad\qquad\square$

Localisation and Changes of Coordinates

Our next result shows that multiplication by a smooth cutoff function defines a bounded linear operator on $H^s(\Omega)$ or on $\widetilde{H}^s(\Omega)$.

Theorem 3.20 Suppose that $\phi \in C^r_{\text{comp}}(\mathbb{R}^n)$ for some integer $r \geq 1$, and let $|s| \leq r$. If $u \in H^s(\Omega)$, then $\phi u \in H^s(\Omega)$ and

$$\|\phi u\|_{H^s(\Omega)} \leq C_r \|\phi\|_{W^r_\infty(\mathbb{R}^n)} \|u\|_{H^s(\Omega)}.$$

The same holds with $H^s(\Omega)$ replaced by $\widetilde{H}^s(\Omega)$.

Proof. Suppose first that $\Omega = \mathbb{R}^n$. In view of Theorem 3.16, the result is clear for $s = r$, and hence by the duality relation (3.22), for $s = -r$. The case $-r < s < r$ then follows by interpolation, using Theorem B.7 (from Appendix B). An alternative proof, leading to a different norm for ϕ, is given in Exercise 3.16.

Now let Ω be any open subset of $\mathbb{R}^n$, and let $u = U|_\Omega$ for some $U \in H^s(\mathbb{R}^n)$. Since $\phi u = (\phi U)|_\Omega$ or, strictly speaking, $\phi|_\Omega u = (\phi U)|_\Omega$, we see that

$$\|\phi u\|_{H^s(\Omega)} \leq \|\phi U\|_{H^s(\mathbb{R}^n)} \leq C_r \|\phi\|_{W^r_\infty(\mathbb{R}^n)} \|U\|_{H^s(\mathbb{R}^n)},$$

and the desired estimate follows after taking the infimum over U. With $H^s(\Omega)$ replaced by $\widetilde{H}^s(\Omega)$, the proof is even easier because $\|\phi u\|_{\widetilde{H}^s(\Omega)} = \|\phi u\|_{H^s(\mathbb{R}^n)}$ and $\|u\|_{\widetilde{H}^s(\Omega)} = \|u\|_{H^s(\mathbb{R}^n)}$. $\qquad\qquad\square$

In conjunction with Theorem 3.20, the following notion is often useful; cf. Exercise 3.19. A *partition of unity* for an open set $S \subseteq \mathbb{R}^n$ is a (finite or infinite) sequence of functions $\phi_1, \phi_2, \ldots$ in $C^\infty(\mathbb{R}^n)$ such that

1. $\phi_j \geq 0$ on $\mathbb{R}^n$ for each j;
2. each point of S has a neighbourhood that intersects $\operatorname{supp}\phi_j$ for only finitely many j;
3. $\sum_{j\geq 1} \phi_j(x) = 1$ for each $x \in S$.

Notice that condition 2 implies that the sum in condition 3 is finite for each $x \in S$. If S is not open, then we say that the ϕ_j form a partition of unity for S if they form a partition of unity for some open neighbourhood of S.

Suppose now that $\mathcal{W}$ is an open cover for S, i.e., $\mathcal{W}$ is a family of open sets for which $S \subseteq \bigcup \mathcal{W}$. We say that a partition of unity $(\phi_j)_{j \geq 1}$ is *subordinate to* $\mathcal{W}$ if for each j there exists $W \in \mathcal{W}$ such that $\operatorname{supp} \phi_j \subseteq W$.

Theorem 3.21 *Given any open cover $\mathcal{W}$ of a set $S \subseteq \mathbb{R}^n$, there exists a partition of unity $(\phi_j)_{j \geq 1}$ for S subordinate to $\mathcal{W}$. Moreover, the ϕ_j can be chosen in such a way that $\operatorname{supp} \phi_j$ is compact for each j.*

Proof. Put $\Omega = \bigcup \mathcal{W}$. We begin by constructing open n-cubes $Q_1, Q_2, \ldots$ with the following properties:

(i) the family $\{Q_j\}_{j \geq 1}$ is an open cover of Ω;
(ii) for each j there exists a set $W \in \mathcal{W}$ such that $Q_j \subseteq W$;
(iii) each point of Ω has a neighbourhood that intersects only finitely many of the Q_j.

Let $\Omega_1 \subsetneq \Omega_2 \subsetneq \Omega_3 \subsetneq \cdots$ be a strictly increasing sequence of bounded open sets whose union is Ω and that satisfy $\overline{\Omega}_j \Subset \Omega_{j+1}$ for each $j \geq 1$. For convenience, we put $\Omega_{-1} = \Omega_0 = \emptyset$, and then define the compact sets $K_j = \overline{\Omega}_j \setminus \Omega_{j-1}$ for $j \geq 1$. Since K_j does not intersect $\overline{\Omega}_{j-2}$, given $x \in K_j$ we can find an open cube $G_{j,x}$ centred at x with $\overline{G}_{j,x} \Subset W \setminus \overline{\Omega}_{j-2}$ for some $W \in \mathcal{W}$. The family $\{G_{j,x} : x \in K_j\}$ is an open cover for K_j, so by compactness we can extract a finite subcover $\mathcal{Q}_j$. After relabelling the cubes, we obtain a countable family $\bigcup_{j \geq 1} \mathcal{Q}_j = \{Q_1, Q_2, \ldots\}$ with the required properties (i)–(iii). (The set Ω_j is disjoint from each cube in $\mathcal{Q}_{j+2} \cup \mathcal{Q}_{j+3} \cup \ldots$, and so intersects only cubes in the finite family $\mathcal{Q}_1 \cup \cdots \cup \mathcal{Q}_{j+1}$.)

For each $j \geq 1$, we can use Exercise 3.6 to construct a function $\psi_j \in C^\infty_{\text{comp}}(\mathbb{R}^n)$ satisfying $\psi_j > 0$ on Q_j, and $\psi_j = 0$ on $\mathbb{R}^n \setminus Q_j$. Property (iii) of the Q_j implies that the sum $\Psi(x) = \sum_{j \geq 1} \psi_j(x)$ defines a function $\Psi \in C^\infty(\Omega)$, and property (i) implies that $\Psi > 0$ on Ω. Hence, we obtain the desired partition of unity by defining $\phi_j(x) = \psi_j(x)/\Psi(x)$ for $x \in \Omega$, and $\phi_j(x) = 0$ otherwise. $\square$

Corollary 3.22 *Given any countable open cover $\{W_1, W_2, \ldots\}$ of a set $S \subseteq \mathbb{R}^n$, there exists a partition of unity $\phi_1, \phi_2, \ldots$ for S having the property that $\operatorname{supp} \phi_j \subseteq W_j$ for each $j \geq 1$.*

Proof. Let $\theta_1, \theta_2, \ldots$ be a partition of unity for S subordinate to the given open cover, define the index sets $I_1 = \{k \geq 1 : \operatorname{supp} \theta_k \Subset W_1\}$ and

$$I_j = \{k \geq 1 : \operatorname{supp} \theta_k \Subset W_j \text{ and } k \notin I_1 \cup \cdots \cup I_{j-1}\} \quad \text{for } j \geq 2,$$

and then put $\phi_j(x) = \sum_{k \in I_j} \theta_k(x)$ for $j \geq 1$. $\qquad \square$

We will now show that the Sobolev spaces on $\mathbb{R}^n$ are invariant under sufficiently regular changes of coordinates; $u \circ \kappa$ denotes the composite function defined by $(u \circ \kappa)(x) = u[\kappa(x)]$.

Theorem 3.23 Suppose that $\kappa : \mathbb{R}^n \to \mathbb{R}^n$ is a bijective mapping and r is a positive integer, such that $\partial^\alpha \kappa$ and $\partial^\alpha \kappa^{-1}$ exist and are (uniformly) Lipschitz on $\mathbb{R}^n$ for $|\alpha| \leq r - 1$. For $1 - r \leq s \leq r$, we have $u \in H^s(\mathbb{R}^n)$ if and only if $u \circ \kappa \in H^s(\mathbb{R}^n)$, in which case

$$\|u \circ \kappa\|_{H^s(\mathbb{R}^n)} \sim \|u\|_{H^s(\mathbb{R}^n)}.$$

Proof. It suffices to show that for $1 - r \leq s \leq r$,

$$\|u \circ \kappa\|_{H^s(\mathbb{R}^n)} \leq C_r \|u\|_{H^s(\mathbb{R}^n)}.$$

If $s = r$, then the estimate follows directly from the chain rule, because $H^r(\mathbb{R}^n) = W^r(\mathbb{R}^n)$. The same estimate holds for $s = 1 - r$ because $H^{1-r}(\mathbb{R}^n) = [H^{r-1}(\mathbb{R}^n)]^*$ and

$$\langle u \circ \kappa, v \rangle = \langle u, (v \circ \kappa^{-1}) |\det(\kappa^{-1})'| \rangle.$$

The case $1 - r < s < r$ then follows by interpolation, using Theorem B.7. $\qquad \square$

Density and Imbedding Theorems

We saw earlier that $\mathcal{D}(\overline{\Omega})$ is dense in $H^s(\Omega)$, but it is easy to find examples where $\mathcal{D}(\overline{\Omega})$ is not dense in $W^s(\Omega)$; see Exercise 3.18. However, we have the following theorem of Meyers and Serrin [64]. The proof relies on a technical lemma.

Lemma 3.24 Let $s \in \mathbb{R}$ and $\epsilon > 0$. For each $u \in H^s(\mathbb{R}^n)$ there exists $v \in \mathcal{D}(\mathbb{R}^n)$ satisfying

$$\|u - v\|_{H^s(\mathbb{R}^n)} < \epsilon \quad \text{and} \quad \operatorname{supp} v \subseteq \{x \in \mathbb{R}^n : \operatorname{dist}(x, \operatorname{supp} u) < \epsilon\}.$$

Proof. See Exercises 3.14 and 3.17. $\qquad \square$

Theorem 3.25 For any open set Ω and any real $s \geq 0$, the set $W^s(\Omega) \cap \mathcal{E}(\Omega)$ is dense in $W^s(\Omega)$.

Proof. Define a strictly increasing sequence of bounded open sets $W_1 \subsetneq W_2 \subsetneq \cdots$ by

$$W_j = \{x \in \Omega : |x| < j \text{ and } \text{dist}(x, \mathbb{R}^n \setminus \Omega) > 1/j\},$$

and choose a partition of unity $\phi_1, \phi_2, \ldots$ as in Corollary 3.22. Let $u \in W^s(\Omega)$ and $\epsilon > 0$. For each j, the function ϕ_j belongs to $\mathcal{D}(\Omega)$, so $\phi_j u \in W^s(\mathbb{R}^n) = H^s(\mathbb{R}^n)$ and we can apply Lemma 3.24 to obtain a sequence $(v_j)_{j=1}^\infty$ of functions in $\mathcal{D}(\Omega)$ satisfying

$$\|\phi_j u - v_j\|_{W^s(\Omega)} \leq \frac{\epsilon}{2^j} \quad \text{and} \quad \text{supp } v_j \subseteq W_{j+1}.$$

(Here, we use the fact that $\|\phi_j u - v_j\|_{W^s(\Omega)} = \|\phi_j u - v_j\|_{W^s(\mathbb{R}^n)}$.) Define $v(x) = \sum_{j=1}^\infty v_j(x)$, and note that this sum is finite for x in any compact subset of Ω, so $v \in \mathcal{E}(\Omega)$ and

$$\|u - v\|_{W^s(\Omega)} = \left\| \sum_{j=1}^\infty (\phi_j u - v_j) \right\|_{W^s(\Omega)} \leq \sum_{j=1}^\infty \frac{\epsilon}{2^j} = \epsilon.$$

$\square$

Next, we prove the Sobolev imbedding theorem, which shows that if s is large enough then the elements of $H^s(\mathbb{R}^n)$ can be thought of as continuous functions, and the elements of $H_0^s(\Omega)$ as functions that vanish on the boundary of Ω.

Theorem 3.26 Suppose $0 < \mu < 1$. If $u \in H^{n/2+\mu}(\mathbb{R}^n)$, then u is (almost everywhere equal to) a Hölder-continuous function. In fact,

$$|u(x)| \leq C\|u\|_{H^{n/2+\mu}(\mathbb{R}^n)}$$

and

$$|u(x) - u(y)| \leq C\|u\|_{H^{n/2+\mu}(\mathbb{R}^n)} |x - y|^\mu$$

for $x, y \in \mathbb{R}^n$.

Proof. By the Fourier inversion formula (Theorem 3.10) and the Cauchy–Schwarz inequality, if $u \in \mathcal{S}(\mathbb{R}^n)$ and $x \in \mathbb{R}^n$, then

$$|u(x)| \leq \int_{\mathbb{R}^n} |\hat{u}(\xi)| \, d\xi \leq C\|u\|_{H^{n/2+\mu}(\mathbb{R}^n)},$$

where

$$C^2 = \int_{\mathbb{R}^n} (1 + |\xi|^2)^{-n/2-\mu} \, d\xi < \infty.$$

Now let $u \in H^{n/2+\mu}(\mathbb{R}^n)$. We choose $u_j \in \mathcal{S}(\mathbb{R}^n)$ such that $u_j \to u$ in $H^{n/2+\mu}(\mathbb{R}^n)$, and observe that, by the estimate above,

$$|u_j(x) - u_k(x)| \leq c\|u_j - u_k\|_{H^{n/2+\mu}(\mathbb{R}^n)}.$$

Therefore, we can define a uniformly continuous function $U : \mathbb{R}^n \to \mathbb{C}$ by $U(x) = \lim_{j \to \infty} u_j(x)$ for $x \in \mathbb{R}^n$. Let $\phi \in \mathcal{D}(\mathbb{R}^n)$. On the one hand, $(u_j, \phi) \to (u, \phi)$ because $u_j \to u$ in $H^{n/2+\mu}(\mathbb{R}^n)$, and on the other hand $(u_j, \phi) \to (U, \phi)$ because $u_j \to U$ uniformly on $\mathbb{R}^n$. We conclude from Theorem 3.7 that, for almost all $x \in \mathbb{R}^n$, $u(x) = U(x)$ and

$$|u(x)| = |U(x)| = \lim_{j \to \infty} |u_j(x)| \leq C \lim_{j \to \infty} \|u_j\|_{H^{n/2+\mu}(\mathbb{R}^n)} = C\|u\|_{H^{n/2+\mu}(\mathbb{R}^n)}.$$

Similarly, we see from (3.25) that $\delta_h u(x) = u(x + h) - u(x)$ is bounded by

$$\int_{\mathbb{R}^n} |\widehat{\delta_h u}(\xi)|\, d\xi \leq M_\mu(h)\|u\|_{H^{n/2+\mu}(\mathbb{R}^n)}$$

where

$$M_\mu(h)^2 = \int_{\mathbb{R}^n} (1 + |\xi|^2)^{-n/2-\mu} |e^{i2\pi\xi \cdot h} - 1|^2\, d\xi.$$

It is clear that $M_\mu(h) \leq C$ for all $h \in \mathbb{R}^n$, and if $0 < |h| \leq 1$ then

$$M_\mu(h)^2 \leq C \int_{|\xi| < 1/|h|} (1 + |\xi|^2)^{-n/2-\mu} |\xi \cdot h|^2\, d\xi$$

$$+ 4 \int_{|\xi| \geq 1/|h|} (1 + |\xi|^2)^{-n/2-\mu}\, d\xi$$

$$\leq C|h|^2 \left(1 + \int_1^{1/|h|} \rho^{1-2\mu}\, d\rho\right) + C \int_{1/|h|}^{\infty} \rho^{-1-2\mu}\, d\rho$$

$$\leq C|h|^2(1 + |h|^{2\mu-2}) + C|h|^{2\mu} \leq C|h|^{2\mu},$$

so $|\delta_h u(x)| \leq C|h|^\mu \|u\|_{H^{n/2+\mu}(\mathbb{R}^n)}$ for all $x \in \mathbb{R}^n$. $\qquad\square$

We shall also prove an important compactness result that originated in a paper of Rellich [84].

Theorem 3.27 *Assume* $-\infty < s < t < \infty$.

(i) *If K is a compact subset of $\mathbb{R}^n$, then the inclusion $H_K^t \subseteq H_K^s$ is compact.*

(ii) *If Ω is a bounded, open subset of $\mathbb{R}^n$, then the inclusion $H^t(\Omega) \subseteq H^s(\Omega)$ is compact.*

Proof. To prove part (i), let $(u_j)_{j=1}^{\infty}$ be a bounded sequence in H_K^t for some compact set $K \Subset \mathbb{R}^n$. We want to show that a subsequence converges in H_K^s. Choose a cutoff function $\chi \in \mathcal{D}(\mathbb{R}^n)$ satisfying $\chi = 1$ on K, so that

$$\hat{u}_j(\xi) = \widehat{\chi u_j}(\xi) = \int_{\mathbb{R}^n} \hat{\chi}(\xi - \eta)\hat{u}_j(\eta)\, d\eta.$$

Using the Cauchy–Schwarz inequality and Peetre's inequality (see Exercise 3.16), we find that

$$(1 + |\xi|^2)^t |\hat{u}_j(\xi)|^2 \leq 2^{|t|} \left(\int_{\mathbb{R}^n} (1 + |\xi - \eta|^2)^{|t|} |\hat{\chi}(\xi - \eta)|^2\, d\eta \right)$$

$$\times \left(\int_{\mathbb{R}^n} (1 + |\eta|^2)^t |\hat{u}_j(\eta)|^2\, d\eta \right)$$

$$= 2^{|t|} \|\chi\|_{H^{|t|}(\mathbb{R}^n)}^2 \|u_j\|_{H^t(\mathbb{R}^n)}^2.$$

Likewise, since $\partial^{\alpha} \hat{u}_j = (\partial^{\alpha} \hat{\chi}) * \hat{u}_j = \hat{\chi}_{\alpha} * \hat{u}_j$ where $\chi_{\alpha}(x) = (-i2\pi x)^{\alpha} \chi(x)$, we have

$$(1 + |\xi|^2)^t |\partial^{\alpha} \hat{u}_j(\xi)|^2 \leq 2^{|t|} \|\chi_{\alpha}\|_{H^{|t|}(\mathbb{R}^n)}^2 \|u_j\|_{H^t(\mathbb{R}^n)}^2.$$

It follows, in particular, that the sequence of Fourier transforms $(\hat{u}_j)_{j=1}^{\infty}$ is uniformly bounded and equicontinuous on any compact subset of $\mathbb{R}^n$. Let K_1, K_2, K_3, ... be an increasing sequence of compact sets with $\bigcup_{l=1}^{\infty} K_l = \mathbb{R}^n$. By the Arzela–Ascoli theorem (Theorem 2.15), there is a subsequence $\hat{u}_j^1$ that converges uniformly on K_1. From this subsequence, we can extract a subsequence $\hat{u}_j^2$ that converges uniformly on K_2. Continuing in this way, we obtain successive subsequences $\hat{u}_j^3$, $\hat{u}_j^4$, ... such that $\hat{u}_j^l$ converges uniformly on K_l. For brevity, we now denote the diagonal subsequence u_j^j by u_j. Thus, the Fourier transforms $\hat{u}_j$ converge uniformly on any compact subset of $\mathbb{R}^n$, and it suffices to show that the functions u_j themselves are Cauchy in H_K^s. Given $\epsilon > 0$, we first choose R large enough so that

$$\int_{|\xi| > R} (1 + |\xi|^2)^s |\hat{u}_j(\xi) - \hat{u}_k(\xi)|^2\, d\xi$$

$$\leq (1 + R^2)^{s-t} \int_{\mathbb{R}^n} (1 + |\xi|^2)^t |\hat{u}_j(\xi) - \hat{u}_k(\xi)|^2\, d\xi$$

$$= \frac{\|u_j - u_k\|_{H^t(\mathbb{R}^n)}^2}{(1 + R^2)^{t-s}} \leq \frac{2(\|u_j\|_{H^t(\mathbb{R}^n)}^2 + \|u_k\|_{H^t(\mathbb{R}^n)}^2)}{(1 + R^2)^{t-s}} < \frac{\epsilon}{2}$$

for all $j, k \geq 1$, and then choose N large enough so that

$$\int_{|\xi| \leq R} (1 + |\xi|^2)^s |\hat{u}_j(\xi) - \hat{u}_k(\xi)|^2 \, d\xi < \frac{\epsilon}{2} \quad \text{for all } j, k \geq N.$$

Hence, $\|u_j - u_k\|^2_{H^s(\mathbb{R}^n)} < \epsilon$ for $j, k \geq N$, completing the proof of part (i).

Suppose now that $(u_j)_{j=1}^\infty$ is a bounded sequence in $H^t(\Omega)$. By hypothesis, Ω is a bounded subset of $\mathbb{R}^n$, so we can find a compact set $K \Subset \mathbb{R}^n$ together with a cutoff function $\chi \in \mathcal{D}_K(\mathbb{R}^n)$ such that $\Omega \subseteq K$ and $\chi = 1$ on Ω. For each j, choose $U_j \in H^t(\mathbb{R}^n)$ with $\|U_j\|_{H^t(\mathbb{R}^n)} = \|u_j\|_{H^t(\Omega)}$. By Theorem 3.20, the sequence $(\chi U_j)_{j=1}^\infty$ is bounded in H^t_K, so by part (i) there is a subsequence, again denoted by $(\chi U_j)_{j=1}^\infty$, that converges in H^s_K. Put $U = \lim_{j \to \infty} \chi U_j \in H^s_K$ and $u = U|_\Omega \in H^s(\Omega)$, and observe that

$$\|u_j - u\|_{H^s(\Omega)} = \|(\chi U_j - U)|_\Omega\|_{H^s(\Omega)} \leq \|\chi U_j - U\|_{H^s_K}.$$

Thus, $u_j \to u$ in $H^s(\Omega)$, proving part (ii). $\qquad\square$

Lipschitz Domains

Denote the boundary of the open set Ω by

$$\Gamma = \partial \Omega = \overline{\Omega} \cap (\mathbb{R}^n \setminus \Omega).$$

Thus far, no use has been made of any regularity assumption on Γ, but henceforth we shall require that, roughly speaking, the boundary of Ω can be represented locally as the graph of a Lipschitz function (using different systems of Cartesian coordinates for different parts of Γ, as necessary). The simplest case occurs when there is a function $\zeta : \mathbb{R}^{n-1} \to \mathbb{R}$ such that

$$\Omega = \{x \in \mathbb{R}^{n-1} : x_n < \zeta(x') \text{ for all } x' = (x_1, \ldots, x_{n-1}) \in \mathbb{R}^{n-1}\}. \quad (3.26)$$

If ζ is Lipschitz, i.e., if there is a constant M such that

$$|\zeta(x') - \zeta(y')| \leq M|x' - y'| \quad \text{for all } x', y' \in \mathbb{R}^{n-1}, \quad (3.27)$$

then we say that Ω is a *Lipschitz hypograph*.

Definition 3.28 The open set Ω is a Lipschitz domain if its boundary Γ is compact and if there exist finite families $\{W_j\}$ and $\{\Omega_j\}$ having the following properties:

(i) *The family $\{W_j\}$ is a finite open cover of Γ, i.e., each W_j is an open subset of $\mathbb{R}^n$, and $\Gamma \subseteq \bigcup_j W_j$.*

(ii) *Each Ω_j can be transformed to a Lipschitz hypograph by a rigid motion, i.e., by a rotation plus a translation.*

(iii) *The set Ω satisfies $W_j \cap \Omega = W_j \cap \Omega_j$ for each j.*

Notice that if Ω is a Lipschitz hypograph as in (3.26), then

$$\Gamma = \{x \in \mathbb{R}^{n-1} : x_n = \zeta(x') \text{ for all } x' = (x_1, \ldots, x_{n-1}) \in \mathbb{R}^{n-1}\}.$$

We also remark that although, in our definition, the boundary of a Lipschitz domain must be compact, the domain itself may be unbounded. In particular, if Ω is a bounded Lipschitz domain, then $\mathbb{R}^n \setminus \overline{\Omega}$ is an unbounded Lipschitz domain.

Sometimes, a different smoothness condition will be needed, so we broaden the above terminology as follows. For any integer $k \geq 0$, we say that the set (3.26) is a C^k *hypograph* if the function $\zeta : \mathbb{R}^{n-1} \to \mathbb{R}$ is C^k, and if $\partial^\alpha \zeta$ is bounded for $|\alpha| \leq k$. In the obvious way, we then define a C^k *domain* by substituting "C^k" for "Lipschitz" throughout Definition 3.28. Likewise, for $0 < \mu \leq 1$, we define a $C^{k,\mu}$ domain by adding the requirement that the kth-order partial derivatives of ζ be Hölder-continuous with exponent μ, i.e.,

$$|\partial^\alpha \zeta(x') - \partial^\alpha \zeta(y')| \leq M|x' - y'|^\mu \qquad \text{for all } x', y' \in \mathbb{R}^{n-1} \text{ and } |\alpha| = k.$$

Hence, a Lipschitz domain is the same thing as a $C^{0,1}$ domain. Notice that in the definition of a C^k or $C^{k,\mu}$ domain, we can assume if we want that ζ has compact support, because Γ is always assumed to be compact.

The class of Lipschitz domains is broad enough to cover most cases that arise in applications of partial differential equations. For instance, if $k \geq 1$ and Γ is a compact, $(n-1)$-dimensional C^k submanifold of $\mathbb{R}^n$, then Ω is a C^k domain and hence also a Lipschitz domain. Furthermore, any polygon in $\mathbb{R}^2$ or polyhedron in $\mathbb{R}^3$ is a Lipschitz domain. One can construct many other examples using the fact that if $\kappa : \mathbb{R}^n \to \mathbb{R}^n$ is a C^1 diffeomorphism and if Ω is a Lipschitz domain, then the set $\kappa(\Omega)$ is again a Lipschitz domain.

Figure 2 shows some examples of open sets that fail to be Lipschitz domains: (i) is disqualified because of the cusp at the point A; (ii) because of the crack BC (a Lipschitz domain cannot be on both sides of its boundary); and (iii) because in any neighbourhood of the point D it is impossible to represent Γ as the graph of a function.

For a Lipschitz domain, in fact even for a C^0 domain, a much stronger density result than Theorem 3.25 holds.

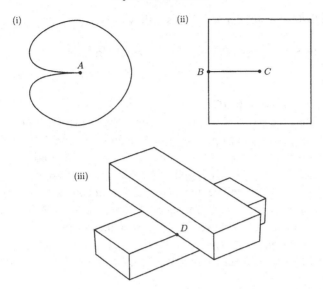

Figure 2. Examples of regions that fail to be Lipschitz domains.

Theorem 3.29 *If Ω is a C^0 domain, then*

(i) $\mathcal{D}(\overline{\Omega})$ is dense in $W^s(\Omega)$ for $s \geq 0$;

(ii) $\mathcal{D}(\Omega)$ is dense in $H^s_{\overline{\Omega}}$, or in other words $\widetilde{H}^s(\Omega) = H^s_{\overline{\Omega}}$, for $s \in \mathbb{R}$.

Proof. Suppose to begin with that Ω is of the form (3.26) for some continuous function $\zeta : \mathbb{R}^{n-1} \to \mathbb{R}$ having compact support.

Let $s \geq 0$, $u \in W^s(\Omega)$ and $\epsilon > 0$. For $\delta > 0$, we define

$$u_\delta(x) = u(x', x_n - \delta) \quad \text{and} \quad \Omega_\delta = \{x \in \mathbb{R}^n : x_n < \zeta(x') + \delta\},$$

so that $u_\delta \in W^s(\Omega_\delta)$. Since $\partial^\alpha u_\delta = (\partial^\alpha u)_\delta$, we can choose δ small enough so that

$$\|u - u_\delta|_\Omega\|_{W^s(\Omega)} < \frac{\epsilon}{2},$$

and then choose a cutoff function $\chi \in \mathcal{E}(\mathbb{R}^n)$ satisfying $\chi = 1$ on Ω and $\chi = 0$ on $\mathbb{R}^n \setminus \Omega_{(\delta/2)}$, so that $\chi u_\delta \in W^s(\mathbb{R}^n)$. Hence, by Theorem 3.16, there exists $V \in \mathcal{D}(\mathbb{R}^n)$ such that

$$\|\chi u_\delta - V\|_{W^s(\mathbb{R}^n)} < \frac{\epsilon}{2},$$

implying that the restriction $v = V|_\Omega$ satisfies

$$\|u - v\|_{W^s(\Omega)} = \|u - u_\delta|_\Omega + (\chi u_\delta - V)|_\Omega\|_{W^s(\Omega)}$$

$$\leq \|u - u_\delta|_\Omega\|_{W^s(\Omega)} + \|\chi u_\delta - V\|_{W^s(\mathbb{R}^n)} < \epsilon.$$

Thus, $\mathcal{D}(\overline{\Omega})$ is dense in $W^s(\Omega)$.

To complete the proof of part (i), we suppose now that Ω is a C^0 domain, and let $\{W_j\}_{j=1}^J$ be a finite open cover of Γ as in Definition 3.28. Define one additional open set $W_0 = \{x \in \Omega : \operatorname{dist}(x, \mathbb{R}^n \setminus \Omega) > \delta\}$, choosing a small enough $\delta > 0$ so that $\overline{\Omega} \subseteq \bigcup_{j=0}^J W_j$. We may assume that W_j is bounded for $1 \leq j \leq J$, but W_0 will be unbounded if Ω is unbounded. Let $(\phi_j)_{j=0}^J$ be a partition of unity for $\overline{\Omega}$ such that $\operatorname{supp}\phi_j \subseteq W_j$ for all j. By the case of a C^0 hypograph considered above, if $1 \leq j \leq J$ then there exists $v_j \in \mathcal{D}(\overline{\Omega})$ such that $\|\phi_j u - v_j\|_{W^s(\Omega)} < \epsilon/(J+1)$. In fact, by Lemma 3.24, the same is true also when $j = 0$, because $\phi_0 u \in W^s(\mathbb{R}^n)$. Put $v = \sum_{j=0}^J v_j \in \mathcal{D}(\overline{\Omega})$; then

$$\|u - v\|_{W^s(\Omega)} = \left\| \sum_{j=0}^J (\phi_j u - v_j) \right\|_{W^s(\Omega)} \leq \sum_{j=0}^J \|\phi_j u - v_j\|_{W^s(\Omega)} < \epsilon.$$

To prove (ii), assume once again that Ω is of the form (3.26), and let $s \in \mathbb{R}$, $u \in \widetilde{H}^s(\Omega)$ and $\epsilon > 0$. This time we put $u_\delta(x) = u(x', x_n + \delta)$, and observe that $u_\delta \in H^s(\mathbb{R}^n)$ and $\operatorname{supp} u_\delta \subseteq \{x \in \mathbb{R}^n : x_n \leq \zeta(x') - \delta\}$. Choose δ small enough to ensure

$$\|u - u_\delta\|_{H^s(\mathbb{R}^n)} < \frac{\epsilon}{2},$$

and then apply Lemma 3.24 to find $v \in \mathcal{D}(\Omega)$ satisfying

$$\|u_\delta - v\|_{H^s(\mathbb{R}^n)} < \frac{\epsilon}{2}.$$

Since $\|u - v\|_{\widetilde{H}^s(\Omega)} = \|(u - u_\delta) + (u_\delta - v)\|_{H^s(\mathbb{R}^n)} < \epsilon$, we see that $\mathcal{D}(\Omega)$ is a dense subspace of $\widetilde{H}^s(\Omega)$. As with part (i), the result carries over to C^0 domains with the help of a partition of unity.   □

Theorems 3.29 and A.4 allow us to apply Theorems 3.14 and 3.18, and hence deduce the following important result.

Theorem 3.30 *If Ω is a Lipschitz domain, then*

(i) $H^s(\Omega)^* = \widetilde{H}^{-s}(\Omega)$ and $\widetilde{H}^s(\Omega)^* = H^{-s}(\Omega)$ for all $s \in \mathbb{R}$;
(ii) $W^s(\Omega) = H^s(\Omega)$ for all $s \geq 0$.

We remarked earlier that $\tilde{H}^s(\Omega) \subseteq H_0^s(\Omega)$. The next two technical lemmas will enable us to establish the reverse inclusion apart from certain exceptional values of s.

Lemma 3.31 *Suppose* $u \in \mathcal{D}(\mathbb{R})$. *If* $0 < s < \frac{1}{2}$, *then*

$$\int_0^\infty x^{-2s} |u(x)|^2 \, dx \leq C_s \int_0^\infty \int_0^\infty \frac{|u(x) - u(y)|^2}{|x - y|^{1+2s}} \, dx \, dy.$$

If, in addition, $u(0) = 0$, *then the inequality holds also for* $\frac{1}{2} < s < 1$.

Proof. Observe that the double integral converges for $0 < s < 1$. For $x > 0$, define

$$v(x) = \frac{1}{x} \int_0^x [u(x) - u(y)] \, dy = u(x) - \frac{1}{x} \int_0^x u(y) \, dy$$

and

$$w(x) = \int_x^\infty \frac{v(y)}{y} \, dy.$$

Since $v'(x) = u'(x) - x^{-1} v(x)$ and $w'(x) = -x^{-1} v(x)$, we see that

$$u'(x) = v'(x) - w'(x) \quad \text{for } x > 0.$$

Furthermore, u has compact support, and both $v(x)$ and $w(x)$ tend to zero as x tends to infinity, so

$$u(x) = v(x) - w(x) \quad \text{for } x > 0.$$

By the Cauchy–Schwarz inequality,

$$|v(x)|^2 \leq \frac{1}{x} \int_0^x |u(y) - u(x)|^2 \, dy,$$

implying

$$\begin{aligned}
\int_0^\infty x^{-2s} |v(x)|^2 \, dx &\leq \int_0^\infty x^{-1-2s} \int_0^x |u(y) - u(x)|^2 \, dy \, dx \\
&= \int_0^\infty \int_y^\infty x^{-1-2s} |u(y) - u(x)|^2 \, dx \, dy \\
&\leq \int_0^\infty \int_0^\infty \frac{|u(x) - u(y)|^2}{|x - y|^{1+2s}} \, dx \, dy
\end{aligned}$$

for $0 < s < 1$. By Exercise 3.20,

$$\int_0^\infty x^{-2s} |w(x)|^2 \, dx \le \frac{1}{\left(\frac{1}{2} - s\right)^2} \int_0^\infty x^{-2s} |v(x)|^2 \, dx \quad \text{for } s < \tfrac{1}{2},$$

and the first part of the lemma follows.

One easily verifies by Taylor expansion that $v(x) \to 0$ as $x \to 0^+$, so $w(x) = v(x) - u(x) \to -u(0)$ as $x \to 0^+$. Thus, if $u(0) = 0$ then $w(0) = 0$, implying that

$$w(x) = w(x) - w(0) = w(x) - \int_0^\infty \frac{v(y)}{y} \, dy = -\int_0^x \frac{v(y)}{y} \, dy.$$

Hence, by Exercise 3.20,

$$\int_0^\infty x^{-2s} |w(x)|^2 \, dx \le \frac{1}{\left(s - \frac{1}{2}\right)^2} \int_0^\infty x^{-2s} |v(x)|^2 \, dx \quad \text{for } s > \tfrac{1}{2},$$

giving the second part of the lemma. □

Lemma 3.32 *If Ω is a Lipschitz domain and $u \in \mathcal{D}(\overline{\Omega})$, then*

$$\int_\Omega \operatorname{dist}(x, \Gamma)^{-2s} |u(x)|^2 \, dx \le C \|u\|_{H^s(\Omega)}^2 \quad \text{for } 0 < s < \tfrac{1}{2}.$$

If, in addition, $u = 0$ on Γ, then this inequality holds also for $\frac{1}{2} < s < 1$.

Proof. We prove the result for Ω a Lipschitz hypograph given by $x_n < \zeta(x')$. If $y \in \Gamma$ and M is a Lipschitz constant for ζ, as in (3.27), then $y_n = \zeta(y')$ and so

$$|\zeta(x') - x_n| = |y_n - x_n + \zeta(x') - \zeta(y')| \le |x_n - y_n| + M|x' - y'|$$
$$\le \sqrt{1 + M^2} |x - y|,$$

implying that

$$\operatorname{dist}(x, \Gamma) \ge \frac{\zeta(x') - x_n}{\sqrt{1 + M^2}} \quad \text{for } x \in \Omega.$$

Using the substitution $x_n = \zeta(x') - t$, followed by Lemma 3.31, we see that

$$\int_\Omega \operatorname{dist}(x, \Gamma)^{-2s} |u(x)|^2 \, dx \le (1 + M^2)^s \int_{x_n < \zeta(x')} [\zeta(x') - x_n]^{-2s} |u(x)|^2 \, dx$$

$$= C \int_{\mathbb{R}^{n-1}} \int_0^\infty t^{-2s} \left| u\big(x', \zeta(x') - t\big) \right|^2 dt \, dx'$$

$$\le C \int_{\mathbb{R}^{n-1}} \int_{y<\zeta(x')} \int_{z<\zeta(x')} \frac{|u(x', y) - u(x', z)|^2}{|y - z|^{1+2s}}$$

$$\times\, dy\, dz\, dx',$$

so if $u = U|_\Omega$ where $U \in \mathcal{D}(\mathbb{R}^n)$, then, arguing as in the proofs of Lemma 3.15 and Theorem 3.16,

$$\int_\Omega \operatorname{dist}(x, \Gamma)^{-2s} |u(x)|^2\, dx \le C \int_{-\infty}^{\infty} \int_{\mathbb{R}^n} \frac{|U(x', x_n + h) - U(x)|^2}{|h|^{1+2s}}\, dx\, dh$$

$$= C \int_{\mathbb{R}^n} |\hat{U}(\xi)|^2 \int_{-\infty}^{\infty} \frac{|e^{i2\pi\xi_n h} - 1|^2}{|h|^{1+2s}}\, dh\, d\xi$$

$$= C \int_{\mathbb{R}^n} |\xi_n|^{2s} |\hat{U}(\xi)|^2\, d\xi \le C \|U\|_{H^s(\mathbb{R}^n)}^2,$$

and the result follows because $\mathcal{D}(\mathbb{R}^n)$ is dense in $H^s(\mathbb{R}^n)$. $\qquad\square$

Notice that the value $s = \frac{1}{2}$ is excluded in the two lemmas above; cf. Exercise 3.22. Also, recall our earlier discussion of the imbedding (3.24).

Theorem 3.33 *Let $s \ge 0$. If Ω is a Lipschitz domain, then*

$$\widetilde{H}^s(\Omega) = \{u \in L_2(\Omega) : \tilde{u} \in H^s(\mathbb{R}^n)\} \subseteq H_0^s(\Omega),$$

where $\tilde{u}$ denotes the extension of u by zero:

$$\tilde{u}(x) = \begin{cases} u(x) & \text{if } x \in \Omega, \\ 0 & \text{if } x \in \mathbb{R}^n \setminus \Omega. \end{cases}$$

In fact,

$$\widetilde{H}^s(\Omega) = H_0^s(\Omega) \quad \text{provided } s \notin \left\{ \tfrac{1}{2}, \tfrac{3}{2}, \tfrac{5}{2}, \ldots \right\}.$$

Proof. For the moment, we think of the elements of $\widetilde{H}^s(\Omega)$ as distributions on $\mathbb{R}^n$. If $u \in \widetilde{H}^s(\Omega)$, then the restriction $v = u|_\Omega$ belongs to $L_2(\Omega)$, and $u = \tilde{v}$ as a distribution on $\mathbb{R}^n$, so $\tilde{v} \in H^s(\mathbb{R}^n)$. Conversely, if $u \in L_2(\Omega)$ and $\tilde{u} \in H^s(\mathbb{R}^n)$, then $\operatorname{supp} \tilde{u} \subseteq \overline{\Omega}$, so $\tilde{u} \in H_{\overline{\Omega}}^s = \widetilde{H}^s(\Omega)$. We have already seen that $\widetilde{H}^s(\Omega) \subseteq H_0^s(\Omega)$.

Now view $\widetilde{H}^s(\Omega)$ as a subspace of $L_2(\Omega)$, and let $u \in \mathcal{D}(\Omega)$. For any integer $r \ge 0$, Theorems 3.16 and 3.30 give

$$\|u\|_{\widetilde{H}^r(\Omega)}^2 = \|\tilde{u}\|_{H^r(\mathbb{R}^n)}^2 \sim \sum_{|\alpha|\le r} \|\partial^\alpha \tilde{u}\|_{L_2(\mathbb{R}^n)}^2 = \sum_{|\alpha|\le r} \|\partial^\alpha u\|_{L_2(\Omega)}^2$$

$$= \|u\|_{W^r(\Omega)}^2 \sim \|u\|_{H^r(\Omega)}^2,$$

so $\tilde{H}^r(\Omega) = H_0^r(\Omega)$. For $s = r + \mu$ with $0 < \mu < 1$,

$$
\begin{aligned}
\|u\|_{\tilde{H}^s(\Omega)}^2 &= \|\tilde{u}\|_{H^s(\mathbb{R}^n)}^2 \sim \sum_{|\alpha|\le r} \|\partial^\alpha \tilde{u}\|_{L_2(\mathbb{R}^n)}^2 \\
&\quad + \sum_{|\alpha|=r} \iint_{\mathbb{R}^n \times \mathbb{R}^n} \frac{|\partial^\alpha \tilde{u}(x) - \partial^\alpha \tilde{u}(y)|^2}{|x-y|^{2\mu+n}} \, dx\, dy \\
&= \sum_{|\alpha|\le r} \|\partial^\alpha u\|_{L_2(\Omega)}^2 + \sum_{|\alpha|=r} \iint_{\Omega \times \Omega} \frac{|\partial^\alpha u(x) - \partial^\alpha u(y)|^2}{|x-y|^{2\mu+n}} \, dx\, dy \\
&\quad + \sum_{|\alpha|=r} \iint_{\Omega \times (\mathbb{R}^n \setminus \Omega)} \frac{|\partial^\alpha u(x)|^2}{|x-y|^{2\mu+n}} \, dx\, dy \\
&\quad + \sum_{|\alpha|=r} \iint_{(\mathbb{R}^n \setminus \Omega) \times \Omega} \frac{|\partial^\alpha u(y)|^2}{|x-y|^{2\mu+n}} \, dx\, dy \\
&= \|u\|_{W^s(\Omega)}^2 + 2 \sum_{|\alpha|=r} \int_\Omega |\partial^\alpha u(x)|^2 w_\mu(x) \, dx,
\end{aligned}
$$

where the weight w_μ is defined by

$$
w_\mu(x) = \int_{\mathbb{R}^n \setminus \Omega} \frac{dy}{|x-y|^{2\mu+n}} \quad \text{for } x \in \Omega.
$$

Introducing polar coordinates about x, we see that

$$
w_\mu(x) \le C \operatorname{dist}(x, \Gamma)^{-2\mu} \quad \text{for } x \in \Omega,
$$

so by Lemma 3.32,

$$
\sum_{|\alpha|=r} \int_\Omega w_\mu(x) |\partial^\alpha u(x)|^2 \, dx \le C_\mu \sum_{|\alpha|=r} \|\partial^\alpha u\|_{H^\mu(\Omega)}^2 \le C_\mu \|u\|_{H^s(\Omega)}^2
$$

provided $s \ne$ integer $+ \frac{1}{2}$. Hence, in this case, $H_0^s(\Omega) \subseteq \tilde{H}^s(\Omega)$. $\square$

Sobolev Spaces on the Boundary

Any Lipschitz domain Ω has a surface measure σ, and an outward unit normal ν that exists σ-almost everywhere on Γ. In fact, by Rademacher's theorem [83], [105, Theorem 11A, p. 272], if $\zeta : \mathbb{R}^{n-1} \to \mathbb{R}$ is Lipschitz, then ζ is Fréchet-differentiable almost everywhere with

$$
\|\operatorname{grad} \zeta\|_{L_\infty(\mathbb{R}^{n-1})} \le M,
$$

where M is any Lipschitz constant for ζ, as in (3.27). If Ω is the Lipschitz hypograph (3.26), then

$$d\sigma_x = \sqrt{1 + |\operatorname{grad}\zeta(x')|^2}\,dx' \quad \text{and} \quad \nu(x) = \frac{(-\operatorname{grad}\zeta(x'), 1)}{\sqrt{1 + |\operatorname{grad}\zeta(x')|^2}}$$

$$\text{for } x \in \Gamma. \quad (3.28)$$

The divergence theorem can then be proved in a straightforward manner.

Theorem 3.34 *If Ω is a Lipschitz domain, and if $F : \mathbb{R}^n \to \mathbb{R}^n$ is a C^1 vector field with compact support, then*

$$\int_\Omega \operatorname{div} F\,dx = \int_\Gamma F \cdot \nu\,d\sigma.$$

Proof. Assume to begin with that Ω is a Lipschitz hypograph (3.26). For $1 \le k \le n - 1$, we define $u_k : \mathbb{R}^{n-1} \to \mathbb{R}$ by

$$u_k(x') = \int_{-\infty}^{\zeta(x')} F_k(x', x_n)\,dx_n.$$

The function u_k is Lipschitz with

$$\partial_k u_k(x') = F_k\big(x', \zeta(x')\big)\partial_k\zeta(x') + \int_{-\infty}^{\zeta(x')} \partial_k F_k(x)\,dx_n,$$

and $\int_{\mathbb{R}^{n-1}} \partial_k u_k(x')\,dx' = 0$ because u_k has compact support. Thus,

$$\int_{x_n < \zeta(x')} \partial_k F_k(x)\,dx = -\int_{\mathbb{R}^{n-1}} F_k\big(x', \zeta(x')\big)\partial_k\zeta(x')\,dx' \quad \text{for } 1 \le k \le n - 1,$$

with

$$\int_{x_n < \zeta(x')} \partial_n F_n(x)\,dx = \int_{\mathbb{R}^{n-1}} F_n\big(x', \zeta(x')\big)\,dx',$$

so that

$$\int_\Omega \operatorname{div} F(x)\,dx = \int_{\mathbb{R}^{n-1}} F\big(x', \zeta(x')\big) \cdot \big(-\operatorname{grad}\zeta(x'), 1\big)\,dx' = \int_\Gamma F \cdot \nu\,d\sigma.$$

Now let Ω be a Lipschitz *domain*, and take a partition of unity $(\phi_j)_{j=0}^J$ for $\overline{\Omega}$ as in the proof of Theorem 3.29. Since $\phi_0 \in \mathcal{D}(\Omega)$, it is obvious that

$$\int_{-\infty}^{\infty} \partial_k(\phi_0 F_k)\,dx_k = 0,$$

so $\int_\Omega \operatorname{div}(\phi_0 F)\,dx = 0$. Hence, from the case of a Lipschitz hypograph treated above,

$$\int_\Omega \operatorname{div} F\,dx = \sum_{j=0}^{J} \int_\Omega \operatorname{div}(\phi_j F)\,dx = \sum_{j=1}^{J} \int_\Gamma \nu \cdot (\phi_j F)\,d\sigma = \int_\Gamma \nu \cdot F\,d\sigma,$$

where we used the fact that $\operatorname{supp}\phi_j \cap \partial(W_j \cap \Omega) \subseteq W_j \cap \Gamma$ for $1 \le j \le J$. $\qquad\square$

If Ω is a Lipschitz hypograph, then we can construct Sobolev spaces on its boundary Γ in terms of Sobolev spaces on $\mathbb{R}^{n-1}$, as follows. For $u \in L_2(\Gamma) = L_2(\Gamma, \sigma)$, we define

$$u_\zeta(x') = u\big(x', \zeta(x')\big) \quad \text{for } x' \in \mathbb{R}^{n-1},$$

put

$$H^s(\Gamma) = \{u \in L_2(\Gamma) : u_\zeta \in H^s(\mathbb{R}^{n-1})\} \quad \text{for } 0 \le s \le 1,$$

and equip this space with the inner product

$$(u, v)_{H^s(\Gamma)} = (u_\zeta, v_\zeta)_{H^s(\mathbb{R}^{n-1})}.$$

Recalling that $d\sigma$ is given by (3.28), we put

$$\|u\|_{H^{-s}(\Gamma)} = \left\|u_\zeta \sqrt{1 + |\operatorname{grad}\zeta|^2}\right\|_{H^{-s}(\mathbb{R}^{n-1})} \quad \text{for } 0 < s \le 1,$$

and then define $H^{-s}(\Gamma)$ to be the completion of $L_2(\Gamma)$ in this norm. It follows that $H^{-s}(\Gamma)$ is a realisation of the dual space of $H^s(\Gamma)$, with

$$\|u\|_{H^{-s}(\Gamma)} \sim \sup_{0 \ne v \in H^s(\Gamma)} \frac{|\langle u, v\rangle_\Gamma|}{\|v\|_{H^s(\Gamma)}} = \sup_{0 \ne v \in H^s(\Gamma)} \frac{|(u, v)_\Gamma|}{\|v\|_{H^s(\Gamma)}} \quad \text{for } |s| \le 1,$$

where

$$\langle u, v\rangle_\Gamma = \int_\Gamma u(x)v(x)\,d\sigma(x) \quad \text{and} \quad (u, v)_\Gamma = \int_\Gamma \overline{u(x)}v(x)\,d\sigma(x).$$

If $\kappa(\Omega)$ is a Lipschitz hypograph for some rigid motion $\kappa : \mathbb{R}^n \to \mathbb{R}^n$, then we define $H^s(\Gamma)$ in the same way except that $u_\zeta(x') = u[\kappa^{-1}(x', \zeta(x'))]$.

Suppose now that Ω is a Lipschitz domain. Using the notation of Definition 3.28, we choose a partition of unity $\{\phi_j\}$ subordinate to the open cover $\{W_j\}$ of Γ, i.e., we choose $\phi_j \in \mathcal{D}(W_j)$ satisfying $\sum_j \phi_j(x) = 1$ for all $x \in \Gamma$. The inner product in $H^s(\Gamma)$ is then defined by

$$(u, v)_{H^s(\Gamma)} = \sum_j (\phi_j u, \phi_j v)_{H^s(\Gamma_j)}, \qquad (3.29)$$

where $\Gamma_j = \partial\Omega_j$. Theorems 3.20 and 3.23 imply that a different choice of $\{W_j\}$, $\{\Omega_j\}$ and $\{\phi_j\}$ would yield the same space $H^s(\Gamma)$ with an equivalent norm, for $|s| \le 1$. If Ω is $C^{k-1,1}$ for $k \ge 0$, then $H^s(\Gamma)$ is well defined for $|s| \le k$.

We shall also require Sobolev spaces defined over only a part of the boundary of Ω. Consider a disjoint union

$$\Gamma = \Gamma_1 \cup \Pi \cup \Gamma_2, \qquad (3.30)$$

where Γ_1 and Γ_2 are disjoint, non-empty, relatively open subsets of Γ, having Π as their common boundary in Γ. When Ω is a Lipschitz hypograph, we call (3.30) a *Lipschitz dissection* of Γ if there is a Lipschitz function $\varrho : \mathbb{R}^{n-2} \to \mathbb{R}$ such that

$$\Gamma_1 = \{x \in \Gamma : x_{n-1} < \varrho(x'')\},$$
$$\Pi = \{x \in \Gamma : x_{n-1} = \varrho(x'')\},$$
$$\Gamma_2 = \{x \in \Gamma : x_{n-1} > \varrho(x'')\},$$

where $x'' = (x_1, \ldots, x_{n-2})$; for $n = 2$, the function ϱ reduces to a constant. In the obvious way, we extend the notion of a Lipschitz dissection to the case when Ω is the image under a rigid motion of a Lipschitz hypograph.

Next, suppose that Ω is a Lipschitz *domain*. We say that (3.30) is a Lipschitz dissection of Γ if, in the notation of Definition 3.28, there are Lipschitz dissections $\partial\Omega_j = \Gamma_{1j} \cup \Pi_j \cup \Gamma_{2j}$ such that

$$W_j \cap \Gamma_1 = W_j \cap \Gamma_{1j}, \quad W_j \cap \Pi = W_j \cap \Pi_j, \quad W_j \cap \Gamma_2 = W_j \cap \Gamma_{2j},$$

for all j. We remark that, in this case, the subsets Γ_1 and Γ_2 are not necessarily connected. Let $\mathcal{D}(\Gamma_1) = \{\phi \in \mathcal{D}(\Gamma) : \operatorname{supp}\phi \subseteq \Gamma_1\}$; by defining

$$H^s(\Gamma_1) = \{U|_{\Gamma_1} : U \in H^s(\Gamma)\},$$
$$\widetilde{H}^s(\Gamma_1) = \text{closure of } \mathcal{D}(\Gamma_1) \text{ in } H^s(\Gamma),$$
$$H_0^s(\Gamma) = \text{closure of } \mathcal{D}(\Gamma_1) \text{ in } H^s(\Gamma_1),$$

the properties of Sobolev spaces on Lipschitz domains in $\mathbb{R}^n$ carry over to Sobolev spaces on Γ_1, subject to the condition that $|s| \le 1$, or $|s| \le k$ if (3.30) is $C^{k-1,1}$ in the obvious sense.

The Trace Operator

In studying boundary value problems, we shall need to make sense of the restriction $u|_\Gamma$ as an element of a Sobolev space on Γ when u belongs to a Sobolev space on Ω. The main idea is contained in the following lemma.

Lemma 3.35 *Define the trace operator* $\gamma : \mathcal{D}(\mathbb{R}^n) \to \mathcal{D}(\mathbb{R}^{n-1})$ *by*

$$\gamma u(x) = u(x', 0) \quad \text{for } x' \in \mathbb{R}^{n-1}.$$

If $s > \frac{1}{2}$, *then* γ *has a unique extension to a bounded linear operator*

$$\gamma : H^s(\mathbb{R}^n) \to H^{s-1/2}(\mathbb{R}^{n-1}).$$

Proof. For $u \in \mathcal{D}(\mathbb{R}^n)$, the Fourier inversion formula (3.15) gives

$$\gamma u(x') = \int_{\mathbb{R}^n} \hat{u}(\xi) e^{i2\pi \xi' \cdot x'} \, d\xi = \int_{\mathbb{R}^{n-1}} \left(\int_{-\infty}^{\infty} \hat{u}(\xi', \xi_n) \, d\xi_n \right) e^{i2\pi \xi' \cdot x'} \, d\xi',$$

and so

$$\widehat{\gamma u}(\xi') = \int_{-\infty}^{\infty} \hat{u}(\xi', \xi_n) \, d\xi_n = \int_{-\infty}^{\infty} (1 + |\xi|^2)^{-s/2} (1 + |\xi|^2)^{s/2} \hat{u}(\xi', \xi_n) \, d\xi_n.$$

Applying the Cauchy–Schwarz inequality, we obtain the bound

$$|\widehat{\gamma u}(\xi')|^2 \le M_s(\xi') \int_{-\infty}^{\infty} (1 + |\xi|^2)^s |\hat{u}(\xi)|^2 \, d\xi_n,$$

where, using the substitution $\xi_n = (1 + |\xi'|^2)^{1/2} t$,

$$M_s(\xi') = \int_{-\infty}^{\infty} \frac{d\xi_n}{(1 + |\xi'|^2 + |\xi_n|^2)^s} = \frac{1}{(1 + |\xi'|^2)^{s-1/2}} \int_{-\infty}^{\infty} \frac{dt}{(1 + t^2)^s}.$$

The integral with respect to t converges because $s > \frac{1}{2}$, so if we write $M_s = M_s(0)$ then

$$(1 + |\xi'|^2)^{s-1/2} |\widehat{\gamma u}(\xi')|^2 \le M_s \int_{-\infty}^{\infty} (1 + |\xi|^2)^s |\hat{u}(\xi)|^2 \, d\xi_n.$$

Integrating over $\xi' \in \mathbb{R}^{n-1}$ gives

$$\|\gamma u\|_{H^{s-1/2}(\mathbb{R}^{n-1})}^2 \le M_s \|u\|_{H^s(\mathbb{R}^n)}^2,$$

and since $\mathcal{D}(\mathbb{R}^n)$ is dense in $H^s(\mathbb{R}^n)$, we obtain a unique continuous extension for γ. $\qquad\qquad\Box$

The lemma above is sharp in the sense that

$$H^{s-1/2}(\mathbb{R}^{n-1}) = \{\gamma u : u \in H^s(\mathbb{R}^n)\} \quad \text{for } s > \tfrac{1}{2},$$

because γ has a continuous right inverse η_0, as we now show.

Lemma 3.36 *For each integer $j \geq 0$, there exists a linear operator*

$$\eta_j : S(\mathbb{R}^{n-1}) \to S(\mathbb{R}^n)$$

satisfying

$$\partial^\alpha(\eta_j u)(x', 0) = \begin{cases} \partial^{\alpha'} u(x') & \text{if } \alpha_n = j, \\ 0 & \text{if } \alpha_n \neq j, \end{cases}$$

for $x' \in \mathbb{R}^{n-1}$, $u \in S(\mathbb{R}^{n-1})$ and any multi-index $\alpha = (\alpha', \alpha_n)$. Moreover, η_j has a unique extension to a bounded linear operator

$$\eta_j : H^{s-j-1/2}(\mathbb{R}^{n-1}) \to H^s(\mathbb{R}^n) \quad \text{for } s \in \mathbb{R}.$$

Proof. Choose a function $\theta_j \in \mathcal{D}(\mathbb{R})$ satisfying $\theta_j(y) = y^j/j!$ for $|y| \leq 1$, and define

$$\eta_j u(x) = \int_{\mathbb{R}^{n-1}} \frac{\hat{u}(\xi')\theta_j[(1 + |\xi'|^2)^{1/2} x_n]}{(1 + |\xi'|^2)^{j/2}} e^{i2\pi \xi' \cdot x'} \, d\xi' \quad \text{for } x \in \mathbb{R}^n.$$

Since $\theta_j^{(k)}(0) = \delta_{jk}$, we see that

$$\partial^\alpha(\eta_j u)(x', 0) = \int_{\mathbb{R}^{n-1}} (i2\pi \xi')^\alpha \hat{u}(\xi')\delta_{j\alpha_n} e^{i2\pi \xi' \cdot x'} \, d\xi' = \partial^{\alpha'} u(x')\delta_{j\alpha_n},$$

as required. The substitution $x_n = (1 + |\xi'|^2)^{-1/2} y$ gives

$$\widehat{\eta_j u}(\xi) = \frac{\hat{u}(\xi')}{(1 + |\xi'|^2)^{j/2}} \int_{-\infty}^{\infty} e^{-i2\pi \xi_n x_n} \theta_j[(1 + |\xi'|^2)^{1/2} x_n] \, dx_n$$

$$= \frac{\hat{u}(\xi')\hat{\theta}_j[(1 + |\xi'|^2)^{-1/2}\xi_n]}{(1 + |\xi'|^2)^{(j+1)/2}},$$

and the substitution $\xi_n = (1 + |\xi'|^2)^{1/2} t$ gives

$$\|\eta_j u\|^2_{H^s(\mathbb{R}^n)} = \int_{\mathbb{R}^{n-1}} \frac{|\hat{u}(\xi')|^2}{(1 + |\xi'|^2)^{j+1}}$$

$$\times \int_{-\infty}^{\infty} (1 + |\xi'|^2 + \xi_n^2)^s |\hat{\theta}_j[(1 + |\xi'|^2)^{-1/2}\xi_n]|^2 \, d\xi_n \, d\xi'$$

$$= C_s \|u\|^2_{H^{s-j-1/2}(\mathbb{R}^{n-1})},$$

where

$$C_s = \int_{-\infty}^{\infty} (1+t^2)^s |\hat{\theta}_j(t)|^2 \, dt < \infty,$$

for all $s \in \mathbb{R}$.   □

For Sobolev spaces on domains, we can now prove the following.

Theorem 3.37 Define the trace operator $\gamma : \mathcal{D}(\overline{\Omega}) \to \mathcal{D}(\Gamma)$ by

$$\gamma u = u|_\Gamma.$$

If Ω is a $C^{k-1,1}$ domain, and if $\frac{1}{2} < s \le k$, then γ has a unique extension to a bounded linear operator

$$\gamma : H^s(\Omega) \to H^{s-1/2}(\Gamma), \tag{3.31}$$

and this extension has a continuous right inverse.

Proof. Since $H^s(\mathbb{R}^n)$ is invariant under a $C^{k-1,1}$ change of coordinates if $1 - k \le s \le k$, one sees, via a partition of unity and a local flattening of the boundary, that if $\frac{1}{2} < s \le k$ then

$$\|\gamma u\|_{H^{s-1/2}(\Gamma)} \le C \|U\|_{H^s(\mathbb{R}^n)} \quad \text{if } u = U|_\Omega \text{ for } U \in \mathcal{D}(\mathbb{R}^n).$$

Hence, $\|\gamma u\|_{H^{s-1/2}(\Gamma)} \le C\|u\|_{H^s(\Omega)}$ for all $u \in \mathcal{D}(\overline{\Omega})$, and we obtain a unique continuous extension because $\mathcal{D}(\overline{\Omega})$ is dense in $H^s(\Omega)$. A right inverse for γ can be pieced together using the same partition of unity, by means of the operator η_0 from Lemma 3.36.   □

The preceding theorem applies, in particular, for $\frac{1}{2} < s \le 1$ if Ω is a Lipschitz domain. For technical reasons, we shall require the following, stronger result of Costabel [14].

Theorem 3.38 If Ω is a Lipschitz domain, then the trace operator (3.31) is bounded for $\frac{1}{2} < s < \frac{3}{2}$.

Proof. It suffices to consider a Lipschitz hypograph (3.26). Let $u \in \mathcal{D}(\mathbb{R}^n)$, and put

$$u_\zeta(x) = u\big(x', \zeta(x') + x_n\big) \quad \text{for } x = (x', x_n) \in \mathbb{R}^n.$$

By definition,

$$\|\gamma u\|_{H^{s-1/2}(\Gamma)} = \|u_\zeta(\cdot, 0)\|_{H^{s-1/2}(\mathbb{R}^{n-1})},$$

and to estimate the right-hand side, we introduce the notation

$$\tilde{u}(x', \xi_n) = \int_{-\infty}^{\infty} e^{-i2\pi \xi_n x_n} u(x', x_n)\, dx_n$$

for the partial Fourier transform of $u(x)$ with respect to x_n. Observe that

$$\tilde{u}_\zeta(x', \xi_n) = e^{i2\pi \xi_n \zeta(x')} \tilde{u}(x', \xi_n),$$

so

$$\|\tilde{u}_\zeta(\cdot, \xi_n)\|_{L_2(\mathbb{R}^{n-1})} = \|\tilde{u}(\cdot, \xi_n)\|_{L_2(\mathbb{R}^{n-1})}$$

and

$$\|\tilde{u}_\zeta(\cdot, \xi_n)\|_{H^1(\mathbb{R}^{n-1})}^2 \le C\|\tilde{u}(\cdot, \xi_n)\|_{H^1(\mathbb{R}^{n-1})}^2 + C\xi_n^2 \|\tilde{u}(\cdot, \xi_n)\|_{L_2(\mathbb{R}^{n-1})}^2.$$

Thus, if we define the anisotropic Sobolev space E^s with norm

$$\|u\|_{E^s}^2 = \int_{-\infty}^{\infty} \left[|\xi_n|^{2s} \|\tilde{u}(\cdot, \xi_n)\|_{L_2(\mathbb{R}^{n-1})}^2 + |\xi_n|^{2s-2} \|\tilde{u}(\cdot, \xi_n)\|_{H^1(\mathbb{R}^{n-1})}^2 \right] d\xi_n$$

$$= \int_{\mathbb{R}^n} a_s(\xi) |\hat{u}(\xi)|^2 \, d\xi,$$

where $a_s(\xi) = |\xi_n|^{2s} + |\xi_n|^{2s-2}(1 + |\xi'|^2) = |\xi_n|^{2s-2}(1 + |\xi|^2)$, then

$$\|u_\zeta\|_{E^s} \le C\|u\|_{E^s} \quad \text{for } s \in \mathbb{R}.$$

It is easy to see that

$$\|u\|_{E^s} \le C\|u\|_{H^s(\mathbb{R}^n)} \quad \text{for } s \ge 1,$$

and we claim

$$\|u(\cdot, 0)\|_{H^{s-1/2}(\mathbb{R}^{n-1})} \le C\|u\|_{E^s} \quad \text{for } \tfrac{1}{2} < s < \tfrac{3}{2}.$$

In fact, the substitution $\xi_n = (1 + |\xi'|^2)^{1/2} t$ yields

$$\int_{-\infty}^{\infty} \frac{d\xi_n}{a_s(\xi)} = C_s (1 + |\xi'|^2)^{-(s-1/2)}, \quad \text{where } C_s = \int_{-\infty}^{\infty} \frac{dt}{t^{2s-2}(1 + t^2)},$$

with $C_s < \infty$ for $\frac{1}{2} < s < \frac{3}{2}$, so, applying the Cauchy–Schwarz inequality,

$$\|u(\cdot, 0)\|^2_{H^{s-1/2}(\mathbb{R}^{n-1})}$$

$$= \int_{\mathbb{R}^{n-1}} (1 + |\xi'|^2)^{s-1/2} \left| \int_{-\infty}^{\infty} \hat{u}(\xi', \xi_n) \, d\xi_n \right|^2 d\xi'$$

$$\leq \int_{\mathbb{R}^{n-1}} (1 + |\xi'|^2)^{s-1/2} \left(\int_{-\infty}^{\infty} \frac{d\xi_n}{a_s(\xi)} \right) \left(\int_{-\infty}^{\infty} a_s(\xi) |\hat{u}(\xi)|^2 \, d\xi_n \right) d\xi'$$

$$= C_s \int_{\mathbb{R}^n} a_s(\xi) |\hat{u}(\xi)|^2 \, d\xi = C_s \|u\|^2_{E^s}.$$

Thus, we see that

$$\|u_\zeta(\cdot, 0)\|_{H^{s-1/2}(\mathbb{R}^{n-1})} \leq C\|u_\zeta\|_{E^s} \leq C\|u\|_{E^s} \leq C\|u\|_{H^s(\mathbb{R}^n)} \quad \text{for } 1 \leq s < \tfrac{3}{2},$$

which, combined with Theorem 3.37, shows that the trace operator (3.31) is bounded for $\frac{1}{2} < s < \frac{3}{2}$. $\qquad\square$

The next lemma is a version of a standard fact [41, p. 47] about distributions supported by a hyperplane, and will allow us to characterise $H^s_0(\Omega)$ using the trace operator. The symbol $\otimes$ means the tensor product of distributions, so, formally, $(v_j \otimes \delta^{(j)})(x) = v_j(x')\delta^{(j)}(x_n)$. In the proof, we use the notation $\mathbb{R}^n_+ = \{x \in \mathbb{R}^n : x_n > 0\}$ for the upper half space.

Lemma 3.39 *Consider the hyperplane $F = \{x \in \mathbb{R}^n : x_n = 0\}$.*

(i) If $s \geq -\frac{1}{2}$, then $H^s_F = \{0\}$.

(ii) If $s < -\frac{1}{2}$, then H^s_F is the set of distributions on $\mathbb{R}^n$ having the form

$$u = \sum_{0 \leq j < -s - 1/2} v_j \otimes \delta^{(j)} \quad \text{with } v_j \in H^{s+j+1/2}(\mathbb{R}^{n-1}). \tag{3.32}$$

Proof. First we show that any u of the form (3.32) belongs to H^s_F. Obviously, $\operatorname{supp} u \subseteq F$, and since $\mathcal{F}_{x \to \xi}\{v_j(x')\delta^{(j)}(x_n)\} = \hat{v}_j(\xi')(i 2\pi \xi_n)^j$, we have

$$\left\| v_j \otimes \delta^{(j)} \right\|^2_{H^s(\mathbb{R}^n)} = \int_{\mathbb{R}^n} (1 + |\xi|^2)^s |\hat{v}_j(\xi')|^2 |2\pi \xi_n|^{2j} \, d\xi.$$

The substitution $\xi_n = (1 + |\xi'|^2)^{1/2} t$ gives

$$\left\| v_j \otimes \delta^{(j)} \right\|^2_{H^s(\mathbb{R}^n)} = C_{j,s} \|v_j\|^2_{H^{s+j+1/2}(\mathbb{R}^{n-1})}, \tag{3.33}$$

where

$$C_{j,s} = \int_{-\infty}^{\infty} (1 + t^2)^s |2\pi t|^{2j} \, dt,$$

and here $C_{j,s} < \infty$ for $s + j < -\frac{1}{2}$, so $u \in H^s_F$.

Next, we show that if $u \in H_F^s$, and if $\phi \in \mathcal{D}(\mathbb{R}^n)$ satisfies

$$\partial_n^j \phi(x', 0) = 0 \quad \text{for } 0 \leq j \leq -s - \tfrac{1}{2},$$

then $\langle u, \phi \rangle = 0$. Indeed, by Exercise 3.22, if we define

$$\phi_\pm(x) = \begin{cases} \phi(x) & \text{if } x \in \mathbb{R}_\pm^n, \\ 0 & \text{otherwise,} \end{cases}$$

then $\phi_\pm \in \tilde{H}^{-s}(\mathbb{R}_\pm^n)$. It follows that there is a sequence $(\phi_m)_{m=1}^\infty$ in $\mathcal{D}(\mathbb{R}^n \setminus F)$ converging to ϕ in $H^{-s}(\mathbb{R}^n)$ as $m \to \infty$. Hence, $\langle u, \phi \rangle = \lim_{m \to \infty} \langle u, \phi_m \rangle = 0$. In particular, $u = 0$ if $u \in H_F^s$ for $s > -\tfrac{1}{2}$.

Suppose now that $u \in H_F^s$ with $-k - \tfrac{3}{2} < s < -k - \tfrac{1}{2}$ for an integer $k \geq 0$, and assume $0 \leq j \leq k$. Let η_j be as in Lemma 3.36, and define $v_j \in \mathcal{D}^*(\mathbb{R}^{n-1})$ by

$$\langle v_j, \phi \rangle = (-1)^j \langle u, \eta_j \phi \rangle \quad \text{for } \phi \in \mathcal{D}(\mathbb{R}^{n-1}).$$

Observe that

$$|\langle v_j, \phi \rangle| \leq \|u\|_{H^s(\mathbb{R}^n)} \|\eta_j \phi\|_{H^{-s}(\mathbb{R}^n)} \leq C\|u\|_{H^s(\mathbb{R}^n)} \|\phi\|_{H^{-s-j-1/2}(\mathbb{R}^{n-1})},$$

so $\|v_j\|_{H^{s+j+1/2}(\mathbb{R}^{n-1})} \leq C\|u\|_{H^s(\mathbb{R}^n)}$, and that

$$\left\langle u - \sum_{j=0}^k v_j \otimes \delta^{(j)}, \phi \right\rangle = \langle u, \rho \rangle \quad \text{for } \phi \in \mathcal{D}(\mathbb{R}^n)$$

where

$$\rho = \phi - \sum_{j=0}^k \eta_j \psi_j \quad \text{with } \psi_j(x') = \partial_n^j \phi(x', 0).$$

Since $\partial_n^l \rho(x', 0) = 0$ for $0 \leq l \leq k$, and since $-s - \tfrac{1}{2} < k + 1$, we have $\langle u, \rho \rangle = 0$ and so (3.32) holds.

It only remains to deal with the case $s = -k - \tfrac{1}{2}$. The argument above shows that if $u \in H_F^{-k-1/2}$ then $u = \sum_{j=0}^k v_j \otimes \delta^{(j)}$ with $v_j \in H^{j-k}(\mathbb{R}^{n-1})$ for $0 \leq j \leq k - 1$, and with $v_k \in H^{-\epsilon}(\mathbb{R}^{n-1})$ for any $\epsilon > 0$. Thus, it suffices to show that $v_k = 0$. In fact, (3.33) implies $v_j \otimes \delta^{(j)} \in H^{-k-1/2}(\mathbb{R}^n)$ for $0 \leq j \leq k - 1$, giving $v_k \otimes \delta^{(k)} \in H^{-k-1/2}(\mathbb{R}^n)$. However,

$$\left\| v_k \otimes \delta^{(k)} \right\|_{H^{-k-1/2-\epsilon}(\mathbb{R}^n)}^2 = C_{k,-k-1/2-\epsilon} \|v_k\|_{H^{-\epsilon}(\mathbb{R}^{n-1})}^2,$$

and $C_{k,-k-1/2-\epsilon} \to \infty$ as $\epsilon \to 0$, giving the desired result. $\qquad \square$

Theorem 3.40 *Assume that Ω is a $C^{k-1,1}$ domain.*

(i) If $0 \le s \le \frac{1}{2}$, then $H_0^s(\Omega) = H^s(\Omega)$.

(ii) If $\frac{1}{2} < s \le k$, then $H_0^s(\Omega) = \{u \in H^s(\Omega) : \gamma(\partial^\alpha u) = 0 \text{ for } |\alpha| < s - \frac{1}{2}\}$.

Proof. It suffices to deal with a half space $\Omega = \mathbb{R}_+^n$. We will use the characterisation of the dense subsets of a Banach space given by Exercise 2.5.

First suppose $0 \le s \le \frac{1}{2}$. If $w \in H^s(\Omega)^* = \widetilde{H}^{-s}(\Omega)$ satisfies $\langle w, \phi \rangle = 0$ for all $\phi \in \mathcal{D}(\Omega)$, then supp w is a subset of the hyperplane $F = \{x \in \mathbb{R}^n : x_n = 0\}$, and we conclude from part (i) of Lemma 3.39 that $w = 0$. Hence, $\mathcal{D}(\Omega)$ is dense in $H^s(\Omega)$, i.e., $H_0^s(\Omega) = H^s(\Omega)$.

Next suppose that $s > \frac{1}{2}$, and let $E = \{u \in H^s(\Omega) : \gamma(\partial^\alpha u) = 0 \text{ for } |\alpha| < s - \frac{1}{2}\}$, noting that this definition makes sense because $\gamma \circ \partial^\alpha : H^s(\Omega) \to H^{s-|\alpha|}(F)$ for $|\alpha| < s - \frac{1}{2}$. Let $\ell \in E^*$ satisfy $\ell(\phi) = 0$ for all $\phi \in \mathcal{D}(\Omega)$. By the Hahn–Banach theorem, there is a $w \in H^s(\Omega)^* = \widetilde{H}^{-s}(\Omega)$ such that $\ell(\phi) = \langle w, \phi \rangle$ for all $\phi \in E$. Since $w \in H_F^{-s}$, part (ii) of Lemma 3.39 shows that $w = \sum_{0 \le j < s - 1/2} v_j \otimes \delta^{(j)}$ with $v_j \in H^{-s+j+1/2}(\mathbb{R}^{n-1})$. Hence, for every $\phi \in E$,

$$\ell(\phi) = \langle w, \phi \rangle = \sum_{0 \le j < s - 1/2} (-1)^j \langle v_j, \gamma(\partial_n^j \phi) \rangle = 0,$$

and so $\mathcal{D}(\Omega)$ is dense in E, proving part (ii). $\qquad \square$

Vector-Valued Functions

Thus far, the present chapter has dealt only with spaces of scalar-valued (generalised) functions. The results obtained extend in a straightforward manner to spaces of vector-valued functions

$$u : \Omega \to \mathbb{C}^m,$$

and this final section does no more than establish some notational conventions.

We denote the space of compactly supported, $\mathbb{C}^m$-valued, C^∞ test functions by

$$\mathcal{D}(\Omega)^m = \mathcal{D}(\Omega; \mathbb{C}^m).$$

The (sequentially) continuous linear functionals on $\mathcal{D}(\Omega)^m$ are then the $\mathbb{C}^m$-valued distributions on Ω, and we view these objects as generalised $\mathbb{C}^m$-valued functions, by writing

$$\langle u, v \rangle_\Omega = \int_\Omega u(x) \cdot v(x) \, dx, \tag{3.34}$$

where the dot denotes the bilinear form on $\mathbb{C}^m$ whose restriction to $\mathbb{R}^m$ coincides with the standard Euclidean inner product, i.e.,

$$u \cdot v = \sum_{j=1}^{m} u_j v_j.$$

The set of all $\mathbb{C}^m$-valued distributions on Ω is denoted by $\mathcal{D}^*(\Omega)^m = \mathcal{D}^*(\Omega; \mathbb{C}^m)$. We think of u as a column vector or $m \times 1$ matrix, and let u^* denote the row vector or $1 \times m$ matrix obtained by transposing the complex conjugate $\bar{u}$. Using matrix multiplication, we may then write the standard unitary inner product in $\mathbb{C}^m$ as

$$u^* v = \bar{u} \cdot v = \sum_{j=1}^{m} \bar{u}_j v_j.$$

In this way, the sesquilinear form associated with the bilinear pairing (3.34) is given by

$$(u, v)_\Omega = \int_\Omega u(x)^* v(x) \, dx.$$

Of course, if u and v are square-integrable functions from Ω to $\mathbb{C}^m$, then $(u, v)_\Omega$ is their inner product in $L_2(\Omega)^m = L_2(\Omega; \mathbb{C}^m)$. The definitions of the vector Sobolev spaces on Ω,

$$W_p^s(\Omega)^m = W_p^s(\Omega; \mathbb{C}^m), \quad H^s(\Omega)^m = H^s(\Omega; \mathbb{C}^m), \quad \widetilde{H}^s(\Omega)^m = \widetilde{H}^s(\Omega; \mathbb{C}^m),$$

should now be obvious. Likewise for the vector Sobolev spaces on Γ. Occasionally, we shall encounter normed spaces of *matrix*-valued functions, such as $L_\infty(\Omega)^{m \times m} = L_\infty(\Omega; \mathbb{C}^{m \times m})$, whose meaning should also be obvious.

Exercises

3.1 Suppose that $u \in L_p(\Omega)$ satisfies

$$\left| \langle u, v \rangle_\Omega \right| \le M \|v\|_{L_{p^*}(\Omega)} \quad \text{for all } v \in L_{p^*}(\Omega).$$

(i) Show that if $1 \le p < \infty$, then $\|u\|_{L_p(\Omega)} \le M$. [Hint: take $v = \text{sign}(u) |u|^{p-1}$.]

(ii) Show that if $p = \infty$, then for every measurable set $E \subseteq \Omega$ with $|E| > 0$, the mean value of $|u|$ over E is bounded by M, i.e.,

$$\frac{1}{|E|} \int_E |u(x)| \, dx \le M.$$

Deduce that $\|u\|_{L_\infty(\Omega)} \le M$. [Hint: take $v = \text{sign}(u) \chi_E$, where χ_E is the characteristic function of E.]

3.2 Prove that convolution is associative:

$$(u * v) * w = u * (v * w) \quad \text{for } u, v, w \in L_1(\mathbb{R}^n).$$

3.3 For $1 \le j \le k$, let $f_j \in L_1(\mathbb{R}^n)$ be a compactly supported, non-negative function satisfying $\|f_j\|_{L_1(\mathbb{R}^n)} = 1$, and let $0 \le \lambda_j \le 1$. Fix $x \in \mathbb{R}^n$, and define

$$g(\lambda) = \left(f_1^{\lambda_1} * \cdots * f_k^{\lambda_k} \right)(x) \quad \text{for } \lambda = (\lambda_1, \ldots, \lambda_k).$$

(Take f_j^0 to be identically 1 outside as well as inside supp f_j.)
  (i) Show that $g(\tilde{e}_j) = 1$ where $\tilde{e}_j \in \mathbb{R}^k$ is the vector with components $(\tilde{e}_j)_l = 1 - \delta_{jl}$.
  (ii) Use the fact that, for any positive a_j and any non-negative λ_j and μ_j,

$$\prod_{j=1}^k a_j^{(1-t)\lambda_j + t\mu_j} = \exp\left((1-t) \sum_{j=1}^k \log a_j^{\lambda_j} + t \sum_{j=1}^k \log a_j^{\mu_j} \right)$$

to show that the function $g : [0, 1]^k \to [0, \infty)$ is convex.
  (iii) Deduce that $g(\lambda) \le 1$ if $\lambda_1 + \cdots + \lambda_k = k - 1$. [Hint: $\lambda = \sum_{j=1}^k (1 - \lambda_j)\tilde{e}_j$.]
  (iv) Hence show that

$$|(u_1 * \cdots * u_k)(x)| \le \|u_1\|_{L_{p_1}(\mathbb{R}^n)} \cdots \|u_k\|_{L_{p_k}(\mathbb{R}^n)} \quad \text{if } \sum_{j=1}^k \frac{1}{p_j} = k - 1.$$

3.4 Show that supp$(u * v) \subseteq$ supp $u +$ supp $v = \{x + y : x \in$ supp u and $y \in$ supp $v\}$.

3.5 Recall the notation (3.7).
  (i) Show that

$$\left(\frac{d}{dt} \right)^k u(x + ty) = u^{(k)}(x + ty; y).$$

[Hint: $k!/\alpha!$ equals the number of permutations of $k = |\alpha|$ objects when there are α_j objects of type j, for $1 \le j \le n$, and it is assumed that objects of different types are distinguishable, but objects of the same type are indistinguishable.]
  (ii) Use integration by parts to verify Taylor's formula for a function of one variable:

$$f(s) = \sum_{j=0}^k \frac{f^{(j)}(0)}{j!} s^j + \frac{s^{k+1}}{k!} \int_0^1 (1-t)^k f^{(k+1)}(ts)\, dt.$$

(iii) By taking $f(s) = u(x + sy)$, derive Taylor's formula (3.8) for a function of n variables.

3.6 Define $f : \mathbb{R} \to \mathbb{R}$ by

$$f(t) = \begin{cases} e^{-1/t} & \text{if } t > 0, \\ 0 & \text{if } t \leq 0. \end{cases}$$

(i) Show that $f^{(j)}(t) \to 0$ as $t \downarrow 0$, for each $j \geq 0$.

(ii) Deduce that $f \in C^{\infty}(\mathbb{R})$.

(iii) Construct a C^{∞} function $g : \mathbb{R} \to \mathbb{R}$ with $g > 0$ on $(-1, 1)$ and with supp $g = [-1, 1]$.

(iv) Construct a C^{∞} function $\psi \in C^{\infty}_{\text{comp}}(\mathbb{R}^n)$ satisfying (3.9).

3.7 Let $\Omega = (0, 1)$, choose any $\phi \in \mathcal{D}(\Omega)$ not identically zero, and define $\phi_j \in \mathcal{D}(\Omega)$ by $\phi_j(x) = \phi(j^{-1}x)$. Show that $\phi_j \to 0$ in $\mathcal{E}(\Omega)$, but not in $\mathcal{D}(\Omega)$.

3.8 Establish that the inclusions $\mathcal{D}(\Omega) \subseteq \mathcal{E}(\Omega)$ and $\mathcal{D}(\mathbb{R}^n) \subseteq \mathcal{S}(\mathbb{R}^n) \subseteq \mathcal{E}(\mathbb{R}^n)$ are continuous with dense image, by proving each of the following statements:

(i) If $\phi_j \to 0$ in $\mathcal{D}(\Omega)$, then $\phi_j \to 0$ in $\mathcal{E}(\Omega)$.

(ii) If $\phi_j \to 0$ in $\mathcal{D}(\mathbb{R}^n)$, then $\phi_j \to 0$ in $\mathcal{S}(\mathbb{R}^n)$.

(iii) If $\phi_j \to 0$ in $\mathcal{S}(\mathbb{R}^n)$, then $\phi_j \to 0$ in $\mathcal{E}(\mathbb{R}^n)$.

(iv) Let $K_1 \subseteq K_2 \subseteq \cdots$ be an increasing sequence of compact sets whose union is Ω, and let $\chi_j \in \mathcal{D}(\Omega)$ satisfy $\chi_j = 1$ on K_j. If $\phi \in \mathcal{E}(\Omega)$, then $\chi_j\phi \to \phi$ in $\mathcal{E}(\Omega)$.

(v) Let $\chi \in \mathcal{D}(\mathbb{R}^n)$ satisfy $\chi(x) = 1$ for $|x| \leq 1$, and define $\chi_j \in \mathcal{D}(\mathbb{R}^n)$ for each positive integer j by $\chi_j(x) = \chi(j^{-1}x)$. If $\phi \in \mathcal{S}(\mathbb{R}^n)$, then $\chi_j\phi \to \phi$ in $\mathcal{S}(\mathbb{R}^n)$.

3.9 Consider a linear functional $\ell : \mathcal{D}(\Omega) \to \mathbb{C}$. Show that ℓ is sequentially continuous (and hence a distribution on Ω) if and only if for each compact set $K \Subset \Omega$ there exists an integer $m \geq 0$ such that

$$|\ell(\phi)| \leq C_{K,m} \sum_{|\alpha| \leq m} \sup_K |\partial^{\alpha}\phi| \quad \text{for all } \phi \in \mathcal{D}_K(\Omega).$$

[Hint: to prove the necessity of the condition, suppose for a contradiction that there is a K for which no such m and $C_{K,m}$ exist, and deduce the existence of a sequence $\phi_j \to 0$ in $\mathcal{D}_K(\Omega)$ such that $\ell(\phi_j) = 1$.]

3.10 Prove from the definition (3.12) that if $u \in \mathcal{D}^*(\mathbb{R}^n)$ and $\phi \in \mathcal{D}(\mathbb{R}^n)$, then $u * \phi$ is C^{∞}, with

$$\partial^{\alpha}(u * \phi) = (\partial^{\alpha}u) * \phi = u * (\partial^{\alpha}\phi).$$

3.11 Show that

$$\mathcal{F}_{x \to \xi}\left\{e^{-\pi|x|^2}\right\} = \prod_{j=1}^{n} \int_{-\infty}^{\infty} e^{-i2\pi\xi_j x_j - \pi x_j^2} \, dx_j = e^{-\pi|\xi|^2}.$$

3.12 For $\psi \in L_1(\mathbb{R}^n)$, show that

$$\mathcal{F}_{x \to \xi}\{\epsilon^{-n}\psi(\epsilon^{-1}x)\} = (\mathcal{F}\psi)(\epsilon\xi) \quad \text{and}$$

$$\mathcal{F}_{\xi \to x}^*\{\psi(\epsilon\xi)\} = \epsilon^{-n}(\mathcal{F}^*\psi)(\epsilon^{-1}x).$$

3.13 Let k be a positive integer. We denote the kth-order forward difference operator by δ_h^k, and then define the kth-order L_2 modulus of continuity by

$$\omega^k(t, u) = \sup_{|h| \le t} \left\| \delta_h^k u \right\|_{L_2(\mathbb{R}^n)} \quad \text{for } t > 0.$$

(i) Adapt the proof of Theorem 3.16 to show that if $0 < s < k$, then

$$\|u\|_{H^s(\mathbb{R}^n)}^2 \sim \|u\|_{L_2(\mathbb{R}^n)}^2 + \int_{\mathbb{R}^n} \frac{\left\| \delta_h^k u \right\|_{L_2(\mathbb{R}^n)}^2}{|h|^{2s+n}} \, dh.$$

(ii) Show that

$$[\omega^k(t, u)]^2 \le C_k \int_{\mathbb{R}^n} \min(|\xi|^{2k} t^{2k}, 1)|\hat{u}(\xi)|^2 \, d\xi.$$

(iii) Deduce that for $0 < s < k$,

$$\int_{\mathbb{R}^n} \frac{\left\| \delta_h^k u \right\|_{L_2(\mathbb{R}^n)}^2}{|h|^{2s+n}} \, dh \sim \int_0^{\infty} \frac{[\omega^k(t, u)]^2}{t^{2s+1}} \, dt$$

$$\sim \sum_{j=-\infty}^{\infty} 2^{2js} [\omega^k(2^{-j}, u)]^2.$$

3.14 Let $\chi_j \in \mathcal{D}(\mathbb{R}^n)$ be as in part (v) of Exercise 3.8, and let $s \in \mathbb{R}$. Show that if $u \in H^s(\mathbb{R}^n)$, then $\chi_j u \to u$ in $H^s(\mathbb{R}^n)$.

3.15 Consider a distribution $u \in \mathcal{D}^*(\Omega)$ with $\mathrm{supp}\, u \subseteq K \Subset \Omega$.

(i) Show that there is a unique $\tilde{u} \in \mathcal{D}^*(\mathbb{R}^n)$ satisfying $\tilde{u} = u$ on Ω, and $\tilde{u} = 0$ on $\mathbb{R}^n \setminus K$. [Hint: use a cutoff function $\chi \in \mathcal{D}(\Omega)$ with $\chi = 1$ on K.]

(ii) Show that if $s \in \mathbb{R}$ and $u \in H^s(\Omega)$, then $\|\tilde{u}\|_{H^s(\mathbb{R}^n)} \le C_{s,K} \|u\|_{H^s(\Omega)}$.

3.16 Give an alternative proof of Theorem 3.20 for $\Omega = \mathbb{R}^n$, as follows.

(i) Prove *Peetre's inequality:*

$$(1+|\xi|^2)^s \leq 2^{|s|}(1+|\xi-\eta|^2)^{|s|}(1+|\eta|^2)^s \quad \text{for } \xi, \eta \in \mathbb{R}^n \text{ and } s \in \mathbb{R}.$$

(ii) Use the relation $\widehat{\phi u} = \hat{\phi} * \hat{u}$ to deduce that if $u \in H^s(\mathbb{R}^n)$, then $\phi u \in H^s(\mathbb{R}^n)$ with $\|\phi u\|_{H^s(\mathbb{R}^n)} \leq C_s \|u\|_{H^s(\mathbb{R}^n)}$, where

$$C_s = 2^{|s|/2} \int_{\mathbb{R}^n} (1+|\xi|^2)^{|s|/2} |\hat{\phi}(\xi)| \, d\xi.$$

3.17 Let ψ and ψ_ϵ be as in (3.9) and (3.10).

(i) Show that if $u \in \mathcal{D}^*(\mathbb{R}^n)$, then the convolution $u_\epsilon = \psi_\epsilon * u$ belongs to $\mathcal{E}(\mathbb{R}^n)$ and

$$\operatorname{supp} u_\epsilon \subseteq \{x \in \mathbb{R}^n : \operatorname{dist}(x, \operatorname{supp} u) \leq \epsilon\}.$$

(ii) Let $s \in \mathbb{R}$, and show that if $u \in H^s(\mathbb{R}^n)$ then

$$\|u_\epsilon\|_{H^s(\mathbb{R}^n)} \leq \|u\|_{H^s(\mathbb{R}^n)} \quad \text{and} \quad \lim_{\epsilon \to 0} \|u_\epsilon - u\|_{H^s(\mathbb{R}^n)} = 0.$$

3.18 Show that if Ω is the crack domain shown in Figure 2(ii), then $\mathcal{D}(\overline{\Omega})$ is not dense in $H^s(\Omega)$ for $s > n/2$.

3.19 Let $u \in \mathcal{D}^*(\Omega)$ and $s \in \mathbb{R}$.

(i) Suppose W is an open set and $\chi \in \mathcal{D}(W)$. Show that if $u \in H^s(W \cap \Omega)$, then $\chi u \in H^s(\Omega)$ and $\|\chi u\|_{H^s(\Omega)} \leq C_{\chi,s} \|u\|_{H^s(W \cap \Omega)}$.

(ii) Suppose $(\phi_j)_{j=0}^J$ is a partition of unity of the type used in the proof of Theorem 3.29. Show that $u \in H^s(\Omega)$ if and only if $\phi_j u \in H^s(W_j \cap \Omega)$ for $0 \leq j \leq J$, in which case

$$\|u\|_{H^s(\Omega)}^2 \sim \sum_{j=0}^J \|\phi_j u\|_{H^s(W_j \cap \Omega)}^2.$$

3.20 Prove the following inequalities, due to Hardy: for $\alpha > 0$ and $1 \leq p < \infty$,

$$\left[\int_0^\infty \left(x^{-\alpha} \int_0^x |f(y)| \frac{dy}{y}\right)^p \frac{dx}{x}\right]^{1/p} \leq \frac{1}{\alpha} \left(\int_0^\infty |y^{-\alpha} f(y)|^p \frac{dy}{y}\right)^{1/p}$$

and

$$\left[\int_0^\infty \left(x^\alpha \int_x^\infty |f(y)| \frac{dy}{y}\right)^p \frac{dx}{x}\right]^{1/p} \leq \frac{1}{\alpha} \left(\int_0^\infty |y^\alpha f(y)|^p \frac{dy}{y}\right)^{1/p}.$$

[Hint: make the substitution $y = xt$ in the inner integral, and then apply Minkowski's inequality, i.e., think of $t \mapsto f(\cdot\, t)$ as a map from $(0, 1)$ or $(1, \infty)$ into a weighted L_p space on $(0, \infty)$.]

3.21 Let $1 \le p < \infty$.

    (i) Show that, for $-\infty < a < b < \infty$ and $f \in L_p(a,b)$,

$$\int_a^b \left(\frac{1}{x-a} \int_a^x |f(t)|\, dt \right)^p dx \le \left(\frac{p}{p-1} \right)^p \int_a^b |f(t)|^p\, dt,$$

$$\int_a^b \left(\frac{1}{b-x} \int_x^b |f(t)|\, dt \right)^p dx \le \left(\frac{p}{p-1} \right)^p \int_a^b |f(t)|^p\, dt.$$

    [Hint: use Exercise 3.20.]

    (ii) Show that, for $u \in \mathcal{D}(\mathbb{R}^2)$,

$$\int_a^b \int_a^b \left| \frac{u(x,x) - u(y,y)}{x-y} \right|^p (\sqrt{2}\, dx)(\sqrt{2}\, dy)$$

$$\le \left(\frac{2p}{p-1} \right)^p \int_a^b \int_a^b \left[|\partial_1 u(x,y)|^p + |\partial_2 u(x,y)|^p \right] dx\, dy.$$

    [Hint: $u(x,x) - u(y,y) = \int_y^x \partial_2 u(x,t)\, dt + \int_y^x \partial_1 u(t,y)\, dt$ for $y < x$.]

    (iii) Let $\gamma u(x') = u(x', 0)$ for $x' \in \mathbb{R}^{n-1}$ and $u \in \mathcal{D}(\mathbb{R}^n)$. Show that

$$\int_{\mathbb{R}^{n-1}} \int_{\mathbb{R}^{n-1}} \frac{|\gamma u(x') - \gamma u(y')|^p}{|x'-y'|^p}\, dx'\, dy' \le C_{p,n} \sum_{j=1}^n \int_{\mathbb{R}^n} |\partial_j u(x)|^p\, dx.$$

    (iv) Show that, for $u \in \mathcal{D}(\mathbb{R}^n)$,

$$\|\gamma u\|_{L_p(\mathbb{R}^{n-1})} \le C \|u\|_{W_p^1(\mathbb{R}^n)}.$$

    [Hint: write $\gamma u(x') = \int_{-\infty}^0 \partial_n(\chi u)(x)\, dx_n$ for a suitable function $\chi(x_n)$.]

    (v) Deduce that $\gamma : W_p^k(\mathbb{R}^n) \to W_p^{k-1/p}(\mathbb{R}^{n-1})$ for each integer $k \ge 0$.

3.22 Consider the half space $\Omega = \{x \in \mathbb{R}^n : x_n < 0\}$. Let $u \in \mathcal{D}(\overline{\Omega})$, and define

$$U(x) = \begin{cases} u(x) & \text{if } x_n < 0, \\ 0 & \text{if } x_n \ge 0. \end{cases}$$

    (i) Show that $|\hat{U}(\xi)| \le C_{k,u}(1 + |\xi'|)^{-k}(1 + |\xi_n|)^{-1}$ for every $k > 0$.

    (ii) Hence show that $U \in H_{\overline{\Omega}}^s = \tilde{H}^s(\Omega)$ for $s < \frac{1}{2}$.

    (iii) Show that if $\partial_n^k u(x', 0) = 0$ for $0 \le k \le j$, then $U \in \tilde{H}^s(\Omega)$ for $s < j + \frac{3}{2}$.

4

Strongly Elliptic Systems

We are now ready to begin our study of boundary value problems for linear elliptic systems of second-order partial differential equations. The first task is to explain how, via the first Green identity, such problems fit into the abstract scheme treated in Chapter 2. We then define the class of *strongly elliptic* operators, and investigate when such operators are coercive. After that, an existence and uniqueness theorem for weak solutions in $H^1(\Omega)^m$ is given, expressed in the form of the Fredholm alternative. Next, we prove regularity of the solution on the interior and up to the boundary, under appropriate assumptions on the data and the domain. We also prove the transmission property, which will be used later to show regularity at the boundary of surface potentials for smooth domains. The final section of the chapter presents some rather technical estimates of Nečas relating the H^1-norm of the trace and the L_2-norm of the conormal derivative. These estimates will allow us to prove some limited regularity of surface potentials for Lipschitz domains.

The First and Second Green Identities

Suppose that Ω is a non-empty open, possibly unbounded subset of $\mathbb{R}^n$, and consider a linear second-order partial differential operator $\mathcal{P}$ of the form

$$\mathcal{P}u = -\sum_{j=1}^{n}\sum_{k=1}^{n} \partial_j (A_{jk}\partial_k u) + \sum_{j=1}^{n} A_j \partial_j u + Au \quad \text{on } \Omega, \qquad (4.1)$$

where the coefficients

$$A_{jk} = \left[a_{pq}^{jk}\right], \quad A_j = \left[a_{pq}^{j}\right], \quad A = [a_{pq}]$$

are functions from Ω into $\mathbb{C}^{m \times m}$, the space of complex $m \times m$ matrices. Thus, $1 \le p \le m$ and $1 \le q \le m$, and $\mathcal{P}$ acts on a (column) vector-valued function

113

$u : \Omega \rightarrow \mathbb{C}^m$ to give a vector-valued function $\mathcal{P}u : \Omega \rightarrow \mathbb{C}^m$, whose components are

$$(\mathcal{P}u)_p = -\sum_{j=1}^{n}\sum_{k=1}^{n}\sum_{q=1}^{m} \partial_j \left(a_{pq}^{jk} \partial_k u_q\right) + \sum_{j=1}^{n}\sum_{q=1}^{m} a_{pq}^j \partial_j u_q + \sum_{q=1}^{m} a_{pq} u_q$$

$$\text{for } 1 \leq p \leq m.$$

We shall see in Lemma 4.1 that $\mathcal{P}$ is naturally associated with a sesquilinear form Φ, defined by

$$\Phi(u, v) = \int_{\Omega} \left(\sum_{j=1}^{n}\sum_{k=1}^{n}(A_{jk}\partial_k u)^* \partial_j v + \sum_{j=1}^{n}(A_j \partial_j u)^* v + (Au)^* v\right) dx. \quad (4.2)$$

(Recall that $*$ denotes the conjugate transpose of a matrix or vector.) It will always be assumed that the coefficients A_{jk}, A_j and A belong to $L_\infty(\Omega)^{m \times m}$, so that Φ is bounded on $H^1(\Omega)^m$:

$$|\Phi(u, v)| \leq C \|u\|_{H^1(\Omega)^m} \|v\|_{H^1(\Omega)^m} \quad \text{for } u, v \in H^1(\Omega)^m.$$

If, in addition, the leading coefficients A_{jk} are Lipschitz, then $\mathcal{P} : H^2(\Omega)^m \rightarrow L_2(\Omega)^m$ is a bounded linear operator. When the lower-order terms are dropped from $\mathcal{P}$, we are left with the *principal part* $\mathcal{P}_0$, which can be written in divergence form as

$$\mathcal{P}_0 u = -\sum_{j=1}^{n} \partial_j \mathcal{B}_j u \quad \text{where } \mathcal{B}_j u = \sum_{k=1}^{n} A_{jk}\partial_k u. \quad (4.3)$$

If Ω is a Lipschitz domain, then the *conormal derivative* $\mathcal{B}_\nu u$ is defined by

$$\mathcal{B}_\nu u = \sum_{j=1}^{n} \nu_j \gamma(\mathcal{B}_j u) \quad \text{on } \Gamma, \quad (4.4)$$

where, as usual, ν is the outward unit normal to Ω, γ is the trace operator for Ω, and $\Gamma = \partial\Omega$ is the boundary of Ω. The conormal derivative arises naturally via the following lemma, known as the *first Green identity*.

Lemma 4.1 *If Ω is a Lipschitz domain, and if the coefficients A_{jk} are Lipschitz, then*

$$\Phi(u, v) = (\mathcal{P}u, v)_\Omega + (\mathcal{B}_\nu u, \gamma v)_\Gamma \quad \text{for } u \in H^2(\Omega)^m \text{ and } v \in H^1(\Omega)^m.$$

Proof. By the divergence theorem (Theorem 3.34), if $w \in C^1_{\text{comp}}(\overline{\Omega})$ then

$$\int_\Omega \partial_j w \, dx = \int_\Gamma \nu_j w \, d\sigma,$$

and one sees with the help of the density and trace results in Theorems 3.29 and 3.38 that this formula holds in fact for any $w \in H^1(\Omega)$. Taking $w = (\mathcal{B}_j u)^* v$, we obtain

$$\int_\Omega \partial_j [(\mathcal{B}_j u)^* v] \, dx = \int_\Gamma \nu_j \gamma [(\mathcal{B}_j u)^* v] \, d\sigma$$
$$\text{for } u \in H^2(\Omega)^m \text{ and } v \in H^1(\Omega)^m,$$

and the result follows after summing over j, because by (4.3),

$$\sum_{j=1}^n \partial_j [(\mathcal{B}_j u)^* v] = -(\mathcal{P}_0 u)^* v + \sum_{j=1}^n (\mathcal{B}_j u)^* \partial_j v. \qquad \square$$

In order to state a dual version of Lemma 4.1, we define

$$\widetilde{\mathcal{B}}_j u = \sum_{k=1}^n A^*_{kj} \partial_k u + A^*_j u,$$

and put

$$\mathcal{P}^* u = -\sum_{j=1}^n \sum_{k=1}^n \partial_j (A^*_{kj} \partial_k u) - \sum_{j=1}^n \partial_j (A^*_j u) + A^* u$$
$$= -\sum_{j=1}^n \partial_j \widetilde{\mathcal{B}}_j u + A^* u \quad \text{on } \Omega,$$

and

$$\widetilde{\mathcal{B}}_\nu u = \sum_{j=1}^n \nu_j \gamma (\widetilde{\mathcal{B}}_j u) \quad \text{on } \Gamma. \qquad (4.5)$$

In fact, by arguing as before, but now with $w = u^* \widetilde{\mathcal{B}}_j v$, one easily verifies the following identity.

Lemma 4.2 *If Ω is a Lipschitz domain, and if the coefficients A_{jk} and A_j are Lipschitz, then*

$$\Phi(u, v) = (u, \mathcal{P}^* v)_\Omega + (\gamma u, \widetilde{\mathcal{B}}_\nu v)_\Gamma \quad \text{for } u \in H^1(\Omega)^m \text{ and } v \in H^2(\Omega)^m.$$

Thus,

$$(\mathcal{P}u, v)_\Omega = \Phi(u, v) = (u, \mathcal{P}^*v)_\Omega \quad \text{for } u \in H^2(\Omega)^m \text{ and } v \in \mathcal{D}(\Omega)^m,$$

because $\gamma v = \widetilde{B}_\nu v = 0$ on Γ. Hence, if $u \in H^1(\Omega)^m$, then we can *define* $\mathcal{P}u$ as a distribution on Ω by

$$(\mathcal{P}u, v)_\Omega = \Phi(u, v) \quad \text{for } v \in \mathcal{D}(\Omega)^m,$$

even if the coefficients A_{jk} are not Lipschitz, but only belong to $L_\infty(\Omega)^{m \times m}$. Likewise, we can define the distribution $\mathcal{P}^*u$ for any $u \in H^1(\Omega)^m$ by

$$(\mathcal{P}^*u, v)_\Omega = \Phi^*(u, v) \quad \text{for } v \in \mathcal{D}(\Omega)^m.$$

The operator $\mathcal{P}^*$ is called the *formal adjoint* of $\mathcal{P}$. Its principal part is given by

$$(\mathcal{P}^*)_0 u = -\sum_{j=1}^n \sum_{k=1}^n \partial_j (A_{kj}^* \partial_k u) = -\sum_{j=1}^n \partial_j \left(\widetilde{B}_j u - A_j^* u \right),$$

which coincides with the formal adjoint of $\mathcal{P}_0$, allowing us to write

$$\mathcal{P}_0^* = (\mathcal{P}^*)_0 = (\mathcal{P}_0)^*.$$

However, the conormal derivative of u relative to $\mathcal{P}^*$ is

$$\sum_{j=1}^n \nu_j \gamma(A_{kj}^* \partial_k u) = \widetilde{B}_\nu u - \sum_{j=1}^n \nu_j \gamma(A_j^* u),$$

which coincides with $\widetilde{B}_\nu u$ if and only if $A_j = 0$ for all j. One says that $\mathcal{P}$ is *formally self-adjoint* if $\mathcal{P}^* = \mathcal{P}$, i.e., if the coefficients of $\mathcal{P}$ satisfy

$$A_{kj}^* = A_{jk}, \quad A_j = 0, \quad A^* = A. \tag{4.6}$$

In this case, $\widetilde{B}_\nu = B_\nu$ and the sesquilinear form (4.2) is Hermitian, i.e.,

$$\Phi(v, u) = \overline{\Phi(u, v)} \quad \text{for } u, v \in H^1(\Omega)^m.$$

The next lemma will allow us to extend the definition of the conormal derivative.

Lemma 4.3 *Suppose that Ω is a Lipschitz domain. If $u \in H^1(\Omega)^m$ and $f \in \widetilde{H}^{-1}(\Omega)^m$ satisfy*

$$\mathcal{P}u = f \quad \text{on } \Omega,$$

then there exists $g \in H^{-1/2}(\Gamma)^m$ such that

$$\Phi(u, v) = (f, v)_\Omega + (g, \gamma v)_\Gamma \quad \text{for } v \in H^1(\Omega)^m.$$

Furthermore, g is uniquely determined by u and f, and we have

$$\|g\|_{H^{-1/2}(\Gamma)^m} \le C\|u\|_{H^1(\Omega)^m} + C\|f\|_{\widetilde{H}^{-1}(\Omega)^m}.$$

Proof. By Theorem 3.37, there exists a bounded linear operator $\eta : H^{1/2}(\Gamma)^m \to H^1(\Omega)^m$ satisfying $\gamma \eta v = v$ for all $v \in H^{1/2}(\Gamma)^m$. Since $[H^1(\Omega)^m]^* = \widetilde{H}^{-1}(\Omega)^m$ and $[H^{1/2}(\Gamma)^m]^* = H^{-1/2}(\Gamma)^m$, we can define $g \in H^{-1/2}(\Gamma)^m$ by

$$(g, w)_\Gamma = \Phi(u, \eta w) - (f, \eta w)_\Omega \quad \text{for } w \in H^{1/2}(\Gamma)^m.$$

Given $v \in H^1(\Omega)^m$, consider the function $v_0 = v - \eta \gamma v$. Since $\gamma v_0 = 0$, we have $v_0 \in H^1_0(\Omega)^m$ by Theorem 3.40, so there is a sequence $(\phi_j)_{j=0}^\infty$ in $\mathcal{D}(\Omega)^m$ that converges to v_0 in $H^1(\Omega)^m$. Hence, using the fact that $\mathcal{P}u = f$ on Ω,

$$\Phi(u, v_0) = \lim_{j \to \infty} \Phi(u, \phi_j) = \lim_{j \to \infty} (f, \phi_j)_\Omega = (f, v_0)_\Omega,$$

and by the definition of g,

$$\begin{aligned}
\Phi(u, v) = \Phi(u, v_0 + \eta \gamma v) &= (f, v_0)_\Omega + (g, \gamma v)_\Gamma + (f, \eta \gamma v)_\Omega \\
&= (f, v)_\Omega + (g, \gamma v)_\Gamma,
\end{aligned}$$

as required. Finally, if g_1 and g_2 both satisfy the conclusions of the lemma, then for the difference $g_1 - g_2 \in H^{-1/2}(\Gamma)^m$ we have $(g_1 - g_2, \gamma v)_\Gamma = 0$ for all $v \in H^1(\Omega)^m$, and therefore $(g_1 - g_2, w)_\Gamma = 0$ for all $w \in H^{1/2}(\Gamma)^m$, implying $g_1 = g_2$. $\square$

Note that g is not uniquely determined by u alone, but depends on the choice of f. The problem is that we could have $\mathcal{P}u = f_1 = f_2$ on Ω, with the difference $f_1 - f_2$ a non-zero distribution on $\mathbb{R}^n$ having support in Γ. However, provided f is clear from the context, we shall write $\mathcal{B}_\nu u = g$ and call this distribution the conormal derivative of u. In particular, if $\mathcal{P}u \in L_2(\Omega)^m$, then we always define $\mathcal{B}_\nu u$ by making the natural choice

$$f = \begin{cases} \mathcal{P}u & \text{on } \Omega, \\ 0 & \text{on } \mathbb{R}^n \setminus \overline{\Omega}, \end{cases}$$

thereby ensuring consistency with the original definition of the conormal derivative. We extend the definition of $\widetilde{\mathcal{B}}_\nu u$ in the same fashion, with the help of Lemma 4.2. The next theorem follows at once; cf. (1.8) and (1.9).

Theorem 4.4 Let Ω be a Lipschitz domain, and suppose that $u, v \in H^1(\Omega)^m$.

(i) If $\mathcal{P}u \in L_2(\Omega)^m$, then the first Green identity holds:

$$\Phi(u, v) = (\mathcal{P}u, v)_\Omega + (\mathcal{B}_\nu u, \gamma v)_\Gamma.$$

(ii) If $\mathcal{P}^* v \in L_2(\Omega)^m$, then

$$\Phi(u, v) = (u, \mathcal{P}^* v)_\Omega + (\gamma u, \widetilde{\mathcal{B}}_\nu v)_\Gamma.$$

(iii) If both $\mathcal{P}u$ and $\mathcal{P}^* v$ belong to $L_2(\Omega)$, then the second Green identity holds:

$$(\mathcal{P}u, v)_\Omega - (u, \mathcal{P}^* v)_\Omega = (\gamma u, \widetilde{\mathcal{B}}_\nu v)_\Gamma - (\mathcal{B}_\nu u, \gamma v)_\Gamma.$$

Strongly Elliptic Operators

Let V be a closed subspace of $H^1(\Omega)^m$, such that V is dense in $L_2(\Omega)^m$. Following the terminology established in Chapter 2, we say that Φ and $\mathcal{P}$ are *coercive* on V if

$$\operatorname{Re} \Phi(u, u) \ge c\|u\|^2_{H^1(\Omega)^m} - C\|u\|^2_{L_2(\Omega)^m} \quad \text{for } u \in V.$$

Obviously, in this context $L_2(\Omega)$ acts as the pivot space for V. When seeking to determine whether or not a given differential operator is coercive, we can ignore the lower-order terms, and consider just the sesquilinear form corresponding to the principal part,

$$\Phi_0(u, v) = \int_\Omega \sum_{j=1}^n \sum_{k=1}^n (A_{jk}\partial_k u)^* \partial_j v \, dx.$$

Lemma 4.5 The differential operator $\mathcal{P}$ is coercive on V if and only if its principal part $\mathcal{P}_0$ is coercive on V.

Proof. For any $\epsilon > 0$, we have

$$|\Phi(u, v) - \Phi_0(u, v)| \le C\|u\|_{H^1(\Omega)^m}\|v\|_{L_2(\Omega)^m}$$
$$\le C\big(\epsilon\|u\|^2_{H^1(\Omega)^m} + \epsilon^{-1}\|v\|^2_{L_2(\Omega)^m}\big),$$

so if $\mathcal{P}_0$ is coercive, i.e., if $\operatorname{Re} \Phi_0(u, u) \ge c\|u\|^2_{H^1(\Omega)^m} - C\|u\|^2_{L_2(\Omega)^m}$, then

$$\operatorname{Re} \Phi(u, u) \ge (c - \epsilon C)\|u\|^2_{H^1(\Omega)^m} - C(1 + \epsilon^{-1})\|u\|^2_{L_2(\Omega)^m},$$

and by choosing ϵ sufficiently small, we see that $\mathcal{P}$ is coercive. The converse is proved in the same way. □

The differential operator $\mathcal{P}$ is said to be *strongly elliptic* on Ω if

$$\text{Re} \sum_{j=1}^{n} \sum_{k=1}^{n} [A_{jk}(x)\xi_k\eta]^*\xi_j\eta \geq c|\xi|^2|\eta|^2 \quad \text{for all } x \in \Omega, \xi \in \mathbb{R}^n \text{ and } \eta \in \mathbb{C}^m.$$
(4.7)

Depending on the subspace V and the regularity of Ω, this purely algebraic condition on the leading coefficients is often necessary and sufficient for $\mathcal{P}$ to be coercive.

Theorem 4.6 *Assume that the coefficients A_{jk} are (bounded and) uniformly continuous on Ω. The differential operator $\mathcal{P}$ is strongly elliptic if and only if it is coercive on $H_0^1(\Omega)^m$.*

Proof. Suppose that $\mathcal{P}$ is strongly elliptic. First we consider the special case when the leading coefficients A_{jk} are constant. Let $u \in H_0^1(\Omega)^m = \widetilde{H}^1(\Omega)^m$, i.e., let $u \in H^1(\mathbb{R}^n)^m$ with supp $\subseteq \overline{\Omega}$. Since $\mathcal{F}_{x \to \xi}\{\partial_j u(x)\} = i2\pi\xi_j\hat{u}(\xi)$, Plancherel's theorem implies that

$$\int_{\mathbb{R}^n} (A_{jk}\partial_k u)^* \partial_j u \, dx = (2\pi)^2 \int_{\mathbb{R}^n} [A_{jk}\xi_k\hat{u}(\xi)]^*\xi_j\hat{u}(\xi) \, d\xi,$$

so, taking $\eta = \hat{u}(\xi)$ in (4.7),

$$\text{Re } \Phi_0(u, u) \geq (2\pi)^2 \int_{\mathbb{R}^n} c|\xi|^2|\hat{u}(\xi)|^2 \, d\xi = c \sum_{j=1}^{n} \int_{\mathbb{R}^n} |\partial_j u|^2 \, dx$$

$$\geq c\|u\|_{H^1(\Omega)^m}^2 - c\|u\|_{L_2(\Omega)^m}^2.$$

By Lemma 4.5, we see that $\mathcal{P}$ is coercive on $H_0^1(\Omega)^m$.

To handle the general case, let $\epsilon > 0$ and choose $\delta > 0$ such that

$$\max_{j,k} |A_{jk}(x) - A_{jk}(y)| < \epsilon \quad \text{for } |x - y| < \delta.$$
(4.8)

Cover $\overline{\Omega}$ with a locally finite family of open balls $B_1, B_2, B_3, \ldots$, each of radius δ. (If Ω is bounded, then the family of balls will be finite.) Since the diameters of the balls are bounded away from zero, we can assume that for each $d > 0$, there is a number N_d such that any given set of diameter less than d intersects at most N_d balls. By Corollary 3.22 and Exercise 4.6, we can find real-valued functions $\phi_1, \phi_2, \phi_3, \ldots$ in $C_{\text{comp}}^\infty(\mathbb{R}^n)$ with $\phi_l \geq 0$ and supp $\phi_l \subseteq B_l$, such that

$$\sum_{l \geq 1} \phi_l(x)^2 = 1, \quad \sum_{l \geq 1} \phi_l(x) \leq C, \quad \text{and} \quad \sum_{l \geq 1} |\partial_j \phi_l(x)| \leq C, \quad \text{for } x \in \overline{\Omega}.$$

Note that the number of non-zero terms in each sum is finite, and is bounded independently of x. Since

$$[A_{jk}\partial_k(\phi_l u)]^*\partial_j(\phi_l v) = \phi_l^2(A_{jk}\partial_k u)^*\partial_j v + (\partial_k\phi_l)\phi_l(A_{jk}u)^*\partial_j v$$
$$+ \phi_l(\partial_j\phi_l)(A_{jk}\partial_k u)^*v + (\partial_k\phi_l)(\partial_j\phi_l)(A_{jk}u)^*v,$$

we have

$$\left|\sum_{l\geq 1}\Phi_0(\phi_l u, \phi_l v)\right| \leq C\|u\|_{H^1(\Omega)^m}\|v\|_{H^1(\Omega)^m}$$

and also

$$\operatorname{Re}\Phi_0(u, u) \geq \sum_{l\geq 1}\operatorname{Re}\Phi_0(\phi_l u, \phi_l u) - C\|u\|_{H^1(\Omega)^m}\|u\|_{L_2(\Omega)^m}.$$

Let Φ_0^l denote the sesquilinear form obtained from Φ_0 by freezing the coefficients $A_{jk}(x)$ at $x = x^l$, the centre of the ball B_l, and observe that

$$\Phi_0(\phi_l u, \phi_l u) - \Phi_0^l(\phi_l u, \phi_l u) = \int_{B_l}\left\{[A_{jk}(x) - A_{jk}(x^l)]\partial_k(\phi_l u)\right\}^*\partial_j(\phi_l u)\,dx.$$

From the special case considered earlier, we know that Φ_0^l is coercive on $H_0^1(\Omega)$, with constants independent of l, and by (4.8),

$$|A_{jk}(x) - A_{jk}(x^l)| < \epsilon \quad \text{for } x \in B_l,$$

so

$$\operatorname{Re}\Phi_0(\phi_l u, \phi_l u) \geq \operatorname{Re}\Phi_0^l(\phi_l u, \phi_l u) - \epsilon\|\phi_l u\|_{H_0^1(\Omega)^m}^2$$
$$\geq (c - \epsilon)\|\phi_l u\|_{H_0^1(\Omega)^m}^2 - C\|\phi_l u\|_{L_2(\Omega)^m}^2,$$

and

$$\operatorname{Re}\Phi_0(u, u) \geq (c - \epsilon)\sum_{l\geq 1}\|\phi_l u\|_{H_0^1(\Omega)^m}^2$$
$$- C\sum_{l\geq 1}\|\phi_l u\|_{L_2(\Omega)^m}^2 - C\|u\|_{H_0^1(\Omega)^m}\|u\|_{L_2(\Omega)^m}.$$

Since the ϕ_l^2 form a partition of unity,

$$\sum_{l\geq 1}\|\phi_l u\|_{L_2(\Omega)}^2 = \int_\Omega\sum_{l\geq 1}\phi_l(x)^2|u(x)|^2\,dx = \|u\|_{L_2(\Omega)^m}^2$$

and

$$\sum_{l \geq 1} \|\partial_j(\phi_l u)\|_{L_2(\Omega)^m}^2 = \sum_{l \geq 1} \|\phi_l \partial_j u + (\partial_j \phi_l) u\|_{L_2(\Omega)^m}^2$$

$$\geq \|\partial_j u\|_{L_2(\Omega)^m}^2 - C\|u\|_{H_0^1(\Omega)^m} \|u\|_{L_2(\Omega)^m}.$$

Using the inequality $ab \leq \frac{1}{2}(\epsilon' a^2 + b^2/\epsilon')$, we see that

$$\operatorname{Re} \Phi_0(u, u) \geq (c - \epsilon - \epsilon')\|u\|_{H_0^1(\Omega)^m}^2 - C\left(1 + \frac{1}{\epsilon'}\right)\|u\|_{L_2(\Omega)^m}^2,$$

so $\mathcal{P}$ is coercive on $H_0^1(\Omega)$.

To prove the converse, take a real-valued cutoff function $\psi \in C_{\text{comp}}^\infty(\mathbb{R}^n)$ satisfying

$$\psi \geq 0 \text{ on } \mathbb{R}^n, \quad \psi = 0 \text{ for } |x| \geq 1, \quad \int_{\mathbb{R}^n} \psi(x)^2 \, dx = 1.$$

Let $x_0 \in \Omega$, and put $\psi_\epsilon(x) = \epsilon^{-n/2}\psi(\epsilon^{-1}(x - x_0))$, so that for ϵ sufficiently small, $\psi_\epsilon \in C_{\text{comp}}^\infty(\Omega)$ with

$$\psi_\epsilon \geq 0 \text{ on } \Omega, \quad \psi = 0 \text{ for } |x - x_0| \geq \epsilon, \quad \int_\Omega \psi_\epsilon(x)^2 \, dx = 1.$$

Thus, $\psi_\epsilon(x)^2$ converges to $\delta(x - x_0)$ as $\epsilon \downarrow 0$. Consider the function

$$u_\epsilon(x) = \psi_\epsilon(x)e^{i\xi \cdot x}\eta.$$

Since

$$\|u_\epsilon\|_{L_2(\Omega)^m} = |\eta|,$$

and since $\partial_j u_\epsilon = (i\psi_\epsilon \xi_j + \partial_j \psi_\epsilon)e^{i\xi \cdot x}\eta$ and $\partial_j \psi_\epsilon = \epsilon^{-1}(\partial_j \psi)_\epsilon$, we have

$$\|\partial_j u_\epsilon\|_{L_2(\Omega)^m}^2 = \left(\xi_j^2 + \epsilon^{-2}\|\partial_j \psi\|_{L_2(\mathbb{R}^n)}^2\right)|\eta|^2.$$

Now,

$$(A_{jk}\partial_k u_\epsilon)^* \partial_j u_\epsilon = \psi_\epsilon^2 (A_{jk}\xi_k \eta)^* \xi_j \eta + i\psi_\epsilon(\partial_k \psi_\epsilon)(A_{jk}\eta)^* \xi_j \eta$$

$$- i\psi_\epsilon(\partial_j \psi_\epsilon)(A_{jk}\xi_k \eta)^* \eta + (\partial_k \psi_\epsilon)(\partial_j \psi_\epsilon)(A_{jk}\eta)^* \eta,$$

so if we define

$$A_{jk}^\epsilon = \int_\Omega \psi_\epsilon(x)^2 A_{jk}(x) \, dx,$$

then

$$\mathrm{Re} \sum_{j=1}^{n} \sum_{k=1}^{n} \left(A_{jk}^{\epsilon}\xi_k\eta\right)^* \xi_j\eta \geq \mathrm{Re}\, \Phi_0(u_\epsilon, u_\epsilon) - C(\epsilon^{-2} + \epsilon^{-1}|\xi|)|\eta|^2.$$

If we now assume that $\mathcal{P}$ is coercive on $H_0^1(\Omega)^m$, then

$$\mathrm{Re}\, \Phi_0(u_\epsilon, u_\epsilon) \geq c\|u_\epsilon\|_{H_0^1(\Omega)^m}^2 - C\|u_\epsilon\|_{L_2(\Omega)^m}^2 \geq c(\epsilon^{-2} + |\xi|^2)|\eta|^2 - C|\eta|^2,$$

implying that

$$\mathrm{Re} \sum_{j=1}^{n} \sum_{k=1}^{n} \left(A_{jk}^{\epsilon}\xi_k\eta\right)^* \xi_j\eta \geq c|\xi|^2|\eta|^2 - C(1 + \epsilon^{-2} + \epsilon^{-1}|\xi|)|\eta|^2.$$

Now replace ξ by $t\xi$ where $t > 0$, divide through by t^2, and send $t \to \infty$ to obtain (4.7) with A_{jk}^{ϵ} in place of A_{jk}. Since $A_{jk}^{\epsilon} \to A_{jk}(x_0)$ as $\epsilon \downarrow 0$, we conclude that $\mathcal{P}$ is strongly elliptic. $\qquad\square$

For scalar problems, i.e., when $m = 1$, the strong ellipticity condition (4.7) simplifies to

$$\mathrm{Re} \sum_{j=1}^{n} \sum_{k=1}^{n} \overline{A_{jk}(x)\xi_k}\,\xi_j \geq c|\xi|^2 \qquad \text{for all } x \in \Omega \text{ and } \xi \in \mathbb{R}^n,$$

and the next result is usually sufficient for establishing that the differential operator is coercive on the whole of $H^1(\Omega)$, not just on $H_0^1(\Omega)$; cf. Exercise 4.1.

Theorem 4.7 Assume that $\mathcal{P}$ has scalar coefficients (i.e., $m = 1$), and that $\mathcal{P}$ is strongly elliptic on Ω. If the leading coefficients satisfy

$$A_{kj} = A_{jk} \quad \text{on } \Omega, \quad \text{for all } j \text{ and } k,$$

then $\mathcal{P}$ is coercive on $H^1(\Omega)$.

Proof. Define $F : \Omega \times \mathbb{C}^n \to \mathbb{C}$ by

$$F(x, \xi) = \sum_{j=1}^{n} \sum_{k=1}^{n} \overline{A_{jk}(x)\xi_k}\,\xi_j.$$

The symmetry condition on the leading coefficients implies that

$$F(x, \zeta + i\eta) = F(x, \zeta) + F(x, \eta) \quad \text{for } \zeta, \eta \in \mathbb{R}^n,$$

and so by strong ellipticity, $\operatorname{Re} F(x, \xi) \geq c|\xi|^2$ for all $\xi \in \mathbb{C}^n$ (not just for $\xi \in \mathbb{R}^n$). Hence, for all $u \in H^1(\Omega)$,

$$\operatorname{Re} \Phi_0(u, u) = \int_\Omega \operatorname{Re} F(x, \operatorname{grad} u) \, dx \geq c \| \operatorname{grad} u \|^2_{L_2(\Omega)}.$$

$\square$

In a similar fashion, when $m > 1$ it is easy to see that $\mathcal{P}$ is coercive on $H^1(\Omega)^m$ if $A_{kj} = A_{jk}^T$ and

$$\operatorname{Re} \sum_{j=1}^n \sum_{k=1}^n [A_{jk}(x)\xi_k]^* \xi_j \geq c \sum_{j=1}^n |\xi_j|^2$$

$$\text{for all } x \in \Omega \text{ and } \xi_1, \dots, \xi_n \in \mathbb{R}^m, \quad (4.9)$$

but this assumption excludes some strongly elliptic operators with important applications; see Exercise 10.3.

Using an approach due to Nečas [72, pp. 187–195], we shall prove a sufficient condition for coercivity on $H^1(\Omega)^m$ in the case when the leading coefficients can be split into sums of Hermitian rank-1 matrices, i.e.,

$$A_{jk} = \sum_{l=1}^L b_{lj} b_{lk}^*,$$

where the b_{lj} are (column) vectors in $\mathbb{C}^m$. It follows that $\mathcal{P}_0$ must be formally self-adjoint, and that

$$\Phi_0(u, v) = \int_\Omega \sum_{l=1}^L \overline{\mathcal{N}_l u} \mathcal{N}_l v \, dx \qquad \text{where } \mathcal{N}_l u = \sum_{j=1}^n b_{lj}^* \partial_j u. \quad (4.10)$$

Note that the first-order differential operator $\mathcal{N}_l$ acts on a vector-valued function u to produce a scalar-valued function $\mathcal{N}_l u$, and that

$$\Phi_0(u, u) = \sum_{l=1}^L \| \mathcal{N}_l u \|^2_{L_2(\Omega)} \geq 0 \quad \text{for } u \in H^1(\Omega)^m.$$

An important example of a strongly elliptic operator of this type is described in Chapter 10; see in particular Theorem 10.2.

The proof of coercivity is based on the following technical lemma, whose proof turns out to be surprisingly difficult.

124   *Strongly Elliptic Systems*

Lemma 4.8 *If Ω is a Lipschitz domain, then for any integers $p \geq 0$ and $q \geq 1$, and for any $u \in \mathcal{D}(\overline{\Omega})$,*

$$\|u\|_{H^{-p}(\Omega)} \leq C\|u\|_{H^{-p-q}(\Omega)} + C\sum_{|\alpha|\leq q}\|\partial^{\alpha}u\|_{H^{-p-q}(\Omega)}.$$

Proof. Since $\mathcal{D}(\Omega)$ is dense in $H^s(\Omega)$ if $s \leq 0$, it suffices to consider $u \in \mathcal{D}(\Omega)$. Plancherel's theorem gives

$$\|u\|_{H^{-p}(\mathbb{R}^n)}^2 \leq C\int_{\mathbb{R}^n}(1+|\xi|^2)^{-p-q}\left(1+\sum_{|\alpha|=q}|\xi^{\alpha}|^2\right)|\hat{u}(\xi)|^2\,d\xi$$

$$\leq C\|u\|_{H^{-p-q}(\mathbb{R}^n)}^2 + C\sum_{|\alpha|=q}\|\partial^{\alpha}u\|_{H^{-p-q}(\mathbb{R}^n)}^2,$$

so by Exercise 3.15,

$$\|u\|_{H^{-p}(\Omega)} \leq C_K\left(\|u\|_{H^{-p-q}(\Omega)} + \sum_{|\alpha|=q}\|\partial^{\alpha}u\|_{H^{-p-q}(\Omega)}\right) \quad \text{if } \operatorname{supp} u \subseteq K \Subset \Omega.$$

In doing away with the dependence on K, we can assume that Ω is a Lipschitz hypograph, given by $x_n < \zeta(x')$. Our strategy is to make a change of variable $x = \kappa(y)$, with $x \in \Omega$ and y in the negative half space $\mathbb{R}^n_-$.

For $\epsilon > 0$, let ψ_ϵ be as in (3.9) and (3.10), introduce the C^∞ function $f(y',\epsilon) = (\psi_\epsilon * \zeta)(y')$, and define

$$\kappa_\epsilon(y) = \big(y', f(y', -\epsilon y_n) + y_n\big) \quad \text{for } y_n < 0.$$

Since $\operatorname{grad}\zeta \in L_\infty(\mathbb{R}^{n-1})$, we find that

$$\left|\left(\frac{\partial}{\partial y_1}\right)^{\alpha_1}\cdots\left(\frac{\partial}{\partial y_{n-1}}\right)^{\alpha_{n-1}}\left(\frac{\partial}{\partial \epsilon}\right)^{\alpha_n}f(y',\epsilon)\right| \leq \frac{C}{\epsilon^{|\alpha|-1}} \quad \text{for } |\alpha| \geq 1. \quad (4.11)$$

Thus, $\partial_n(\kappa_\epsilon)_n(y) = 1 - \epsilon\partial_n f(y', -\epsilon y_n) = 1 + O(\epsilon)$, and we now fix ϵ small enough so that

$$c \leq \partial_n(\kappa_\epsilon)_n(y) \leq C \quad \text{for } y_n < 0,$$

and write $\kappa = \kappa_\epsilon$. In this way, $\kappa_n(y)$ is a strictly increasing function of $y_n \in (-\infty, 0)$, with $\kappa_n(y) \uparrow \zeta(y')$ as $y_n \uparrow 0$, and so $\kappa : \mathbb{R}^n_- \to \Omega$ is a C^∞ diffeomorphism, and it can be shown using (4.11) that

$$|\partial^{\alpha}\kappa(y)| \leq \frac{C}{|y_n|^{|\alpha|-1}} \quad \text{for } |\alpha| \geq 1.$$

In the substitution $x = \kappa(y)$, we have $x' = y'$, so the Jacobian is simply

$$\det D\kappa(y) = \partial_n \kappa_n(y).$$

Also, by differentiating the equation $x_n = f(x', -\epsilon y_n) + y_n$ with respect to x_j, one sees that

$$\left| \frac{\partial y_n}{\partial x_j} \right| = \left| \frac{-\partial_j f(x', -\epsilon y_n)}{1 - \epsilon \partial_n f(x', -\epsilon y_n)} \right| \le C \quad \text{for } 1 \le j \le n - 1,$$

and by differentiating with respect to x_n, one sees that

$$\left| \frac{\partial y_n}{\partial x_n} \right| = \frac{1}{|1 - \epsilon \partial_n f(x', -\epsilon y_n)|} \le C.$$

The higher-order partial derivatives of κ^{-1} can be estimated in a similar fashion, giving

$$\partial^\alpha \kappa^{-1}(x) \le \frac{C}{|y_n|^{|\alpha|-1}} \quad \text{for } |\alpha| \ge 1.$$

Let us now show that

$$\|u \circ \kappa\|_{H^{-p-1}(\mathbb{R}^n_-)} \le C \|u\|_{H^{-p-1}(\Omega)},$$
$$\|\partial_j(u \circ \kappa)\|_{H^{-p-1}(\mathbb{R}^n_-)} \le C \|\partial_j u\|_{H^{-p-1}(\Omega)} + C \|\partial_n u\|_{H^{-p-1}(\Omega)}. \tag{4.12}$$

For $\phi \in \mathcal{D}(\mathbb{R}^n_-)$,

$$\left| (u \circ \kappa, \phi)_{\mathbb{R}^n_-} \right| = \left| \int_\Omega u(x)(\phi \circ \kappa^{-1})(x) \det D\kappa^{-1}(x)\, dx \right|$$
$$\le C \|u\|_{H^{-p-1}(\Omega)} \|(\phi \circ \kappa^{-1}) \det D\kappa^{-1}\|_{\tilde{H}^{p+1}(\Omega)},$$

and since

$$\partial_j(u \circ \kappa)(y) = \begin{cases} \partial_j u(x) + \partial_n u(x)\partial_j \kappa_n(y) & \text{for } 1 \le j \le n - 1, \\ \partial_n u(x)\partial_n \kappa_n(y) & \text{for } j = n, \end{cases}$$

we have

$$\left| (\partial_j(u \circ \kappa), \phi)_{\mathbb{R}^n_-} \right| \le C \|\partial_j u\|_{H^{-p-1}(\Omega)} \|(\phi \circ \kappa^{-1}) \det D\kappa^{-1}\|_{\tilde{H}^{p+1}(\Omega)}$$
$$+ C \|\partial_n u\|_{H^{-p-1}(\Omega)} \|(\phi \partial_j \kappa_n) \circ \kappa^{-1} \det D\kappa^{-1}\|_{\tilde{H}^{p+1}(\Omega)},$$

where only the second term on the right is present if $j = n$. With the help of Theorem 3.33 and Exercise 4.3, we find that both

$$\|(\phi \circ \kappa^{-1}) \det D\kappa^{-1}\|_{\tilde{H}^{p+1}(\Omega)} \quad \text{and} \quad \|(\phi \partial_j \kappa_n) \circ \kappa^{-1} \det D\kappa^{-1}\|_{\tilde{H}^{p+1}(\Omega)}$$

are bounded by

$$C \sum_{|\alpha| \le p+1} \int_{y_n < 0} |y_n|^{2(|\alpha|-p-1)} |\partial^\alpha \phi(y)|^2 \, dy \le C \|\phi\|^2_{\widetilde{H}^{p+1}(\mathbb{R}^n_-)},$$

implying that (4.12) holds. Also, if $\phi \in \mathcal{D}(\Omega)$ then

$$|(u, \phi)_\Omega| = \left| \left(u \circ \kappa, (\phi \circ \kappa) \det D\kappa \right)_{\mathbb{R}^n_-} \right|$$
$$\le C \|u \circ \kappa\|_{H^{-p}(\mathbb{R}^n_-)} \|(\phi \circ \kappa) \det D\kappa\|_{\widetilde{H}^p(\mathbb{R}^n_-)}$$

and

$$\|(\phi \circ \kappa) \det D\kappa\|_{\widetilde{H}^p(\mathbb{R}^n_-)} \le C \sum_{|\alpha| \le p} \int_\Omega |\zeta(x') - x_n|^{2(|\alpha|-p)} |\partial^\alpha \phi(x)|^2 \, dx$$
$$\le C \|\phi\|^2_{H^p_0(\Omega)},$$

so $\|u\|_{H^{-p}(\Omega)} \le C \|u \circ \kappa\|_{H^{-p}(\mathbb{R}^n_-)}$. Hence, using the Seeley extension operator E from Exercise A.3, we have

$$\|u\|_{H^{-p}(\Omega)} \le C \|E(u \circ \kappa)\|_{H^{-p}(\mathbb{R}^n)}$$
$$\le C \|E(u \circ \kappa)\|_{H^{-p-1}(\mathbb{R}^n)} + C \sum_{j=1}^n \|\partial_j E(u \circ \kappa)\|_{H^{-p-1}(\mathbb{R}^n)}$$
$$\le C \|u \circ \kappa\|_{H^{-p-1}(\mathbb{R}^n_-)} + C \sum_{j=1}^n \|\partial_j (u \circ \kappa)\|_{H^{-p-1}(\mathbb{R}^n_-)}$$
$$\le C \|u\|_{H^{-p-1}(\Omega)} + C \sum_{j=1}^n \|\partial_j u\|_{H^{-p-1}(\Omega)},$$

which proves the result when $q = 1$. The general case now follows by induction on q. □

Theorem 4.9 Assume that Ω is a Lipschitz domain and that Φ_0 has the form (4.10), and put

$$N_l(\xi) = \sum_{j=1}^n b^*_{lj} \xi_j \quad \text{for } 1 \le l \le L.$$

The operator $\mathcal{P}$ is coercive on $H^1(\Omega)^m$ if, for $1 \le r \le m$, there is an integer $q_r \ge 1$ such that, for every multi-index α with $|\alpha| = q_r + 1$, there exist polynomials $Q_1, \dots, Q_L$, each homogeneous of degree q_r (with scalar coefficients), satisfying

$$\sum_{l=1}^L Q_l(\xi) N_l(\xi)^* = \xi^\alpha e_r \quad \text{for all } \xi \in \mathbb{R}^n,$$

where e_r is the rth standard basis vector in $\mathbb{C}^m$. *(Note that Q_l depends on r and α.)*

Proof. Let $u \in \mathcal{D}(\overline{\Omega})^m$ and $\phi \in \mathcal{D}(\Omega)$, and let $\mathcal{Q}_l$ be the partial differential operator corresponding to $Q_l(\xi)$ in the same way that $\mathcal{N}_l$ corresponds to $N_l(\xi)$; thus, under Fourier transformation,

$$\mathcal{F}_{x \to \xi}\{\mathcal{N}_l u(x)\} = N_l(\mathrm{i}2\pi\xi)\hat{u}(\xi) \quad \text{and} \quad \mathcal{F}_{x \to \xi}\{\mathcal{Q}_l u(x)\} = Q_l(\mathrm{i}2\pi\xi)\hat{u}(\xi).$$

Suppose β and δ are multi-indices with $|\beta| = 1$ and $|\delta| = q_r$. Applying Plancherel's theorem, and taking $\alpha = \beta + \delta$ and $u_r = e_r^* u$, we have

$$(\partial^\beta u_r, \partial^\delta \phi)_\Omega = (\partial^\beta u_r, \partial^\delta \phi) = \int_{\mathbb{R}^n} \mathrm{i}^{q_r-1}(2\pi\xi)^\alpha \overline{\hat{u}_r(\xi)}\hat{\phi}(\xi)\,d\xi$$

$$= \sum_{l=1}^L \int_{\mathbb{R}^n} \overline{N_l(\mathrm{i}2\pi\xi)\hat{u}(\xi)}\, Q_l(\mathrm{i}2\pi\xi)\hat{\phi}(\xi)\,d\xi$$

$$= \sum_{l=1}^L (\mathcal{N}_l u, \mathcal{Q}_l \phi) = \sum_{l=1}^L (\mathcal{N}_l u, \mathcal{Q}_l \phi)_\Omega,$$

so

$$|(\partial^\delta \partial^\beta u_r, \phi)_\Omega| \le C \sum_{l=1}^L \|\mathcal{N}_l u\|_{L_2(\Omega)} \|\mathcal{Q}_l \phi\|_{L_2(\Omega)}$$

$$\le C \sum_{l=1}^L \|\mathcal{N}_l u\|_{L_2(\Omega)} \|\phi\|_{\widetilde{H}^{q_r}(\Omega)^m},$$

implying that

$$\|\partial^\delta \partial^\beta u_r\|_{H^{-q_r}(\Omega)}^2 \le C \sum_{l=1}^L \|\mathcal{N}_l u\|_{L_2(\Omega)}^2 = C\Phi_0(u, u).$$

Thus, by applying Lemma 4.8 with $p = 0$,

$$\|\partial^\beta u_r\|_{L_2(\Omega)}^2 \le C\|\partial^\beta u_r\|_{H^{-q_r}(\Omega)}^2 + C \sum_{|\alpha| \le q_r} \|\partial^\alpha \partial^\beta u_r\|_{H^{-q_r}(\Omega)}^2$$

$$\le C \sum_{|\alpha| < q_r} \|\partial^\alpha \partial^\beta u_r\|_{H^{-q_r}(\Omega)}^2 + C \sum_{|\delta| = q_r} \|\partial^\delta \partial^\beta u_r\|_{H^{-q_r}(\Omega)}^2$$

$$\le C\|u_r\|_{L_2(\Omega)}^2 + C\Phi_0(u, u),$$

and we have only to sum over $|\beta| = 1$ and $1 \le r \le m$. $\qquad\square$

Boundary Value Problems

In this section, we shall see how the Fredholm alternative allows us to answer fundamental questions of existence and uniqueness for the solutions of elliptic boundary value problems. Suppose that the boundary of the domain Ω has a Lipschitz dissection

$$\Gamma = \Gamma_D \cup \Pi \cup \Gamma_N,$$

and that boundary conditions of Dirichlet and Neumann type are specified on Γ_D and Γ_N, respectively. Thus, our task is to find $u \in H^1(\Omega)^m$ satisfying

$$\begin{aligned}
\mathcal{P}u &= f \text{ on } \Omega, \\
\gamma u &= g_D \text{ on } \Gamma_D, \\
\mathcal{B}_\nu u &= g_N \text{ on } \Gamma_N,
\end{aligned} \tag{4.13}$$

for given f, g_D and g_N. Using our *definitions* of $\mathcal{P}u$ and $\mathcal{B}_\nu u$ as distributions, we see that (4.13) is equivalent to

$$\gamma u = g_D \text{ on } \Gamma_D \text{ and } \Phi(u, v) = (f, v)_\Omega + (g_N, \gamma v)_{\Gamma_N} \quad \text{for } v \in H_D^1(\Omega)^m, \tag{4.14}$$

where

$$H_D^1(\Omega)^m = \{v \in H^1(\Omega)^m : \gamma v = 0 \text{ on } \Gamma_D\}.$$

Recalling Theorem 2.27, we expect that if there exist non-trivial solutions to the homogeneous problem

$$\begin{aligned}
\mathcal{P}u &= 0 \quad \text{on } \Omega, \\
\gamma u &= 0 \quad \text{on } \Gamma_D, \\
\mathcal{B}_\nu u &= 0 \quad \text{on } \Gamma_N,
\end{aligned} \tag{4.15}$$

then we ought to consider also the homogeneous *adjoint* problem,

$$\begin{aligned}
\mathcal{P}^*v &= 0 \quad \text{on } \Omega, \\
\gamma v &= 0 \quad \text{on } \Gamma_D, \\
\widetilde{\mathcal{B}}_\nu v &= 0 \quad \text{on } \Gamma_N.
\end{aligned} \tag{4.16}$$

Theorem 4.10 Assume that Ω is a bounded Lipschitz domain, and that $\mathcal{P}$ is coercive on $H_D^1(\Omega)^m$. Let $f \in \widetilde{H}^{-1}(\Omega)^m$, $g_D \in H^{1/2}(\Gamma_D)^m$ and $g_N \in H^{-1/2}(\Gamma_N)^m$, and let W denote the set of solutions in $H^1(\Omega)^m$ to the homogeneous problem (4.15). There are two mutually exclusive possibilities:

(i) The homogeneous problem has only the trivial solution, i.e., $W = \{0\}$. In this case, the homogeneous adjoint problem (4.16) also has only the trivial solution in $H^1(\Omega)^m$, and for the inhomogeneous problem (4.13) we get a unique solution $u \in H^1(\Omega)^m$. Moreover,

$$\|u\|_{H^1(\Omega)^m} \leq C \|f\|_{\tilde{H}^{-1}(\Omega)^m} + C \|g_D\|_{H^{1/2}(\Gamma_D)^m} + C \|g_N\|_{H^{-1/2}(\Gamma_N)^m}.$$

(ii) The homogeneous problem has exactly p linearly independent solutions, i.e., $\dim W = p$, for some finite $p \geq 1$. In this case, the homogeneous adjoint problem (4.16) also has exactly p linearly independent solutions, say $v_1, \ldots, v_p \in H^1(\Omega)^m$, and the inhomogeneous problem (4.13) is solvable in $H^1(\Omega)^m$ if and only if

$$(v_j, f)_\Omega + (\gamma v_j, g_N)_{\Gamma_N} = (\tilde{\mathcal{B}}_\nu v_j, g_D)_{\Gamma_D} \quad for \; 1 \leq j \leq p.$$

When the data satisfy these p conditions, the solution set is $u + W$, where u is any particular solution. Moreover

$$\|u + W\|_{H^1(\Omega)^m/W} \leq C \|f\|_{\tilde{H}^{-1}(\Omega)^m} + C \|g_D\|_{H^{1/2}(\Gamma_D)^m} + C \|g_N\|_{H^{-1/2}(\Gamma_N)^m}.$$

Proof. Put $V = H_D^1(\Omega)^m$ and $H = L_2(\Omega)^m$, and consider the operator $A : V \to V^*$ determined by Φ in the usual way. We note that since Ω is bounded, Theorem 3.27 applies, so the inclusion $V \subseteq H$ is compact. Hence, by Theorem 2.34, A is Fredholm with index 0. Furthermore, each distribution $f_0 \in \tilde{H}^{-1}(\Omega)$ gives rise to a unique functional $F_0 \in V^*$, defined by $(F_0, v) = (f_0, v)_\Omega$ for $v \in V$ (although different f_0's can give the same F_0 if $\Gamma_D \neq \emptyset$). Thus, for $u_0 \in V$ and $f_0 \in \tilde{H}^{-1}(\Omega)$, the equation $Au_0 = F_0$ is equivalent to

$$\gamma u_0 = 0 \text{ on } \Gamma_D \text{ and } \Phi(u_0, v) = (f_0, v)_\Omega \quad \text{for } v \in V. \tag{4.17}$$

To handle the inhomogeneous Dirichlet condition in (4.13), we write $u = u_0 + \eta E g_D$, where $E : H^{1/2}(\Gamma_D)^m \to H^{1/2}(\Gamma)^m$ and $\eta : H^{1/2}(\Gamma)^m \to H^1(\Omega)^m$ are extension operators constructed with the help of Theorems A.4 and 3.37. In this way, $\gamma u = g_D$ on Γ_D if and only if $\gamma u_0 = 0$ on Γ_D. Choose any extension $\tilde{g}_N \in H^{-1/2}(\Gamma)^m$ of g_N, and define $f_0 \in \tilde{H}^{-1}(\Omega)^m$ by

$$(f_0, v)_\Omega = (f, v)_\Omega - \Phi(\eta E g_D, v) + (\tilde{g}_N, \gamma v)_\Gamma \quad \text{for } v \in H^1(\Omega)^m,$$

so that (4.14) is equivalent to (4.17), or in other words, so that (4.13) is equivalent to the equation $Au_0 = F_0$. We remark that u_0 can be thought of as a solution of

$$\mathcal{P}u_0 = f_1 \quad \text{on } \Omega,$$
$$\gamma u_0 = 0 \quad \text{on } \Gamma_D,$$
$$\mathcal{B}_\nu u_0 = g_1 \quad \text{on } \Gamma_N,$$

where $f_1 = f - \mathcal{P}\eta E g_D$ and $g_1 = g_N - \mathcal{B}_\nu \eta E g_D$. Notice that for $v \in V$,

$$|(F_0, v)| = |(f_0, v)_\Omega| \leq C\|f\|_{\widetilde{H}^{-1}(\Omega)^m}\|v\|_{H^1(\Omega)^m} + C\|\eta E g_D\|_{H^1(\Omega)^m}\|v\|_{H^1(\Omega)^m}$$
$$+ C\|g_N\|_{H^{-1/2}(\Gamma_N)^m}\|\gamma v\|_{\widetilde{H}^{1/2}(\Gamma_N)^m}$$
$$\leq C\big(\|f\|_{\widetilde{H}^{-1}(\Omega)^m} + \|g_D\|_{H^{1/2}(\Gamma_D)^m} + \|g_N\|_{H^{-1/2}(\Gamma_N)^m}\big)\|v\|_{H^1(\Omega)^m}$$

so

$$\|F_0\|_{V^*} \leq C\|f\|_{\widetilde{H}^{-1}(\Omega)^m} + C\|g_D\|_{H^{1/2}(\Gamma_D)^m} + C\|g_N\|_{H^{-1/2}(\Gamma_N)^m}.$$

The homogeneous problem (4.15) is equivalent to $Au = 0$, so $W = \ker A$. Likewise, the homogeneous adjoint problem (4.16) is equivalent to

$$\gamma v = 0 \text{ on } \Gamma_D \quad \text{and} \quad \Phi(u, v) = 0 \quad \text{for } u \in V,$$

which in turn is equivalent to $A^*v = 0$. Thus, in case (i) there exists a unique $u_0 = A^{-1}F_0$, and hence also a unique $u = u_0 + \eta E g_D$, with

$$\|u\|_{H^1(\Omega)^m} \leq \|u_0\|_V + \|\eta E g_D\|_{H^1(\Omega)^m} \leq C\|F_0\|_{V^*} + C\|g_D\|_{H^{1/2}(\Gamma_D)^m}.$$

In case (ii), the equation $Au_0 = F_0$ is solvable if and only if $(F_0, v_j) = 0$ for $1 \leq j \leq p$. Since the first Green identity gives

$$\Phi(\eta E g_D, v_j) = (\eta E g_D, \mathcal{P}^* v_j)_\Omega + (\gamma \eta E g_D, \widetilde{\mathcal{B}}_\nu v_j)_\Gamma = (g_D, \widetilde{\mathcal{B}}_\nu v_j)_{\Gamma_D},$$

and since $(\tilde{g}_N, \gamma v_j)_\Gamma = (g_N, \gamma v_j)_{\Gamma_N}$, it follows that (4.13) is solvable if and only if

$$(f, v_j)_\Omega - (g_D, \widetilde{\mathcal{B}}_\nu v_j)_{\Gamma_D} + (g_N, \gamma v_j)_{\Gamma_N} = 0 \quad \text{for } 1 \leq j \leq p.$$

When these p conditions are satisfied, $\|u_0 + W\|_{V/W} \leq C\|F_0\|_{V^*}$, and by choosing $w \in W$ such that $\|u_0 + w\|_V = \|u_0 + W\|_{V/W}$, we have

$$\|u + w\|_{H^1(\Omega)^m} \leq \|u_0 + w\|_V + \|\eta E g_D\|_{H^1(\Omega)^m} \leq C\|F_0\|_{V^*} + C\|g_D\|_{H^{1/2}(\Gamma_D)^m},$$

giving the required estimate for $\|u + W\|_{H^1(\Omega)^m/W}$. $\qquad\square$

Note that for a pure Dirichlet problem, i.e., when $\Gamma = \Gamma_D$, any strongly elliptic operator with leading coefficients in $W_\infty^1(\Omega)^{m \times m}$ is coercive on $V = H_D^1(\Omega)^m = H_0^1(\Omega)^m$ by Theorem 4.6. Also, we may take $f \in V^* = H^{-1}(\Omega)^m$, because no use is made of the conormal derivative. For a pure Neumann problem, i.e., when $\Gamma = \Gamma_N$, the proof of the theorem above is greatly simplified because $u = u_0 \in V = H^1(\Omega)^m$ and $F_0 = f_0 \in V^* = \widetilde{H}^{-1}(\Omega)^m$.

Next in importance after the Dirichlet and Neumann problems is the *third boundary value problem*:

$$\begin{aligned}
\mathcal{P}u &= f \quad \text{on } \Omega, \\
\mathcal{B}_\nu u + B\gamma u &= g \quad \text{on } \Gamma.
\end{aligned} \tag{4.18}$$

Here, the coefficient B is a known function (matrix-valued, if $m > 1$). The third boundary value problem reduces to a pure Neumann problem if B is identically zero. In fact, for non-zero B, (4.18) is only a compact perturbation of the Neumann problem, and we can state the Fredholm alternative in terms of the homogeneous problem

$$\begin{aligned}
\mathcal{P}u &= 0 \quad \text{on } \Omega, \\
\mathcal{B}_\nu u + B\gamma u &= 0 \quad \text{on } \Gamma,
\end{aligned} \tag{4.19}$$

and the homogeneous adjoint problem

$$\begin{aligned}
\mathcal{P}^*v &= 0 \quad \text{on } \Omega, \\
\widetilde{\mathcal{B}}_\nu v + B^*\gamma v &= 0 \quad \text{on } \Gamma,
\end{aligned} \tag{4.20}$$

as follows.

Theorem 4.11 Assume that Ω is a bounded Lipschitz domain, that the differential operator $\mathcal{P}$ is coercive on $H^1(\Omega)^m$, and that $B \in L_\infty(\Gamma)^{m \times m}$. Let $f \in \widetilde{H}^{-1}(\Omega)^m$ and $g \in H^{-1/2}(\Gamma)^m$, and let W denote the set of solutions in $H^1(\Omega)^m$ to the homogeneous problem (4.19). There are two mutually exclusive possibilities.

(i) The homogeneous problem has only the trivial solution, i.e., $W = \{0\}$. In this case, the homogeneous adjoint problem (4.20) also has only the trivial solution in $H^1(\Omega)^m$, and for the inhomogeneous problem (4.18) we get a unique solution $u \in H^1(\Omega)^m$. Moreover,

$$\|u\|_{H^1(\Omega)^m} \leq C\|f\|_{\widetilde{H}^{-1}(\Omega)^m} + C\|g\|_{H^{-1/2}(\Gamma)^m}.$$

(ii) *The homogeneous problem has exactly p linearly independent solutions,
i.e., dim $W = p$, for some finite $p \geq 1$. In this case, the homogeneous adjoint problem (4.20) also has exactly p linearly independent solutions, say
$v_1, \ldots, v_p \in H^1(\Omega)^m$, and the inhomogeneous problem (4.18) is solvable
in $H^1(\Omega)^m$ if and only if*

$$(v_j, f)_\Omega + (\gamma v_j, g)_\Gamma = 0 \quad \text{for } 1 \leq j \leq p.$$

*When the data satisfy these p conditions, the solution set is $u + W$, where
u is any particular solution. Moreover*

$$\|u + W\|_{H^1(\Omega)^m/W} \leq C\|f\|_{\widetilde{H}^{-1}(\Omega)^m} + C\|g\|_{H^{-1/2}(\Gamma)^m}.$$

Proof. Put $(Au, v)_\Omega = \Phi(u, v)$ and $(Ku, v)_\Omega = (B\gamma u, \gamma v)_\Gamma$ for $u, v \in H^1(\Omega)^m$. The operator $A : H^1(\Omega)^m \to \widetilde{H}^{-1}(\Omega)^m$ is bounded, and for any $0 < \epsilon < \frac{1}{2}$,

$$|(Ku, v)_\Omega| \leq C\|\gamma u\|_{L_2(\Gamma)^m}\|v\|_{L_2(\Gamma)^m} \leq C\|u\|_{H^{1/2+\epsilon}(\Omega)^m}\|v\|_{H^1(\Omega)^m},$$

so $K : H^{1/2+\epsilon}(\Omega)^m \to \widetilde{H}^{-1}(\Omega)^m$ is bounded and hence $K : H^1(\Omega)^m \to \widetilde{H}^{-1}(\Omega)^m$ is compact. Since A is coercive on $H^1(\Omega)^m$, it follows that $A + K : H^1(\Omega)^m \to \widetilde{H}^{-1}(\Omega)^m$ is a Fredholm operator with index 0. By the first Green
identity, a function $u \in H^1(\Omega)^m$ is a solution of (4.18) if and only if

$$\big((A + K)u, v\big)_\Omega = (f, v)_\Omega + (\mathcal{B}_v u, \gamma v)_\Gamma + (B\gamma u, \gamma v)_\Gamma,$$

so $(A + K)u = F$ where $(F, v)_\Omega = (f, v)_\Omega + (g, \gamma v)_\Gamma$. The estimate

$$|(g, \gamma v)_\Gamma| \leq C\|g\|_{H^{-1/2}(\Gamma)^m}\|\gamma v\|_{H^{1/2}(\Gamma)^m} \leq C\|g\|_{H^{-1/2}(\Gamma)^m}\|v\|_{H^1(\Omega)^m}$$

shows that $F \in \widetilde{H}^{-1}(\Omega)^m$, so the results follow directly from Theorem 2.27
after noting that, by the dual version of the first Green identity (Lemma 4.2),
the adjoint of $A + K$ is given by

$$\big(u, (A + K)^*v\big)_\Omega = \Phi(u, v) + (B\gamma u, \gamma v)_\Gamma$$
$$= (u, \mathcal{P}^*v)_\Omega + (\gamma u, \widetilde{B}_v v + B^*\gamma v)_\Gamma,$$

assuming in the usual way that $\mathcal{P}^*v$ is a specified distribution in $\widetilde{H}^{-1}(\Omega)^m$. $\square$

We conclude this section with a result on the spectral properties of elliptic
operators.

Theorem 4.12 *Assume that Ω is a bounded Lipschitz domain, that $\mathcal{P}$ is coercive on $H_D^1(\Omega)^m$, and that $B \in L_\infty(\Gamma)^{m \times m}$. If $\mathcal{P}$ is formally self-adjoint, and if*

$B^* = B$, *then there exist sequences of functions* $\phi_1, \phi_2, \phi_3, \ldots$ *in* $H^1(\Omega)^m$, *and of real numbers* $\lambda_1, \lambda_2, \lambda_3, \ldots$, *having the following properties:*

(i) *Each* ϕ_j *is an eigenfunction of* $\mathcal{P}$ *with eigenvalue* λ_j:

$$\mathcal{P}\phi_j = \lambda_j \phi_j \quad \text{on } \Omega,$$
$$\gamma\phi_j = 0 \qquad \text{on } \Gamma_{\mathrm{D}},$$
$$\mathcal{B}_\nu\phi_j + B\gamma\phi_j = 0 \qquad \text{on } \Gamma_{\mathrm{N}}.$$

(ii) *The eigenfunctions* $\phi_1, \phi_2, \phi_3, \ldots$ *form a complete orthonormal system in* $L_2(\Omega)^m$.

(iii) *The eigenvalues satisfy* $-C < \lambda_1 \leq \lambda_2 \leq \lambda_3 \leq \cdots$ *with* $\lambda_j \to \infty$ *as* $j \to \infty$.

(iv) *For* $u, v \in H_{\mathrm{D}}^1(\Omega)^m$,

$$\Phi(u, v) = \sum_{j=1}^{\infty} \lambda_j (u, \phi_j)_\Omega (\phi_j, v)_\Omega \quad \text{and}$$

$$\|u\|_{H_{\mathrm{D}}^1(\Omega)^m}^2 \sim \sum_{j=1}^{\infty} (\lambda_j + C) |(\phi_j, u)_\Omega|^2.$$

Proof. Put $H = L_2(\Omega)^m$ and $V = H_{\mathrm{D}}^1(\Omega)^m$, and consider the operator $A : V \to V^*$ defined by

$$(Au, v)_\Omega = \Phi(u, v) + C(u, v)_\Omega + (B\gamma u, \gamma v)_{\Gamma_{\mathrm{N}}} \quad \text{for } u, v \in V.$$

Since A is self-adjoint, as well as positive and bounded below on V for C sufficiently large, the desired results follow at once from Theorem 2.37 and Corollary 2.38. $\qquad\square$

Regularity of Solutions

A key feature of the elliptic equation $\mathcal{P}u = f$ is that, away from the boundary, the solution u is smoother than the right-hand side f. We shall prove this fact, and also a result on regularity up to the boundary, using a method introduced by Nirenberg [78] based on estimates of the lth partial difference quotient,

$$\triangle_{l,h}u(x) = \frac{u(x + he_l) - u(x)}{h} \quad \text{for } 1 \leq l \leq n \text{ and } h \in \mathbb{R}. \quad (4.21)$$

(Here, e_l denotes that lth standard basis vector for $\mathbb{R}^n$.) The method of proof relies on the fact that ∂_j commutes with $\triangle_{l,h}$ for any j and l, and also on the next lemma, which allows one to deduce L_2-estimates for $\partial_l u$ from uniform L_2-estimates for $\triangle_{l,h} u$.

Lemma 4.13 *Let* $1 \le l \le n$, *and for brevity write* $\|u\| = \|u\|_{L_2(\mathbb{R}^n)^m}$.

(i) *If* $\partial_l u \in L_2(\mathbb{R}^n)^m$, *then* $\|\triangle_{l,h} u\| \le \|\partial_l u\|$ *for all* $h \in \mathbb{R}$, *and* $\|\triangle_{l,h} u - \partial_l u\| \to 0$ *as* $h \to 0$.

(ii) *If there is a constant* M *such that* $\|\triangle_{l,h} u\| \le M$ *for all* h *sufficiently small, then* $\partial_l u \in L_2(\mathbb{R}^n)^m$ *and* $\|\partial_l u\| \le M$.

Proof. To prove part (i), suppose that $\partial_l u \in L_2(\mathbb{R}^n)^m$. We have

$$\triangle_{l,h} u(x) = \int_0^1 \partial_l u(x + the_l)\, dt,$$

so the Cauchy–Schwarz inequality implies that

$$|\triangle_{l,h} u(x)|^2 \le \int_0^1 |\partial_l u(x + the_l)|^2\, dt.$$

By integrating with respect to x, reversing the order of integration, and making the substitution $y = x + the_l$, we obtain the estimate $\|\triangle_{l,h} u\|^2 \le \|\partial_l u\|^2$. It is clear that $\triangle_{l,h} u \to \partial_l u$ in $L_2(\mathbb{R}^n)^m$ if u is smooth with compact support, so for each $\epsilon > 0$ we choose $u_\epsilon \in \mathcal{D}(\mathbb{R}^n)^m$ such that $\|\partial_l u_\epsilon - \partial_l u\| < \epsilon$. From the estimate

$$\|\triangle_{l,h} u - \partial_l u\| \le \|\triangle_{l,h}(u - u_\epsilon)\| + \|\triangle_{l,h} u_\epsilon - \partial_l u_\epsilon\| + \|\partial_l u_\epsilon - \partial_l u\|$$

$$\le \|\triangle_{l,h} u_\epsilon - \partial_l u_\epsilon\| + 2\|\partial_l u_\epsilon - \partial_l u\| < \|\triangle_{l,h} u_\epsilon - \partial_l u_\epsilon\| + 2\epsilon,$$

we have $\limsup_{h \to 0} \|\triangle_{l,h} u - \partial_l u\| \le 2\epsilon$, showing that $\triangle_{l,h} u \to \partial_l u$ in $L_2(\mathbb{R}^n)^m$.

To prove part (ii), assume that $\|\triangle_{l,h} u\| \le M$. By Theorem 2.31, there is a sequence $h_j \to 0$ and a function $v \in L_2(\mathbb{R}^n)^m$ such that $\triangle_{l,h_j} u \rightharpoonup v$ in $L_2(\mathbb{R}^n)^m$, i.e.,

$$\lim_{j \to \infty} (\triangle_{l,h_j} u, \phi) = (v, \phi) \quad \text{for each } \phi \in L_2(\mathbb{R}^n)^m.$$

If $\phi \in \mathcal{D}(\mathbb{R}^n)^m$, then

$$-(u, \partial_l \phi) = -\lim_{j \to \infty} (u, \triangle_{l,-h_j} \phi) = \lim_{j \to \infty} (\triangle_{l,h_j} u, \phi) = (v, \phi),$$

so $\partial_l u = v \in L_2(\mathbb{R}^n)^m$. Part (i) now implies that $\triangle_{l,h} u \to \partial_l u$ in $L_2(\mathbb{R}^n)^m$, and thus $\|\partial_l u\| = \lim_{h \to 0} \|\triangle_{l,h} u\| \le M$. $\qquad\square$

In stating the next lemma, we use the summation convention, i.e., we sum any repeated indices from 1 to n, except where indicated otherwise. The proof involves only routine calculations, and is left to the reader.

Lemma 4.14 *Let* $[\mathcal{P}, \mathcal{Q}] = \mathcal{P}\mathcal{Q} - \mathcal{Q}\mathcal{P}$ *denote the commutator of* $\mathcal{P}$ *with* $\mathcal{Q}$.

(i) *The commutator of* $\mathcal{P}$ *with any pointwise multiplication operator* $\chi : u \mapsto \chi u$ *is a differential operator of order* 1 *(not 2):*

$$[\mathcal{P}, \chi]u = (\partial_j \chi)(A_j u - A_{jk}\partial_k u) - \partial_j\big[(\partial_k \chi)A_{jk}u\big].$$

(ii) *The commutator of* $\mathcal{P}$ *with* ∂_l^r *is a differential operator of order* $r + 1$ *(not* $r + 2$):

$$[\mathcal{P}, \partial_l^r]u = \partial_j \sum_{p=0}^{r-1} \binom{r}{p}(\partial_l^{r-p} A_{jk})\partial_l^p \partial_k u$$
$$- \sum_{p=1}^{r-1} \binom{r}{p}\big[(\partial_l^{r-p} A_j)\partial_l^p \partial_j u + (\partial_l^{r-p} A)\partial_l^p u\big].$$

(iii) *The commutator of* $\triangle_{l,h}$ *with any pointwise multiplication operator* $u \mapsto \chi u$ *is given by*

$$[\triangle_{l,h}, \chi]u = (\triangle_{l,h}\chi)u(\cdot + he_l).$$

We require two other properties of $\triangle_{l,h}$; again, the proof is left as an exercise for the reader.

Lemma 4.15 *Assume that* supp $u \subseteq \Omega \cap (\Omega - he_l)$ *and* supp $v \subseteq \Omega \cap (\Omega + he_l)$.

(i) *If* u, $v \in L_2(\Omega)^m$, *then* $(\triangle_{l,h}u, v)_\Omega = -(u, \triangle_{l,-h}v)_\Omega$.
(ii) *If* u, $v \in H^1(\Omega)^m$ *and if the coefficients of* $\mathcal{P}$ *satisfy* $|\triangle_{l,h}A_{jk}| \leq C$ *and* $|\triangle_{l,h}A_j| \leq C$ *on* Ω, *then*

$$|\Phi(\triangle_{l,h}u, v) + \Phi(u, \triangle_{l,-h}v)| \leq C\|u\|_{H^1(\Omega)^m}\|v\|_{H^1(\Omega)^m}.$$

We now prove regularity in the interior.

Theorem 4.16 *Let* Ω_1 *and* Ω_2 *be bounded, open subsets of* $\mathbb{R}^n$, *such that* $\overline{\Omega}_1 \Subset \Omega_2$, *and assume that* $\mathcal{P}$ *is strongly elliptic on* Ω_2. *For any integer* $r \geq 0$, *if* $u \in H^1(\Omega_2)^m$ *and* $f \in H^r(\Omega_2)^m$ *satisfy*

$$\mathcal{P}u = f \quad on \ \Omega_2,$$

and if the coefficients of $\mathcal{P}$ *belong to* $C^{r,1}(\overline{\Omega}_2)^{m \times m}$, *then* $u \in H^{r+2}(\Omega_1)^m$ *and*

$$\|u\|_{H^{r+2}(\Omega_1)^m} \leq C\|u\|_{H^1(\Omega_2)^m} + C\|f\|_{H^r(\Omega_2)^m}.$$

Proof. Choose a real-valued cutoff function $\chi \in \mathcal{D}(\Omega_2)$ with $\chi = 1$ on Ω_1. Part (i) of Lemma 4.14 gives

$$\mathcal{P}(\chi u) = f_1 \quad \text{with} \quad \|f_1\|_{L_2(\Omega_2)^m} \leq \|\chi f\|_{L_2(\Omega_2)} + C\|u\|_{H^1(\Omega_2)^m}, \quad (4.22)$$

so

$$\Phi_{\Omega_2}(\chi u, v) = (f_1, v)_{\Omega_2} \quad \text{for } v \in H_0^1(\Omega_2)^m.$$

Thus, by part (ii) of Lemma 4.15 followed by part (i) of Lemma 4.13,

$$
\begin{aligned}
\left|\Phi_{\Omega_2}\big(\triangle_{l,h}(\chi u), v\big)\right| &\leq |\Phi_{\Omega_2}(\chi u, \triangle_{l,-h}v)| + C\|\chi u\|_{H^1(\Omega_2)^m}\|v\|_{H^1(\Omega_2)^m} \\
&= |(f_1, \triangle_{l,-h}v)_{\Omega_2}| + C\|\chi u\|_{H^1(\Omega_2)^m}\|v\|_{H^1(\Omega_2)^m} \\
&\leq \big(\|f_1\|_{L_2(\Omega_2)^m} + \|\chi u\|_{H^1(\Omega_2)^m}\big)\|v\|_{H^1(\Omega_2)^m}
\end{aligned}
$$

if h is sufficiently small and v has compact support in Ω_2. The operator $\mathcal{P}$ is coercive on $H_0^1(\Omega_2)^m$ by Theorem 4.6, so by taking $v = \triangle_{l,h}(\chi u)$ we obtain the estimate

$$
\begin{aligned}
\|\triangle_{l,h}(\chi u)\|^2_{H^1(\Omega_2)^m} &\leq C\|\triangle_{l,h}(\chi u)\|^2_{L_2(\Omega_2)^m} \\
&\quad + \big(\|f_1\|_{L_2(\Omega_2)^m} + \|\chi u\|_{H^1(\Omega_2)^m}\big)\|\triangle_{l,h}(\chi u)\|_{H^1(\Omega_2)^m}.
\end{aligned}
$$

Applying the inequality $ab \leq \frac{1}{2}(\epsilon a^2 + \epsilon^{-1}b^2)$ with ϵ sufficiently small, we conclude that

$$
\begin{aligned}
\|\triangle_{l,h}(\chi u)\|^2_{H^1(\Omega_2)^m} &\leq C\|\triangle_{l,h}(\chi u)\|^2_{L_2(\Omega_2)^m} + C\|f_1\|^2_{L_2(\Omega_2)^m} + C\|\chi u\|^2_{H^1(\Omega_2)^m} \\
&\leq C\|u\|^2_{H^1(\Omega_2)^m} + C\|f\|^2_{L_2(\Omega_2)^m}. \quad (4.23)
\end{aligned}
$$

Part (ii) of Lemma 4.13 now gives the result in the case $r = 0$.

Proceeding by induction on r, we let $r \geq 1$ and choose an open set Ω_1' satisfying $\overline{\Omega}_1 \Subset \Omega_1' \subset \overline{\Omega_1'} \Subset \Omega_2$. By the induction hypothesis,

$$\|u\|_{H^{r+1}(\Omega_1')^m} \leq C\|u\|_{H^1(\Omega_2)^m} + C\|f\|_{H^{r-1}(\Omega_2)^m},$$

and by part (ii) of Lemma 4.14,

$$\mathcal{P}(\partial_l^r u) = f_2 \quad \text{on } \Omega_1', \quad (4.24)$$

with

$$\|f_2\|_{L_2(\Omega_1')^m} \le \|\partial_l^r f\|_{L_2(\Omega_1')^m} + C\|u\|_{H^{r+1}(\Omega_1')^m} \le \|f\|_{H^r(\Omega_2)^m} + C\|u\|_{H^1(\Omega_2)^m}.$$

Since $f_2 \in L_2(\Omega_1')^m$, we can apply the result for $r = 0$, and deduce that

$$\|\partial_l^r u\|_{H^2(\Omega_1)^m} \le C\|\partial_l^r u\|_{H^1(\Omega_1')^m} + C\|f_2\|_{L_2(\Omega_1')^m}$$
$$\le C\|u\|_{H^1(\Omega_2)^m} + C\|f\|_{H^r(\Omega_2)^m},$$

so the induction goes through. □

Next, we consider regularity up to the boundary. As well as assuming conditions on f and the coefficients of $\mathcal{P}$, we require extra smoothness for the trace or conormal derivative of u (i.e., for the boundary data), and also for the boundary itself. In the proof, we estimate the tangential derivatives of u using difference quotients as above, but another trick is needed to deal with the normal derivatives. Hence the next lemma.

Lemma 4.17 If $\mathcal{P}$ is strongly elliptic, then each diagonal leading coefficient has a uniformly bounded inverse, or equivalently, $|\eta| \le C|A_{jj}\eta|$ (no sum over j) for $\eta \in \mathbb{C}^m$.

Proof. Take $\xi_j = \delta_{jl}$ in (4.7), and get $\mathrm{Re}(A_{ll}\eta)^*\eta \ge c|\eta|^2$. □

Refer to Figure 3 for an illustration of the sets used in the next theorem.

Theorem 4.18 Let G_1 and G_2 be open subsets of $\mathbb{R}^n$, such that $\overline{G}_1 \Subset G_2$ and G_1 intersects the boundary Γ of a Lipschitz domain Ω. Put

$$\Omega_1 = G_1 \cap \Omega, \quad \Omega_2 = G_2 \cap \Omega, \quad \Gamma_2 = G_2 \cap \Gamma,$$

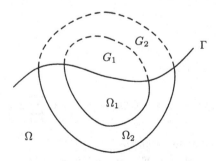

Figure 3. The sets used in Theorem 4.18.

and suppose, for an integer $r \geq 0$, that Γ_2 is $C^{r+1,1}$. Assume further that $\mathcal{P}$ is strongly elliptic on Ω_2, that $u \in H^1(\Omega_2)^m$ and $f \in H^r(\Omega_2)^m$ satisfy

$$\mathcal{P}u = f \quad on \ \Omega_2,$$

and that the coefficients of $\mathcal{P}$ belong to $C^{r,1}(\overline{\Omega}_2)^{m \times m}$.

(i) If $\gamma u \in H^{r+3/2}(\Gamma_2)^m$, then $u \in H^{r+2}(\Omega_1)^m$ and

$$\|u\|_{H^{r+2}(\Omega_1)^m} \leq C\|u\|_{H^1(\Omega_2)^m} + C\|\gamma u\|_{H^{r+3/2}(\Gamma_2)^m} + C\|f\|_{H^r(\Omega_2)^m}.$$

(ii) If $\mathcal{P}$ is coercive on $H^1(\Omega)^m$, and if $\mathcal{B}_\nu u \in H^{r+1/2}(\Gamma_2)^m$, then $u \in H^{r+2}(\Omega_1)^m$ and

$$\|u\|_{H^{r+2}(\Omega_1)^m} \leq C\|u\|_{H^1(\Omega_2)^m} + C\|\mathcal{B}_\nu u\|_{H^{r+1/2}(\Gamma_2)^m} + C\|f\|_{H^r(\Omega_2)^m}.$$

Proof. After a $C^{r+1,1}$ change of coordinates, we can assume that Ω is the half space $x_n < 0$ (see Exercise 4.2).

To prove part (i) in the case $r = 0$, we assume first that $\gamma u = 0$ on Γ_2. Choose a cutoff function $\chi \in \mathcal{D}(G_2)$ satisfying $\chi = 1$ on G_1; then $\chi u \in H_0^1(\Omega_2)$, and by arguing as in the first part of the proof of Theorem 4.16 we see that

$$\|\partial_l(\chi u)\|_{H^1(\Omega_2)^m} \leq C\|u\|_{H^1(\Omega_2)^m} + C\|f\|_{L_2(\Omega_2)^m} \quad \text{for } 1 \leq l \leq n-1.$$

We cannot estimate $\partial_n(\chi u)$ in this way, because $\Delta_{l,h}(\chi u)$ might not belong to $H_0^1(\Omega_2)$ if $l = n$. However, the partial differential equation can be rewritten as

$$-A_{nn}\partial_n^2 u = f_3, \tag{4.25}$$

where

$$f_3 = f + (\partial_n A_{nn})\partial_n u + \sum_{(j,k)\neq(n,n)} \partial_j(A_{jk}\partial_k u) - \sum_{j=1}^n A_j \partial_j u - Au,$$

so by Lemma 4.17,

$$\left\|\partial_n^2 u\right\|_{L_2(\Omega_1)^m} \leq C\|f_3\|_{L_2(\Omega_1)^m}$$

$$\leq C\|f\|_{L_2(\Omega_1)^m} + C\|u\|_{H^1(\Omega_1)^m} + C\sum_{l=1}^{n-1}\|\partial_l u\|_{H^1(\Omega_1)^m}, \tag{4.26}$$

giving the desired estimate for $\|u\|_{H^2(\Omega_1)^m}$.

For non-zero $\gamma u \in H^{3/2}(\Gamma_2)^m$, we use Lemma 3.36 to find $w \in H^2(\Omega_2)^m$ satisfying $\gamma w = \gamma u$ on Γ_2, and $\|w\|_{H^2(\Omega_2)^m} \leq C\|\gamma u\|_{H^{3/2}(\Gamma_2)^m}$. The difference $u - w \in H^1(\Omega_2)^m$ satisfies $\mathcal{P}(u - w) = f - \mathcal{P}w$ on Ω_2, and $\gamma(u - w) = 0$

on Γ_2, so by the argument above,

$$\|u - w\|_{H^2(\Omega_1)^m} \le C\|u - w\|_{H^1(\Omega_2)^m} + C\|f - \mathcal{P}w\|_{L_2(\Omega_2)^m},$$

and therefore

$$\|u\|_{H^2(\Omega_1)^m} \le C\|u\|_{H^1(\Omega_2)^m} + C\|f\|_{L_2(\Omega_2)^m} + C\|w\|_{H^2(\Omega_2)^m},$$

completing the proof of part (i) for $r = 0$.

Proceeding by induction on r, we let $r \ge 1$ and choose an open set G_1' satisfying $\overline{G_1} \Subset G_1' \subset \overline{G_1'} \Subset G_2$. By the induction hypothesis,

$$\|u\|_{H^{r+1}(\Omega_1')^m} \le C\|u\|_{H^1(\Omega_2)^m} + C\|\gamma u\|_{H^{r+1/2}(\Gamma_2)^m} + C\|f\|_{H^{r-1}(\Omega_2)^m},$$

where $\Omega_1' = G_1' \cap \Omega$. If $1 \le l \le n - 1$, then $\gamma(\partial_l u) = \partial_l(\gamma u)$, so it follows from (4.24) and the result for $r = 0$ proved above that

$$\left\|\partial_l^r u\right\|_{H^2(\Omega_1)^m} \le C\left\|\partial_l^r u\right\|_{H^1(\Omega_1')^m} + C\left\|\partial_l^r \gamma u\right\|_{H^{3/2}(\Gamma_2)^m} + C\|f_2\|_{L_2(\Omega_1')^m}.$$

Thus,

$$\left\|\partial_l^r u\right\|_{H^2(\Omega_1)^m} \le C\|u\|_{H^1(\Omega_2)^m} + C\|\gamma u\|_{H^{r+3/2}(\Gamma_2)^m} + C\|f\|_{H^r(\Omega_2)^m}$$
$$\text{for } 1 \le l \le n - 1,$$

and to estimate $\partial_n^{r+2} u$ we apply ∂_n^r to both sides of (4.25) and obtain

$$-A_{nn}\partial_n^{r+2} u = \partial_n^r f_3 + \sum_{p=0}^{r-1} \binom{r}{p}\left(\partial_n^{r-p} A_{nn}\right)\partial_n^{p+2} u.$$

Hence,

$$\left\|\partial_n^{r+2} u\right\|_{L_2(\Omega_1)^m} \le C\|f\|_{H^r(\Omega_1)^m} + C\|u\|_{H^{r+1}(\Omega_1)^m} + C\sum_{l=1}^{n-1} \|\partial_l u\|_{H^{r+1}(\Omega_1)^m},$$
$$(4.27)$$

and the induction goes through.

Turning to part (ii), the first Green identity applied to the equation $\mathcal{P}(\chi u) = f_1$ gives

$$\Phi(\chi u, v) = (f_1, v)_{\Omega_2} + \left(\mathcal{B}_\nu(\chi u), \gamma v\right)_{\Gamma_2} \quad \text{for all } v \in H^1(\Omega_2)^m. \quad (4.28)$$

Here, f_1 is the same as in (4.22) from the proof of interior regularity, but now χ does not vanish on the boundary, so we need coercivity on $H^1(\Omega)^m$, not just on $H_0^1(\Omega)^m$. If $1 \le l \le n - 1$, then we can repeat the argument leading to (4.23). The only change is the presence of an extra term involving the conormal derivative of χu. In this way, we arrive at the following estimate for the tangential difference quotients:

$$\|\triangle_{l,h}(\chi u)\|_{H^1(\Omega_2)}^2 \le C\|u\|_{H^1(\Omega_2)^m}^2 + C\|f\|_{L_2(\Omega_2)^m}^2 + C\left|\left(\mathcal{B}_\nu(\chi u), \gamma \triangle_{l,-h} v\right)_{\Gamma_2}\right|,$$

where $v = \triangle_{l,h}(\chi u)$. Let $\Gamma_1' = G_1' \cap \Gamma_2$. We can assume that $\operatorname{supp} \chi \Subset G_1'$, and then, since $[\mathcal{B}_\nu, \chi]u = \nu_j \gamma[(\partial_k \chi) A_{jk} u]$,

$$\|\mathcal{B}_\nu(\chi u)\|_{\tilde{H}^{1/2}(\Gamma_1')^m} \le \|\chi \mathcal{B}_\nu u\|_{\tilde{H}^{1/2}(\Gamma_1')^m} + \|\nu_j \gamma[(\partial_k \chi) A_{jk} u]\|_{\tilde{H}^{1/2}(\Gamma_1')^m}$$
$$\le C\|\mathcal{B}_\nu u\|_{H^{1/2}(\Gamma_2)^m} + C\|\gamma u\|_{H^{1/2}(\Gamma_2)^m}.$$

By Exercise 4.4,

$$\|\gamma \triangle_{l,-h} v\|_{H^{-1/2}(\Gamma_1')^m} = \|\triangle_{l,-h} \gamma v\|_{H^{-1/2}(\Gamma_1')^m} \le \|\partial_l \gamma v\|_{H^{-1/2}(\Gamma_1')^m}$$
$$\le C\|\gamma v\|_{H^{1/2}(\Gamma_1')^m} \le C\|v\|_{H^1(\Omega_2)^m},$$

so

$$\left|\left(\mathcal{B}_\nu(\chi u), \gamma \triangle_{l,-h} v\right)_{\Gamma_2}\right| = \left|\left(\mathcal{B}_\nu(\chi u), \gamma \triangle_{l,-h} v\right)_{\Gamma_1'}\right|$$
$$\le \|\mathcal{B}_\nu(\chi u)\|_{\tilde{H}^{1/2}(\Gamma_1')^m} \|\gamma \triangle_{l,-h} v\|_{H^{-1/2}(\Gamma_1')^m} \tag{4.29}$$
$$\le C\left(\|\mathcal{B}_\nu u\|_{H^{1/2}(\Gamma_2)^m} + \|u\|_{H^1(\Omega_2)^m}\right)\|\triangle_{l,h}(\chi u)\|_{H^1(\Omega_2)^m},$$

and thus,

$$\|\triangle_{l,h}(\chi u)\|_{H^1(\Omega_2)^m} \le C\|u\|_{H^1(\Omega_2)^m} + C\|\mathcal{B}_\nu u\|_{H^{1/2}(\Gamma_2)^m} + C\|f\|_{L_2(\Omega_2)^m}$$
$$\text{for } 1 \le l \le n-1.$$

After applying Lemma 4.13(ii), and estimating $\partial_n^2 u$ using (4.26), part (ii) follows for $r = 0$.

For $r \ge 1$, we make the induction hypothesis that

$$\|u\|_{H^{r+1}(\Omega_1')^m} \le C\|u\|_{H^1(\Omega_2)^m} + C\|\mathcal{B}_\nu u\|_{H^{r-1/2}(\Gamma_2)^m} + C\|f\|_{H^{r-1}(\Omega_2)^m}.$$

Since

$$\mathcal{B}_\nu\left(\partial_l^r u\right) = \partial_l^r(\mathcal{B}_\nu u) - \gamma \sum_{p=0}^{r-1} \binom{r}{p}\left(\partial_l^{r-p} A_{nk}\right)\left(\partial_l^p \partial_k u\right),$$

we see that if $1 \le l \le n-1$, then

$$\left\|\mathcal{B}_\nu\left(\partial_l^r u\right)\right\|_{H^{1/2}(\Gamma_1')^m} \le \|\mathcal{B}_\nu u\|_{H^{r+1/2}(\Gamma_1')^m} + C\|u\|_{H^{r+1}(\Omega_1')^m}$$

and so, with f_2 as in (4.24),

$$\left\|\partial_l^r u\right\|_{H^2(\Omega_1)^m} \le C\left\|\partial_l^r u\right\|_{H^1(\Omega_1')^m} + C\left\|\mathcal{B}_\nu\left(\partial_l^r u\right)\right\|_{H^{1/2}(\Gamma_1')^m} + C\|f_2\|_{L_2(\Omega_1')^m}.$$

Thus,

$$\left\|\partial_l^r u\right\|_{H^2(\Omega_1)^m} \le C\|u\|_{H^1(\Omega_2)^m} + C\|\mathcal{B}_\nu u\|_{H^{r+1/2}(\Gamma_2)^m} + C\|f\|_{H^r(\Omega_2)^m}$$
$$\text{for } 1 \le l \le n-1,$$

and after combining this estimate with (4.27), the induction goes through. $\quad\square$

The Transmission Property

In our later study of surface potentials and boundary integral equations, we will often consider two equations simultaneously, both involving the same operator $\mathcal{P}$, but with one over the interior domain $\Omega^- = \Omega$ and the other over the complementary exterior domain $\Omega^+ = \mathbb{R}^n \setminus (\Omega^- \cup \Gamma)$. We then have the disjoint union

$$\mathbb{R}^n = \Omega^+ \cup \Gamma \cup \Omega^-,$$

with $\Gamma = \partial\Omega^+ = \partial\Omega^-$ the common boundary of the two domains, as depicted in Figure 1 of Chapter 1. The vector ν will always denote the outward unit normal to Ω^-, and therefore the *in*ward unit normal to Ω^+. (If one thinks of Ω^+ as locally the half space $x_n > 0$, then $\nu = e_n$.) We denote the one-sided trace operator for $\Omega^\pm$ by $\gamma^\pm$, and the sesquilinear form and conormal derivative for $\Omega^\pm$ by $\Phi^\pm = \Phi_{\Omega^\pm}$ and $\mathcal{B}_\nu^\pm$, respectively. On account of our convention regarding the meaning of ν, the first Green identity (Lemma 4.1) takes the form

$$\Phi^\pm(u, v) = (\mathcal{P}u, v)_{\Omega^\pm} \mp (\mathcal{B}_\nu^\pm u, \gamma^\pm v)_\Gamma \quad \text{for } u \in H^2(\Omega^\pm)^m \text{ and}$$
$$v \in H^1(\Omega^\pm)^m, \quad (4.30)$$

and dually (Lemma 4.2)

$$\Phi^\pm(u, v) = (u, \mathcal{P}^*v)_{\Omega^\pm} \mp (\gamma^\pm u, \widetilde{\mathcal{B}}_\nu^\pm v)_\Gamma \quad \text{for } u \in H^1(\Omega^\pm)^m \text{ and}$$
$$v \in H^2(\Omega^\pm)^m. \quad (4.31)$$

As explained in the discussion following Lemma 4.3, we extend the definitions of $\mathcal{B}_\nu^\pm$ and $\widetilde{\mathcal{B}}_\nu^\pm$ to make these identities valid for $u, v \in H^1(\Omega^\pm)^m$ whenever in the former case $\mathcal{P}u$, or in the latter case $\mathcal{P}^*v$, belongs to $\widetilde{H}^{-1}(\Omega^\pm)^m$.

In this setting, it is convenient to think of u as a function defined on the whole of $\mathbb{R}^n$, but possibly having jumps in its trace and conormal derivatives across the surface Γ. We denote these jumps by

$$[u]_\Gamma = \gamma^+ u - \gamma^- u, \quad [\mathcal{B}_\nu u]_\Gamma = \mathcal{B}_\nu^+ u - \mathcal{B}_\nu^- u, \quad [\widetilde{\mathcal{B}}_\nu u]_\Gamma = \widetilde{\mathcal{B}}_\nu^+ u - \widetilde{\mathcal{B}}_\nu^- u,$$

provided both of the corresponding one-sided quantities make sense. When the jump vanishes, the $+$ or $-$ superscript is redundant and will often be dropped, i.e., we shall write

$$\gamma u = \gamma^+ u = \gamma^- u \quad \text{if } [u]_\Gamma = 0,$$

and similarly for the conormal derivatives. For brevity, we sometimes write

$$u^\pm = u|_{\Omega^\pm}$$

to emphasise that we are considering the pieces of u separately.

Lemma 4.19 *Let $f^\pm \in \widetilde{H}^{-1}(\Omega^\pm)^m$, define $f = f^+ + f^- \in H^{-1}(\mathbb{R}^n)^m$, and suppose that $u \in L_2(\mathbb{R}^n)^m$ with $u^\pm \in H^1(\Omega^\pm)^m$. If*

$$\mathcal{P}u^\pm = f^\pm \quad \text{on } \Omega^\pm, \tag{4.32}$$

then

$$\Phi^+(u^+, v) + \Phi^-(u^-, v) = (f, v) - ([\mathcal{B}_\nu u]_\Gamma, \gamma v)_\Gamma \quad \text{for } v \in H^1(\mathbb{R}^n)^m, \tag{4.33}$$

and

$$(\mathcal{P}u, \phi) = (f, \phi) + ([u]_\Gamma, \widetilde{\mathcal{B}}_\nu \phi)_\Gamma - ([\mathcal{B}_\nu u]_\Gamma, \gamma \phi)_\Gamma \quad \text{for } \phi \in \mathcal{D}(\mathbb{R}^n)^m. \tag{4.34}$$

Thus, $\mathcal{P}u = f$ on $\mathbb{R}^n$ if and only if $[u]_\Gamma = [\mathcal{B}_\nu u]_\Gamma = 0$.

Proof. Equation (4.33) follows at once from (4.30). The definition of $\mathcal{P}u$ as a distribution on $\mathbb{R}^n$ gives

$$(\mathcal{P}u, \phi) = (u, \mathcal{P}^*\phi) = (u^+, \mathcal{P}^*\phi)_{\Omega^+} + (u^-, \mathcal{P}^*\phi)_{\Omega^-},$$

and by (4.31),

$$(u^\pm, \mathcal{P}^*\phi)_{\Omega^\pm} = \Phi^\pm(u^\pm, \phi) \pm (\gamma^\pm u, \widetilde{\mathcal{B}}_\nu \phi)_\Gamma,$$

so (4.34) follows from (4.33). $\square$

We can now prove the *transmission property* for (4.32): if the jumps in u and its conormal derivative are smooth, and if the right hand sides f^+ and f^- are smooth up to the boundary, then the restrictions u^+ and u^- must be smooth up to the boundary. Jumps are also allowed in the coefficients of $\mathcal{P}$, although we shall not use this fact in later chapters. The proof again uses Nirenberg's method of difference quotients. Results like the following theorem, but involving more general types of transmission conditions on Γ, appear in work by Schechter [90] and Roĭtberg and Šefteĺ [89].

Theorem 4.20 Let G_1 and G_2 be bounded open subsets of $\mathbb{R}^n$ such that $\overline{G}_1 \Subset G_2$ and G_1 intersects Γ, and put

$$\Omega_j^{\pm} = G_j \cap \Omega^{\pm} \text{ and } \Gamma_j = G_j \cap \Gamma \qquad \text{for } j = 1, 2.$$

Suppose, for an integer $r \geq 0$, that Γ_2 is $C^{r+1,1}$, and consider the two equations

$$\mathcal{P}u^{\pm} = f^{\pm} \quad \text{on } \Omega_2^{\pm},$$

where $\mathcal{P}$ is strongly elliptic on G_2 with coefficients in $C^{r,1}\big(\overline{\Omega_2^{\pm}}\big)^{m \times m}$. If $u \in L_2(G_2)^m$ satisfies

$$u^{\pm} \in H^1(\Omega_2^{\pm})^m, \quad [u]_{\Gamma} \in H^{r+3/2}(\Gamma_2)^m, \quad [\mathcal{B}_\nu u]_{\Gamma} \in H^{r+1/2}(\Gamma_2)^m,$$

and if $f^{\pm} \in H^r(\Omega_2^{\pm})^m$, then $u^{\pm} \in H^{r+2}(\Omega_1^{\pm})^m$ and

$$\|u^+\|_{H^{r+2}(\Omega_1^+)^m} + \|u^-\|_{H^{r+2}(\Omega_1^-)^m}$$

$$\leq C\|u^+\|_{H^1(\Omega_2^+)^m} + C\|u^-\|_{H^1(\Omega_2^-)^m}$$

$$+ C\big\|[u]_{\Gamma_2}\big\|_{H^{r+3/2}(\Gamma_2)^m} + C\big\|[\mathcal{B}_\nu u]_{\Gamma_2}\big\|_{H^{r+1/2}(\Gamma_2)^m}$$

$$+ C\|f^+\|_{H^r(\Omega_2^+)^m} + C\|f^-\|_{H^r(\Omega_2^-)^m}.$$

Proof. As in the proof of Theorem 4.16, it suffices to consider the case when Ω^+ is the half space $x_n > 0$. Suppose to begin with that $[u]_{\Gamma} = 0$ on Γ_2, so that $u \in H^1(G_2)^m$ by Exercise 4.5. We fix a cutoff function $\chi \in \mathcal{D}(G_2)$ such that $\chi = 1$ on G_1, then, as with (4.22),

$$\mathcal{P}(\chi u^{\pm}) = f_1^{\pm} \quad \text{on } \Omega_2^{\pm},$$

where the functions f_1^+ and f_1^- satisfy

$$\|f_1^{\pm}\|_{L_2(\Omega_2^{\pm})^m} \leq C\|\chi f\|_{L_2(\Omega_2^{\pm})^m} + C\|u^{\pm}\|_{H^1(\Omega_2^{\pm})^m}.$$

By Lemma 4.19,

$$\Phi_{G_2}(\chi u, v) = (f_1, v)_{G_2} - \big([\mathcal{B}_\nu(\chi u)]_{\Gamma}, \gamma v\big)_{\Gamma_2} \quad \text{for } v \in H_0^1(G_2)^m,$$

an equation eerily like (4.28). Repeating the argument from the proof of interior regularity, but with an extra term involving $[\mathcal{B}_\nu(\chi u)]_\Gamma$, we see that

$$\|\Delta_{l,h}(\chi u)\|^2_{H^1(G_2)^m} \leq C\|u\|^2_{H^1(G_2)^m} + C\|f_1\|^2_{L_2(G_2)^m}$$
$$+ C\left|\left([\mathcal{B}_\nu(\chi u)]_\Gamma, \gamma\Delta_{l,-h}v\right)_{\Gamma_2}\right|,$$

where $v = \Delta_{l,h}(\chi u)$; cf. (4.23). If $1 \leq l \leq n-1$, then the argument leading to (4.29) shows that

$$\left|\left([\mathcal{B}_\nu(\chi u)]_\Gamma, \gamma\Delta_{l,-h}v\right)_{\Gamma_2}\right| \leq \left(\|[\mathcal{B}_\nu u]_\Gamma\|_{H^{1/2}(\Gamma_2)^m} + \|u\|_{H^1(G_2)^m}\right)$$
$$\times \|\Delta_{l,h}(\chi u)\|_{H^1(G_2)^m},$$

and therefore

$$\|\partial_l(\chi u)\|_{H^1(G_2)^m} \leq C\|u\|_{H^1(G_2)^m} + C\|[\mathcal{B}_\nu u]_\Gamma\|_{H^{1/2}(\Gamma_2)^m} + C\|f_1\|_{L_2(G_2)^m}.$$

Of course, $\partial_n(\chi u)$ is generally not in $H^1(G_2)^m$, because $[\partial_n u]_\Gamma \neq 0$, but as in (4.25),

$$-A_{nn}\partial^2_n u^\pm = f_3^\pm \quad \text{on } \Omega_2^\pm,$$

with

$$\|f_3^\pm\|_{L_2(\Omega_1^\pm)^m} \leq C\|f^\pm\|_{L_2(\Omega_1^\pm)^m} + C\|u\|_{H^1(\Omega_1^\pm)^m} + C\sum_{l=1}^{n-1}\|\partial_l u\|_{H^1(\Omega_1^\pm)^m}.$$

Hence, $u^\pm \in H^2(\Omega_1^\pm)^m$ and the desired estimate holds for $r = 0$.

For non-zero $[u]_\Gamma \in H^{3/2}(\Gamma_2)^m$, we use the extension operator η_0 of Lemma 3.36 to construct $w \in H^2(G_2)^m$ satisfying

$$\gamma w = [u]_\Gamma \quad \text{on } \Gamma_2 \quad \text{and} \quad \|\mathcal{B}_\nu w\|_{H^{1/2}(\Gamma_2)^m} + \|w\|_{H^2(G_2)^m} \leq C\|[u]_\Gamma\|_{H^{3/2}(\Gamma_2)^m}.$$

Consider the function

$$u_1 = \begin{cases} u^+ & \text{on } \Omega_2^+, \\ u^- + w & \text{on } \Omega_2^-, \end{cases}$$

Since

$$\mathcal{P}u_1^\pm = \begin{cases} f^+ & \text{on } \Omega_2^+, \\ f^- + \mathcal{P}w & \text{on } \Omega_2^-, \end{cases}$$

with $[u_1]_\Gamma = [u]_\Gamma - w = 0$ on Γ_2, and $[\mathcal{B}_\nu u_1]_\Gamma = [\mathcal{B}_\nu u]_\Gamma - \mathcal{B}_\nu w \in H^{1/2}(\Gamma_2)^m$, the preceding argument applies, showing that $u_1^\pm \in H^2(\Omega_1^\pm)^m$.

Therefore, $u^{\pm} \in H^2(\Omega_1^{\pm})^m$ and

$$
\begin{aligned}
\|u^+\|_{H^2(\Omega_1^+)^m} &+ \|u^-\|_{H^2(\Omega_1^-)^m} \\
&\leq \|u_1^+\|_{H^2(\Omega_1^+)^m} + \|u_1^-\|_{H^2(\Omega_1^-)^m} + \|w\|_{H^2(\Omega_1^-)^m} \\
&\leq C\|u_1\|_{H^1(G_2)^m} + C\|[\mathcal{B}_\nu u_1]_\Gamma\|_{H^{1/2}(\Gamma_2)^m} + C\|f^+\|_{L_2(\Omega_2^+)^m} \\
&\quad + C\|f^- + \mathcal{P}w\|_{L_2(\Omega_2^-)^m} + \|w\|_{H^2(\Omega_1^-)^m},
\end{aligned}
$$

which, because

$$
\begin{aligned}
\|w\|_{H^1(\Omega_2^-)^m} + \|\mathcal{B}_\nu w\|_{H^{1/2}(\Gamma_2)^m} + \|\mathcal{P}w\|_{L_2(\Omega_2^-)^m} &\leq C\|w\|_{H^2(\Omega_2^-)^m} \\
&\leq C\|[u]_\Gamma\|_{H^{3/2}(\Gamma_2)^m},
\end{aligned}
$$

shows that the desired estimate for the case $r = 0$ holds also when $[u]_\Gamma \neq 0$. An inductive argument like the one used for part (ii) of Theorem 4.18 takes care of $r \geq 1$. $\qquad\square$

Estimates for the Steklov–Poincaré Operator

Consider the semi-homogeneous Dirichlet problem,

$$
\begin{aligned}
\mathcal{P}u &= 0 \quad \text{on } \Omega, \\
\gamma u &= g \quad \text{on } \Gamma
\end{aligned}
\tag{4.35}
$$

and the adjoint problem

$$
\begin{aligned}
\mathcal{P}^* v &= 0 \quad \text{on } \Omega, \\
\gamma v &= \phi \quad \text{on } \Gamma.
\end{aligned}
\tag{4.36}
$$

If the fully homogeneous problem has only the trivial solution in $H^1(\Omega)^m$, i.e., if $g = 0$ implies $u = 0$, then under the usual assumptions we are able to define solution operators

$$
\mathcal{U} : g \mapsto u \quad \text{and} \quad \mathcal{V} : \phi \mapsto v,
$$

with

$$
\mathcal{U} : H^{1/2}(\Gamma)^m \to H^1(\Omega)^m \quad \text{and} \quad \mathcal{V} : H^{1/2}(\Gamma)^m \to H^1(\Omega)^m.
\tag{4.37}
$$

We can also form the *Steklov–Poincaré operators* $\mathcal{B}_\nu \mathcal{U} : g \mapsto \mathcal{B}_\nu u$ and $\widetilde{\mathcal{B}}_\nu \mathcal{V} : \phi \mapsto \widetilde{\mathcal{B}}_\nu v$, that satisfy

$$
\mathcal{B}_\nu \mathcal{U} : H^{1/2}(\Gamma)^m \to H^{-1/2}(\Gamma)^m \quad \text{and} \quad \widetilde{\mathcal{B}}_\nu \mathcal{V} : H^{1/2}(\Gamma)^m \to H^{-1/2}(\Gamma)^m.
\tag{4.38}
$$

The purpose of this section is to prove that, under certain conditions, $\mathcal{B}_\nu \mathcal{U}$ and $\tilde{\mathcal{B}}_\nu \mathcal{V}$ are also bounded from $H^1(\Gamma)^m$ to $L_2(\Gamma)^m$, a fact that will be used later in our study of surface potentials and boundary integral operators; see Theorem 6.12. Notice that

$$(\mathcal{B}_\nu \mathcal{U} g, \phi)_\Gamma = \Phi(\mathcal{U} g, \mathcal{V} \phi) = (g, \tilde{\mathcal{B}}_\nu \mathcal{V} \phi)_\Gamma, \qquad (4.39)$$

so $(\mathcal{B}_\nu \mathcal{U})^* = \tilde{\mathcal{B}}_\nu \mathcal{V}$.

If Ω is at least $C^{1,1}$, then the regularity estimates of Theorem 4.18 apply, and we can extend (4.37) and (4.38) as follows.

Theorem 4.21 Assume that Ω is a bounded, $C^{r+1,1}$ domain, for some integer $r \geq 0$, and that $\mathcal{P}$ is strongly elliptic on Ω, with coefficients in $C^{r,1}(\overline{\Omega})^{m \times m}$. If the Dirichlet problem (4.35) has only the trivial solution in $H^1(\Omega)^m$ when $g = 0$, then the solution operator has the mapping property

$$\mathcal{U} : H^{s+1/2}(\Gamma)^m \to H^{s+1}(\Omega)^m \quad for\ 0 \leq s \leq r+1,$$

and the Steklov–Poincaré operator has the mapping property

$$\mathcal{B}_\nu \mathcal{U} : H^{s+1/2}(\Gamma)^m \to H^{s-1/2}(\Gamma)^m \quad for\ -r-1 \leq s \leq r+1.$$

Proof. The case $s = 0$ is covered already in (4.37) and (4.38). Part (i) of Theorem 4.18 shows that $\mathcal{U} : H^{r+3/2}(\Gamma)^m \to H^{r+2}(\Omega)^m$, and thus $\mathcal{B}_\nu \mathcal{U} : H^{r+3/2}(\Gamma)^m \to H^{r+1/2}(\Gamma)^m$, which means that the result holds for $s = r + 1$. Boundedness for the range $0 \leq s \leq r + 1$ now follows by interpolation, i.e., by Theorems B.8 and B.11. The same arguments apply to $\mathcal{V}$ and $\tilde{\mathcal{B}}_\nu \mathcal{V}$, so, in view of (4.39), we can extend $\mathcal{B}_\nu \mathcal{U}$ in a unique way to a linear operator that is bounded for $-r - 1 \leq s \leq 0$. $\qquad \square$

Our task is much harder when Ω is permitted to be Lipschitz. In this case, we will estimate $\|\mathcal{B}_\nu u\|_{L_2(\Omega)^m}$ with the help of the following integral identity. Here, for the sake of brevity, we use the summation convention, i.e., we sum any repeated indices from 1 to n.

Lemma 4.22 Assume that Ω is Lipschitz, and that the leading coefficients A_{jk} belong to $W_\infty^1(\Omega)^{m \times m}$. For any real-valued functions $h_1, h_2, \ldots, h_n \in W_\infty^1(\Omega)$, and for any $u, v \in H^2(\Omega)^m$,

$$\int_\Gamma \nu_l \gamma \left\{ \left[(h_l A_{jk} - h_j A_{lk} - h_k A_{jl}) \partial_k u \right]^* \partial_j v \right\} dx$$

$$= \int_\Omega \left\{ (D_{jk} \partial_k u)^* \partial_j v + (\mathcal{P}_0 u)^* (h_j \partial_j v) + (h_k \partial_k u)^* (\mathcal{P}_0^* v) \right\} dx,$$

where

$$D_{jk} = \partial_l(h_l A_{jk}) - (\partial_l h_j) A_{lk} - (\partial_l h_k) A_{jl}.$$

Proof. By the divergence theorem, it suffices to show that

$$\partial_l\{\big[(h_l A_{jk} - h_j A_{lk} - h_k A_{jl})\partial_k u\big]^* \partial_j v\}$$
$$= (D_{jk}\partial_k u)^* \partial_j v + (\mathcal{P}_0 u)^*(h_j \partial_j v) + (h_k \partial_k u)^*(\mathcal{P}_0^* v).$$

In fact, the left-hand side expands to a sum of five terms,

$$\big[\partial_l(h_l A_{jk} - h_j A_{lk} - h_k A_{jl})(\partial_k u)\big]^* \partial_j v$$
$$+ \big[(h_l A_{jk} - h_k A_{jl})\partial_l \partial_k u\big]^* \partial_j v + \big[-h_j A_{lk}\partial_l \partial_k u\big]^* \partial_j v$$
$$+ \big[(h_l A_{jk} - h_j A_{lk})\partial_k u\big]^* \partial_l \partial_j v + \big[-h_k A_{jl}\partial_k u\big]^* \partial_l \partial_j v.$$

The second term vanishes because its factor (...) is skew-symmetric in l and k, and likewise the fourth term vanishes because its factor (...) is skew-symmetric in l and j. The third term equals

$$h_j\big[\mathcal{P}_0 u + (\partial_l A_{lk})\partial_k u\big]^* \partial_j v = (\mathcal{P}_0 u)^*(h_j \partial_j v) + \big[h_j(\partial_l A_{lk})\partial_k u\big]^* \partial_j v,$$

and the fifth term equals

$$-(h_k \partial_k u)^* A_{jl}^* \partial_l \partial_j v = (h_k \partial_k u)^*\big[\mathcal{P}_0^* v + (\partial_l A_{jl}^*)\partial_j v\big]$$
$$= (h_k \partial_k u)^*(\mathcal{P}_0^* v) + \big[h_k(\partial_l A_{jl})\partial_k u\big]^* \partial_j v,$$

so we get the desired right hand side with

$$D_{jk} = \partial_l(h_l A_{jk} - h_j A_{lk} - h_k A_{jl}) + h_j \partial_l A_{lk} + h_k \partial_l A_{jl}. \qquad \square$$

Rellich [85] used a special case of the above identity to obtain an integral representation for the Dirichlet eigenvalues of the Laplacian. Subsequently, Payne and Weinberger [81] generalised the Rellich identity to handle second-order elliptic systems with variable coefficients, and used it to bound the errors in certain approximations to $\Phi(u, u)$ and pointwise values of u, when u is the solution to a Dirichlet or Neumann problem. In what follows, we use the arguments of Nečas [72, Chapitre 5].

We will use certain first-order partial differential operators of the form

$$\mathcal{Q}u = Q_k \gamma(\partial_k u) \quad \text{with } Q_k \in L_\infty(\Gamma)^{m \times m}.$$

If $\nu_k Q_k = 0$ on Γ, then such a $\mathcal{Q}$ is said to be *tangential* to Γ.

Lemma 4.23 Assume that Ω is a Lipschitz domain, and let $u \in C^1_{\mathrm{comp}}(\overline{\Omega})^m$.

(i) If $\mathcal{Q}$ is a first-order, tangential differential operator, then

$$\|\mathcal{Q}u\|_{L_2(\Gamma)^m} \le C\|\gamma u\|_{H^1(\Gamma)^m}.$$

(ii) The normal and conormal derivatives of u satisfy

$$\|\mathcal{B}_\nu u\|_{L_2(\Gamma)^m} \le C\|\partial u/\partial \nu\|_{L_2(\Gamma)^m} + C\|\gamma u\|_{H^1(\Gamma)^m},$$

and, when $\mathcal{P}$ is strongly elliptic on Ω,

$$\|\partial u/\partial \nu\|_{L_2(\Gamma)^m} \le C\|\mathcal{B}_\nu u\|_{L_2(\Gamma)^m} + C\|\gamma u\|_{H^1(\Gamma)^m}.$$

(iii) For $1 \le j \le n$,

$$\|\gamma \partial_j u\|_{L_2(\Gamma)^m} \le \|\partial u/\partial \nu\|_{L_2(\Gamma)^m} + C\|\gamma u\|_{H^1(\Gamma)^m}.$$

Proof. It suffices to deal with the case when Ω is a Lipschitz hypograph given by $x_n < \zeta(x')$. From the formula for the unit normal given in (3.28), we see that the tangency condition $\nu_k \mathcal{Q}_k = 0$ is equivalent to

$$\mathcal{Q}_n(x', \zeta(x')) = \sum_{l=1}^{n-1} \mathcal{Q}_l(x', \zeta(x'))\partial_l \zeta(x'),$$

and so

$$
\begin{aligned}
\mathcal{Q}u(x', \zeta(x')) &= \sum_{l=1}^{n-1} \mathcal{Q}_l(x', \zeta(x'))\big[\partial_l u(x', \zeta(x')) + \partial_n u(x', \zeta(x'))\partial_l \zeta(x')\big] \\
&= \sum_{l=1}^{n-1} \mathcal{Q}_l(x', \zeta(x'))\partial_l u_\zeta(x'),
\end{aligned}
$$

where $u_\zeta(x') = u(x', \zeta(x'))$. Part (i) follows at once.

Next, consider the identity

$$(\nu_j A_{jk}\nu_k)\frac{\partial u}{\partial \nu} = \mathcal{B}_\nu u + \mathcal{Q}u, \qquad \text{where } \mathcal{Q}u = \nu_j(A_{jl}\nu_l\nu_k - A_{jk})\gamma(\partial_k u).$$

The first half of part (ii) is immediate, because the differential operator $\mathcal{Q}$ is tangential to Γ. The second part follows because the condition (4.7) for strong ellipticity implies that the $m \times m$ matrix $\nu_j A_{jk}\nu_k$ is uniformly positive-definite, so $\|(\nu_j A_{jk}\nu_k)\partial u/\partial \nu\|_{L_2(\Gamma)^m} \ge c\|\partial u/\partial \nu\|_{L_2(\Gamma)^m}$. Finally, to prove part (iii), we observe that

$$\gamma(\partial_j u) = \nu_j \frac{\partial u}{\partial \nu} + Q_j u, \qquad \text{where } Q_j u = (\delta_{jk} - \nu_j \nu_k)\gamma(\partial_k u),$$

and the differential operator Q_j is tangential to Γ. $\qquad\qquad\qquad$ □

Concerning the next theorem, we remark that only part (i) is actually used later.

Theorem 4.24 Assume that Ω is Lipschitz, that $\mathcal{P}$ is strongly elliptic on Ω, and that the leading coefficients A_{jk} are Lipschitz and satisfy

$$A_{kj}^* = A_{jk}.$$

(Thus, the principal part of $\mathcal{P}$ is formally self-adjoint.) Let $u \in H^1(\Omega)^m$ and $f \in L_2(\Omega)^m$ satisfy

$$\mathcal{P}u = f \quad on\ \Omega.$$

(i) If $\gamma u \in H^1(\Gamma)^m$, then $\mathcal{B}_\nu u \in L_2(\Omega)^m$, and

$$\|\mathcal{B}_\nu u\|_{L_2(\Gamma)^m} \le C\|\gamma u\|_{H^1(\Gamma)^m} + C\|u\|_{H^1(\Omega)^m} + C\|f\|_{L_2(\Omega)^m}.$$

(ii) If $\mathcal{P}$ is a scalar operator (i.e, if $m = 1$), if $A_{kj} = A_{jk}$ (so the leading coefficients are purely real), and if $\mathcal{B}_\nu u \in L_2(\Gamma)$, then $\gamma u \in H^1(\Gamma)$, and

$$\|\gamma u\|_{H^1(\Gamma)} \le C\|\mathcal{B}_\nu u\|_{L_2(\Gamma)} + C\|u\|_{H^1(\Omega)} + C\|f\|_{L_2(\Omega)}.$$

Proof. Suppose in the first instance that $u \in H^2(\Omega)^m$. Taking $v = u$ in Lemma 4.22, and noting that $\mathcal{P}_0^* = \mathcal{P}_0$, we obtain

$$\int_\Gamma \nu_l \gamma \big\{\big[(h_l A_{jk} - h_j A_{lk} - h_k A_{jl})\partial_k u\big]^* \partial_j u\big\}\, dx$$

$$= \int_\Omega \big\{(D_{jk}\partial_k u)^* \partial_j u + 2\,\mathrm{Re}\,\big[(\mathcal{P}_0 u)^*(h_j \partial_j u)\big]\big\}\, dx. \qquad (4.40)$$

We split the integrand on the left-hand side into a sum of three terms,

$$(Q_j u)^*(\partial_j u) + (\partial_k u)^* Q_k' u - \nu_l\big[(h_l A_{jk})\partial_k u\big]^* \partial_j u,$$

where Q_j and Q_k' are tangential differential operators, defined by

$$Q_j u = \nu_l (h_l A_{jk} - h_k A_{jl})\partial_k u \quad \text{and} \quad Q_k' u = \nu_l (h_l A_{jk}^* - h_j A_{lk}^*)\partial_j u.$$

The matrix $A_\nu = \nu_j A_{jk} \nu_k \in L_\infty(\Gamma)^{m \times m}$ is Hermitian and uniformly positive-definite on Γ, because $\mathcal{P}$ is strongly elliptic. Thus, $A_\nu^{-1} \in L_\infty(\Gamma)^{m \times m}$ exists and is uniformly positive-definite. A short calculation shows that

$$(A_{jk}\partial_k u)^* \partial_j u = \left(A_\nu^{-1}\mathcal{B}_\nu u\right)^* \mathcal{B}_\nu u + (\mathcal{Q}_j'' u)^* \partial_j u,$$

where $\mathcal{Q}_j'' u = (A_{jk} - \nu_p A_{jp} A_\nu^{-1} A_{qk}\nu_q)\partial_k u$. The operator $\mathcal{Q}_j''$ is also tangential, so if we choose the h_l such that

$$h_l \nu_l \geq c, \tag{4.41}$$

then

$$
\begin{aligned}
\|\mathcal{B}_\nu u\|_{L_2(\Gamma)^m}^2 &\leq C \int_\Gamma h_l \nu_l \left(A_\nu^{-1}\mathcal{B}_\nu u\right)^* \mathcal{B}_\nu u \, d\sigma \\
&\leq C \|\gamma \partial_j u\|_{L_2(\Gamma)^m} \left(\|\mathcal{Q}_j u\|_{L_2(\Gamma)^m} + \|\mathcal{Q}_j' u\|_{L_2(\Gamma)^m} + \|\mathcal{Q}_j'' u\|_{L_2(\Gamma)^m} \right) \\
&\quad + C \|u\|_{H^1(\Omega)^m}^2 + C \|\mathcal{P}_0 u\|_{L_2(\Omega)^m} \|u\|_{H^1(\Omega)^m}.
\end{aligned}
$$

By Lemma 4.23,

$$
\begin{aligned}
\|\gamma \partial_j u\|_{L_2(\Gamma)^m} &\leq C \|\partial u/\partial \nu\|_{L_2(\Gamma)^m} + C \|\gamma u\|_{H^1(\Gamma)^m} \\
&\leq C \|\mathcal{B}_\nu u\|_{L_2(\Gamma)^m} + C \|\gamma u\|_{H^1(\Gamma)^m},
\end{aligned}
$$

and so

$$
\begin{aligned}
\|\mathcal{B}_\nu u\|_{L_2(\Gamma)^m}^2 &\leq C \|\mathcal{B}_\nu u\|_{L_2(\Gamma)^m} \|\gamma u\|_{H^1(\Gamma)^m} + C \|\gamma u\|_{H^1(\Gamma)^m}^2 \\
&\quad + C \|u\|_{H^1(\Omega)^m}^2 + C \|\mathcal{P}_0 u\|_{L_2(\Omega)^m} \|u\|_{H^1(\Omega)^m}.
\end{aligned}
$$

Since $\|\mathcal{P}_0 u\|_{L_2(\Omega)^m} \leq \|f\|_{L_2(\Omega)^m} + C \|u\|_{H^1(\Omega)^m}$, the estimate in part (i) follows.

Now drop the requirement that $u \in H^2(\Omega)^m$, i.e., allow $u \in H^1(\Omega)^m$, but assume $\gamma u \in H^1(\Gamma)^m$. It suffices to consider Ω of the form $x_n < \zeta(x')$, where ζ is Lipschitz with compact support. By Theorem 4.6, the operator $\mathcal{P}$ is coercive on $H_0^1(\Omega)^m$, so for λ sufficiently large,

$$\Phi(u, u) + \lambda \|u\|_{L_2(\Omega)^m}^2 \geq c \|u\|_{H_0^1(\Omega)^m}^2 \quad \text{for } u \in H_0^1(\Omega)^m.$$

Choose a sequence $\zeta_r \in C^\infty(\mathbb{R}^{n-1})$ such that

1. $\zeta_r \to \zeta$ in $L_\infty(\mathbb{R}^{n-1})$, and $\operatorname{grad} \zeta_r \to \operatorname{grad} \zeta$ in $L_p(\mathbb{R}^{n-1})$ for $1 \leq p < \infty$;
2. $\zeta_r \leq \zeta$ on $\mathbb{R}^{n-1}$, and $\|\operatorname{grad} \zeta_r\|_{L_\infty(\mathbb{R}^{n-1})} \leq C$, for all r;
3. $\zeta_r(x') = \zeta(x') = 0$ for $|x'|$ sufficiently large.

We let Ω_r denote the set of points $x \in \mathbb{R}^n$ satisfying $x_n < \zeta_r(x')$, and put $\Gamma_r = \partial\Omega_r$. Obviously, $\Omega_r \subseteq \Omega$. Define $g : \Omega \to \mathbb{C}^m$ by

$$g(x) = \gamma u(x', \zeta(x')),$$

so that $g(x)$ is independent of x_n. We easily verify that

$$\|g\|_{H^1(\Omega)^m} \le C\|\gamma u\|_{H^1(\Gamma)^m} \quad \text{and} \quad \|\gamma_r g\|_{H^1(\Gamma_r)^m} \le C\|\gamma u\|_{H^1(\Gamma)^m},$$

where γ_r is the trace operator for Γ_r, and where, in the second estimate, the constant C is independent of r. The operator $\mathcal{P} + \lambda$ is positive and bounded below on $H_0^1(\Omega)^m$, and hence also on $H_0^1(\Omega_r)^m$, so there is a unique solution $u_r \in H^1(\Omega_r)^m$ to the Dirichlet problem

$$(\mathcal{P} + \lambda)u_r = f + \lambda u \quad \text{on } \Omega_r,$$

$$\gamma_r u_r = \gamma_r g \quad \text{on } \Gamma_r.$$

Moreover,

$$\|u_r\|_{H^1(\Omega_r)^m} \le C\|f + \lambda u\|_{\tilde{H}^{-1}(\Omega_r)^m} + C\|\gamma_r g\|_{H^1(\Gamma_r)^m},$$

and (as one sees from the proof of Lemma 2.32) the constants are independent of r. We extend u_r to Ω by defining $u_r = g$ on $\Omega \setminus \Omega_r$, and observe that because $\gamma_r u_r = \gamma_r g$,

$$\|u_r\|_{H^1(\Omega)^m} \le \|u_r\|_{H^1(\Omega_r)^m} + \|g\|_{H^1(\Omega\setminus\Omega_r)^m}$$

$$\le C\|\gamma u\|_{H^1(\Gamma)^m} + C\|f + \lambda u\|_{L_2(\Omega)^m}.$$

Since Γ_r is smooth, we have $u_r \in H^2(\Omega_r)^m$ by Theorem 4.18, and so the argument in the first part of the proof shows that

$$\|\mathcal{B}_\nu u_r\|_{L_2(\Gamma_r)^m} \le C\|\gamma_r u_r\|_{H^1(\Gamma_r)^m} + C\|u_r\|_{H^1(\Omega_r)^m} + C\|f + \lambda u\|_{L_2(\Omega_r)^m}$$

$$\le C\|\gamma u\|_{H^1(\Gamma)^m} + C\|u\|_{H^1(\Omega)^m} + C\|f\|_{L_2(\Omega)^m}, \tag{4.42}$$

with the constants again independent of r.

We claim that u_r converges to u in $H^1(\Omega)^m$. Let Φ_r denote the sesquilinear form on Ω_r, and apply the first Green identity to obtain

$$\Phi_r(u_r, v) + \lambda(u_r, v)_{\Omega_r} = (f + \lambda u, v)_{\Omega_r} + (\mathcal{B}_\nu u_r, \gamma_r v)_{\Gamma_r} \quad \text{for } v \in H^1(\Omega_r)^m, \tag{4.43}$$

and

$$\Phi(u, v) + \lambda(u, v)_\Omega = (f + \lambda u, v)_\Omega + (\mathcal{B}_\nu u, \gamma v)_\Gamma \quad \text{for } v \in H^1(\Omega)^m.$$

Hence, if $v \in H_0^1(\Omega)^m$, then

$$
\begin{aligned}
\Phi(u_r - u, v) + \lambda(u_r - u, v)_\Omega &= \left[\Phi_r(u_r, v) + \lambda(u_r, v)_{\Omega_r}\right] \\
&\quad - \left[\Phi(u, v) + \lambda(u, v)_\Omega\right] \\
&\quad + \left[\Phi(u_r, v) - \Phi_r(u_r, v) + \lambda(u_r, v)_{\Omega \setminus \Omega_r}\right] \\
&= (\mathcal{B}_\nu u_r, \gamma_r v)_{\Gamma_r} - (f + \lambda u, v)_{\Omega \setminus \Omega_r} \\
&\quad + \left[\Phi(u_r, v) - \Phi_r(u_r, v) + \lambda(u_r, v)_{\Omega \setminus \Omega_r}\right],
\end{aligned}
$$

so by taking $v = u_r - u$ and remembering that $u_r = g$ on $\Omega \setminus \Omega_r$,

$$
\begin{aligned}
\|u_r - u\|^2_{H_0^1(\Omega)^m} &\leq C \|\mathcal{B}_\nu u_r\|_{L_2(\Gamma_r)^m} \|\gamma_r(u_r - u)\|_{L_2(\Gamma_r)^m} \\
&\quad + C\left(\|f + \lambda u\|_{L_2(\Omega \setminus \Omega_r)^m} + \|g\|_{H^1(\Omega \setminus \Omega_r)^m}\right) \|u_r - u\|_{H_0^1(\Omega)^m}.
\end{aligned}
$$

Define

$$
w_r(x') = \sqrt{1 + |\text{grad}\,\zeta_r(x')|^2} \quad \text{and} \quad w(x') = \sqrt{1 + |\text{grad}\,\zeta(x')|^2},
$$

and note that $1 \leq w_r(x') \leq C$, $1 \leq w(x') \leq C$, and $w_r \to w$ in $L_p(\mathbb{R}^{n-1})$ for $1 \leq p < \infty$. We have

$$
\begin{aligned}
\|\gamma_r(u_r - u)\|^2_{L_2(\Gamma_r)^m} &= \int_{\mathbb{R}^{n-1}} \left|u(x', \zeta(x')) - u(x', \zeta_r(x'))\right|^2 w_r(x')\, dx' \\
&\leq C \int_{\mathbb{R}^{n-1}} [\zeta(x') - \zeta_r(x')] \int_{\zeta_r(x')}^{\zeta(x')} |\partial_n u(x', x_n)|^2\, dx_n\, dx' \\
&\leq C \|\zeta - \zeta_r\|_{L_\infty(\mathbb{R}^{n-1})} \|u\|^2_{H^1(\Omega \setminus \Omega_r)^m}, \qquad (4.44)
\end{aligned}
$$

which, in combination with (4.42), implies that $\|u_r - u\|_{H_0^1(\Omega)^m} \to 0$, as claimed.

The sequence of functions $\psi_r(x') = \mathcal{B}_\nu u_r(x', \zeta_r(x'))$ is bounded in $L_2(\mathbb{R}^{n-1})^m$ because

$$
\|\psi_r\|^2_{L_2(\mathbb{R}^{n-1})^m} \leq \int_{\mathbb{R}^{n-1}} \left|\mathcal{B}_\nu u_r(x', \zeta_r(x'))\right|^2 w_r(x')\, dx' = \|\mathcal{B}_\nu u_r\|^2_{L_2(\Gamma_r)^m},
$$

so by Theorem 2.31, after passing to a subsequence we can assume that ψ_r converges weakly to a function ψ in $L_2(\mathbb{R}^{n-1})^m$, i.e., $(\psi_r, v) \to (\psi, v)$ for each $v \in L_2(\mathbb{R}^{n-1})^m$. Define $\tilde{\psi}(x', \zeta(x')) = \psi(x')$, and let $v \in \mathcal{D}(\overline{\Omega})^m$, say. We have

$$
\begin{aligned}
(\mathcal{B}_\nu u_r, \gamma_r v)_{\Gamma_r} &= \int_{\mathbb{R}^{n-1}} \psi_r(x')^* v(x', \zeta_r(x')) w_r(x')\, dx' \\
&\to \int_{\mathbb{R}^{n-1}} \psi(x')^* v(x', \zeta(x')) w(x')\, dx' = (\tilde{\psi}, \gamma v)_\Gamma,
\end{aligned}
$$

and therefore, sending $r \to \infty$ in (4.43) gives

$$\Phi(u, v) = (f, v)_\Omega + (\tilde{\psi}, \gamma v)_\Gamma \quad \text{for } v \in \mathcal{D}(\overline{\Omega})^m.$$

Hence, $\mathcal{B}_\nu u = \tilde{\psi} \in L_2(\Gamma)^m$, and by Exercise 2.11, the estimate of part (i) follows from the uniform estimate (4.42) for $\mathcal{B}_\nu u_r$.

To prove (ii), assume $m = 1$, and suppose once again that $u \in H^2(\Omega)$ and that the h_l satisfy (4.41). This time, we write the integrand on the left-hand side of (4.40) as

$$\nu_l h_l \overline{A_{jk} \partial_k u} \partial_j u - \overline{\mathcal{B}_\nu u} h_j \partial_j u - h_k \overline{\partial_k u} \mathcal{B}_\nu u,$$

and hence obtain, using strong ellipticity,

$$\sum_{j=1}^n \|\gamma \partial_j u\|_{L_2(\Gamma)}^2 \leq C \int_\Gamma h_l \nu_l \gamma \{\overline{A_{jk} \partial_k u} \, \partial_j u\} \, d\sigma$$

$$\leq C \|\mathcal{B}_\nu u\|_{L_2(\Gamma)} \sum_{j=1}^n \|\gamma \partial_j u\|_{L_2(\Gamma)}$$

$$+ C \|u\|_{H^1(\Omega)}^2 + C \|\mathcal{P}_0 u\|_{L_2(\Omega)} \|u\|_{H^1(\Omega)}.$$

Thus,

$$\sum_{j=1}^n \|\gamma \partial_j u\|_{L_2(\Gamma)}^2 \leq C \|\mathcal{B}_\nu u\|_{L_2(\Gamma)}^2 + C \|u\|_{H^1(\Omega)}^2 + C \|f\|_{L_2(\Omega)}^2,$$

and the estimate of part (ii) follows.

Next, we allow $u \in H^1(\Omega)$ and assume $\mathcal{B}_\nu u \in L_2(\Gamma)$. Define $g_r \in L_2(\Gamma_r)$ by

$$g_r\big(x', \zeta_r(x')\big) = \mathcal{B}_\nu u\big(x', \zeta(x')\big) w(x') / w_r(x'),$$

and let $u_r \in H^1(\Omega_r)$ be the solution of the Neumann problem (with λ sufficiently large, as before)

$$(\mathcal{P} + \lambda) u_r = f + \lambda u \quad \text{on } \Omega_r,$$

$$\mathcal{B}_\nu u_r = g_r \quad \text{on } \Gamma_r.$$

We easily verify that $\|g_r\|_{L_2(\Gamma_r)} \leq C \|\mathcal{B}_\nu u\|_{L_2(\Gamma)}$, and by Theorem 4.7,

$$\Phi_r(v, v) + \lambda(v, v)_{\Omega_r} \geq c \|v\|_{H^1(\Omega_r)}^2 \quad \text{for } v \in H^1(\Omega_r),$$

with the constant independent of r in both cases. Thus, we have the uniform bounds

$$\|u_r\|_{H^1(\Omega_r)} \leq C\|f + \lambda u\|_{\tilde{H}^{-1}(\Omega_r)} + C\|g_r\|_{H^{-1/2}(\Gamma_r)}$$
$$\leq C\|\mathcal{B}_\nu u\|_{L_2(\Gamma)} + C\|u\|_{H^1(\Omega)} + C\|f\|_{L_2(\Omega)},$$

and, since $u_r \in H^2(\Omega_r)$ by Theorem 4.18, our earlier argument gives

$$\sum_{j=1}^{n} \|\gamma \partial_j u_r\|_{L_2(\Gamma_r)} \leq C\|g_r\|_{L_2(\Gamma_r)} + C\|u_r\|_{H^1(\Omega_r)} + C\|f\|_{L_2(\Omega_r)}$$

$$\leq C\|\mathcal{B}_\nu u\|_{L_2(\Gamma)} + C\|u_r\|_{H^1(\Omega_r)} + C\|f\|_{L_2(\Omega)}. \quad (4.45)$$

Using the method in the proof of Theorem A.4, we can extend u_r to a function in $H^1(\mathbb{R}^n)$ in such a way that $\|u_r\|_{H^1(\Omega \setminus \Omega_r)} \to 0$. To show that u_r converges to u in $H^1(\Omega)$, we observe that for $v \in H^1(\Omega)$,

$$\Phi(u_r - u, v) + \lambda(u_r - u, v)_\Omega = \left[\Phi_r(u_r, v) + \lambda(u_r, v)_{\Omega_r}\right]$$
$$- \left[\Phi(u, v) + \lambda(u, v)_\Omega\right]$$
$$+ \left[\Phi(u_r, v) - \Phi_r(u_r, v) + \lambda(u_r, v)_{\Omega \setminus \Omega_r}\right]$$
$$= (g_r, \gamma_r v)_{\Gamma_r} - (\mathcal{B}_\nu u, \gamma v)_\Gamma - (f + \lambda u, v)_{\Omega \setminus \Omega_r}$$
$$+ \Phi_{\Omega \setminus \Omega_r}(u_r, v) + \lambda(u_r, v)_{\Omega \setminus \Omega_r}.$$

Since

$$|(g_r, \gamma_r v)_{\Gamma_r} - (\mathcal{B}_\nu u, \gamma v)_\Gamma|^2$$
$$= \left| \int_{\mathbb{R}^{n-1}} \overline{\mathcal{B}_\nu u(x', \zeta(x'))} \left[v(x', \zeta_r(x')) - v(x', \zeta(x'))\right] w(x') \, dx' \right|^2$$
$$\leq C\|\mathcal{B}_\nu u\|_{L_2(\Gamma)}^2 \|\zeta_r - \zeta\|_{L_\infty(\mathbb{R}^{n-1})}^2 \|v\|_{H^1(\Omega \setminus \Omega_r)}^2,$$

we see by taking $v = u_r - u$ and again applying Theorem 4.7, this time over Ω instead of Ω_r, that

$$\|u_r - u\|_{H^1(\Omega)} \leq C\|\zeta_r - \zeta\|_{L_\infty(\mathbb{R}^{n-1})}^{1/2} \|\mathcal{B}_\nu u\|_{L_2(\Gamma)}$$
$$+ C\|f + \lambda u\|_{L_2(\Omega \setminus \Omega_r)} + C\|u_r\|_{H^1(\Omega \setminus \Omega_r)},$$

and so $u_r \to u$ in $H^1(\Omega)$. Define $\psi_r(x') = (\gamma_r u_r)(x', \zeta_r(x'))$ and $\psi(x') = (\gamma u)(x', \zeta(x'))$. By writing $\psi_r(x') - \psi(x') = \left[u_r(x', \zeta_r(x')) - u_r(x', \zeta(x'))\right] + (u_r - u)(x', \zeta(x'))$, and estimating the first term in a manner similar to (4.44), we obtain

$$|(\psi_r - \psi, v)| \leq \left[\|\zeta_r - \zeta\|_{L_\infty(\mathbb{R}^{n-1})}^{1/2} \|u_r\|_{H^1(\Omega \setminus \Omega_r)} + \|\gamma(u_r - u)\|_{L_2(\Gamma)}\right] \|v\|_{L_2(\mathbb{R}^{n-1})},$$

and therefore $\psi_r \to \psi$ in $L_2(\mathbb{R}^{n-1})$. Finally, by (4.45),

$$\|\psi_r\|_{H^1(\mathbb{R}^{n-1})} \le C\|u_r\|_{H^1(\Gamma_r)} \le C\|\mathcal{B}_\nu u\|_{L_2(\Gamma)} + C\|u\|_{H^1(\Omega)} + C\|f\|_{L_2(\Omega)},$$

so Exercise 2.11 shows that $\psi \in H^1(\mathbb{R}^{n-1})$ with $\|\psi\|_{H^1(\mathbb{R}^{n-1})} \le \limsup$ $\|\psi_r\|_{H^1(\mathbb{R}^{n-1})}$, implying that $u \in H^1(\Gamma)$, and that the estimate of part (ii) holds. $\qquad\square$

It is now a simple matter to obtain the desired mapping properties for the Steklov–Poincaré operators. We can also introduce a meaningful concept of a solution $u = \mathcal{U}g \in L_2(\Omega)^m$ for $g \in L_2(\Gamma)^m$ (but see also the sharper mapping property for $\mathcal{U}$ proved in Theorem 6.12).

Theorem 4.25 *Assume that Ω is a Lipschitz domain, that $\mathcal{P}$ is strongly elliptic on Ω, and that the coefficients A_{jk} and A_j are Lipschitz (not just L_∞). If*

$$A_{kj}^* = A_{jk},$$

and if the solution operators (4.37) exist, then the Steklov–Poincaré operators satisfy

$$\mathcal{B}_\nu \mathcal{U} : H^{s+1/2}(\Gamma)^m \to H^{s-1/2}(\Gamma)^m \quad \text{and} \quad \widetilde{\mathcal{B}}_\nu \mathcal{V} : H^{s+1/2}(\Gamma)^m \to H^{s-1/2}(\Gamma)^m$$

for $-\frac{1}{2} \le s \le \frac{1}{2}$. Furthermore, the solution operators have bounded extensions

$$\mathcal{U} : L_2(\Gamma)^m \to L_2(\Omega)^m \quad \text{and} \quad \mathcal{V} : L_2(\Gamma)^m \to L_2(\Omega)^m.$$

Proof. For $s = 0$, the first part of the theorem was established in (4.38). We obtain the result for $s = \frac{1}{2}$ by applying part (i) of Theorem 4.24 to (4.35) and (4.36). The case $s = -\frac{1}{2}$ then follows by duality, using (4.39). Finally, interpolation gives the complete range $-\frac{1}{2} \le s \le \frac{1}{2}$.

In the second part of the theorem, it suffices to consider $\mathcal{U}$. Let $g \in H^{1/2}(\Gamma)^m$ and $f \in L_2(\Omega)^m$, and put $u = \mathcal{U}g \in H^1(\Omega)^m$. Our assumptions imply the existence of a unique $w \in H^1(\Omega)^m$ such that

$$\mathcal{P}^* w = f \quad \text{on } \Omega,$$
$$\gamma w = 0 \quad \text{on } \Gamma.$$

The first Green identity gives, on the one hand,

$$\Phi(u, w) = (\mathcal{P}u, w)_\Omega + (\mathcal{B}_\nu u, \gamma w)_\Gamma = 0,$$

and on the other hand,

$$\Phi(u, w) = (u, \mathcal{P}^* w)_\Omega + (\gamma u, \widetilde{B}_\nu w)_\Gamma = (u, f)_\Omega + (g, \widetilde{B}_\nu w)_\Gamma,$$

so

$$|(u, f)_\Omega| = \left|(g, \widetilde{B}_\nu w)_\Gamma\right| \le \|g\|_{L_2(\Gamma)^m} \|\widetilde{B}_\nu w\|_{L_2(\Gamma)^m}.$$

By part (i) of Theorem 4.24,

$$\|\widetilde{B}_\nu w\|_{L_2(\Gamma)^m} \le C\|\gamma w\|_{H^1(\Gamma)^m} + C\|w\|_{H^1(\Omega)^m} + C\|f\|_{L_2(\Omega)^m} \le C\|f\|_{L_2(\Omega)^m},$$

and hence $|(u, f)_\Omega| \le C\|g\|_{L_2(\Gamma)^m} \|f\|_{L_2(\Omega)^m}$, implying that $\|\mathcal{U}g\|_{L_2(\Omega)^m} = \|u\|_{L_2(\Omega)^m} \le C\|g\|_{L_2(\Gamma)^m}$. $\qquad\square$

Exercises

4.1 Show that $\mathcal{P}u = -\partial_j(\tilde{A}_{jk}\partial_k u) + \tilde{A}_j \partial_j u + Au$, where $\tilde{A}_{jk} = \frac{1}{2}(A_{jk} + A_{kj}) = \tilde{A}_{kj}$ and $\tilde{A}_j = A_j + \frac{1}{2}\partial_k(A_{jk} - A_{kj})$.

4.2 Let $\kappa : \overline{\Omega^\kappa} \to \overline{\Omega}$ be a C^1 diffeomorphism. We denote the Jacobian of the coordinate transformation $x = \kappa(y)$ by

$$J(y) = |\det \kappa'(y)| = \left|\frac{\partial(x_1, \ldots, x_n)}{\partial(y_1, \ldots, y_n)}\right|,$$

and write $u^\kappa = u \circ \kappa$ and $v^\kappa = v \circ \kappa$, so that $u(x) = u^\kappa(y)$ and $v(x) = v^\kappa(y)$.

(i) Let Φ be the sesquilinear form (4.2). Show that $\Phi(u, v) = \Phi^\kappa(u^\kappa, v^\kappa)$, where Φ^κ is the sesquilinear form with coefficients

$$A^\kappa_{jk}(y) = \left(\frac{\partial y_j}{\partial x_r} A_{rs}(x) \frac{\partial y_k}{\partial x_s}\right) J(y),$$

$$A^\kappa_j(y) = \left(A_r(x)\frac{\partial y_j}{\partial x_r}\right) J(y), \qquad A^\kappa(y) = A(x)J(y).$$

(ii) Let $\mathcal{P}^\kappa$ denote the differential operator with the coefficients in part (i). Show that $\mathcal{P}u = f$ on Ω if and only if $\mathcal{P}^\kappa u^\kappa = f^\kappa$ on Ω^κ, where $f^\kappa(y) = f(x)J(y)$.

(iii) Show that $\mathcal{P}$ is strongly elliptic (respectively, coercive) on Ω if and only if $\mathcal{P}^\kappa$ is strongly elliptic (respectively, coercive) on Ω^κ.

4.3 Show that if $f(0) = 0$ and $\alpha > \frac{1}{2}$, then

$$\left(\int_0^\infty |t^{-\alpha} f(t)|^2 dt\right)^{1/2} \le \frac{1}{\alpha - \frac{1}{2}} \left(\int_0^\infty |t^{1-\alpha} f'(t)|^2 dt\right)^{1/2}.$$

[Hint: use Exercise 3.20.]

4.4 Show for all $s \in \mathbb{R}$ that if $\partial_l u \in H^s(\mathbb{R}^n)^m$, then the difference quotient (4.21) satisfies $\|\Delta_{l,h} u\|_{H^s(\mathbb{R}^n)^m} \leq \|\partial_l u\|_{H^s(\mathbb{R}^n)^m}$. [Hint: $|e^{i\theta} - 1| \leq |\theta|$ for $\theta \in \mathbb{R}$.]

4.5 Suppose that $u \in L_2(\mathbb{R}^n)$ and $u^{\pm} = u|_{\mathbb{R}^n_{\pm}} \in H^1(\mathbb{R}^n_{\pm})$. Show that

$$(\partial_j u, \phi)$$

$$= (\partial_j u^+, \phi)_{\mathbb{R}^n_+} + (\partial_j u^-, \phi)_{\mathbb{R}^n_-} + \begin{cases} 0 & \text{if } 1 \leq j \leq n-1, \\ ([u]_\Gamma, \gamma\phi)_\Gamma & \text{if } j = n, \end{cases}$$

where $\Gamma = \mathbb{R}^{n-1} \times \{0\}$. Deduce that $u \in H^1(\mathbb{R}^n)$ if and only if $[u]_\Gamma = 0$.

4.6 Show that in Corollary 3.22 we can choose the functions ϕ_j for the partition of unity in such a way that $\phi_j(x) = \psi_j(x)^2$, where $\psi_j \in C^\infty_{\text{comp}}(\mathbb{R}^n)$. [Hint: start by considering g in Exercise 3.6.]

4.7 Consider the classical *cooling-off problem* [19, p. 56]:

$$\frac{\partial u}{\partial t} - a\triangle u = 0 \quad \text{on } \Omega, \quad \text{for } t > 0,$$

$$\frac{\partial u}{\partial \nu} + b\gamma u = 0 \quad \text{on } \Gamma, \quad \text{for } t > 0,$$

$$u = u_0 \quad \text{on } \Omega, \quad \text{when } t = 0,$$

where a and b are positive constants. Let $\phi_1, \phi_2, \ldots$ and $\lambda_1, \lambda_2, \ldots$ be the eigenfunctions and eigenvalues of the stationary problem, as in Theorem 4.12, i.e.,

$$-a\triangle \phi_j = \lambda_j \phi_j \quad \text{on } \Omega,$$

$$\frac{\partial \phi_j}{\partial \nu} + b\gamma \phi_j = 0 \quad \text{on } \Gamma,$$

and derive the series representation

$$u(x, t) = \sum_{j=1}^{\infty} e^{-\lambda_j t} (\phi_j, u_0)_\Omega \phi_j(x) \quad \text{for } x \in \Omega \text{ and } t > 0.$$

In what sense does this sum converge?

5
Homogeneous Distributions

For $a \in \mathbb{C}$, a function $u : \mathbb{R}^n \setminus \{0\} \to \mathbb{C}$ is said to be *homogeneous of degree a* if

$$u(tx) = t^a u(x) \qquad \text{for all } t > 0 \text{ and } x \in \mathbb{R}^n \setminus \{0\}. \tag{5.1}$$

To extend this concept to distributions, we introduce the linear operator M_t, defined by

$$M_t u(x) = u(tx) \qquad \text{for } 0 \ne t \in \mathbb{R} \text{ and } x \in \mathbb{R}^n,$$

and observe that for every $u \in L_{1,\text{loc}}(\mathbb{R}^n)$,

$$\langle M_t u, \phi \rangle = |t|^{-n} \langle u, M_{1/t} \phi \rangle \qquad \text{for } t \ne 0 \text{ and } \phi \in \mathcal{D}(\mathbb{R}^n). \tag{5.2}$$

If $u \in \mathcal{D}^*(\mathbb{R}^n)$ then (5.2) serves to *define* $M_t u \in \mathcal{D}^*(\mathbb{R}^n)$, because $M_{1/t} : \mathcal{D}(\mathbb{R}^n) \to \mathcal{D}(\mathbb{R}^n)$ is continuous and linear. We then say that $u \in \mathcal{D}^*(\mathbb{R}^n)$ is homogeneous of degree a on $\mathbb{R}^n$ if $M_t u = t^a u$ on $\mathbb{R}^n$ in the sense of distributions, for all $t > 0$.

This chapter develops the theory of homogeneous distributions, using Hadamard's notion of a finite-part integral to extend homogeneous functions on $\mathbb{R}^n \setminus \{0\}$ to distributions on $\mathbb{R}^n$. We consider in some detail the Fourier transform of, and the change of variables formula for, such finite-part extensions. A technique used several times is to reduce the general n-dimensional case to a one-dimensional problem by transforming to polar coordinates. Most of the material that follows can be found in standard texts such as Schwartz [92], Gel'fand and Shilov [27], and Hörmander [41], but the final two sections – dealing with finite-part integrals on surfaces – include some less well-known results from the thesis of Kieser [48].

The results of this chapter will be applied later in our study of fundamental solutions of elliptic partial differential operators, and of boundary integral operators with non-integrable kernels.

Finite-Part Integrals

We begin our study of homogeneous distributions by focusing on the simplest example, namely, the one-dimensional, homogeneous function

$$x_+^a = \begin{cases} x^a & \text{if } x > 0, \\ 0 & \text{if } x < 0. \end{cases}$$

If $\operatorname{Re} a > -1$, then x_+^a is locally integrable on $\mathbb{R}$, and is obviously homogeneous of degree a as a distribution on $\mathbb{R}$, not just as a function on $\mathbb{R} \setminus \{0\}$. To deal with the interesting case $\operatorname{Re} a \leq -1$, we use the following concept, introduced by Hadamard [37] in the context of Cauchy's problem for hyperbolic equations.

Definition 5.1 Let $a_1, \ldots, a_N$ and $b_1, \ldots, b_{N+2}$ be complex numbers, with

$$\operatorname{Re} a_j \geq 0 \quad \text{and} \quad a_j \neq 0 \quad \text{for } 1 \leq j \leq N, \tag{5.3}$$

and

$$a_j \neq a_k \quad \text{whenever } j \neq k. \tag{5.4}$$

If a function g satisfies

$$g(\epsilon) = \sum_{j=1}^{N} \frac{b_j}{\epsilon^{a_j}} + b_{N+1} \log \epsilon + b_{N+2} + o(1) \quad \text{as } \epsilon \downarrow 0,$$

then the term b_{N+2} is called the finite part of $g(\epsilon)$ as $\epsilon \downarrow 0$, and we write

$$\operatorname*{f.p.}_{\epsilon \downarrow 0} g(\epsilon) = b_{N+2}.$$

When no singular terms are present, i.e., when $b_1 = \cdots = b_{N+1} = 0$, the finite part is just the limit of $g(\epsilon)$ as $\epsilon \downarrow 0$. The next lemma shows that, when it exists, the finite part is unique.

Lemma 5.2 Let $a_1, \ldots, a_N$ and $b_1, \ldots, b_{N+2}$ be complex numbers, and assume that the a_j satisfy (5.3) and (5.4). If

$$\lim_{\epsilon \downarrow 0} \left(\sum_{j=1}^{N} \frac{b_j}{\epsilon^{a_j}} + b_{N+1} \log \epsilon + b_{N+2} \right) = 0,$$

then $b_j = 0$ for $1 \leq j \leq N + 2$.

Proof. Let $\mu = \max_{1 \leq j \leq N} \operatorname{Re} a_j$, and suppose to begin with that $\mu = 0$. In this case, (5.3) implies that all of the a_j are purely imaginary, say $a_j = i\lambda_j$

with $\lambda_j \in \mathbb{R} \setminus \{0\}$, so $|\epsilon^{-a_j}| = |\exp(-i\lambda_j \log \epsilon)| = 1$ for $1 \le j \le N$, and we see at once that $b_{N+1} = 0$. Moreover, there exist a sequence $(\epsilon_m)_{m=0}^\infty$ and real numbers $\theta_1, \ldots, \theta_N$ such that $\epsilon_m \downarrow 0$ and $\lim_{m\to\infty} \epsilon_m^{a_j} = e^{i\theta_j}$ for $1 \le j \le N$. Thus, putting $\epsilon = \epsilon_m e^x$ and sending $m \to \infty$, we see that

$$\sum_{j=1}^N b_j e^{-i(\lambda_j x + \theta_j)} + b_{N+2} = 0 \quad \text{for } x \in \mathbb{R}. \tag{5.5}$$

By (5.4), we may assume without loss of generality that $\lambda_1 < \lambda_2 < \cdots < \lambda_N$, and by analytic continuation we may replace x with ix in (5.5) to obtain

$$\sum_{j=1}^N b_j e^{\lambda_j x - i\theta_j} + b_{N+2} = 0 \quad \text{for } x \in \mathbb{R}.$$

If $\lambda_N > 0$, then we divide through by $e^{\lambda_N x}$ and send $x \to \infty$, to conclude that $b_N = 0$. Otherwise, λ_1 must be negative, so we divide through by $e^{\lambda_1 x}$ and send $x \to -\infty$ to conclude that $b_1 = 0$. Repeating this argument, one sees that $b_j = 0$ for $1 \le j \le N$, and then (5.5) reduces to $b_{N+2} = 0$.

Now suppose $\mu > 0$. Since

$$\lim_{\epsilon \downarrow 0} \epsilon^\mu \left(\sum_{j=1}^N \frac{b_j}{\epsilon^{a_j}} + b_{N+1} \log \epsilon + b_{N+2} \right) = \lim_{\epsilon \downarrow 0} \sum_{\operatorname{Re} a_j = \mu} \frac{b_j}{\epsilon^{a_j - \mu}},$$

it follows as above that $b_j = 0$ if $\operatorname{Re} a_j = \mu$, and repetition of this argument shows that $b_j = 0$ if $\operatorname{Re} a_j > 0$, which gets us back to the case when $\mu = 0$. $\square$

For $\epsilon > 0$ and $\phi \in \mathcal{S}(\mathbb{R})$, we define the integral

$$H_{a,\epsilon}(\phi) = \int_\epsilon^\infty x^a \phi(x)\, dx$$

and its finite part,

$$H_a(\phi) = \operatorname*{f.p.}_{\epsilon \downarrow 0} H_{a,\epsilon}(\phi).$$

The following lemma establishes the existence of $H_a(\phi)$, and shows that we can define a temperate distribution f.p. $x_+^a \in \mathcal{S}^*(\mathbb{R})$, called the *finite-part extension* of x_+^a, by putting

$$\langle \text{f.p.}\, x_+^a, \phi(x) \rangle = H_a(\phi) \quad \text{for } \phi \in \mathcal{S}(\mathbb{R}).$$

Obviously, if x is restricted to $\mathbb{R} \setminus \{0\}$, then f.p. x_+^a coincides with x_+^a.

Lemma 5.3 For any integer $k \geq 1$ and any $\phi \in \mathcal{S}(\mathbb{R})$,

$$H_a(\phi) = \frac{(-1)^k H_{a+k}(\phi^{(k)})}{(a+1)(a+2)\cdots(a+k)} \quad \text{if Re } a > -k-1 \quad \text{and}$$
$$a \neq -1, -2, \ldots, -k,$$

but

$$H_{-1}(\phi) = -\int_0^\infty \phi'(x) \log x \, dx$$

and

$$H_{-k-1}(\phi) = \left(\frac{1}{k!}\sum_{j=1}^k \frac{1}{j}\right)\phi^{(k)}(0) + \frac{1}{k!}H_{-1}(\phi^{(k)}).$$

Proof. Integration by parts gives

$$H_{a,\epsilon}(\phi) = -\frac{\epsilon^{a+1}\phi(\epsilon)}{a+1} - \frac{H_{a+1,\epsilon}(\phi')}{a+1} \quad \text{if } a \neq -1.$$

Suppose first that Re $a > -k-1$ and $a \neq -1, -2, \ldots, -k$. By Taylor expansion,

$$\epsilon^{a+1}\phi(\epsilon) = \sum_{j=0}^{k-1} \frac{\phi^{(j)}(0)}{j!}\epsilon^{j+a+1} + O(\epsilon^{\text{Re } a+k+1}),$$

so if $H_{a+1}(\phi')$ exists, then

$$H_a(\phi) = \frac{-H_{a+1}(\phi')}{a+1},$$

and the first part of the lemma follows by induction.

Next, integration by parts and Taylor expansion give

$$H_{-1,\epsilon}(\phi) = -\phi(\epsilon)\log\epsilon - \int_\epsilon^\infty \phi'(x)\log x \, dx$$
$$= -\phi(0)\log\epsilon - \int_0^\infty \phi'(x)\log x \, dx + O(\epsilon\log\epsilon),$$

implying the formula for $H_{-1}(\phi)$. Finally, when $a = -k-1$,

$$H_{-k-1,\epsilon}(\phi) = -\frac{\epsilon^{-k}\phi(\epsilon)}{k} + \frac{H_{-k,\epsilon}(\phi')}{k},$$

and by Taylor expansion,

$$\epsilon^{-k}\phi(\epsilon) = \sum_{j=1}^{k-1} \frac{\phi^{(j)}(0)}{j!}\epsilon^{j-k} + \frac{\phi^{(k)}(0)}{k!} + O(\epsilon),$$

so if $H_{-k}(\phi')$ exists then

$$H_{-k-1}(\phi) = \frac{\phi^{(k)}(0)}{k \times k!} + \frac{1}{k}H_{-k}(\phi').$$

The formula for $H_{-k-1}(\phi)$ follows by induction. $\qquad\square$

The next lemma shows that the distribution f.p. x_+^a is homogeneous of degree a on $\mathbb{R}$, *except* when a is a negative integer.

Lemma 5.4 *If* $\phi \in \mathcal{S}(\mathbb{R})$ *and* $t > 0$, *then*

$$t^{-1}H_a\big(M_{1/t}\phi\big) = t^a H_a(\phi) \quad \text{for } a \neq -1, -2, -3, \ldots,$$

but

$$t^{-1}H_{-k-1}\big(M_{1/t}\phi\big) = t^{-k-1}H_{-k-1}(\phi) + (t^{-k-1}\log t)\frac{\phi^{(k)}(0)}{k!}$$

$$\text{for any integer } k \geq 0.$$

Proof. If $\operatorname{Re} a > -1$, then it suffices to make the substitution $x = ty$ to obtain

$$t^{-1}H_a\big(M_{1/t}\phi\big) = t^{-1}\int_0^\infty x^a \phi(x/t)\,dx = t^a \int_0^\infty y^a \phi(y)\,dy = t^a H_a(\phi);$$

cf. (5.2). Now suppose that $\operatorname{Re} a > -k-1$ and $a \neq -1, -2, \ldots, -k$. For brevity, write $b_k = (-1)^k/[(a+1)\cdots(a+k)]$; then by Lemma 5.3,

$$t^{-1}H_a\big(M_{1/t}\phi\big) = t^{-1}b_k H_{a+k}\big((M_{1/t}\phi)^{(k)}\big) = t^{-1}b_k H_{a+k}\big(t^{-k}M_{1/t}\phi^{(k)}\big)$$
$$= b_k t^{-k-1}H_{a+k}\big(M_{1/t}\phi^{(k)}\big) = b_k t^a H_{a+k}\big(\phi^{(k)}\big) = t^a H_a(\phi).$$

However, when $a = -1$, we have

$$t^{-1}H_{-1}\big(M_{1/t}\phi\big) = -t^{-1}\int_0^\infty \big(M_{1/t}\phi\big)'(x)\log x\,dx$$

$$= -t^{-2}\int_0^\infty \phi'(t^{-1}x)\log x\,dx,$$

and the substitution $x = ty$ gives

$$t^{-1} H_{-1}(M_{1/t}\phi) = -t^{-1} \int_0^\infty \phi'(y) \log(ty)\, dy = t^{-1} H_{-1}(\phi) + \phi(0)t^{-1} \log t.$$

In general, if we let $c_k = (1/k!) \sum_{j=1}^k 1/j$, then

$$
\begin{aligned}
t^{-1} H_{-k-1}(M_{1/t}\phi) &= t^{-1} c_k (M_{1/t}\phi)^{(k)}(0) + \frac{t^{-1}}{k!} H_{-1}\big((M_{1/t}\phi)^{(k)}\big) \\
&= c_k t^{-k-1} \phi^{(k)}(0) + \frac{t^{-k-1}}{k!} H_{-1}(M_{1/t}\phi^{(k)}) \\
&= t^{-k-1}\left(c_k \phi^{(k)}(0) + \frac{1}{k!} H_{-1}(\phi^{(k)})\right) + \frac{\phi^{(k)}(0)}{k!} t^{-k-1} \log t,
\end{aligned}
$$

giving the second formula in the lemma. $\qquad\square$

We shall also have use for the homogeneous function

$$x_-^a = (-x)_+^a = \begin{cases} 0 & \text{if } x > 0, \\ |x|^a & \text{if } x < 0, \end{cases}$$

and its finite-part extension,

$$
\begin{aligned}
\langle \text{f.p.}\, x_-^a, \phi(x) \rangle &= \text{f.p.} \int_{-\infty}^{-\epsilon} |x|^a \phi(x)\, dx \\
&= \text{f.p.} \int_\epsilon^\infty x^a \phi(-x)\, dx = H_a(M_{-1}\phi).
\end{aligned}
$$

It is easy to verify that

$$\text{f.p.}(-x)_\pm^a = \text{f.p.}\, x_\mp^a, \tag{5.6}$$

and that f.p. x_-^a is homogeneous of degree a on $\mathbb{R}$, except when a is a negative integer. Indeed, Lemma 5.4 shows that if $t > 0$, then

$$\text{f.p.}(tx)_\pm^a = t^a\, \text{f.p.}\, x_\pm^a \quad \text{for } a \neq -1, -2, -3, \ldots,$$

but

$$\text{f.p.}(tx)_\pm^{-k-1} = t^{-k-1}\, \text{f.p.}\, x_\pm^{-k-1} + (\mp 1)^k \frac{t^{-k-1} \log t}{k!} \delta^{(k)}(x)$$

$$\text{for any integer } k \geq 0. \tag{5.7}$$

This loss of homogeneity does not occur in the case of the function x^{-k-1}.

Indeed, the finite-part extension

$$\langle \text{f.p.}\, x^{-k-1}, \phi(x) \rangle = \underset{\epsilon \downarrow 0}{\text{f.p.}} \int_{|x| > \epsilon} x^{-k-1} \phi(x)\, dx$$

$$= H_{-k-1}(\phi) + (-1)^{k+1} H_{-k-1}(M_{-1}\phi),$$

satisfies

$$\text{f.p.}\, x^{-k-1} = \text{f.p.}\, x_+^{-k-1} + (-1)^{k+1}\, \text{f.p.}\, x_-^{-k-1} \tag{5.8}$$

and so, by (5.7) and (5.6),

$$\text{f.p.}\,(tx)^{-k-1} = t^{-k-1}\, \text{f.p.}\, x^{-k-1} \quad \text{if } 0 \neq t \in \mathbb{R}. \tag{5.9}$$

One can formally integrate by parts $k + 1$ times to express $\langle \text{f.p.}\, x^{-k-1}, \phi(x) \rangle$ as a convergent integral.

Lemma 5.5 For $\phi \in S(\mathbb{R})$ and for any integer $k \geq 0$,

$$\langle \text{f.p.}\, x^{-k-1}, \phi(x) \rangle = \frac{-1}{k!} \int_{-\infty}^{\infty} \phi^{(k+1)}(x) \log |x|\, dx.$$

Proof. By (5.8) and Lemma 5.3,

$$\langle \text{f.p.}\, x^{-k-1}, \phi(x) \rangle = H_{-k-1}(\phi) + (-1)^{k+1} H_{-k-1}(M_{-1}\phi)$$

$$= \left(\frac{1}{k!} \sum_{j=1}^{k} \frac{1}{j} \right) \left[\phi^{(k)}(0) + (-1)^{k+1}(M_{-1}\phi)^{(k)}(0) \right]$$

$$+ \frac{1}{k!} H_{-1}\big(\phi^{(k)} + (-1)^{k+1}(M_{-1}\phi)^{(k)} \big),$$

and thus, because $(M_{-1}\phi)^{(k)} = (-1)^k M_{-1}\phi^{(k)}$,

$$\langle \text{f.p.}\, x^{-k-1}, \phi(x) \rangle = \frac{1}{k!} H_{-1}\big(\phi^{(k)} - M_{-1}\phi^{(k)} \big)$$

$$= \frac{-1}{k!} \int_0^{\infty} \big[\phi^{(k+1)}(x) + \phi^{(k+1)}(-x) \big] \log x\, dx,$$

giving the desired formula. □

Later, when studying the Fourier transforms of f.p. $x_\pm^a$ and f.p. x^{-k-1}, we shall encounter the distribution $(x \pm i0)^a$, defined by

$$\langle (x \pm i0)^a, \phi(x) \rangle = \lim_{y \downarrow 0} \int_{-\infty}^{\infty} (x \pm iy)^a \phi(x)\, dx, \tag{5.10}$$

with the branch of $z^a = \exp(a \log z)$ chosen so that $-\pi < \arg z \leq \pi$.

Lemma 5.6 The formula (5.10) defines a temperate distribution $(x \pm i0)^a \in$ $\mathcal{S}^*(\mathbb{R})$, *given by*

$$(x \pm i0)^a = \text{f.p.} \, x_+^a + e^{\pm i\pi a} \, \text{f.p.} \, x_-^a \quad \text{for } a \neq -1, -2, -3, \ldots,$$

and

$$(x \pm i0)^{-k-1} = \text{f.p.} \, x^{-k-1} \pm i\pi \frac{(-1)^{k+1}}{k!} \delta^{(k)}(x) \quad \text{for any integer } k \geq 0.$$

Proof. If $\text{Re} \, a > -1$, then we may apply the dominated convergence theorem to obtain the first formula. If $-k - 1 < \text{Re} \, a < -k$ for some integer $k \geq 0$, so that $\text{Re}(a + k) > -1$, then integration by parts gives

$$\int_{-\infty}^{\infty} (x \pm iy)^a \phi(x) \, dx = \int_{-\infty}^{\infty} \frac{(-1)^k (x \pm iy)^{a+k}}{(a+1)(a+2)\cdots(a+k)} \phi^{(k)}(x) \, dx,$$

so in the limit as $y \downarrow 0$, we have

$$\langle (x \pm i0)^a, \phi(x) \rangle = \left\langle \frac{(-1)^k \big(x_+^{a+k} + e^{\pm i\pi(a+k)} x_-^{a+k}\big)}{(a+1)(a+2)\cdots(a+k)}, \phi^{(k)} \right\rangle$$

$$= \left\langle \frac{d^k}{dx^k} \frac{x_+^{a+k} + (-1)^k e^{\pm i\pi a} x_-^{a+k}}{(a+1)(a+2)\cdots(a+k)}, \phi(x) \right\rangle,$$

and we have only to apply Exercise 5.3. However, when $a = -k - 1$,

$$\int_{-\infty}^{\infty} (x \pm iy)^{-k-1} \phi(x) \, dx = \frac{-1}{k!} \int_{-\infty}^{\infty} \log(x \pm iy) \, \phi^{(k+1)}(x) \, dx,$$

and since $-\pi < \arg(x \pm iy) \leq \pi$,

$$\lim_{y \downarrow 0} \log(x \pm iy) = \begin{cases} \log x & \text{if } x > 0, \\ \log|x| \pm i\pi & \text{if } x < 0. \end{cases}$$

Thus, by Lemma 5.5,

$$\langle (x \pm i0)^{-k-1}, \phi(x) \rangle = \frac{-1}{k!} \int_{-\infty}^{\infty} \log|x| \, \phi^{(k+1)}(x) \, dx$$

$$- \frac{1}{k!} \int_{-\infty}^{0} (\pm i\pi) \, \phi^{(k+1)}(x) \, dx$$

$$= \langle \text{f.p.} \, x^{-k-1}, \phi(x) \rangle - \frac{1}{k!} (\pm i\pi) \, \phi^{(k)}(0),$$

which yields the desired expression for $(x \pm i0)^{-k-1}$. $\qquad\square$

Extension from $\mathbb{R}^n \setminus \{0\}$ to $\mathbb{R}^n$

If $u \in L_{1,\mathrm{loc}}(\mathbb{R}^n \setminus \{0\})$, then we can try to define the finite-part extension f.p. u as a distribution on $\mathbb{R}^n$ by writing

$$\langle \mathrm{f.p.}\, u, \phi \rangle = \mathrm{f.p.}_{\epsilon \downarrow 0} \int_{|x| > \epsilon} u(x)\phi(x)\, dx \quad \text{for } \phi \in \mathcal{D}(\mathbb{R}^n).$$

In the special case when the finite part of the integral is just a limit, we speak of the *principal value* p.v. u, i.e.,

$$\langle \mathrm{p.v.}\, u, \phi \rangle = \lim_{\epsilon \downarrow 0} \int_{|x| > \epsilon} u(x)\phi(x)\, dx.$$

Suppose now that $u \in C^\infty(\mathbb{R}^n \setminus \{0\})$ is homogeneous of degree a. By introducing polar coordinates $\rho = |x|$ and $\omega = x/\rho$, so that $x = \rho\omega$ and $dx = \rho^{n-1}\, d\rho\, d\omega$, we find that

$$\int_{|x| > \epsilon} u(x)\phi(x)\, dx = \int_{|\omega| = 1} u(\omega) \int_{\rho > \epsilon} \rho^{a+n-1}\phi(\rho\omega)\, d\rho\, d\omega.$$

This formula prompts us to define the linear operator R_a by

$$R_a\phi(x) = \mathrm{f.p.}_{\epsilon \downarrow 0} \int_\epsilon^\infty \rho^{a+n-1}\phi(\rho x)\, d\rho = \langle \mathrm{f.p.}\, \rho_+^{a+n-1}, \phi(\rho x) \rangle$$

$$\text{for } x \in \mathbb{R}^n \setminus \{0\}, \quad (5.11)$$

so that

$$\langle \mathrm{f.p.}\, u, \phi \rangle = \int_{|\omega| = 1} u(\omega) R_a\phi(\omega)\, d\omega \quad \text{for } \phi \in \mathcal{S}(\mathbb{R}^n). \quad (5.12)$$

Here, to justify taking the finite part inside the integral with respect to ω, it suffices to check that the $o(1)$ term in the expansion of $\int_\epsilon^\infty \rho^{a-n+1}\phi(\rho\omega)\, d\rho$ tends to zero *uniformly* for $|\omega| = 1$. As a consequence of Lemma 5.4, we are able to prove the following.

Theorem 5.7 *Suppose that $u \in C^\infty(\mathbb{R}^n \setminus 0)$ is a homogeneous function of degree a.*

(i) *If $a \neq -n, -n-1, -n-2, \ldots$, then f.p. u is the unique extension of u to a homogeneous distribution on $\mathbb{R}^n$.*

(ii) *If $a = -n - k$ for some integer $k \geq 0$, then a homogeneous extension* *exists if and only if u satisfies the orthogonality condition*

$$\int_{|\omega|=1} \omega^\alpha u(\omega) \, d\omega = 0 \quad \text{whenever } |\alpha| = k. \tag{5.13}$$

In this case, the homogeneous extensions of u consist of all distributions of the form

$$\text{f.p. } u + \sum_{|\alpha|=k} c_\alpha \partial^\alpha \delta,$$

with arbitrary coefficients $c_\alpha \in \mathbb{C}$.

Proof. Consider

$$\langle M_t \text{ f.p. } u, \phi \rangle = t^{-n} \langle \text{f.p. } u, M_{1/t}\phi \rangle = \int_{|\omega|=1} u(\omega) t^{-n} R_a M_{1/t}\phi(\omega) \, d\omega.$$

Since $R_a M_{1/t}\phi(\omega) = R_a\phi(t^{-1}\omega)$, we see from Exercise 5.4 that for a as in part (i),

$$\langle M_t \text{ f.p. } u, \phi \rangle = \int_{|\omega|=1} u(\omega) t^{-n} (t^{-1})^{-a-n} R_a\phi(\omega) \, d\omega = \langle t^a \text{ f.p. } u, \phi \rangle.$$

However, if $a = -n - k$, then

$$\langle M_t \text{ f.p. } u, \phi \rangle$$

$$= \int_{|\omega|=1} u(\omega) t^{-n} \left((t^{-1})^k R_{-n-k}\phi(\omega) - (t^{-1})^k \log(t^{-1}) \sum_{|\alpha|=k} \frac{\partial^\alpha \phi(0)}{\alpha!} \omega^\alpha \right) d\omega$$

$$= \langle t^{-n-k} \text{ f.p. } u, \phi \rangle + t^{-n-k} \log t \sum_{|\alpha|=k} \frac{\partial^\alpha \phi(0)}{\alpha!} \int_{|\omega|=1} \omega^\alpha u(\omega) \, d\omega,$$

so f.p. u is homogeneous on $\mathbb{R}^n$ if and only if (5.13) holds.

To settle the question of uniqueness, and to show the necessity of the orthogonality condition for existence if $a = -n - k$, let $\dot{u} \in \mathcal{D}^*(\mathbb{R}^n)$ be any extension of u. Since $\dot{u} - \text{f.p. } u = 0$ on $\mathbb{R}^n \setminus 0$, Theorem 3.9 implies that

$$\dot{u} - \text{f.p. } u = \sum_\alpha c_\alpha \partial^\alpha \delta,$$

where the sum is finite. The result follows because $\partial^\alpha \delta$ is homogeneous of degree $-n - |\alpha|$; see Exercises 5.1 and 5.2. □

The next theorem complements the one above, and introduces a particularly important class of homogeneous functions satisfying the orthogonality condition (5.13).

Theorem 5.8 Suppose that $u \in C^{\infty}(\mathbb{R}^n \setminus \{0\})$ is a homogeneous function of degree $-n - k$, for some integer $k \geq 0$. If u has parity opposite to k, i.e., if

$$u(-x) = (-1)^{k+1} u(x) \quad for\ x \in \mathbb{R}^n \setminus \{0\}, \tag{5.14}$$

then

$$\langle \mathrm{f.p.}\, u, \phi \rangle = \frac{1}{2} \int_{|\omega|=1} u(\omega) \langle \mathrm{f.p.}\, \rho^{-k-1}, \phi(\rho\omega) \rangle\, d\omega \quad for\ \phi \in \mathcal{S}(\mathbb{R}^n),$$

and f.p. u *is the unique extension of u to a homogeneous distribution with parity opposite to k.*

Proof. The parity condition (5.14) implies that

$$\langle \mathrm{f.p.}\, u, \phi \rangle = \int_{|\omega|=1} u(-\omega) R_{-n-k}\phi(-\omega)\, d\omega$$
$$= \int_{|\omega|=1} (-1)^{k+1} u(\omega) R_{-n-k}\phi(-\omega)\, d\omega,$$

and by (5.6),

$$R_{-n-k}\phi(-\omega) = \langle \mathrm{f.p.}\, \rho_+^{-k-1}, \phi(-\rho\omega) \rangle = \langle \mathrm{f.p.}\, \rho_-^{-k-1}, \phi(\rho\omega) \rangle,$$

so we have

$$\langle \mathrm{f.p.}\, u, \phi \rangle = \frac{1}{2} \int_{|\omega|=1} u(\omega) \left[R_{-n-k}\phi(\omega) + (-1)^{k+1} R_{-n-k}(-\omega) \right] d\omega$$
$$= \frac{1}{2} \int_{|\omega|=1} u(\omega) \langle \mathrm{f.p.}\, \rho_+^{-k-1} + (-1)^{k+1} \mathrm{f.p.}\, \rho_-^{-k-1}, \phi \rangle\, d\omega,$$

giving the desired formula; recall (5.8). The homogeneity of f.p. u follows from Theorem 5.7 because the parity condition (5.14) implies the orthogonality condition (5.13). Alternatively, one sees from (5.9) that

$$M_t\, \mathrm{f.p.}\, u = |t|^{-n} t^{-k}\, \mathrm{f.p.}\, u \quad \text{on } \mathbb{R}^n \text{ for } t \in \mathbb{R} \setminus \{0\},$$

and in particular, f.p. u has parity opposite to k. Finally, if $|\alpha| = k$ then $\partial^\alpha \delta$ has the same parity as k, so f.p. u is the *only* homogeneous extension of u having the opposite parity to k. $\qquad \square$

The uniqueness results in Theorems 5.7 and 5.8 yield a simple proof of the following fact.

Theorem 5.9 *Suppose that $u \in C^\infty(\mathbb{R}^n \setminus \{0\})$ is a homogeneous function of degree a, and assume, if $a = -n - k$ for some integer $k \geq 0$, that u has parity opposite to k. Then for any multi-index α,*

$$\partial^\alpha \text{ f.p. } u = \text{f.p.}(\partial^\alpha u) \quad \text{on } \mathbb{R}^n.$$

Proof. By Exercise 5.2, if $a \neq -1, -2, -3, \ldots$, then ∂^α f.p. u and f.p.$(\partial^\alpha u)$ are homogeneous extensions of $\partial^\alpha u$ with degree $a - |\alpha|$, and must therefore coincide. If $a = -n - k$ but u has parity opposite to k, then

$$\partial^\alpha u(-x) = (-1)^{k+|\alpha|+1} \partial^\alpha u(x) \quad \text{for } 0 \neq x \in \mathbb{R}^n,$$

so $\partial^\alpha u$ is homogeneous of degree $-n - (k + |\alpha|)$ and has parity opposite to $k + |\alpha|$. Thus, ∂^α f.p. u and f.p.$(\partial^\alpha u)$ are again homogeneous extensions of $\partial^\alpha u$, and both have parity opposite to k, so they must coincide. $\square$

Fourier Transforms

Our aim in this section is to compute the Fourier transform of the finite-part extension of a homogeneous function. Following the pattern of previous sections, the one-dimensional distributions f.p. $x_\pm^a$ and f.p. x^{-k-1} will be treated first. In order to state the next lemma, we require the gamma function,

$$\Gamma(a) = \int_0^\infty x^{a-1} e^{-x} \, dx \quad \text{for Re } a > 0.$$

In the usual way, Γ is extended by means of the identity

$$\Gamma(a + 1) = a\Gamma(a)$$

to a meromorphic function on $\mathbb{C}$ having simple poles at $0, -1, -2, \ldots$, and satisfying $\Gamma(k + 1) = k!$ for any integer $k \geq 0$.

Lemma 5.10 *Let $\Pi_a^\pm(\xi) = \mathcal{F}_{x \to \xi}\{\text{f.p. } x_\pm^a\}$. If $a \neq -1, -2, -3, \ldots$, then*

$$\Pi_a^\pm(\xi) = \frac{\Gamma(a + 1)}{(\pm i2\pi)^{a+1}} (\xi \mp i0)^{-a-1},$$

but for any integer $k \geq 0$,

$$\Pi_{-k-1}^\pm(\xi) = -\frac{(\mp i2\pi\xi)^k}{k!} \left(\log 2\pi|\xi| \pm \frac{i\pi}{2} \text{ sign}(\xi) - \Gamma'(1) - \sum_{j=1}^k \frac{1}{j} \right).$$

(When $k = 0$, the empty sum over j is interpreted as zero.)

Proof. To begin with, suppose that $\operatorname{Re} a > -1$. For any $\eta > 0$, the function $e^{-2\pi\eta|x|}x_\pm^a$ belongs to $L_1(\mathbb{R})$, and

$$\mathcal{F}_{x\to\xi}\left\{e^{-2\pi\eta|x|}x_\pm^a\right\} = \int_0^\infty e^{-2\pi(\eta\pm i\xi)x}x^a\,dx.$$

Making the substitution $z = 2\pi(\eta \pm i\xi)x$, and then applying Cauchy's theorem to shift the contour of integration back to the positive real axis, we find that

$$\mathcal{F}_{x\to\xi}\left\{e^{-2\pi\eta|x|}x_\pm^a\right\} = \left[2\pi(\eta \pm i\xi)\right]^{-a-1}\int_0^\infty e^{-z}z^a\,dz$$

$$= \frac{1}{(\pm i2\pi)^{a+1}}\Gamma(a+1)(\xi \mp i\eta)^{-a-1}.$$

Sending $\eta \downarrow 0$, we obtain the first part of the lemma for $\operatorname{Re} a > -1$, and the case $\operatorname{Re} a < -1$ then follows by analytic continuation; see Exercise 5.6.

For the second part of the lemma, consider

$$H_{-k-1,\epsilon}(e^{-i2\pi\xi\cdot}) = \int_\epsilon^\infty e^{-i2\pi\xi x}x^{-k-1}\,dx,$$

where $\int_\epsilon^\infty$ is interpreted as $\lim_{r\to\infty}\int_\epsilon^r$ if $k = 0$. Suppose $\xi \in \mathbb{R}^\pm$, and make the substitution $z = i2\pi\xi x = \pm i2\pi|\xi|x$ to obtain

$$H_{-k-1,\epsilon}(e^{-i2\pi\xi\cdot}) = (i2\pi\xi)^k\int_{\pm i2\pi|\xi|\epsilon}^{\pm i\infty} e^{-z}z^{-k-1}\,dz.$$

Applying Cauchy's theorem (and Jordan's lemma, if $k = 0$), we see that

$$\int_{\pm i2\pi|\xi|\epsilon}^{\pm i\infty} e^{-z}z^{-k-1}\,dz = \int_{2\pi|\xi|\epsilon}^\infty e^{-z}z^{-k-1}\,dz - \int_{C_\epsilon^\pm} e^{-z}z^{-k-1}\,dz,$$

where $C_\epsilon^\pm = \{2\pi|\xi|\epsilon e^{\pm i\omega} : 0 < \omega < \pi/2\}$. Since

$$\int_{C_\epsilon^\pm} e^{-z}z^{-k-1}\,dz = \pm\frac{1}{4}\int_{|z|=2\pi|\xi|\epsilon} e^{-z}z^{-k-1}\,dz$$

$$= \pm\frac{i\pi}{2}\operatorname*{res}_{z=0} e^{-z}z^{-k-1} = \pm\frac{i\pi}{2}\frac{(-1)^k}{k!},$$

it follows that, with $\varphi(x) = e^{-x}$,

$$H_{-k-1,\epsilon}(e^{-i2\pi\xi\cdot}) = (i2\pi\xi)^k\left(H_{-k-1,2\pi|\xi|\epsilon}(\varphi) \mp \frac{i\pi}{2}\frac{(-1)^k}{k!}\right).$$

By Exercise 5.7 together with Lemma 5.3, we have

$$
\text{f.p.}_{\epsilon \downarrow 0} H_{-k-1, 2\pi|\xi|\epsilon}(\varphi) = H_{-k-1}(\varphi) - \frac{\varphi^{(k)}(0)}{k!} \log 2\pi|\xi|
$$

$$
= \frac{\varphi^{(k)}(0)}{k!} \left(\sum_{j=1}^{k} \frac{1}{j} - \log 2\pi|\xi| \right) + \frac{1}{k!} H_{-1}(\varphi^{(k)}),
$$

$$
= \frac{(-1)^k}{k!} \left(\sum_{j=1}^{k} \frac{1}{j} - \log 2\pi|\xi| + \int_0^\infty e^{-x} \log x \, dx \right),
$$

and thus

$$
H_{-k-1}(e^{-i2\pi\xi \cdot}) = -\frac{(-i2\pi\xi)^k}{k!} \left(\log 2\pi|\xi| + \frac{i\pi}{2} \operatorname{sign}(\xi) - \Gamma'(1) - \sum_{j=1}^{k} \frac{1}{j} \right).
$$

For any $\phi \in \mathcal{S}(\mathbb{R})$,

$$
\langle \Pi_{-k-1}^+, \phi \rangle = \langle \text{f.p.} \, x_+^{-k-1}, \hat\phi(x) \rangle = \text{f.p.}_{\epsilon \downarrow 0} \int_\epsilon^\infty x^{-k-1} \int_{-\infty}^\infty e^{-i2\pi\xi x} \phi(\xi) \, d\xi \, dx
$$

$$
= \text{f.p.}_{\epsilon \downarrow 0} \int_{-\infty}^\infty H_{-k-1,\epsilon}(e^{-i2\pi\xi \cdot}) \phi(\xi) \, d\xi
$$

$$
= \int_{-\infty}^\infty H_{-k-1}(e^{-i2\pi\xi \cdot}) \phi(\xi) \, d\xi,
$$

where the final step is justified because the $o(1)$ term in the expansion of $H_{-k-1,\epsilon}(e^{-i2\pi\xi \cdot})$ can be bounded by $f(\epsilon)g(\xi)$, with $f(\epsilon) = o(1)$ as $\epsilon \downarrow 0$, and $g(\xi)$ having only polynomial growth as $|\xi| \to \infty$. The formula for $\Pi_{-k-1}^+(\xi)$ is now established, and the one for $\Pi_{-k-1}^-(\xi)$ then follows by (5.6) because the Fourier transform commutes with M_{-1}; see Exercise 5.8. $\qquad\square$

As an immediate consequence of Lemmas 5.6 and 5.10, we obtain the formulae below; see also Exercise 5.10.

Corollary 5.11 For any integer k, let $\Pi_k(\xi) = \mathcal{F}_{x \to \xi}\{\text{f.p.} \, x^k\}$. If $k \geq 0$, then

$$
\Pi_k(\xi) = \frac{1}{(-i2\pi)^k} \delta^{(k)}(\xi) \quad and \quad \Pi_{-k-1}(\xi) = \frac{(-i2\pi)^{k+1}}{2k!} \xi^k \operatorname{sign}(\xi).
$$

Turning to the general, n-dimensional setting, we require the following technical lemma.

Lemma 5.12 *Let $a \in \mathbb{C}$ and $u \in \mathcal{D}^*(\mathbb{R}^n)$. If u is homogeneous of degree a on $\mathbb{R}^n \setminus \{0\}$, i.e., if $M_t u = t^a u$ on $\mathbb{R}^n \setminus \{0\}$ for all $t > 0$, then $u \in \mathcal{S}^*(\mathbb{R}^n)$. If, in addition, u is C^∞ on $\mathbb{R}^n \setminus \{0\}$, then $\hat{u}$ is C^∞ on $\mathbb{R}^n \setminus \{0\}$.*

Proof. Let $\psi \in C^\infty_{\text{comp}}(\mathbb{R}^n \setminus \{0\})$ be as in Exercise 5.12, and define $\chi \in C^\infty_{\text{comp}}(\mathbb{R}^n)$ by

$$\chi(x) = 1 - \int_0^1 \psi(tx) \, \frac{dt}{t} \quad \text{for } x \in \mathbb{R}^n.$$

Put $u_0 = \chi u$ and $u_1 = (1 - \chi)u$; then $u_0 \in \mathcal{S}^*$ and $\hat{u}_0 \in C^\infty(\mathbb{R}^n)$ because u_0 has compact support, so it suffices to consider u_1. We have

$$\langle u_1, \phi \rangle = \left\langle \int_0^1 \psi(tx) \, \frac{dt}{t} u(x), \phi(x) \right\rangle = \int_0^1 \langle u, \phi M_t \psi \rangle \, \frac{dt}{t} \quad \text{for } \phi \in \mathcal{D}(\mathbb{R}^n),$$

and, because u is homogeneous of degree a on $\mathbb{R}^n \setminus \{0\}$,

$$\langle u, \phi M_t \psi \rangle = \langle u, M_t (\psi M_{1/t} \phi) \rangle = \langle t^{-n} M_{1/t} u, \psi M_{1/t} \phi \rangle$$
$$= t^{-n-a} \langle u, \psi M_{1/t} \phi \rangle.$$

Let $K = \operatorname{supp} \psi$; then there is an integer k such that

$$|\langle u, \psi M_{1/t} \phi \rangle| \le C \sum_{|\alpha| \le k} \max_K |\partial^\alpha (\psi M_{1/t} \phi)| \le C \sum_{|\alpha| \le k} t^{-|\alpha|} \max_{x \in K} |\partial^\alpha \phi(t^{-1} x)|.$$

Choose an integer $r > n + \operatorname{Re} a$; then because $0 \notin K$,

$$t^{-|\alpha|} \max_{x \in K} |\partial^\alpha \phi(t^{-1} x)| \le C t^{-|\alpha|} \max_{x \in K} \sum_{|\beta| = r + |\alpha|} |x^\beta \partial^\alpha \phi(t^{-1} x)|$$
$$\le C t^r \sum_{|\beta| = r + |\alpha|} \max_{x \in K} |(t^{-1} x)^\beta \partial^\alpha \phi(t^{-1} x)|,$$

so

$$|\langle u_1, \phi \rangle| \le C \sum_{|\alpha| \le k} \sum_{|\beta| = r + |\alpha|} \sup_{y \in \mathbb{R}^n} |y^\beta \partial^\alpha \phi(y)| \int_0^1 t^{r - n - \operatorname{Re} a} \, \frac{dt}{t},$$

showing that $u_1 \in \mathcal{S}^*(\mathbb{R}^n)$.

Now make the additional assumption that u is C^∞ on $\mathbb{R}^n \setminus \{0\}$. Since $\chi = 1$ on a neighbourhood of 0, we see that u_1 is C^∞ on $\mathbb{R}^n$, and

$$|\partial^\alpha u_1(x)| \le C(1 + |x|)^{\operatorname{Re} a - |\alpha|} \quad \text{for } x \in \mathbb{R}^n.$$

Hence, if $|\beta| > n + |\alpha| + \operatorname{Re} a$, then the function $\partial_x^\beta[(-i2\pi x)^\alpha u_1(x)]$ belongs to $L_1(\mathbb{R}^n)$, and we deduce from (3.17) that $(i2\pi\xi)^\beta\partial^\alpha\hat{u}_1(\xi)$ is continuous on $\mathbb{R}^n$. □

Lemma 5.12 shows that the Fourier transform of a homogeneous distribution always exists as a temperate distribution. Furthermore, the following holds.

Theorem 5.13 Let $a \in \mathbb{C}$. If u is a homogeneous distribution of degree a on $\mathbb{R}^n$, then its Fourier transform $\hat{u}$ is a homogeneous distribution of degree $-a - n$ on $\mathbb{R}^n$.

Proof. For $t > 0$ and $\phi \in \mathcal{D}(\mathbb{R}^n)$,

$$\langle M_t\hat{u}, \phi\rangle = \langle M_t\mathcal{F}u, \phi\rangle = t^{-n}\langle u, \mathcal{F}M_{1/t}\phi\rangle,$$

and by Exercise 5.8,

$$t^{-n}\langle u, \mathcal{F}M_{1/t}\phi\rangle = t^{-n}\langle u, t^n M_t\mathcal{F}\phi\rangle = t^{-n}\langle M_{1/t}u, \mathcal{F}\phi\rangle$$
$$= \langle t^{-n-a}u, \mathcal{F}\phi\rangle = \langle t^{-n-a}\mathcal{F}u, \phi\rangle,$$

so $M_t\hat{u} = t^{-n-a}\hat{u}$. □

If $u \in C^\infty(\mathbb{R}^n \setminus \{0\})$ is homogeneous of degree a, then by Lemma 5.12 the finite-part extension f.p. u is a temperate distribution on $\mathbb{R}^n$, and, recalling (5.12), the Fourier transform of f.p. u is given by

$$\langle \mathcal{F}\text{ f.p. } u, \phi\rangle = \langle \text{f.p. } u, \hat{\phi}\rangle = \int_{|\omega|=1} u(\omega)R_a\hat{\phi}(\omega)\,d\omega \quad \text{for } \phi \in \mathcal{S}(\mathbb{R}^n).$$

$$(5.15)$$

We can express $R_a\hat{\phi}$ in terms of the one-dimensional Fourier transform in Lemma 5.10.

Lemma 5.14 If $x \in \mathbb{R}^n \setminus \{0\}$, then

$$R_a\hat{\phi}(x) = \langle \Pi_{a+n-1}^+(\xi \cdot x), \phi(\xi)\rangle \quad \text{for } \phi \in \mathcal{S}(\mathbb{R}^n).$$

Proof. By (5.11), we have $R_a\hat{\phi}(x) = \langle \text{f.p. } \rho_+^{a+n-1}, \hat{\phi}(\rho x)\rangle$. To express the function $\rho \mapsto \hat{\phi}(\rho x)$ as a one-dimensional Fourier transform, we make the substitution $\xi = \xi_\perp + tx/|x|^2$, where $t = \xi \cdot x$ and thus $\xi_\perp$ is the orthogonal projection of ξ onto the hyperplane normal to x. In this way,

$$\hat{\phi}(\rho x) = \int_{-\infty}^\infty \int_{x\cdot\xi_\perp=0} e^{-i2\pi\rho t}\phi\left(\xi_\perp + \frac{t}{|x|^2}x\right)d\xi_\perp \frac{dt}{|x|} = \hat{\phi}_x(\rho),$$

where

$$\phi_x(t) = \frac{1}{|x|} \int_{x \cdot \xi_\perp = 0} \phi\left(\xi_\perp + \frac{t}{|x|^2}x\right) d\xi_\perp,$$

so

$$R_a\hat{\phi}(\rho x) = \langle \text{f.p. } \rho_+^{a+n-1}, \hat{\phi}_x(\rho) \rangle = \langle \Pi_{a+n-1}^+, \phi_x \rangle.$$

Now see Exercise 5.11.   □

Together, (5.15) and Lemma 5.14 show that the Fourier transform of f.p. u is given by

$$\langle \mathcal{F} \text{ f.p. } u, \phi \rangle = \int_{|\omega|=1} u(\omega) \langle \Pi_{a+n-1}^+(\xi \cdot \omega), \phi(\xi) \rangle \, d\omega \quad \text{for } \phi \in \mathcal{S}(\mathbb{R}^n),$$
$$(5.16)$$

and similarly, the *inverse* Fourier transform of f.p. u is given by

$$\langle \mathcal{F}^* \text{ f.p. } u, \phi \rangle = \int_{|\omega|=1} u(\omega) \langle \Pi_{a+n-1}^+(-x \cdot \omega), \phi(x) \rangle \, d\omega \quad \text{for } \phi \in \mathcal{S}(\mathbb{R}^n).$$
$$(5.17)$$

If $\operatorname{Re} a < 1 - n$ so that $\Pi_{a+n-1}^+ \in L_{1,\text{loc}}(\mathbb{R})$, then we may write

$$\mathcal{F}_{x\to\xi}\{\text{f.p. } u(x)\} = \int_{|\omega|=1} \Pi_{a+n-1}^+(\xi \cdot \omega) u(\omega) \, d\omega$$

and

$$\mathcal{F}_{\xi\to x}^*\{\text{f.p. } u(\xi)\} = \int_{|\omega|=1} \Pi_{a+n-1}^+(-x \cdot \omega) u(\omega) \, d\omega; \qquad (5.18)$$

see Exercise 5.12 for alternative formulae that do not require any restriction on a.

Change of Variables

We wish to investigate the effect of a change of variable $x = \kappa(y)$, where $\kappa : \Omega^\kappa \to \Omega$ is a C^∞ diffeomorphism satisfying

$$\kappa(0) = 0,$$

and Ω^κ and Ω are open neighbourhoods of 0 in $\mathbb{R}^n$. For any $\epsilon > 0$, let

$$B_\epsilon = \{y \in \mathbb{R}^n : |y| < \epsilon\}$$

denote the open ball in $\mathbb{R}^n$ with radius ϵ and centre 0, so that if $u \in L_{1,\text{loc}}(\Omega \setminus \{0\})$ and $\phi \in \mathcal{D}(\Omega)$, then

$$\int_{\kappa(\Omega^\kappa \setminus B_\epsilon)} u(x)\phi(x)\,dx = \int_{\Omega^\kappa \setminus B_\epsilon} u\big(\kappa(y)\big)\phi\big(\kappa(y)\big)|\det \kappa'(y)|\,dy. \quad (5.19)$$

Theorem 5.15 below implies the existence of the finite part of the right-hand side of (5.19) when u is homogeneous, or, more generally, when u is the sum of finitely many homogeneous functions and a remainder term that is integrable on Ω.

Theorem 5.15 If $u \in C^\infty(\mathbb{R}^n \setminus \{0\})$ is homogeneous of degree a, then there exist $\epsilon > 0$ and functions $w_0, w_1, w_2, \ldots$ and $R_0, R_1, R_2, \ldots$ with the following properties:

(i) For every $m \geq 0$, the composite function $u \circ \kappa$ admits the expansion

$$u\big(\kappa(y)\big) = \sum_{j=0}^m w_j(y) + R_m(y) \quad \text{for } 0 < |y| \leq \epsilon.$$

(ii) For every $j \geq 0$, the function w_j is C^∞ on $\mathbb{R}^n \setminus \{0\}$ and homogeneous of degree $a + j$:

$$w_j(ty) = t^{a+j} w_j(y) \quad \text{for } t > 0 \text{ and } y \in \mathbb{R}^n \setminus \{0\}.$$

(iii) For every $m \geq 0$, the function R_m is C^∞ on $\overline{B}_\epsilon \setminus \{0\}$ and, for every multi-index α,

$$|\partial^\alpha R_m(y)| \leq C_{m,\alpha}|y|^{a+m+1-|\alpha|} \quad \text{for } 0 < |y| \leq \epsilon.$$

If, in addition, $a = -n - k$ for some integer $k \geq 0$, and u satisfies the parity condition (5.14), then for all $j \geq 0$ the parity of the function w_j is opposite to that of $k - j$, i.e.,

$$w_j(-x) = (-1)^{k-j+1} w_j(x) \quad \text{for } x \in \mathbb{R}^n \setminus \{0\}.$$

Proof. Since κ is a diffeomorphism, the derivative $\kappa'(0) : \mathbb{R}^n \to \mathbb{R}^n$ is an isomorphism, and since $\kappa(0) = 0$,

$$\kappa(y) = \kappa'(0)y + O(|y|^2) \quad \text{as } y \to 0.$$

Letting $h(y) = \kappa(y) - \kappa'(0)y$, we see that there exists $\epsilon > 0$ such that

$$c|y| \leq |\kappa'(0)y + th(y)| \leq C|y| \quad \text{for } 0 < |y| \leq \epsilon \text{ and } 0 \leq t \leq 1.$$

Thus, by Taylor expansion of u about $\kappa'(0)y$,

$$u\big(\kappa(y)\big) = \sum_{j=0}^{m} \frac{1}{j!} u^{(j)}\big(\kappa'(0)y; h(y)\big) + R_{m,1}(y),$$

where

$$R_{m,1}(y) = \frac{1}{m!} \int_0^1 (1-t)^m u^{(m+1)}\big(\kappa'(0)y + th(y); h(y)\big)\, dt.$$

In turn, Taylor expansion of h about 0 allows us to write

$$h(y) = \sum_{r=2}^{m} h_r(y) + R_{m,2}(y),$$

where

$$h_r(y) = \frac{1}{r!} \kappa^{(r)}(0; y) \quad \text{and} \quad R_{m,2}(y) = \frac{1}{m!} \int_0^1 (1-t)^m \kappa^{(m+1)}(ty; y)\, dt.$$

Hence,

$$u^{(p)}\big(\kappa'(0)y; h(y)\big) = \sum_{r_1=2}^{m} \cdots \sum_{r_p=2}^{m} u^{(p)}\big(\kappa'(0)y; h_{r_1}(y), \ldots, h_{r_p}(y)\big) + \widetilde{R}_{m,p}(y),$$

and we see that

$$w_j(y) = \sum \frac{1}{p!} u^{(p)}\big(\kappa'(0)y; h_{r_1}(y), \ldots, h_{r_p}(y)\big),$$

where the sum is over all $p \geq 0$, $r_1 \geq 2, \ldots, r_p \geq 2$ satisfying

$$r_1 + \cdots + r_p - p = j.$$

(Notice that $j \geq p \geq 0$.) Since $\partial^\alpha u$ is homogeneous of degree $a - |\alpha|$, and h_r is homogeneous of degree r, it follows that w_j is homogeneous of degree $a - p + r_1 + \cdots + r_p = a + j$. The estimates for $R_m(y)$ follow from the bounds

$$|\partial^\alpha R_{m,1}(y)| \leq C|y|^{a+m+1-|\alpha|} \quad \text{and} \quad |\partial^\alpha R_{m,2}(y)| \leq C \min\big(1, |y|^{m+1-|\alpha|}\big)$$

for $0 < |y| \leq \epsilon$. Finally, if $a = -n - k$ and $u(-x) = (-1)^{k+1}u(x)$, then $\partial^\alpha u(-x) = (-1)^{k+1-|\alpha|}\partial^\alpha u(x)$ and so the term w_j is homogeneous of degree $-n - (k - j)$, with

$$w_j(-y) = \sum \frac{1}{p!} u^{(p)}\big(-\kappa'(0)y; h_{r_1}(-y), \ldots, h_{r_p}(-y)\big)$$

$$= (-1)^{k+1-p+r_1+\cdots+r_p} w_j(y)$$

because $h_r(-y) = (-1)^r h_r(y)$. $\qquad\qquad\qquad\qquad\qquad\qquad\qquad\square$

Now consider the left-hand side of (5.19). Since $\kappa(\Omega^\kappa \setminus B_\epsilon) = \Omega \setminus \kappa(B_\epsilon)$, the question arises as to whether

$$\text{f.p.} \int_{\Omega \setminus \kappa(B_\epsilon)} u(x)\phi(x)\,dx = \text{f.p.} \int_{\Omega \setminus B_\epsilon} u(x)\phi(x)\,dx.$$
$$\epsilon \downarrow 0 \qquad\qquad\qquad\qquad \epsilon \downarrow 0$$

In fact, Exercise 5.7 shows that these two finite-part integrals can differ, even if κ is *linear*, when a is an integer less than or equal to $-n$. Once again, we seek first to understand the one-dimensional case.

Lemma 5.16 Suppose $n = 1$. If $a \neq -1, -2, -3, \ldots$, then

$$\text{f.p.} \int_{\kappa(\epsilon)}^{\infty} x^a \phi(x)\,dx = \langle \text{f.p.}\, x_+^a, \phi(x)\rangle \qquad (5.20)$$
$$\epsilon \downarrow 0$$

and for any integer $k \geq 0$,

$$\text{f.p.} \int_{\mathbb{R} \setminus \kappa(B_\epsilon)} x^{-k-1} \phi(x)\,dx = \langle \text{f.p.}\, x^{-k-1}, \phi(x)\rangle \qquad (5.21)$$
$$\epsilon \downarrow 0$$

Proof. The case $\text{Re}\, a > -1$ is trivial, so suppose that $\text{Re}\, a > -k - 1$ and $a \neq -1, -2, \ldots, -k$ for some integer $k \geq 1$. Integration by parts gives

$$H_{a,\kappa(\epsilon)}(\phi) = -\frac{[\kappa(\epsilon)]^{a+1}\phi(\kappa(\epsilon))}{a+1} - \frac{H_{a+1,\kappa(\epsilon)}(\phi')}{a+1},$$

and by Taylor expansion,

$$[\kappa(\epsilon)]^{a+1}\phi(\kappa(\epsilon)) = \sum_{j=0}^{k-1} \frac{\phi^{(j)}(0)}{j!}[\kappa(\epsilon)]^{a+j+1} + O(\epsilon^{a+k+1})$$

and

$$[\kappa(\epsilon)]^{a+j+1} = \sum_{l=j}^{k-1} c_{jl}\epsilon^{a+l+1} + O(\epsilon^{a+k+1})$$

for some $c_{jl} \in \mathbb{R}$. Thus,

$$\text{f.p.}[\kappa(\epsilon)]^{a+1}\phi(\kappa(\epsilon)) = 0,$$
$$\epsilon \downarrow 0$$

and we have

$$\text{f.p.}\, H_{a,\kappa(\epsilon)}(\phi) = \frac{-1}{a+1}\,\text{f.p.}\, H_{a+1,\kappa(\epsilon)}(\phi'),$$
$$\epsilon \downarrow 0 \qquad\qquad\qquad \epsilon \downarrow 0$$

provided the right-hand side exists. Induction on k yields

$$
\text{f.p.}_{\epsilon \downarrow 0} H_{a,\kappa(\epsilon)}(\phi) = \text{f.p.}_{\epsilon \downarrow 0} \frac{(-1)^k H_{a+k,\kappa(\epsilon)}(\phi^{(k)})}{(a+1)(a+2)\cdots(a+k)}
$$

$$
= \frac{(-1)^k H_{a+k}(\phi^{(k)})}{(a+1)(a+2)\cdots(a+k)},
$$

which, by Lemma 5.3, shows that (5.20) holds.

To prove (5.21), let

$$
J_k(\phi) = \text{f.p.}_{\epsilon \downarrow 0} J_{k,\epsilon}(\phi), \qquad \text{where} \quad J_{k,\epsilon}(\phi) = \int_{\mathbb{R} \backslash \kappa(B_\epsilon)} x^{-k-1} \phi(x)\, dx.
$$

If we can show that

$$
J_k(\phi) = \frac{-1}{k!} \int_{-\infty}^{\infty} \phi^{(k+1)}(x) \log |x|\, dx, \tag{5.22}
$$

then (5.21) will follow by Lemma 5.5. Integration by parts gives

$$
J_{0,\epsilon}(\phi) = \phi\big(\kappa(-\epsilon)\big) \log |\kappa(-\epsilon)| - \phi\big(\kappa(\epsilon)\big) \log |\kappa(\epsilon)|
$$

$$
- \int_{\mathbb{R} \backslash \kappa(B_\epsilon)} \phi'(x) \log |x|\, dx,
$$

and by Taylor expansion,

$$
\phi\big(\kappa(\pm\epsilon)\big) \log |\kappa(\pm\epsilon)| = \phi(0) \log |\kappa'(0)\epsilon| + O\left(\epsilon \log \frac{1}{\epsilon}\right),
$$

so (5.22) holds for $k = 0$. If $k > 0$, then

$$
J_{k,\epsilon}(\phi) = \frac{\phi\big(\kappa(\epsilon)\big)}{k[\kappa(\epsilon)]^k} - \frac{\phi\big(\kappa(-\epsilon)\big)}{k[\kappa(-\epsilon)]^k} + \frac{J_{k-1,\epsilon}(\phi')}{k}
$$

$$
= \frac{J_{k-1,\epsilon}(\phi')}{k} + \frac{1}{k} \sum_{j=0}^{k-1} \frac{\phi^{(j)}(0)}{j!} \left(\frac{1}{[\kappa(\epsilon)]^{k-j}} - \frac{1}{[\kappa(-\epsilon)]^{k-j}} \right) + O(\epsilon).
$$

Given any integer $m \geq 1$, we can define a C^∞ function $f : \Omega \to \mathbb{R}$ by

$$
f(y) = \begin{cases} \left(\dfrac{y}{\kappa(y)} \right)^m & \text{if } y \in \Omega \backslash 0, \\[2ex] \dfrac{1}{[\kappa'(0)]^m} & \text{if } y = 0, \end{cases}
$$

and so

$$\underset{\epsilon \downarrow 0}{\text{f.p.}} \frac{1}{[\kappa(\pm \epsilon)]^m} = \underset{\epsilon \downarrow 0}{\text{f.p.}} \frac{f(\pm \epsilon)}{(\pm \epsilon)^m} = \frac{f^{(m)}(0)}{m!}.$$

Thus,

$$J_k(\phi) = \frac{J_{k-1}(\phi')}{k},$$

and (5.22) follows by induction on k. $\qquad\square$

For ϵ sufficiently small, the set $\kappa(B_\epsilon)$ is approximately ellipsoidal and can be described using the function g in the next lemma. Recall that $\mathbb{S}^{n-1} = \{\omega \in \mathbb{R}^n : |\omega| = 1\}$ denotes the unit sphere in $\mathbb{R}^n$.

Lemma 5.17 There is an $\epsilon_0 > 0$ and a C^∞ function

$$g : (-\epsilon_0, \epsilon_0) \times \mathbb{S}^{n-1} \to \mathbb{R},$$

such that

$$\kappa(B_\epsilon) \cap \{\rho\omega : 0 < \rho < \infty\} = \{\rho\omega : 0 < \rho < g(\epsilon, \omega)\}$$
$$\textit{for } 0 < \epsilon < \epsilon_0 \textit{ and } |\omega| = 1. \qquad (5.23)$$

Moreover, g satisfies

$$g(\epsilon, \omega) = -g(-\epsilon, -\omega) \quad and \quad \left| \partial_\epsilon^j g(\epsilon, \omega) \right| \le C_j, \qquad (5.24)$$

for $|\epsilon| < \epsilon_0$, $|\omega| = 1$ and $j \ge 0$.

Proof. Choose $\rho_0 > 0$ such that $t\omega \in \Omega$ for $|t| \le \rho_0$ and $|\omega| = 1$, and define

$$f(t, \omega) = \text{sign}(t)|\kappa^{-1}(t\omega)|.$$

Since $\kappa^{-1}(0) = 0$, Taylor's theorem gives

$$\kappa^{-1}(t\omega) = tL\omega + t^2 M(t, \omega),$$

where

$$L = (\kappa^{-1})'(0) \quad and \quad M(t, \omega) = \int_0^1 (1-s)(\kappa^{-1})^{(2)}(st\omega; \omega)\, ds.$$

Thus,

$$f(t, \omega) = t|L\omega + tM(t, \omega)|,$$

and, because $L\omega \neq 0$ for $|\omega| = 1$, there is a $\rho_1 \in (0, \rho_0)$ such that $f(t, \omega)$ is C^∞ for $|t| \leq \rho_1$ and $|\omega| = 1$. The uniform bound $\partial_t f(0, \omega) = |L\omega| \geq c > 0$ means that we can apply the implicit function theorem and define g by

$$f(t, \omega) = \epsilon \iff t = g(\epsilon, \omega),$$

for $|\epsilon| < \epsilon_0$, with $\epsilon_0 > 0$ independent of ω. The relation (5.23) follows because, for $|x| < \epsilon_0$,

$$\kappa(x) = \rho\omega \iff \rho = g(|x|, \omega),$$

and because $g(\epsilon, \omega)$ is a monotonically increasing function of ϵ. Finally, (5.24) holds because $f(t, \omega) = -f(-t, -\omega)$ and because $|\partial_t^j f(t, \omega)| \leq C_j$ for $|t| \leq \rho_1$ and $|\omega| = 1$. $\square$

We are at last in a position to prove the main result for this section; see Kieser [48, Satz 2.2.12] for the original proof using the calculus of pseudodifferential operators.

Theorem 5.18 *Let* $\kappa : \Omega^\kappa \to \Omega$ *be a diffeomorphism with* $\kappa(0) = 0$, *as above, and suppose that* $u \in C^\infty(\mathbb{R}^n \setminus 0)$ *is homogeneous of degree* a. *If* $a \neq -n$, $-n - 1, -n - 2, \ldots$, *or if* $a = -n - k$ $(k \geq 0)$ *but* u *satisfies the parity condition (5.14), then*

$$\text{f.p.} \int_{\epsilon \downarrow 0}_{\mathbb{R}^n \setminus \kappa(B_\epsilon)} u(x)\phi(x)\, dx = \langle \text{f.p.}\, u, \phi \rangle \quad \text{for } \phi \in \mathcal{S}(\mathbb{R}^n),$$

and hence

$$\text{f.p.} \int_{\epsilon \downarrow 0}_{\Omega \setminus B_\epsilon} u(x)\phi(x)\, dx = \text{f.p.} \int_{\epsilon \downarrow 0}_{\Omega^\kappa \setminus B_\epsilon} u(\kappa(y))\phi(\kappa(y))|\det \kappa'(y)|\, dy$$

$$\text{for } \phi \in \mathcal{D}(\Omega).$$

Proof. Using the notation in Lemma 5.17 and making the substitution $x = \rho\omega$, we have

$$\int_{\mathbb{R}^n \setminus \kappa(B_\epsilon)} u(x)\phi(x)\, dx = \int_{|\omega|=1} u(\omega) \int_{g(\epsilon, \omega)}^\infty \rho^{a+n-1}\phi(\rho\omega)\, d\rho\, d\omega$$

$$\text{for } 0 < \epsilon < \epsilon_0.$$

If $a \neq -n - k$, then by Lemma 5.16,

$$\text{f.p.} \int_{\epsilon \downarrow 0}_{g(\epsilon, \omega)}^\infty \rho^{a+n-1}\phi(\rho\omega)\, d\rho = \langle \text{f.p.}\, \rho_+^{a+n-1}, \phi(\rho\omega) \rangle = R_a\phi(\omega),$$

and the result follows at once from (5.12). If $a = -n - k$ and u satisfies (5.14), then

$$\int_{|\omega|=1} u(\omega) \int_{g(\epsilon,\omega)}^{\infty} \rho^{-k-1} \phi(\rho\omega) \, d\rho \, d\omega$$

$$= \int_{|\omega|=1} u(-\omega) \int_{g(\epsilon,-\omega)}^{\infty} \rho^{-k-1} \phi(-\rho\omega) \, d\rho \, d\omega$$

$$= \int_{|\omega|=1} u(\omega) \int_{-\infty}^{g(-\epsilon,\omega)} \rho^{-k-1} \phi(\rho\omega) \, d\rho \, d\omega$$

$$= \frac{1}{2} \int_{|\omega|=1} u(\omega) \int_{\mathbb{R}\setminus\kappa_\omega(B_\epsilon)} \rho^{-k-1} \phi(\rho\omega) \, d\rho \, d\omega,$$

where κ_ω is the one-dimensional diffeomorphism defined by $\kappa_\omega(\epsilon) = g(\epsilon, \omega)$. Hence, it suffices to apply Theorem 5.8 and Lemma 5.16. $\square$

Finite-Part Integrals on Surfaces

Although we shall avoid making explicit use of the results in this section they nevertheless give some insight into the nature of the surface potentials and boundary integral operators encountered in the later chapters.

Consider a C^∞ surface of the form

$$\Gamma = \{x : x_n = \zeta(x'), x' \in \mathbb{R}^{n-1}\}, \tag{5.25}$$

and assume that the origin has been chosen so that $\zeta(0) = 0$. By (3.28), if $u \in L_{1,\text{loc}}(\Gamma)$ and $\phi \in \mathcal{D}(\Gamma)$, then

$$\int_\Gamma u(x)\phi(x) \, d\sigma_x = \int_{\mathbb{R}^{n-1}} u(x', \zeta(x'))\phi(x', \zeta(x'))\sqrt{1 + |\text{grad}\zeta(x')|^2} \, dx'.$$

For finite-part integrals on Γ, we have the following result of Kieser [48, Satz 4.3.9].

Theorem 5.19 Let Γ be a C^∞ surface passing though the origin, as above, and suppose that $u \in C^\infty(\mathbb{R}^n \setminus \{0\})$ is homogeneous of degree a. If $a \neq -n + 1$, $-n, -n - 1, \ldots$, or if $a = -n + 1 - k$ ($k \geq 0$) but u satisfies $u(-x) = (-1)^{k+1}u(x)$, then

$$\text{f.p.} \int_{\epsilon \downarrow 0} \int_{\Gamma \setminus B_\epsilon} u(x)\phi(x) \, d\sigma_x$$

$$= \text{f.p.} \int_{\epsilon \downarrow 0} \int_{|x'|>\epsilon} u(x', \zeta(x'))\phi(x', \zeta(x'))\sqrt{1 + |\text{grad}\zeta(x')|^2} \, dx'$$

for $\phi \in \mathcal{D}(\Gamma)$.

Proof. We introduce polar coordinates in $\mathbb{R}^{n-1}$, writing

$$x' = r\omega, \quad r = |x'|, \quad \omega = x'/r \in \mathbb{S}^{n-2}.$$

For $x \in \Gamma$, we have $|x|^2 = r^2 + \zeta(r\omega)^2$, and so

$$x \in \Gamma \setminus B_\epsilon \iff r\sqrt{1 + [\zeta(r\omega)/r]^2} > \epsilon.$$

Since $\zeta(0) = 0$, there exists $\epsilon_0 > 0$ and a C^∞ function $g : (-\epsilon_0, \epsilon_0) \times \mathbb{S}^{n-2} \to \mathbb{R}$ such that

$$\Gamma \setminus B_\epsilon = \left\{ \left(x', \zeta(x') \right) : x' = r\omega, r > g(\epsilon, \omega) \text{ and } \omega \in \mathbb{S}^{n-2} \right\} \quad \text{for } 0 < \epsilon < \epsilon_0,$$

with $g(\epsilon, \omega) = -g(-\epsilon, -\omega)$; cf. Lemma 5.17. The result follows by arguing as in the proof of Theorem 5.18, remembering that now the integral is over $\mathbb{R}^{n-1}$ instead of $\mathbb{R}^n$. $\qquad\square$

Finite-part integrals on surfaces arise as boundary values of functions of the form

$$f(x) = \int_\Gamma u(x - y)\psi(y)\, d\sigma_y \quad \text{for } x \in \mathbb{R}^n \setminus \Gamma, \tag{5.26}$$

where the integral is divergent if $x \in \Gamma$. We shall consider u of the following type.

Assumption 5.20 The distribution u is of the form

$$u(x) = \mathcal{F}^*_{\xi \to x}\{v(\xi)\} \quad \text{with} \quad v(\xi) = \frac{p(\xi)}{q(\xi)},$$

where p and q are homogeneous polynomials satisfying

$$(\text{degree of } p) - (\text{degree of } q) = j - 1 \quad \text{for some } j \geq -n + 2,$$

and in addition

$$q(\xi) \neq 0 \quad \text{for all } \xi \in \mathbb{R}^n \setminus \{0\}.$$

Thus, $v \in C^\infty(\mathbb{R}^n \setminus \{0\})$ is homogeneous of degree $j - 1$, and so is locally integrable on $\mathbb{R}^n$ because $j - 1 \geq -n + 1$. By Theorem 5.13, we see that u is a homogeneous distribution on $\mathbb{R}^n$ of degree $-n + 1 - j$, and by Lemma 5.12, u is C^∞ on $\mathbb{R}^n \setminus \{0\}$. Also, by Exercise 5.8,

$$v(-\xi) = (-1)^{j-1}v(\xi) \quad \text{and} \quad u(-x) = (-1)^{j-1}u(x). \tag{5.27}$$

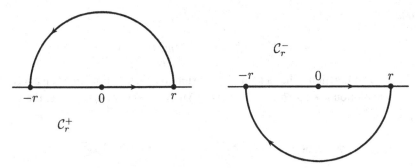

Figure 4. Integration contours in the definition (5.28) of $\int^{\pm}$.

To begin with, we investigate the simplest case, when $\Gamma = \mathbb{R}^{n-1}$ and so

$$f(x) = \int_{\mathbb{R}^{n-1}} u(x' - y', x_n)\psi(y')\,dy' \quad \text{for } x_n \neq 0.$$

The next lemma gives an alternative representation for f in terms of the Fourier transform of ψ. We denote the upper and lower complex half planes by

$$\mathbb{C}^+ = \{z \in \mathbb{C} : \operatorname{Im} z > 0\} \quad \text{and} \quad \mathbb{C}^- = \{z \in \mathbb{C} : \operatorname{Im} z < 0\},$$

put

$$\mathbb{R}^n_{\pm} = \{x \in \mathbb{R}^n : x_n \in \mathbb{R}^{\pm}\},$$

and let $\mathcal{C}_r^+$ and $\mathcal{C}_r^-$ be the closed, semicircular contours shown in Figure 4. If $w(z)$ is continuous for z in the closed half plane $\mathbb{C}^{\pm} \cup \mathbb{R}$, and analytic for $|z| > r_0$ and $z \in \mathbb{C}^{\pm}$, then we denote the integral of w over $\mathcal{C}_r^{\pm}$ by

$$\int^{\pm} w(z)\,dz = \int_{\mathcal{C}_r^{\pm}} w(z)\,dz \quad \text{for } r \geq r_0. \tag{5.28}$$

By Cauchy's theorem, this integral is independent of r.

Lemma 5.21 If u satisfies Assumption 5.20, and if $\psi \in \mathcal{S}(\mathbb{R}^{n-1})$, then

$$f(x) = \int_{\mathbb{R}^{n-1}} m_{\pm}(\xi', x_n)\hat{\psi}(\xi')e^{i2\pi\xi' \cdot x'}\,d\xi' \quad \text{for } x \in \mathbb{R}^n_{\pm},$$

where

$$m_{\pm}(\xi', x_n) = \int^{\pm} v(\xi', \xi_n)e^{i2\pi\xi_n x_n}\,d\xi_n.$$

Furthermore,

$$m_{\pm}(t\xi', t^{-1}x_n) = t^j m_{\pm}(\xi', x_n) \quad \text{for } t > 0 \text{ and } \xi' \neq 0,$$

and

$$m_+(-\xi', -x_n) = (-1)^{j-1} m_-(\xi', x_n).$$

Proof. The function f has a natural extension to a distribution on $\mathbb{R}^n$, namely the convolution $u * (\psi \otimes \delta)$, where $(\psi \otimes \delta)(y) = \psi(y')\delta(y_n)$. Hence, for all $\phi \in \mathcal{S}(\mathbb{R}^n)$,

$$\langle f, \phi \rangle = \langle u * (\psi \otimes \delta), \mathcal{F}\mathcal{F}^*\phi \rangle = \langle \mathcal{F}[u * (\psi \otimes \delta)], \mathcal{F}^*\phi \rangle$$
$$= \int_{\mathbb{R}^n} v(\xi)\hat{\psi}(\xi')\check{\phi}(\xi)\, d\xi,$$

where $\check{\phi} = \mathcal{F}^*\phi$. Suppose now that $\mathrm{supp}\,\phi \subseteq \mathbb{R}_\pm^n$. In this case, for each ξ' the function $\check{\phi}(\xi', \cdot)$ is continuous on $\mathbb{C}^\pm \cup \mathbb{R}$, is analytic on $\mathbb{C}^\pm$, and satisfies bounds of the form

$$|\check{\phi}(\xi)| \le C_{M,N}(1 + |\xi'|)^{-M}(1 + |\xi_n|)^{-N} \quad \text{for } \xi' \in \mathbb{R}^{n-1} \text{ and } \xi_n \in \mathbb{C}^\pm \cup \mathbb{R}.$$

Hence,

$$\langle f, \phi \rangle = \int_{\mathbb{R}^{n-1}} \hat{\psi}(\xi') \int_{-\infty}^\infty v(\xi)\check{\phi}(\xi)\, d\xi_n\, d\xi',$$

and to shift the contour of integration in the ξ_n-plane, we consider the poles of $v(\xi)$, putting $Z(\xi') = \{\xi_n \in \mathbb{C} : q(\xi', \xi_n) = 0\}$. Since q is homogeneous, we have $Z(t\xi') = t Z(\xi')$, and since the coefficients of the polynomial $q(\xi', \cdot)$ are continuous functions of ξ',

$$Z(\xi') \subseteq B_r \quad \text{for } r > c|\xi'|.$$

The formula for $f(x)$ now follows, because

$$\int_{-\infty}^\infty v(\xi)\check{\phi}(\xi)\, d\xi_n = \int^\pm v(\xi)\check{\phi}(\xi)\, d\xi_n = \int^\pm \int_{\mathbb{R}^n} v(\xi) e^{i2\pi\xi \cdot x} \phi(x)\, dx\, d\xi_n$$
$$= \int_{\mathbb{R}^n} \phi(x) e^{i2\pi\xi' \cdot x'} m_\pm(\xi', x_n)\, dx.$$

Finally, since $v(t\xi) = t^{j-1} v(\xi)$ the substitution $z = t\xi_n$ gives

$$m_\pm(t\xi', t^{-1}x_n) = \int^\pm v(t\xi', z) e^{i2\pi z(t^{-1}x_n)}\, dz$$
$$= \int^\pm v(t\xi) e^{i2\pi\xi_n x_n} t\, d\xi_n = t^j m_\pm(\xi, x_n),$$

and since $Z(-\xi') = -Z(\xi')$, the substitution $z = -\xi_n$ gives

$$
m_+(-\xi', -x_n) = \int_{C_r^+} v(-\xi', z)e^{i2\pi z(-x_n)}\, dz
$$

$$
= \int_{-C_r^-} v(-\xi)e^{i2\pi \xi_n x_n}\, (-d\xi_n),
$$

which equals $(-1)^{j-1} m_-(\xi', x_n)$ by (5.27). $\qquad\qquad\qquad\square$

Since $m_\pm(\xi', x_n)$ is a C^∞ function of x_n, we see that the restrictions $f|_{\mathbb{R}_+^n}$ and $f|_{\mathbb{R}_-^n}$ can be extended to C^∞ functions on $\mathbb{R}^n$. We now consider the one-sided boundary values of f on the hyperplane $x_n = 0$, given by

$$
f_\pm(x') = \lim_{z \to (x', 0), z \in \mathbb{R}_\pm^n} f(z) = \int_{\mathbb{R}^{n-1}} m_\pm(\xi', 0)\hat\psi(\xi')e^{i2\pi\xi' \cdot x'}\, d\xi'
$$
$$
\text{for } x' \in \mathbb{R}^{n-1}.
$$

Theorem 5.22 *Suppose that u satisfies Assumption 5.20, and that $\psi \in \mathcal{S}(\mathbb{R}^{n-1})$.*

(i) If $j \le -1$, then

$$
f_+(x') = f_-(x') = \int_{\mathbb{R}^{n-1}} u(x' - y', 0)\psi(y')\, dy'. \qquad (5.29)
$$

(ii) If $j \ge 0$, then

$$
f_+(x') + f_-(x') = 2\,\mathrm{f.p.} \int_{\epsilon\downarrow 0} \int_{|x'-y'|>\epsilon} u(x' - y', 0)\psi(y')\, dy',
$$

and the jump in f across the hyperplane $x_n = 0$ has the form

$$
f_+(x') - f_-(x') = \sum_{|\alpha|=j} c_\alpha \partial^\alpha \psi(x'),
$$

for some coefficients $c_\alpha \in \mathbb{C}$.

Proof. Using the sum and difference of m_+ and m_-, we define

$$
u_s(x') = \mathcal{F}^*_{\xi'\to x'}\{m_s(\xi')\}, \qquad \text{where} \quad m_s(\xi') = m_+(\xi', 0) + m_-(\xi', 0),
$$

and

$$
u_d(x') = \mathcal{F}^*_{\xi'\to x'}\{m_d(\xi')\}, \qquad \text{where} \quad m_d(\xi') = m_+(\xi', 0) - m_-(\xi', 0),
$$

so that

$$f_+ + f_- = u_s * \psi \quad \text{and} \quad f_+ - f_- = u_d * \psi.$$

By Lemma 5.21, m_s and m_d are homogeneous of degree j, and satisfy

$$m_s(-\xi') = (-1)^{j-1} m_s(\xi') \quad \text{and} \quad m_d(-\xi') = (-1)^j m_d(\xi'),$$

so u_s and u_d are homogeneous distributions on $\mathbb{R}^n$ of degree $-j - (n-1)$, and satisfy

$$u_s(-x') = (-1)^{j-1} u_s(x') \quad \text{and} \quad u_d(-x') = (-1)^j u_d(x').$$

Since $u(x - y)$ is C^∞ for $x \neq y$, it is easy to see that if $x' \notin \operatorname{supp} \psi$, then (5.29) holds (even if $j \geq 0$), and in particular $f_+(x') - f_-(x') = 0$. Therefore $\operatorname{supp} u_d(x' - \cdot) \subseteq \{x'\}$. If $j \leq -1$, then $u_d \in L_{1,\mathrm{loc}}(\mathbb{R}^{n-1})$ so $u_d = 0$. If $j \geq 0$, then, with the help of Theorem 3.9 and Exercise 5.1, we deduce from the homogeneity of u_d that

$$u_d * \psi = \sum_{|\alpha|=j} c_\alpha \partial^\alpha \psi,$$

for some $c_\alpha \in \mathbb{C}$. Furthermore, $u_s(x' - y') = 2u(x' - y', 0)$ for $x' \neq y'$, because (5.29) holds when $x' \notin \operatorname{supp} \psi$, so the homogeneous distributions u_s and $2\,\mathrm{f.p.}\,u(\cdot, 0)$ are equal on $\mathbb{R}^{n-1} \setminus \{0\}$. Since both have degree $-(n-1) - j$ and parity $j - 1$, we see by Theorem 5.8 (applied in $\mathbb{R}^{n-1}$, not $\mathbb{R}^n$) that in fact $u_s = 2\,\mathrm{f.p.}\,u(\cdot, 0)$ as distributions on $\mathbb{R}^{n-1}$.   □

Suppose now that Γ is the graph of a C^∞ function $\zeta : \mathbb{R}^{n-1} \to \mathbb{R}$, as in (5.25). We denote the epigraph and hypograph of ζ by

$$\Omega^+ = \{x \in \mathbb{R}^n : x_n > \zeta(x')\} \quad \text{and} \quad \Omega^- = \{x \in \mathbb{R}^n : x_n < \zeta(x')\},$$

and denote the boundary values of the function f defined in (5.26) by

$$f_\pm(x) = \lim_{z \to x, z \in \Omega^\pm} \int_\Gamma u(z - y)\psi(y)\,d\sigma_y \quad \text{for } x \in \Gamma.$$

It is possible to generalise Theorem 5.22 as follows.

Theorem 5.23 *As before, let Γ be the C^∞ surface (5.25), with $\zeta(0) = 0$. If u satisfies Assumption 5.20, and if $\psi \in \mathcal{D}(\Gamma)$, then the restrictions $f|_{\Omega^+}$ and $f|_{\Omega^-}$ can be extended to C^∞ functions on $\mathbb{R}^n$. Moreover, for $x = (x', \zeta(x')) \in \Gamma$ we have*

(i) if $j \leq -1$, then

$$f_+(x) = f_-(x) = \int_\Gamma u(x - y)\psi(y)\,d\sigma_y;$$

(ii) if $j \geq 0$, then

$$f_+(x) + f_-(x) = 2 \, \text{f.p.} \int_{\substack{\epsilon \downarrow 0 \\ \Gamma \setminus B_\epsilon(x)}} u(x - y)\psi(y) \, d\sigma_y$$

and, for some coefficients $b_\alpha \in C^\infty(\mathbb{R}^{n-1})$,

$$f_+(x) - f_-(x) = \sum_{|\alpha| \leq j} b_\alpha(x') \partial_{x'}^\alpha \psi(x', \zeta(x')).$$

We shall not prove this result, because techniques from the theory of pseudo-differential operators would be required; cf. Kieser [48, Satz 4.3.6] or Chazarain and Piriou [10, p. 280]. In all subsequent proofs, we avoid using Theorem 5.23. It does, however, help to account for some of our results in Chapter 7.

Exercises

5.1 Show that the Dirac distribution $\delta \in \mathcal{D}^*(\mathbb{R}^n)$ is homogeneous of degree $-n$.

5.2 Show that if $u \in C^\infty(\mathbb{R}^n \setminus \{0\})$ is homogeneous of degree $a \in \mathbb{C}$ as a function on $\mathbb{R}^n \setminus \{0\}$, then for any multi-index α the partial derivative $\partial^\alpha u$ is homogeneous of degree $a - |\alpha|$ on $\mathbb{R}^n \setminus \{0\}$. Show further that if $u \in \mathcal{D}^*(\mathbb{R}^n)$ is homogeneous of degree a as a distribution on $\mathbb{R}^n$, then $\partial^\alpha u$ is homogeneous of degree $a - |\alpha|$ on $\mathbb{R}^n$.

5.3 Show that

$$\frac{d}{dx} \, \text{f.p.} \, x_\pm^a = \pm a \, \text{f.p.} \, x_\pm^{a-1} \quad \text{for } a \neq -1, -2, -3, \ldots,$$

but

$$\frac{d}{dx} \, \text{f.p.} \, x_\pm^{-k} = \mp k \, \text{f.p.} \, x_\pm^{-k-1} \pm \frac{(\mp 1)^k}{k!} \delta^{(k)}(x) \quad \text{for any integer } k \geq 1.$$

Deduce that

$$\frac{d}{dx} \, \text{f.p.} \, x^{-k} = -k \, \text{f.p.} \, x^{-k-1}$$

5.4 Recall the definition (5.11) of $R_a\phi$. Show that if $\phi \in \mathcal{D}(\mathbb{R}^n)$, $t > 0$ and $x \in \mathbb{R}^n \setminus \{0\}$, then

$$R_a\phi(tx) = t^{-a-n} R_a\phi(x) \quad \text{for } a \neq -n, -n-1, -n-2, \ldots,$$

but

$$R_{-n-k}\phi(tx) = t^k R_{-n-k}\phi(x) - t^k \log t \sum_{|\alpha|=k} \frac{\partial^\alpha \phi(0)}{\alpha!} x^\alpha$$

for any integer $k \geq 0$.

5.5 Show that if u is a homogeneous function in $C^\infty(\mathbb{R}^n \setminus \{0\})$, and if $\psi \in C^\infty(\mathbb{R}^n)$, then ψ f.p. $u = $ f.p.(ψu).

5.6 Show that for each $\phi \in S(\mathbb{R})$, the function $a \mapsto H_a(\phi)$ is analytic for $a \neq -1, -2, -3, \ldots$, with all simple poles, and residues

$$\operatorname*{res}_{a=-k-1} H_a(\phi) = \frac{\phi^{(k)}(0)}{k!}.$$

5.7 Show that if $\lambda > 0$ and $\phi \in S(\mathbb{R})$, then

$$\operatorname*{f.p.}_{\epsilon \downarrow 0} H_{-k-1,\lambda\epsilon}(\phi) = H_{-k-1}(\phi) - \frac{\phi^{(k)}(0)}{k!} \log \lambda$$

for any integer $k \geq 0$.

5.8 Show that for $t \neq 0$,

$$M_t \mathcal{F} = |t|^{-n} \mathcal{F} M_{1/t}, \quad M_t \mathcal{F}^* = |t|^{-n} \mathcal{F}^* M_{1/t},$$
$$\mathcal{F} M_t = |t|^{-n} M_{1/t} \mathcal{F}, \quad \mathcal{F}^* M_t = |t|^{-n} M_{1/t} \mathcal{F}^*.$$

5.9 Show that

$$\operatorname*{f.p.}_{\epsilon \downarrow 0} \int_\epsilon^\infty x^{a-1} e^{-x} \, dx = \Gamma(a) \quad \text{for } a \in \mathbb{C} \setminus \{0, -1, -2, \ldots\}.$$

5.10 Prove Corollary 5.11 directly, i.e., without using Lemma 5.10. For the first part, use (3.17), and for the second part, show that

$$H_{-k-1}(e^{-i2\pi\xi \cdot}) = \frac{(-i2\pi)^{k+1}}{2k!} \xi^k \operatorname{sign}(\xi) \quad \text{for any integer } k \geq 0.$$

5.11 Show that if $u \in L_{1,\text{loc}}(\mathbb{R})$ and $\phi \in \mathcal{D}(\mathbb{R}^n)$, then

$$\int_{\mathbb{R}^n} u(a \cdot x)\phi(x) \, dx = \int_{-\infty}^\infty u(t)\phi_a(t) \, dt,$$

where

$$\phi_a(t) = \frac{1}{|a|} \int_{a \cdot x_\perp = 0} \phi\left(x_\perp + \frac{t}{|a|^2} a\right) dx_\perp,$$

and $dx_\perp$ is the surface element on the $(n-1)$-dimensional hyperplane normal to a. Show further that $\phi_a \in \mathcal{D}(\mathbb{R})$, and that if u is a distribution on $\mathbb{R}$, then $u(a \cdot x)$ makes sense as a distribution on $\mathbb{R}^n$.

5.12 Derive alternative formulae to (5.12), (5.16) and (5.17), as follows.

(i) Show that there exists $f \in C^\infty_{\text{comp}}(0, \infty)$ satisfying

$$\int_0^\infty f(t) \frac{dt}{t} = 1 \quad \text{and} \quad \int_0^\infty f(t) \log t \frac{dt}{t} = 0.$$

[Hint: look for f in the form $f(t) = Cg(\lambda t)$ for appropriate constants $C > 0$ and $\lambda > 0$.]

(ii) Deduce that the cutoff function $\psi(x) = f(|x|)$ belongs to $C^\infty_{\text{comp}}(\mathbb{R}^n \setminus \{0\})$, and satisfies

$$\int_0^\infty \psi(tx) \frac{dt}{t} = 1 \quad \text{and} \quad \int_0^\infty \psi(tx) \log t \frac{dt}{t} = -\log|x|$$
$$\text{for } x \in \mathbb{R}^n \setminus \{0\}.$$

(iii) Show that if $u \in C^\infty(\mathbb{R}^n \setminus \{0\})$ is homogeneous of degree a, then

$$\langle \text{f.p.}\, u, \phi \rangle = \int_{\mathbb{R}^n} u(x) \psi(x) R_a \phi(x)\, dx \quad \text{for } \phi \in \mathcal{S}(\mathbb{R}^n).$$

(iv) Deduce that

$$\mathcal{F}_{x \to \xi}\{\text{f.p.}\, u(x)\} = \langle \Pi^+_{a+n-1}(\xi \cdot x), \psi(x)u(x) \rangle$$

and

$$\mathcal{F}^*_{\xi \to x}\{\text{f.p.}\, u(\xi)\} = \langle \Pi^+_{a+n-1}(-\xi \cdot x), \psi(\xi)u(\xi) \rangle.$$

5.13 Show that if $u \in C^\infty(\mathbb{R}^n \setminus \{0\})$ is homogeneous of degree $-n - k$ for some integer $k \geq 0$, and if u satisfies the parity condition (5.14), then

$$\mathcal{F}^*_{\xi \to x}\{\text{f.p.}\, u(\xi)\} = \frac{1}{2} \int_{|\omega|=1} \Pi_{-k-1}(-x \cdot \omega) u(\omega)\, d\omega$$
$$= \frac{(i2\pi)^{k+1}}{4k!} \int_{|\omega|=1} \text{sign}(x \cdot \omega)(x \cdot \omega)^k u(\omega)\, d\omega.$$

5.14 Suppose that $K \in C^\infty(\mathbb{R}^n \setminus \{0\})$ is homogeneous of degree $-n$, and that

$$\int_{|\omega|=1} K(\omega)\, d\omega = 0.$$

Define

$$\mathcal{K}_\epsilon u(x) = \int_{|y-x|>\epsilon} K(x-y)u(y)\,dy \quad \text{for } x \in \mathbb{R}^n,$$

whenever this integral exists and is finite, and put

$$\mathcal{K}u(x) = \lim_{\epsilon \downarrow 0} \mathcal{K}_\epsilon u(x),$$

whenever this limit exists.

(i) Show that if u is Hölder-continuous and has compact support, then

$$\mathcal{K}u(x) = \int_{|x-y|<\rho_x} K(x-y)[u(y) - u(x)]\,dy$$

where $\rho_x > 0$ is any number such that $u(y) = 0$ for $|x-y| \geq \rho_x$.

(ii) Show, with the assumptions of (i), that $\mathcal{K}_\epsilon u \to \mathcal{K}u$ uniformly on compact subsets of $\mathbb{R}^n$.

(iii) Show that p.v. K exists and is a homogeneous distribution of degree $-n$ on $\mathbb{R}^n$. [Hint: see Theorem 5.7.]

(iv) Show that $\mathcal{K}u = (\text{p.v. } K) * u$.

(v) Deduce that $\|\mathcal{K}u\|_{H^s(\mathbb{R}^n)} \leq C\|u\|_{H^s(\mathbb{R}^n)}$ for $-\infty < s < \infty$, and that $|\mathcal{K}u|_\mu \leq C|u|_\mu$ for $0 < \mu < 1$. [Hint: use Theorem 5.13 and Lemma 3.15.]

6

Surface Potentials

Following the notation of Chapter 4, we consider a second-order partial differential operator

$$\mathcal{P}u = -\sum_{j=1}^{n}\sum_{k=1}^{n} \partial_j (A_{jk}\partial_k u) + \sum_{j=1}^{n} A_j \partial_j u + Au.$$

From this point onwards, we shall always assume that $\mathcal{P}$ has C^∞ coefficients and is strongly elliptic on $\mathbb{R}^n$. Thus, A_{jk}, A_j and A are (bounded) C^∞ functions from $\mathbb{R}^n$ into $\mathbb{C}^{m \times m}$, with the leading coefficients satisfying

$$\mathrm{Re}\sum_{j=1}^{n}\sum_{k=1}^{n}[A_{jk}(x)\xi_k\eta]^*\xi_j\eta \geq c|\xi|^2|\eta|^2 \qquad \text{for } x, \xi \in \mathbb{R}^n \text{ and } \eta \in \mathbb{C}^m.$$

$$(6.1)$$

In this and subsequent chapters, we shall develop integral equation methods for solving boundary value problems involving $\mathcal{P}$. Such methods require a two-sided inverse for $\mathcal{P}$, or, more precisely, they require a linear operator $\mathcal{G}$ with the property that

$$\mathcal{P}\mathcal{G}u = u = \mathcal{G}\mathcal{P}u \quad \text{for } u \in \mathcal{E}^*(\mathbb{R}^n)^m. \tag{6.2}$$

Since $\mathcal{P}$ is a partial *differential* operator, it is natural to seek $\mathcal{G}$ in the form of an *integral* operator:

$$\mathcal{G}u(x) = \int_{\mathbb{R}^n} G(x, y)u(y)\,dy \quad \text{for } x \in \mathbb{R}^n. \tag{6.3}$$

The kernel G is said to be a *fundamental solution* for $\mathcal{P}$, and the same term is also applied to the operator $\mathcal{G}$, although we shall sometimes refer to the latter as a *volume potential*. We shall also work with a kind of approximate fundamental solution, known as a *parametrix*, that is generally easier to construct.

The plan of the chapter is as follows. The first two sections set out the main properties of parametrices and fundamental solutions, emphasising the simplest case when $\mathcal{P}$ has constant coefficients. Next, we prove the third Green identity, in which the single- and double-layer potentials arise. Following the approach of Costabel [14], we then prove the jump relations and mapping properties of these surface potentials for the case of a Lipschitz domain. The final section of the chapter establishes some relations between the surface potentials associated with $\mathcal{P}$ and those associated with $\mathcal{P}^*$.

Parametrices

A *smoothing operator* on $\mathbb{R}^n$ is an integral operator

$$\mathcal{K}u(x) = \int_{\mathbb{R}^n} K(x, y)u(y)\, dy \quad \text{for } x \in \mathbb{R}^n,$$

whose kernel K is C^∞ from $\mathbb{R}^n \times \mathbb{R}^n$ into $\mathbb{C}^{m \times m}$; it is easy to see that any such $\mathcal{K}$ satisfies

$$\mathcal{K} : \mathcal{E}^*(\mathbb{R}^n)^m \to \mathcal{E}(\mathbb{R}^n)^m.$$

Conversely, it can be shown that *every* continuous linear operator from $\mathcal{E}^*(\mathbb{R}^n)^m$ into $\mathcal{E}(\mathbb{R}^n)^m$ has a C^∞ kernel; see [10, p. 28].

A linear operator $\mathcal{G} : \mathcal{E}^*(\mathbb{R}^n)^m \to \mathcal{D}^*(\mathbb{R}^n)^m$ is called a parametrix for $\mathcal{P}$ if there exist smoothing operators $\mathcal{K}_1$ and $\mathcal{K}_2$ such that

$$\mathcal{P}\mathcal{G}u = u - \mathcal{K}_1 u \text{ and } \mathcal{G}\mathcal{P}u = u - \mathcal{K}_2 u \quad \text{for } u \in \mathcal{E}^*(\mathbb{R}^n)^m. \tag{6.4}$$

Roughly speaking, a parametrix allows us to invert $\mathcal{P}$ modulo smooth functions. Later, we shall write $\mathcal{G}$ as an integral operator as in (6.3), and refer also to its kernel $G(x, y)$ as a parametrix for $\mathcal{P}$.

When $\mathcal{P}$ has constant coefficients, we can easily construct a parametrix via the Fourier transform. Indeed, let $P_0(\xi)$ and $P(\xi)$ denote the polynomials corresponding to $\mathcal{P}_0$ and $\mathcal{P}$, respectively, i.e.,

$$P_0(\xi) = (2\pi)^2 \sum_{j=1}^n \sum_{k=1}^n \xi_j A_{jk} \xi_k$$

and

$$P(\xi) = P_0(\xi) + \mathrm{i}2\pi \sum_{j=1}^n A_j \xi_j + A.$$

For any $u \in \mathcal{S}^*(\mathbb{R}^n)^m$,

$$\mathcal{F}_{x \to \xi}\{\mathcal{P}_0 u(x)\} = P_0(\xi)\hat{u}(\xi) \quad \text{and} \quad \mathcal{F}_{x \to \xi}\{\mathcal{P}u(x)\} = P(\xi)\hat{u}(\xi),$$

and the strong ellipticity condition (6.1) can be written as

$$\operatorname{Re} \eta^* P_0(\xi)\eta \geq c|\xi|^2|\eta|^2 \quad \text{for } \xi \in \mathbb{R}^n \text{ and } \eta \in \mathbb{C}^m. \tag{6.5}$$

Thus, if $b = P_0(\xi)\eta$ then $c|\xi|^2|\eta|^2 \leq \operatorname{Re} \eta^* b \leq |\eta||b|$, giving $c|\xi|^2|\eta| \leq |b|$. It follows that the ℓ_2 matrix norm of $P_0(\xi)^{-1}$ satisfies the bound

$$|P_0(\xi)^{-1}| \leq \frac{1}{c|\xi|^2} \quad \text{for } 0 \neq \xi \in \mathbb{R}^n, \tag{6.6}$$

so we can find $\rho_0 > 0$ such that

$$\left| P_0(\xi)^{-1}(i2\pi) \sum_{j=1}^{n} A_j \xi_j \right| \leq \frac{1}{4} \quad \text{and} \quad |P_0(\xi)^{-1} A| \leq \frac{1}{4} \quad \text{for } |\xi| > \rho_0,$$

and hence

$$|P(\xi)^{-1}| \leq 2|P_0(\xi)^{-1}| \leq \frac{C}{|\xi|^2} \quad \text{for } |\xi| > \rho_0. \tag{6.7}$$

Fix a cutoff function $\chi \in C_{\text{comp}}^{\infty}(\mathbb{R}^n)$ satisfying

$$\chi(\xi) = 1 \quad \text{for } |\xi| < 2\rho_0,$$

and define

$$\mathcal{G}u(x) = \mathcal{F}_{\xi \to x}^*\{[1 - \chi(\xi)]P(\xi)^{-1}\hat{u}(\xi)\}. \tag{6.8}$$

We observe that $\mathcal{G}$ is an integral operator as in (6.3) with kernel

$$G(x, y) = \mathsf{G}(x - y), \quad \text{where } \mathsf{G}(z) = \mathcal{F}_{\xi \to z}^*\{[1 - \chi(\xi)]P(\xi)^{-1}\}. \tag{6.9}$$

Theorem 6.1 *If the strongly elliptic operator* $\mathcal{P}$ *has constant coefficients, then the formula (6.8) defines a parametrix for* $\mathcal{P}$, *and moreover*

$$\mathcal{G} : H^{s-1}(\mathbb{R}^n)^m \to H^{s+1}(\mathbb{R}^n)^m \quad \text{for } -\infty < s < \infty.$$

Proof. It is easy to see that

$$\mathcal{G} : \mathcal{S}(\mathbb{R}^n)^m \to \mathcal{S}(\mathbb{R}^n)^m \quad \text{and} \quad \mathcal{G} : \mathcal{S}^*(\mathbb{R}^n)^m \to \mathcal{S}^*(\mathbb{R}^n)^m,$$

and since $\mathcal{F}\{\mathcal{P}\mathcal{G}u\} = u - \chi u = \mathcal{F}\{\mathcal{G}\mathcal{P}u\}$ the condition (6.4) is satisfied with $\mathcal{K}_1 u = \mathcal{K}_2 u = \mathcal{F}^*\{\chi\hat{u}\}$, or equivalently, with

$$K_1(x, y) = K_2(x, y) = (\mathcal{F}^*\chi)(x - y).$$

This kernel is C^∞ because χ has compact support. Also, the estimate (6.7) implies that

$$\begin{aligned}
\|\mathcal{G}u\|^2_{H^{s+1}(\mathbb{R}^n)^m} &= \int_{\mathbb{R}^n} (1 + |\xi|^2)^{s+1} \left| [1 - \chi(\xi)] P(\xi)^{-1} \hat{u}(\xi) \right|^2 d\xi \\
&\leq C \int_{\mathbb{R}^n} (1 + |\xi|^2)^{s+1} \left| [1 - \chi(\xi)] |\xi|^{-2} \hat{u}(\xi) \right|^2 d\xi \\
&\leq C \|u\|^2_{H^{s-1}(\mathbb{R}^n)^m},
\end{aligned}$$

proving the desired mapping property.   □

In the general case when $\mathcal{P}$ is permitted to have non-constant coefficients, a parametrix $\mathcal{G}$ can be constructed using the symbol calculus from the theory of pseudodifferential operators; see Chazarain and Piriou [10, pp. 221–224]. The mapping property of Theorem 6.1 remains valid locally, i.e., given any fixed cutoff functions $\chi_1, \chi_2 \in C^\infty_{\text{comp}}(\mathbb{R}^n)$,

$$\chi_1 \mathcal{G} \chi_2 : H^{s-1}(\mathbb{R}^n)^m \to H^{s+1}(\mathbb{R}^n)^m \quad \text{for } -\infty < s < \infty. \tag{6.10}$$

The next lemma will help us to describe the behaviour of the kernel $G(x, y)$.

Lemma 6.2 Suppose that $v \in C^\infty(\mathbb{R}^n \setminus \{0\})$ is homogeneous of degree $-j$ for some integer $j \geq 1$. If

$$u(x) = \mathcal{F}^*_{\xi \to x}\{\text{f.p. } v(\xi)\},$$

then the distribution u is locally integrable on $\mathbb{R}^n$, and is C^∞ on $\mathbb{R}^n \setminus \{0\}$. Moreover,

(i) if $1 \leq j \leq n - 1$, then u is homogeneous of degree $j - n$;
(ii) if $j \geq n$, then

$$u(x) = u_1(x) + u_2(x) \log|x|,$$

where u_1 and u_2 are homogeneous of degree $j - n$ and C^∞ on $\mathbb{R}^n \setminus \{0\}$, with u_2 a polynomial.

Proof. Part (i) follows at once from Lemma 5.12 and Theorem 5.13, because v is locally integrable on $\mathbb{R}^n$. If $j \geq n$, then by (5.18),

$$u(x) = \int_{|\omega|=1} \Pi^+_{-j+n-1}(-x \cdot \omega)v(\omega)\,d\omega,$$

and part (ii) follows from Lemma 5.10. $\qquad\qquad\qquad\qquad\qquad\qquad\square$

We state the next theorem for the general case, but give a proof only for $\mathcal{P}$ having constant coefficients.

Theorem 6.3 *Assume that $\mathcal{P}$ is strongly elliptic with C^∞ coefficients on $\mathbb{R}^n$. There exists a parametrix $\mathcal{G}$ for $\mathcal{P}$ whose kernel admits an expansion of the form*

$$G(x, y) = \sum_{j=0}^{N} \mathsf{G}_j(x, x - y) + R_N(x, y), \tag{6.11}$$

for each $N \geq 0$, where the functions $\mathsf{G}_0, \mathsf{G}_1, \mathsf{G}_2, \ldots$ and $R_0, R_1, R_2, \ldots$ have the following properties:

(i) *For each $j \geq 0$, the function G_j is C^∞ on $\mathbb{R}^n \times (\mathbb{R}^n \setminus \{0\})$, and has the same parity as j in its second argument, i.e., $\mathsf{G}_j(x, -z) = (-1)^j \mathsf{G}_j(x, z)$.*

(ii) *If $0 \leq j \leq n - 3$, then $\mathsf{G}_j(x, z)$ is homogeneous in z of degree $2 - n + j$.*

(iii) *If $j \geq n - 2$, then G_j has the form*

$$\mathsf{G}_j(x, z) = \mathsf{G}_{j1}(x, z) + \mathsf{G}_{j2}(x, z) \log |z|,$$

where $\mathsf{G}_{j1}(x, z)$ and $\mathsf{G}_{j2}(x, z)$ are homogeneous in z of degree $2 - n + j$, with $\mathsf{G}_{j2}(x, z)$ a polynomial in z.

(iv) *If $0 \leq N + 1 \leq n - 3$, then*

$$\partial^\alpha R_N(x, y) = O\big(|x - y|^{2-n+(N+1)-|\alpha|}\big) \quad \text{as } |x - y| \to 0, \text{ for } |\alpha| \geq 0.$$

(v) *If $N + 1 \geq n - 2$, then R_N is C^{2-n+N} on $\mathbb{R}^n$, and*

$$\partial^\alpha R_N(x, y) = O\big(|x - y|^{2-n+(N+1)-|\alpha|} \log |x - y|\big)$$

as $|x - y| \to 0$, for $|\alpha| \geq 2 - n + (N + 1)$.

Proof. As mentioned above, we assume $\mathcal{P}$ has constant coefficients, and consider G given by (6.9). The choice of ρ_0 ensures that there exists an expansion of the form

$$P(\xi)^{-1} = \sum_{j=0}^{\infty} V_j(\xi) \quad \text{for } |\xi| > \rho_0,$$

with $V_j \in C^{\infty}(\mathbb{R}^n \setminus \{0\})^{m \times m}$ rational and homogeneous of degree $-2 - j$. We define

$$G_j(z) = \mathcal{F}^*_{\xi \to z}\{\text{f.p. } V_j(\xi)\}$$

and apply Lemma 6.2 to obtain (ii) and (iii). The expansion (6.11) serves to define R_N, and we see that $G_j(-z) = (-1)^j G_j(z)$ because $V_j(-\xi) = (-1)^j V_j(\xi)$.

Write $R_N(x, y) = R_{N,1}(x - y) + R_{N,2}(x - y) + R_{N,3}(x - y)$, where

$$\widehat{R}_{N,1} = -\chi \sum_{j=0}^{N+1} \text{f.p. } V_j, \qquad \widehat{R}_{N,2} = \text{f.p. } V_{N+1}, \qquad \widehat{R}_{N,3} = (1-\chi) \sum_{j=N+2}^{\infty} V_j.$$

We see that $R_{N,1}$ is C^{∞} on $\mathbb{R}^n$ because $\widehat{R}_{N,1}$ has compact support. Parts (ii) and (iii) imply that $R_{N,2}(x, y) = R_{N,2}(x - y) = G_{N+1}(x - y)$ satisfies conditions (iv)–(v). To deal with the remaining term $R_{N,3}(x, y) = R_{N,3}(x - y)$, we use the bounds

$$\left| \mathcal{F}_{x \to \xi}\left\{ (-i2\pi z)^{\beta} \bar{\partial}^{\alpha} R_{N,3}(z) \right\} \right| = \left| \partial_{\xi}^{\beta} (i2\pi \xi)^{\alpha} \widehat{R}_{N,3}(\xi) \right|$$
$$\leq C(1 + |\xi|)^{-2 - (N+2) + |\alpha| - |\beta|}.$$

Indeed, taking $\beta = 0$, we see that $\mathcal{F}\{\partial^{\alpha} R_{N,3}\} \in L_1(\mathbb{R}^n)^{m \times m}$ if $-2 - (N+2) + |\alpha| \leq -n - 1$, and thus $\partial^{\alpha} R_{N,3}$ is continuous on $\mathbb{R}^n$ if $|\alpha| \leq 3 - n + N$. Furthermore, by summing over $|\beta| = r \geq 0$, we see that $|z|^r \partial^{\alpha} R_{N,3}(z)$ is bounded for $z \in \mathbb{R}^n$ if $-2 - (N+2) + |\alpha| - r = -n - 1$, i.e., if $-r = 2 - n + (N+1) - |\alpha| \leq 0$, and thus $|\partial^{\alpha} R_{N,3}(z)| \leq C|z|^{2-n+(N+1)-|\alpha|}$ if $|\alpha| \geq 2 - n + (N+1)$. $\qquad \square$

Notice in particular that the parametrix $G(x, y)$ in Theorem 6.3 is C^{∞} for $x \neq y$. We can use this fact, together with the mapping property (6.10), to extend the interior regularity result of Theorem 4.16.

Theorem 6.4 *Let Ω_1 and Ω_2 be open subsets of $\mathbb{R}^n$, such that $\overline{\Omega}_1 \Subset \Omega_2$, and assume that $\mathcal{P}$ is strongly elliptic with C^{∞} coefficients. For s, $t \in \mathbb{R}$, if $u \in H^t(\Omega_2)^m$ and $f \in H^s(\Omega_2)^m$ satisfy*

$$\mathcal{P}u = f \quad \text{on } \Omega_2,$$

then $u \in H^{s+2}(\Omega_1)^m$ and

$$\|u\|_{H^{s+2}(\Omega_1)^m} \leq C\|u\|_{H^t(\Omega_2)^m} + C\|f\|_{H^s(\Omega_2)^m}.$$

Proof. Choose an open set $\Omega \subseteq \Omega_2$ such that $\overline{\Omega}_1 \Subset \Omega$ and $\overline{\Omega} \Subset \Omega_2$, and then choose a cutoff function $\chi_1 \in C^\infty_{\mathrm{comp}}(\Omega_2)$ such that $\chi_1 = 1$ on Ω. We have

$$\chi_1 u - \mathcal{K}_2(\chi_1 u) = \mathcal{G}\mathcal{P}(\chi_1 u) = \mathcal{G}(\chi_1 f) + \mathcal{G}[\mathcal{P}\chi_1 u - \chi_1 \mathcal{P}u],$$

and thus

$$\|u\|_{H^{s+2}(\Omega_1)^m} = \left\|\mathcal{K}_2(\chi_1 u) + \mathcal{G}(\chi_1 f) + \mathcal{G}[\mathcal{P}\chi_1 u - \chi_1 \mathcal{P}u]\right\|_{H^{s+2}(\Omega_1)^m}.$$

Since $\mathcal{K}_2$ is a smoothing operator, $\|\mathcal{K}_2(\chi_1 u)\|_{H^{s+2}(\Omega_1)^m} \leq C\|u\|_{H^t(\Omega_2)^m}$, and it follows from (6.10) that $\|\mathcal{G}(\chi_1 f)\|_{H^{s+2}(\Omega_1)^m} = \|(\chi_1 \mathcal{G}\chi_1)f\|_{H^{s+2}(\Omega_1)^m} \leq C\|f\|_{H^s(\Omega_2)^m}$. Finally, since $\mathcal{P}\chi_1 u - \chi_1 \mathcal{P}u = 0$ on Ω, and since $G(x, y)$ is C^∞ for $x \neq y$, we have

$$\|\mathcal{G}[\mathcal{P}\chi_1 u - \chi_1 \mathcal{P}u]\|_{H^{s+2}(\Omega_1)^m} \leq C\|u\|_{H^t(\Omega_2)^m}.$$

$\square$

One interesting consequence of the above result is that the parametrix is unique modulo smoothing operators.

Corollary 6.5 If $\mathcal{G}_1$ and $\mathcal{G}_2$ are parametrices for $\mathcal{P}$, then $\mathcal{G}_1 - \mathcal{G}_2$ is a smoothing operator, and hence $G_1 - G_2$ is C^∞ on $\mathbb{R}^n \times \mathbb{R}^n$.

In particular, it follows that the mapping properties (6.10), and the expansion in Theorem 6.3, are valid for *every* parametrix of $\mathcal{P}$. Two further consequences are now obvious.

Lemma 6.6 If $\mathcal{G}$ is a parametrix for $\mathcal{P}$, then $\mathcal{G} : \mathcal{D}(\mathbb{R}^n)^m \to \mathcal{E}(\mathbb{R}^n)^m$.

Theorem 6.7 The operator $\mathcal{G}$ is a parametrix for $\mathcal{P}$ if and only if $\mathcal{G}^*$ is a parametrix for $\mathcal{P}^*$.

Fundamental Solutions

A parametrix $\mathcal{G}$ (or its kernel G) is said to be a *fundamental solution* for $\mathcal{P}$ if (6.4) holds with $\mathcal{K}_1 = 0 = \mathcal{K}_2$, i.e., if

$$\mathcal{P}\mathcal{G}u = u = \mathcal{G}\mathcal{P}u \quad \text{for } u \in \mathcal{E}^*(\mathbb{R}^n)^m.$$

When $\mathcal{P}$ has constant coefficients, it is natural to seek a convolution kernel

$$G(x, y) = G(x - y)$$

with $G \in \mathcal{S}^*(\mathbb{R}^n)^{m \times m}$. Indeed, taking Fourier transforms we see that such a G is a fundamental solution if and only if

$$P(\xi)\widehat{G}(\xi) = I,$$

or equivalently,

$$\mathcal{P}G = \delta \quad \text{on } \mathbb{R}^n. \tag{6.12}$$

When the polynomial $P(\xi)$ is homogeneous, i.e., when $P(\xi) = P_0(\xi)$, then we can easily construct a fundamental solution as follows.

Theorem 6.8 *Assume that $\mathcal{P}$ has constant coefficients and no lower-order terms ($A_j = 0$ and $A = 0$).*

(i) *If $n = 2$, then the formula*

$$G(z) = \mathcal{F}^*_{\xi \to z}\{\text{f.p. } P(\xi)^{-1}\} = \int_{|\omega|=1} \left[\Gamma'(1) - \log 2\pi |\omega \cdot z| \right] P(\omega)^{-1} \, d\omega$$

defines a fundamental solution for $\mathcal{P}$.

(ii) *If $n \geq 3$, then $G(z) = \mathcal{F}^*_{\xi \to z}\{P(\xi)^{-1}\}$ defines a fundamental solution for $\mathcal{P}$, and G is homogeneous of degree $2 - n$.*

(iii) *If $n = 3$, then G in (ii) has the integral representation*

$$G(z) = \frac{1}{2|z|} \int_{\mathbb{S}_z^\perp} P(\omega_\perp)^{-1} \, d\omega_\perp,$$

where $\mathbb{S}_z^\perp = \{\omega_\perp \in \mathbb{S}^2 : \omega_\perp \cdot z = 0\}$ is the unit circle in the plane normal to z, and $d\omega_\perp$ is the element of arc length on $\mathbb{S}_z^\perp$.

Proof. We see from (6.6) that $P(\xi)$ is invertible for $\xi \neq 0$. Thus, $P(\xi)^{-1}$ is C^∞ and homogeneous of degree -2 for $\xi \in \mathbb{R}^n \setminus \{0\}$.

Suppose $n = 2$. We know from Lemma 5.12 that f.p. $P(\xi)^{-1}$ is a temperate distribution, so $G(z)$ is well defined as the inverse Fourier transform of f.p. $P(\xi)^{-1}$, and by Exercise 5.5,

$$\mathcal{F}_{z \to \xi}\{\mathcal{P}G(z)\} = P(\xi) \text{ f.p. } P(\xi)^{-1} = I = \mathcal{F}_{z \to \xi}\{\delta(z)\},$$

implying that G is a fundamental solution for $\mathcal{P}$. Since $P(-\omega) = P(\omega)$, we see from (5.18) that

$$G(z) = \frac{1}{2} \int_{|\omega|=1} \left[\Pi^+_{-1}(-\omega \cdot z) + \Pi^+_{-1}(\omega \cdot z) \right] P(\omega)^{-1} \, d\omega,$$

and, by Lemma 5.10,

$$\Pi^+_{-1}(\xi) = \Gamma'(1) - \log 2\pi |\xi| - \frac{i\pi}{2} \operatorname{sign}(\xi),$$

giving the integral representation for $G(z)$, and completing the proof of part (i).

Part (ii) is clear from Theorem 5.13, because for $n \geq 3$ the function $P(\xi)^{-1}$ is locally integrable on $\mathbb{R}^n$, and thus homogeneous of degree -2 as a distribution on $\mathbb{R}^n$.

To prove (iii), we note that by Exercise 5.12, if $n = 3$ then

$$G(z) = \langle \Pi^+_0(-\xi \cdot z), \psi(\xi) P(\xi)^{-1} \rangle \quad \text{for } z \in \mathbb{R}^3 \setminus \{0\}.$$

Moreover, since $P(-\xi) = P(\xi)$, and since we can choose ψ so that $\psi(-\xi) = \psi(\xi)$,

$$G(z) = \tfrac{1}{2} \langle \Pi^+_0(-\xi \cdot z) + \Pi^+_0(\xi \cdot z), \psi(\xi) P(\xi)^{-1} \rangle.$$

Observe that $\Pi^+_0(-\xi \cdot z) = \Pi^-_0(\xi \cdot z)$ and $x^0_+ + x^0_- = 1$, so $\Pi^+_0 + \Pi^-_0 = \mathcal{F}\{1\} = \delta$, and therefore, if we define $\phi(\xi) = \psi(\xi) P(\xi)^{-1}$ and $\phi_z(t)$ as in the proof of Lemma 5.14, then, by Exercise 5.11,

$$G(z) = \tfrac{1}{2} \langle \delta(\xi \cdot z), \psi(\xi) P(\xi)^{-1} \rangle = \tfrac{1}{2} \langle \delta, \phi_z \rangle = \tfrac{1}{2} \phi_z(0)$$

$$= \frac{1}{2|z|} \int_{z \cdot \xi_\perp = 0} \psi(\xi_\perp) P(\xi_\perp)^{-1} \, d\xi_\perp.$$

Introducing polar coordinates in the plane normal to z, i.e., putting $\xi_\perp = \rho \omega_\perp$ with $\rho = |\xi_\perp|$ and $\omega_\perp = \xi_\perp / \rho \in \mathbb{S}^\perp_z$, we find that

$$\int_{\xi_\perp \cdot z = 0} \psi(\xi_\perp) P(\xi_\perp)^{-1} d\xi_\perp = \int_{\rho > 0} \int_{\omega_\perp \in \mathbb{S}^\perp_z} \psi(\rho \omega_\perp) P(\rho \omega_\perp)^{-1} \, \rho \, d\rho \, d\omega_\perp$$

$$= \int_{\mathbb{S}^\perp_z} \left(\int_0^\infty \psi(\rho \omega_\perp) \frac{d\rho}{\rho} \right) P(\omega_\perp)^{-1} \, d\omega_\perp,$$

which yields the desired formula for $G(z)$. $\qquad\square$

Of course, in part (i) of Theorem 6.8 we can simply take

$$G(z) = \int_{\mathbb{S}^1} \left(\log \frac{1}{|\omega \cdot z|} \right) P(\omega)^{-1} \, d\omega, \qquad (6.13)$$

because $\mathcal{P} = \mathcal{P}_0$ annihilates constants.

We shall not prove any other general existence result for fundamental solutions, although Chapter 9 treats a particular example with $A \neq 0$. Dieudonné [19, pp. 253–256] discusses the history of existence proofs for fundamental solutions of general classes of partial differential operators. Gel'fand and Shilov [27, p. 122] give a reasonably simple proof for scalar elliptic operators of arbitrary order with constant coefficients, and Hörmander [41, Theorem 7.3.10] considers arbitrary (not necessarily elliptic) partial differential operators with constant coefficients. Miranda [67, Theorem 19, VIII] treats second-order elliptic equations with variable coefficients.

The Third Green Identity

Let us recall the notation used in our discussion of the transmission property (Theorem 4.20). The set Ω^- is a bounded Lipschitz domain in $\mathbb{R}^n$, Ω^+ is the complementary, *un*bounded Lipschitz domain, $\Gamma = \partial\Omega^+ = \partial\Omega^-$, and we have the sesquilinear forms $\Phi^\pm = \Phi_{\Omega^\pm}$, defined by

$$\Phi^\pm(u, v) = \int_{\Omega^\pm} \left(\sum_{j=1}^n \sum_{k=1}^n (A_{jk}\partial_k u)^* \partial_j v + \sum_{j=1}^n (A_j \partial_j u)^* v + (Au)^* v \right) dx;$$

cf. (4.2). The one-sided trace operators for Ω^+ and Ω^- are denoted by γ^+ and γ^-, respectively, so that

$$\gamma^\pm u = (U^\pm)|_\Gamma \quad \text{when } u = U^\pm|_{\Omega^\pm} \text{ for some } U^\pm \in \mathcal{D}(\mathbb{R}^n)^m.$$

In the usual way, we extend $\gamma^\pm$ to a bounded linear operator

$$\gamma^\pm : H^s(\Omega^\pm)^m \to H^{s-1/2}(\Gamma)^m \quad \text{for } \tfrac{1}{2} < s < \tfrac{3}{2}.$$

The one-sided conormal derivatives of a function $u \in H^2(\Omega^\pm)^m$ are defined by

$$\mathcal{B}_\nu^\pm u = \sum_{j=1}^n \nu_j \gamma^\pm \left(\sum_{k=1}^n A_{jk}\partial_k u \right) \quad \text{and} \quad \widetilde{\mathcal{B}}_\nu^\pm = \sum_{j=1}^n \nu_j \gamma^\pm \left(\sum_{k=1}^n A_{kj}^*\partial_k u + A_j^* u \right),$$

with the usual generalisation via the first Green identity, as in Lemma 4.3. Remember also our convention that the unit normal ν points out of Ω^- and into Ω^+.

When u is defined on the whole of $\mathbb{R}^n$, we sometimes write $u^{\pm} = u|_{\Omega^{\pm}}$ for its restriction to $\Omega^{\pm}$. To avoid redundant $+$ and $-$ signs, we write the one-sided traces as $\gamma^+ u$ and $\gamma^- u$ instead of $\gamma^+ u^+$ and $\gamma^- u^-$, and similarly for the one-sided conormal derivatives. The jumps in these quantities are denoted by

$$[u]_\Gamma = \gamma^+ u - \gamma^- u, \quad [\mathcal{B}_\nu u]_\Gamma = \mathcal{B}_\nu^+ u - \mathcal{B}_\nu^- u, \quad [\widetilde{\mathcal{B}}_\nu u]_\Gamma = \widetilde{\mathcal{B}}_\nu^+ u - \widetilde{\mathcal{B}}_\nu^- u,$$

and we often indicate that a jump vanishes by dropping the $+$ or $-$ superscript; for instance, we write

$$\gamma u = \gamma^+ u = \gamma^- u \quad \text{if } [u]_\Gamma = 0.$$

The first new symbols are γ^*, the adjoint of the two-sided trace operator, and $\widetilde{\mathcal{B}}_\nu^*$, defined by

$$(\gamma^* \psi, \phi) = (\psi, \gamma \phi)_\Gamma \text{ and } (\widetilde{\mathcal{B}}_\nu^* \psi, \phi) = (\psi, \widetilde{\mathcal{B}}_\nu \phi)_\Gamma \qquad \text{for } \phi \in \mathcal{E}(\mathbb{R}^n)^m. \tag{6.14}$$

By Theorem 3.38, $\gamma \phi \in H^{1-\epsilon}(\Gamma)^m$ for $0 < \epsilon \le 2$, so $\gamma^* \psi$ makes sense as a distribution on $\mathbb{R}^n$ for any $\psi \in H^{\epsilon-1}(\Gamma)^m$. Similarly, since $\widetilde{\mathcal{B}}_\nu \phi \in L_\infty(\Gamma)^m$, we see that $\widetilde{\mathcal{B}}_\nu^* \psi$ makes sense as a distribution on $\mathbb{R}^n$ for any $\psi \in L_1(\Gamma)^m$. Obviously,

$$\operatorname{supp} \gamma^* \psi \subseteq \operatorname{supp} \psi \subseteq \Gamma \quad \text{and} \quad \operatorname{supp} \widetilde{\mathcal{B}}_\nu^* \psi \subseteq \operatorname{supp} \psi \subseteq \Gamma.$$

Using γ^* and $\widetilde{\mathcal{B}}_\nu^*$, we can restate Lemma 4.19 as follows.

Lemma 6.9 *Let* $f^{\pm} \in \widetilde{H}^{-1}(\Omega^{\pm})^m$ *and put* $f = f^+ + f^- \in H^{-1}(\mathbb{R}^n)^m$, *and suppose that* $u \in L_2(\mathbb{R}^n)^m$ *with* $u^{\pm} \in H^1(\Omega^{\pm})^m$. *If*

$$\mathcal{P} u^{\pm} = f^{\pm} \quad \text{on } \Omega^{\pm},$$

then

$$\mathcal{P} u = f + \widetilde{\mathcal{B}}_\nu^* [u]_\Gamma - \gamma^* [\mathcal{B}_\nu u]_\Gamma \quad \text{on } \mathbb{R}^n. \tag{6.15}$$

Now let $\mathcal{G}$ be a parametrix for $\mathcal{P}$. Thus, there are smoothing operators $\mathcal{K}_1$ and $\mathcal{K}_2$ such that

$$\mathcal{P} \mathcal{G} u = u - \mathcal{K}_1 u \text{ and } \mathcal{G} \mathcal{P} u = u - \mathcal{K}_2 u \qquad \text{for } u \in \mathcal{E}^*(\mathbb{R}^n)^m,$$

and of course $\mathcal{K}_1 = 0 = \mathcal{K}_2$ if G is a fundamental solution. Motivated by Lemma 6.9, we define the *single-layer potential* SL and the *double-layer potential* DL by

$$\mathrm{SL} = \mathcal{G}\gamma^* \quad \text{and} \quad \mathrm{DL} = \mathcal{G}\widetilde{\mathcal{B}}_\nu^*.$$

Applying $\mathcal{G}$ to both sides of (6.15), the *third Green identity* follows immediately.

Theorem 6.10 If, in addition to the hypotheses of Lemma 6.9, the function u has compact support in $\mathbb{R}^n$ (and thus also f has compact support in $\mathbb{R}^n$), then

$$u = \mathcal{G}f + \mathrm{DL}[u]_\Gamma - \mathrm{SL}[\mathcal{B}_\nu u]_\Gamma + \mathcal{K}_2 u \quad \text{on } \mathbb{R}^n.$$

The definitions above mean that

$$(\mathrm{SL}\,\psi, \phi) = (\psi, \gamma\mathcal{G}^*\phi)_\Gamma \quad \text{and} \quad (\mathrm{DL}\,\psi, \phi) = (\psi, \widetilde{\mathcal{B}}_\nu\mathcal{G}^*\phi)_\Gamma \quad \text{for } \phi \in \mathcal{E}(\mathbb{R}^n)^m,$$

so by considering test functions with $\operatorname{supp}\phi \Subset \mathbb{R}^n \setminus \Gamma$, and recalling from Theorem 6.3 and Corollary 6.5 that $G(x, y)$ is C^∞ for $x \neq y$, one obtains the integral representations

$$\mathrm{SL}\,\psi(x) = \int_\Gamma G(x, y)\psi(y)\,d\sigma_y \tag{6.16}$$

and

$$\mathrm{DL}\,\psi(x) = \int_\Gamma \left[\widetilde{\mathcal{B}}_{\nu,y}G(x, y)^*\right]^*\psi(y)\,d\sigma_y, \tag{6.17}$$

for $x \in \mathbb{R}^n \setminus \Gamma$. Notice also that

$$\mathcal{P}\,\mathrm{SL}\,\psi = \gamma^*\psi - \mathcal{K}_1\gamma^*\psi \quad \text{and} \quad \mathcal{P}\,\mathrm{DL}\,\psi = \widetilde{\mathcal{B}}_\nu^*\psi - \mathcal{K}_1\widetilde{\mathcal{B}}_\nu^*\psi \quad \text{on } \mathbb{R}^n, \tag{6.18}$$

and hence

$$\mathcal{P}\,\mathrm{SL}\,\psi = -\mathcal{K}_1\gamma^*\psi \quad \text{and} \quad \mathcal{P}\,\mathrm{DL}\,\psi = -\mathcal{K}_1\widetilde{\mathcal{B}}_\nu^*\psi \quad \text{on } \Omega^\pm. \tag{6.19}$$

In particular, if G is a fundamental solution, then $\mathcal{P}\,\mathrm{SL}\,\psi = 0 = \mathcal{P}\,\mathrm{DL}\,\psi$ on $\Omega^\pm$.

Jump Relations and Mapping Properties

The surface potentials $\mathrm{SL}\psi$ and $\mathrm{DL}\psi$ are C^∞ on $\Omega^\pm$ because $G(x, y)$ is C^∞ for $x \neq y$. We shall now investigate their behaviour at the boundary Γ. The results in the next two theorems, for general Lipschitz domains, are from the paper of Costabel [14].

Theorem 6.11 Fix a cutoff function $\chi \in C_{\text{comp}}^\infty(\mathbb{R}^n)$. *The single-layer potential* SL *and the double-layer potential* DL *give rise to bounded linear operators*

$$\chi \text{SL} : H^{-1/2}(\Gamma)^m \to H^1(\mathbb{R}^n)^m, \qquad \chi \, \text{DL} : H^{1/2}(\Gamma)^m \to H^1(\Omega^\pm)^m,$$

$$\gamma \text{SL} : H^{-1/2}(\Gamma)^m \to H^{1/2}(\Gamma)^m, \qquad \gamma^\pm \, \text{DL} : H^{1/2}(\Gamma)^m \to H^{1/2}(\Gamma)^m,$$

$$\mathcal{B}_\nu^\pm \, \text{SL} : H^{-1/2}(\Gamma)^m \to H^{-1/2}(\Gamma)^m, \qquad \mathcal{B}_\nu \, \text{DL} : H^{1/2}(\Gamma)^m \to H^{-1/2}(\Gamma)^m,$$

and satisfy the jump relations

$$[\text{SL} \, \psi]_\Gamma = 0 \ \text{ and } \ [\mathcal{B}_\nu \, \text{SL} \, \psi]_\Gamma = -\psi \qquad \text{for } \psi \in H^{-1/2}(\Gamma)^m,$$

and

$$[\text{DL} \, \psi]_\Gamma = \psi \ \text{ and } \ [\mathcal{B}_\nu \, \text{DL} \, \psi]_\Gamma = 0 \qquad \text{for } \psi \in H^{1/2}(\Gamma)^m.$$

Proof. Choose a second cutoff function $\chi_1 \in \mathcal{D}(\mathbb{R}^n)$, satisfying $\chi_1 = 1$ on a neighbourhood of $\Omega^- \cup \Gamma$. For $\psi \in \mathcal{D}(\Gamma)^m$ and $\phi \in \mathcal{D}(\mathbb{R}^n)^m$,

$$(\chi \text{SL} \, \psi, \phi) = (\chi \mathcal{G} \chi_1 \gamma^* \psi, \phi) = \big(\psi, \gamma(\chi_1 \mathcal{G}^* \chi)\phi\big)_\Gamma,$$

and by Theorem 3.38 and (6.10),

$$\gamma : H^1(\mathbb{R}^n)^m \to H^{1/2}(\Gamma)^m \quad \text{and} \quad \chi_1 \mathcal{G}^* \chi : H^{-1}(\mathbb{R}^n)^m \to H^1(\mathbb{R}^n)^m,$$
$$\tag{6.20}$$

so $\|\gamma(\chi_1 \mathcal{G}^* \chi)\phi\|_{H^{1/2}(\mathbb{R}^n)^m} \le C \|\phi\|_{H^{-1}(\mathbb{R}^n)^m}$. Hence,

$$|(\chi \text{SL} \, \psi, \phi)| \le C \|\psi\|_{H^{-1/2}(\Gamma)^m} \|\phi\|_{H^{-1}(\mathbb{R}^n)^m},$$

implying that $\|\chi \text{SL} \, \psi\|_{H^1(\mathbb{R}^n)^m} \le C \|\psi\|_{H^{-1/2}(\Gamma)^m}$. This inequality proves the mapping property for χSL. The mapping property for γSL and the first jump relation $[\text{SL}\psi]_\Gamma = 0$ follow at once because $\gamma : H^1(\mathbb{R}^n)^m \to H^{1/2}(\Gamma)^m$; cf. Exercise 4.5.

The mapping property for $\mathcal{B}_\nu^- \, \text{SL}$ is proved using the first Green identity. In fact, since $\mathcal{P} \, \text{SL} \, \psi = -\mathcal{K}_1 \gamma^* \psi$ on Ω^-, we see by applying Lemma 4.3 that

$$\|\mathcal{B}_\nu^- \, \text{SL} \, \psi\|_{H^{-1/2}(\Gamma)^m} \le C \| \text{SL} \, \psi\|_{H^1(\Omega^-)^m} + C \|\mathcal{K}_1 \gamma^* \psi\|_{\tilde{H}^{-1}(\Omega^-)^m}.$$

If $\phi \in H^1(\Omega^-)^m$, then

$$|(\mathcal{K}_1 \gamma^* \psi, \phi)_{\Omega^-}| = |(\psi, \gamma \mathcal{K}_1 \phi)_\Gamma| \le C \|\psi\|_{H^{-1/2}(\Gamma)^m} \|\mathcal{K}_1 \phi\|_{H^1(\Omega^-)^m}$$
$$\le C \|\psi\|_{H^{-1/2}(\Gamma)^m} \|\phi\|_{H^1(\Omega^-)^m},$$

so $\|\mathcal{K}_1\gamma^*\psi\|_{\tilde{H}^{-1}(\Omega^-)^m} \leq C\|\psi\|_{H^{-1/2}(\Gamma)^m}$. Hence, using the mapping property of SL just proved, $\|\mathcal{B}_\nu^- \mathrm{SL}\,\psi\|_{H^{-1/2}(\Gamma)^m} \leq C\|\psi\|_{H^{-1/2}(\Gamma)^m}$. Essentially the same argument, applied over Ω^+, proves the mapping property for $\mathcal{B}_\nu^+ \mathrm{SL}$. The only complication is that the cutoff function χ_1 is needed to ensure that $\chi_1 \mathrm{SL}\,\psi \in H^1(\Omega^+)^m$; the details are left to the reader.

The jump $[\mathcal{B}_\nu \mathrm{SL}\,\psi]_\Gamma$ is found by applying the formula (6.15) to the function $u = \chi_1 \mathrm{SL}\,\psi$. Indeed, since $[u]_\Gamma = 0$ and $\mathcal{B}_\nu^\pm u = \mathcal{B}_\nu^\pm \mathrm{SL}\,\psi$, if Ω' is an open neighbourhood of $\Omega^- \cup \Gamma$ on which $\chi_1 = 1$, then

$$\mathcal{P}u = -\mathcal{K}_1\gamma^*\psi - \gamma^*[\mathcal{B}_\nu \mathrm{SL}\,\psi]_\Gamma \quad \text{on } \Omega',$$

whereas by (6.18),

$$\mathcal{P}u = \gamma^*\psi - \mathcal{K}_1\gamma^*\psi \quad \text{on } \Omega'.$$

Thus, $\gamma^*\big(\psi + [\mathcal{B}_\nu \mathrm{SL}\,\psi]_\Gamma\big) = 0$, or in other words,

$$(\psi + [\mathcal{B}_\nu \mathrm{SL}\,\psi]_\Gamma, \gamma\phi)_\Gamma = 0 \quad \text{for all } \phi \in \mathcal{D}(\mathbb{R}^n)^m,$$

and so $\psi + [\mathcal{B}_\nu \mathrm{SL}\,\psi]_\Gamma = 0$. All properties of the single-layer potential have now been established.

To handle the double-layer potential, we choose $\lambda > 0$ large enough so that $\mathcal{P} + \lambda$ is positive and bounded below on $H_0^1(\Omega^-)^m$. (Such a λ exists by Theorem 4.6.) Thus, the Dirichlet problem

$$\begin{aligned}(\mathcal{P} + \lambda)u &= 0 \quad \text{on } \Omega^-, \\ \gamma^- u &= g \quad \text{on } \Gamma\end{aligned} \tag{6.21}$$

has a unique solution $u \in H^1(\Omega^-)^m$ for each $g \in H^{1/2}(\Gamma)^m$, and we can define the solution operator $\mathcal{U} : g \mapsto u$. Recall that this operator was discussed in the final section of Chapter 4. Let $\psi \in \mathcal{D}(\Gamma)^m$ and define $u \in L_2(\mathbb{R}^n)^m$ by

$$u = \begin{cases} \mathcal{U}\psi & \text{on } \Omega^-, \\ 0 & \text{on } \Omega^+. \end{cases}$$

Since $\mathcal{P}u = -\lambda u$ on $\Omega^\pm$, the third Green identity (Theorem 6.10) gives

$$u = -\lambda \mathcal{G}u + \mathrm{DL}[u]_\Gamma - \mathrm{SL}[\mathcal{B}_\nu u]_\Gamma + \mathcal{K}_2 u \quad \text{on } \mathbb{R}^n,$$

and since $\gamma^+ u = \mathcal{B}_\nu^+ u = 0$,

$$[u]_\Gamma = -\gamma^- \mathcal{U}\psi = -\psi \quad \text{and} \quad [\mathcal{B}_\nu u]_\Gamma = -\mathcal{B}_\nu \mathcal{U}\psi,$$

so

$$\mathrm{DL}\,\psi = \mathrm{SL}\,\mathcal{B}_\nu^- \mathcal{U}\psi - u - \lambda\mathcal{G}u + \mathcal{K}_2 u \quad \text{on } \mathbb{R}^n. \tag{6.22}$$

The mapping property of $\chi\,\mathrm{SL}$ and the mapping property (4.38) of $\mathcal{B}_\nu^- \mathcal{U}$ imply that

$$\|\chi\,\mathrm{SL}\,\mathcal{B}_\nu^- \mathcal{U}\psi\|_{H^1(\mathbb{R}^n)^m} \le C\|\mathcal{B}_\nu^- \mathcal{U}\psi\|_{H^{-1/2}(\Gamma)^m} \le C\|\psi\|_{H^{1/2}(\Gamma)^m},$$

and by (6.10),

$$\|\chi\mathcal{G}u\|_{H^2(\mathbb{R}^n)^m} = \|(\chi\mathcal{G}\chi_1)u\|_{H^2(\mathbb{R}^n)^m} \le C\|u\|_{L_2(\mathbb{R}^n)^m} = C\|\mathcal{U}\psi\|_{L_2(\Omega^-)^m}.$$

Thus, the mapping property (4.37) of $\mathcal{U}$ gives $\|\chi\,\mathrm{DL}\,\psi\|_{H^1(\Omega^\pm)^m} \le C\|\psi\|_{H^{1/2}(\Gamma)^m}$. The mapping property of $\gamma^\pm\,\mathrm{DL}$ follows at once from (6.20), and we obtain the mapping property of $\mathcal{B}_\nu^\pm\,\mathrm{DL}$ by again using (6.22):

$$\begin{aligned}
\|\mathcal{B}_\nu^\pm\,\mathrm{DL}\,\psi\|_{H^{-1/2}(\Gamma)^m} &\le \|\mathcal{B}_\nu^\pm\,\mathrm{SL}\,\mathcal{B}_\nu^- \mathcal{U}\psi\|_{H^{-1/2}(\Gamma)^m} + \|\mathcal{B}_\nu^\pm u\|_{H^{-1/2}(\Gamma)^m} \\
&\quad + \lambda\|\mathcal{B}_\nu^\pm\mathcal{G}u\|_{H^{-1/2}(\Gamma)^m} + \|\mathcal{B}_\nu\mathcal{K}_2 u\|_{H^{-1/2}(\Gamma)^m} \\
&\le C\|\mathcal{B}_\nu^- \mathcal{U}\psi\|_{H^{-1/2}(\Gamma)^m} + C\|(\chi_1\mathcal{G}\chi_1)u\|_{H^1(\mathbb{R}^n)^m} \\
&\quad + \|u\|_{L_2(\mathbb{R}^n)^m} \\
&\le C\|\psi\|_{H^{1/2}(\Gamma)^m} + C\|\mathcal{U}\psi\|_{H^1(\Omega^-)^m} \\
&\le C\|\psi\|_{H^{1/2}(\Gamma)^m}.
\end{aligned}$$

Finally, $[\mathcal{G}u]_\Gamma = [\mathcal{B}_\nu\mathcal{G}u]_\Gamma = 0$ because $\mathcal{G}u \in H^2(\mathbb{R}^n)$, and

$$[\mathrm{SL}\,\mathcal{B}_\nu^- \mathcal{U}\psi]_\Gamma = 0 \quad \text{and} \quad [\mathcal{B}_\nu\,\mathrm{SL}\,\mathcal{B}_\nu^- \mathcal{U}\psi]_\Gamma = -\mathcal{B}_\nu^- \mathcal{U}\psi,$$

so by (6.22),

$$\begin{aligned}
[\mathrm{DL}\,\psi]_\Gamma = -[u]_\Gamma = \psi \quad \text{and} \quad [\mathcal{B}_\nu\,\mathrm{DL}\,\psi]_\Gamma &= -\mathcal{B}_\nu^- \mathcal{U}\psi - [\mathcal{B}_\nu u]_\Gamma \\
&= -\mathcal{B}_\nu^+ u = 0,
\end{aligned}$$

proving the jump relations for DL. $\qquad\square$

Next, we show that the mapping properties of the surface potentials can be extended to a range of Sobolev spaces. Note, however, that the results below are not quite the best possible; see the discussion following the proof.

Theorem 6.12 *Fix a cutoff function* $\chi \in C^\infty_{\mathrm{comp}}(\mathbb{R}^n)$, *and assume that* $-\frac{1}{2} < s < \frac{1}{2}$.

(i) For the single-layer potential, we have

$$\chi SL : H^{s-1/2}(\Gamma)^m \to H^{s+1}(\mathbb{R}^n)^m,$$
$$\gamma SL : H^{s-1/2}(\Gamma)^m \to H^{s+1/2}(\Gamma)^m.$$

(ii) If $\mathcal{P}$ satisfies the hypotheses of Theorem 4.25, then the solution operator for the Dirichlet problem (6.21) satisfies

$$\mathcal{U} : H^{s+1/2}(\Gamma)^m \to H^{s+1}(\Omega^-)^m,$$

and for the single- and double-layer potentials we have

$$\mathcal{B}_v^\pm SL : H^{s-1/2}(\Gamma)^m \to H^{s-1/2}(\Gamma)^m,$$
$$\chi DL : H^{s+1/2}(\Gamma)^m \to H^{s+1}(\Omega^\pm)^m,$$
$$\gamma^\pm DL : H^{s+1/2}(\Gamma)^m \to H^{s+1/2}(\Gamma)^m,$$
$$\mathcal{B}_v DL : H^{s+1/2}(\Gamma)^m \to H^{s-1/2}(\Gamma)^m.$$

Proof. We prove (i) by generalising the corresponding part in the proof of Theorem 6.11. Using, instead of (6.20),

$$\gamma : H^{-s+1}(\mathbb{R}^n)^m \to H^{-s+1/2}(\Gamma)^m \quad \text{and}$$
$$\chi_1 \mathcal{G}^* \chi : H^{-s-1}(\mathbb{R}^n)^m \to H^{-s+1}(\Gamma)^m, \tag{6.23}$$

we have $\|\gamma(\chi_1 \mathcal{G}\chi)\phi\|_{H^{-s+1/2}(\Gamma)^m} \le C\|\phi\|_{H^{-s-1}(\mathbb{R}^n)^m}$, and hence

$$|(\chi SL\, \chi_1\psi, \phi)| = \left|(\psi, \gamma(\chi_1 \mathcal{G}^*\chi)\phi)_\Gamma\right| \le C\|\psi\|_{H^{-1/2}(\Gamma)^m}\|\phi\|_{H^{-s-1}(\mathbb{R}^n)^m}.$$

To prove (ii), let $\mathcal{V} : \phi \mapsto v$ be the solution operator for the dual problem to (6.21):

$$(\mathcal{P}^* + \lambda)v = 0 \quad \text{on } \Omega^-,$$
$$\gamma^- v = \phi \quad \text{on } \Gamma.$$

By Theorem 4.25,

$$\mathcal{B}_v\mathcal{U} : H^{s+1/2}(\Gamma)^m \to H^{s-1/2}(\Gamma)^m \quad \text{and} \quad \tilde{\mathcal{B}}_v\mathcal{V} : H^{s+1/2}(\Gamma)^m \to H^{s-1/2}(\Gamma)^m, \tag{6.24}$$

and also

$$\mathcal{U} : L_2(\Gamma)^m \to L_2(\Omega^-)^m. \tag{6.25}$$

We choose a number ρ large enough so that $|x| < \rho/2$ for all $x \in \Omega^-$, and put

$$\Omega_\rho^+ = \{x \in \Omega^+ : |x| < \rho\} \quad \text{and} \quad \Gamma_\rho = \{x \in \mathbb{R}^n : |x| = \rho\}.$$

We can assume λ is large enough so that $\mathcal{P} + \lambda$ is positive and bounded below on $H_0^1(\Omega_\rho^+)^m$ as well as on $H^1(\Omega^-)^m$. Hence, there is a bounded linear operator

$$\mathcal{U}_\rho^+ : H^{1/2}(\Gamma)^m \to H^1(\Omega_\rho^+)^m$$

defined by $\mathcal{U}_\rho^+ g = u$, where u is the unique solution of the Dirichlet problem

$$(\mathcal{P} + \lambda)u = 0 \quad \text{on } \Omega_\rho^+,$$
$$\gamma^+ u = g \quad \text{on } \Gamma,$$
$$\gamma_\rho^+ u = 0 \quad \text{on } \Gamma_\rho,$$

and γ_ρ^+ is the trace map from $H^1(\Omega_\rho^+)^m$ onto $H^{1/2}(\Gamma_\rho)^m$. Now let $g \in \mathcal{D}(\Gamma)^m$, and define $w \in L_2(\mathbb{R}^n)^m$ by

$$w = \begin{cases} \mathcal{U}g & \text{on } \Omega^-, \\ \mathcal{U}_\rho^+ g & \text{on } \Omega_\rho^+, \\ 0 & \text{on } \Omega^+ \setminus \Omega_\rho^+. \end{cases}$$

Define ν on Γ_ρ to be the *in*ward unit normal to Ω_ρ^+, and let SL_ρ denote the single-layer potential on the (disconnected) surface $\partial \Omega_\rho^+ = \Gamma \cup \Gamma_\rho$. Since $[w]_{\Gamma \cup \Gamma_\rho} = 0$ and

$$\mathcal{P}w = \begin{cases} -\lambda w & \text{on } \Omega^-, \\ -\lambda w & \text{on } \Omega_\rho^+, \\ 0 & \text{on } \Omega^+ \setminus \Omega_\rho^+, \end{cases}$$

the third Green identity (Theorem 6.10) implies that

$$w = \mathcal{G}(-\lambda w) - \mathrm{SL}_\rho[\mathcal{B}_\nu w]_{\Gamma \cup \Gamma_\rho} + \mathcal{K}_2 w \quad \text{on } \mathbb{R}^n.$$

Using (6.25), we see that

$$\|\mathcal{G}(-\lambda w) + \mathcal{K}_2 w\|_{H^{s+1}(\Omega^-)^m} \le C \|\mathcal{G}w\|_{H^2(\Omega^-)^m} + C \|w\|_{L_2(\Omega^-)^m}$$
$$\le C \|w\|_{L_2(\Omega^- \cup \Omega_\rho^+)^m}$$
$$\le C \|\mathcal{U}g\|_{L_2(\Omega^-)^m} + C \|\mathcal{U}_\rho^+ g\|_{L_2(\Omega_\rho^+)^m}$$
$$\le C \|g\|_{L_2(\Gamma)^m} \le C \|g\|_{H^{s+1/2}(\Gamma)^m}$$

and by part (i) and (6.24),

$$\|\mathrm{SL}_\rho[\mathcal{B}_\nu w]_{\Gamma \cup \Gamma_\rho}\|_{H^{s+1}(\Omega^-)^m} \le C\|[\mathcal{B}_\nu w]_{\Gamma \cup \Gamma_\rho}\|_{H^{s-1/2}(\Gamma \cup \Gamma_\rho)^m}$$
$$\le C\|\mathcal{B}_\nu^- \mathcal{U}g\|_{H^{s-1/2}(\Gamma)^m} + C\|\mathcal{B}_\nu^+ \mathcal{U}_\rho^+ g\|_{H^{s-1/2}(\Gamma)^m}$$
$$+ C\|\mathcal{B}_\nu \mathcal{U}_\rho^+ g\|_{H^{s-1/2}(\Gamma_\rho)^m}$$
$$\le C\|g\|_{H^{s+1/2}(\Gamma)^m}.$$

Hence, $\|\mathcal{U}g\|_{H^{s+1}(\Omega^-)^m} = \|w\|_{H^{s+1}(\Omega^-)^m} \le C\|g\|_{H^{s+1/2}(\Gamma)^m}$.

Next, consider the operator $\mathcal{B}_\nu^- \mathrm{SL}$, and let $\psi, \phi \in \mathcal{D}(\Gamma)^m$. By the second Green identity (Theorem 4.4),

$$\big((\mathcal{P} + \lambda)\,\mathrm{SL}\,\psi, \mathcal{V}\phi\big)_{\Omega^-} - \big(\mathrm{SL}\,\psi, (\mathcal{P}^* + \lambda)\mathcal{V}\phi\big)_{\Omega^-}$$
$$= (\gamma^- \,\mathrm{SL}\,\psi, \widetilde{\mathcal{B}}_\nu^- \mathcal{V}\phi)_\Gamma - (\mathcal{B}_\nu^- \,\mathrm{SL}\,\psi, \gamma^- \mathcal{V}\phi)_\Gamma,$$

and since $\mathcal{P}\,\mathrm{SL}\,\psi = -\mathcal{K}_1 \gamma^* \psi$ and $(\mathcal{P}^* + \lambda)\mathcal{V}\phi = 0$ on Ω^-, and $\gamma^- \mathcal{V}\phi = \phi$ on Γ, we see that

$$(\mathcal{B}_\nu^- \,\mathrm{SL}\,\psi, \phi)_\Gamma = (\gamma \,\mathrm{SL}\,\psi, \widetilde{\mathcal{B}}_\nu^- \mathcal{V}\phi)_\Gamma + (\psi, \gamma \mathcal{K}_1^* \mathcal{V}\phi)_\Gamma - \lambda(\mathrm{SL}\psi, \mathcal{V}\phi)_{\Omega^-}.$$

Therefore,

$$|(\mathcal{B}_\nu^- \,\mathrm{SL}\,\psi, \phi)_\Gamma| \le C\|\gamma \,\mathrm{SL}\,\psi\|_{H^{s+1/2}(\Gamma)^m} \|\widetilde{\mathcal{B}}_\nu^- \mathcal{V}\phi\|_{H^{-s-1/2}(\Gamma)^m}$$
$$+ C\|\psi\|_{H^{s-1/2}(\Gamma)^m} \|\gamma \mathcal{K}_1^* \mathcal{V}\phi\|_{H^{-s+1/2}(\Gamma)^m}$$
$$+ C\|\,\mathrm{SL}\,\psi\|_{L_2(\Omega^-)^m} \|\mathcal{V}\phi\|_{L_2(\Omega^-)^m}$$
$$\le C\|\psi\|_{H^{s-1/2}(\Gamma)^m} \big(\|\phi\|_{H^{-s+1/2}(\Gamma)^m} + \|\mathcal{V}\phi\|_{H^{-s+1}(\Omega^-)^m}\big),$$

and since $\|\mathcal{V}\phi\|_{H^{-s+1}(\Omega^-)^m} \le C\|\phi\|_{H^{-s+1/2}(\Gamma)^m}$, the mapping property of $\mathcal{B}_\nu^- \mathrm{SL}$ follows. We can handle $\mathcal{B}_\nu^+ \mathrm{SL}$ in essentially the same way.

The results for the double-layer potential now follow from (6.22). Indeed,

$$\|\gamma^\pm \mathrm{DL}\,\psi\|_{H^{s+1/2}(\Gamma)^m} \le C\|\chi\,\mathrm{DL}\,\psi\|_{H^{s+1}(\Omega^\pm)^m}$$
$$\le C\|\chi \mathrm{SL}\,\mathcal{B}_\nu^- \mathcal{U}\psi\|_{H^{s+1}(\Omega^\pm)^m} + C\|\chi u\|_{H^{s+1}(\Omega^\pm)^m}$$
$$+ C\|\chi(\mathcal{K}_2 u - \lambda \mathcal{G}u)\|_{H^2(\mathbb{R}^n)^m}$$
$$\le C\|\mathcal{B}_\nu^- \mathcal{U}\psi\|_{H^{s-1/2}(\Gamma)^m} + C\|\mathcal{U}\psi\|_{H^{s+1}(\Omega^-)^m} + C\|u\|_{L_2(\mathbb{R}^n)^m}$$
$$\le C\|\psi\|_{H^{s+1/2}(\Gamma)^m} + C\|\mathcal{U}\psi\|_{H^{s+1}(\Omega^-)^m} \le C\|\psi\|_{H^{s+1/2}(\Gamma)^m},$$

and

$$\|\mathcal{B}_\nu^\pm \mathrm{DL}\,\psi\|_{H^{s-1/2}(\Gamma)^m}$$
$$\le C\|\mathcal{B}_\nu^\pm \mathrm{SL}\,\mathcal{B}_\nu^- \mathcal{U}\psi\|_{H^{s-1/2}(\Gamma)^m} + \|\mathcal{B}_\nu^\pm u\|_{H^{s-1/2}(\Gamma)^m}$$

$$+ \lambda \|\mathcal{B}_\nu \mathcal{G} u\|_{L_2(\Gamma)^m} + \|\mathcal{B}_\nu \mathcal{K}_2 u\|_{L_2(\Gamma)^m}$$

$$\leq C \|\mathcal{B}_\nu^- \mathcal{U} \psi\|_{H^{s-1/2}(\Gamma)^m} + C \|(\chi_1 \mathcal{G} \chi_1) u\|_{H^2(\mathbb{R}^n)^m} + \|u\|_{L_2(\mathbb{R}^n)^m}$$

$$\leq C \|\psi\|_{H^{s+1/2}(\Gamma)^m} + C \|u\|_{L_2(\mathbb{R}^n)^m} \leq C \|\psi\|_{H^{s+1/2}(\Gamma)^m},$$

proving the mapping property for $\mathcal{B}_\nu \, \mathrm{DL} = \mathcal{B}_\nu^\pm \, \mathrm{DL}$. $\qquad\qquad\square$

The proof of Theorem 6.12 breaks down if $s = \pm \frac{1}{2}$. In particular, the trace operator no longer satisfies the mapping property in (6.23); cf. Theorem 3.38. Nevertheless, it turns out that all of the conclusions of Theorem 6.12 are valid for $-\frac{1}{2} \leq s \leq \frac{1}{2}$, even when the principal part of $\mathcal{P}$ is not formally self-adjoint, as required in Theorem 4.25. However, the proofs for the cases $s = \pm \frac{1}{2}$ are difficult, and rely on techniques from harmonic analysis. Of course, the estimates for $s = \pm \frac{1}{2}$, in combination with the interpolation property of the Sobolev spaces, yield a proof of the case $-\frac{1}{2} < s < \frac{1}{2}$ independent of the one given above.

A detailed discussion of harmonic analysis techniques for elliptic equations on nonsmooth domains is beyond the scope of this book. Nevertheless, in view of the importance of the mapping properties for $s = \pm \frac{1}{2}$, a few pointers to the literature may be appropriate. Two useful early papers are the survey by Jerison and Kenig [43] on the Dirichlet and Neumann problems for the Laplacian on a Lipschitz domain, and the study by Fabes, Jodeit and Riviére [21] of the classical method of surface potentials for the Laplacian on a C^1 domain. A key ingredient in [21] is the fact, proved by Calderón [9], that if Γ is C^1 (or even Lipschitz but with a sufficiently small Lipschitz constant), then the Cauchy integral defines a bounded linear operator on $L_2(\Gamma)$. Subsequently, Coifman, McIntosh and Meyer [11] extended Calderon's result to the case when Γ is Lipschitz (without restriction on the size of the Lipschitz constant), after which Verchota [102] was able to extend the results of [21] to Lipschitz domains. Later, other elliptic equations were treated, including the Stokes system [22] and the equations of linear elasticity [17]; see also the survey paper of Kenig [47]. Mitrea, Mitrea and Taylor [68], [69] have only recently treated general strongly elliptic systems. Further historical and bibliographical details up to 1991 appear in a monograph by Kenig [47].

In most of these papers, not only the boundedness of the various boundary operators is of concern, but also their *invertibility*, a question we shall address in Chapter 7. Also, estimates of the surface potentials in Sobolev norms on the *domain* do not appear explicitly in most cases. Instead, bounds are proved for the nontangential maximal functions of the potentials and their derivatives, from which the Sobolev estimates follow; see Jerison and Kenig [43, pp. 62–63], [46, p. 145] and [44, Theorem 4.1].

When Γ is smooth locally, we can use the transmission property to show that the mapping properties in Theorem 6.12 hold also for $s > \frac{1}{2}$. Recall Figure 3 of Chapter 4.

Theorem 6.13 *Let G_1 and G_2 be bounded open subsets of $\mathbb{R}^n$, such that $\overline{G}_1 \Subset G_2$ and G_1 intersects Γ, and put*

$$\Omega_j^\pm = G_j \cap \Omega^\pm \quad and \quad \Gamma_j = G_j \cap \Omega^\pm \qquad for\ j = 1, 2.$$

Suppose, for some integer $r \geq 0$, that Γ_2 is $C^{r+1,1}$, and let

$$\psi \in H^{-r-2}(\Gamma)^m \quad and \quad -\tfrac{1}{2} < s \leq r + 1.$$

(i) If $\psi|_{\Gamma_2} \in H^{s-1/2}(\Gamma_2)^m$, then $\mathrm{SL}\,\psi \in H^{s+1}(\Omega_1^\pm)^m$ and

$$\| \mathrm{SL}\,\psi \|_{H^{s+1}(\Omega_1^\pm)^m} \leq C \|\psi\|_{H^{-r-2}(\Gamma)^m} + C \|\psi\|_{H^{s-1/2}(\Gamma_2)^m}.$$

(ii) If $\psi|_{\Gamma_2} \in H^{s+1/2}(\Gamma_2)^m$, then $\mathrm{DL}\,\psi \in H^{s+1}(\Omega_1^\pm)^m$ and

$$\| \mathrm{DL}\,\psi \|_{H^{s+1}(\Omega_1^\pm)^m} \leq C \|\psi\|_{H^{-r-2}(\Gamma)^m} + C \|\psi\|_{H^{s+1/2}(\Gamma_2)^m}.$$

Proof. Since the fundamental solution $G(x, y)$ is C^∞ for $x \neq y$, we can assume without loss of generality that supp $\psi \Subset \overline{\Gamma}_{3/2}$, where $\Gamma_{3/2} = G_{3/2} \cap \Gamma$ and $\overline{G}_1 \Subset G_{3/2} \subseteq \overline{G}_{3/2} \Subset G_2$. For $-\frac{1}{2} < s < \frac{1}{2}$, we can repeat the proof of Theorem 6.12, noting that the mapping properties (6.24) now hold by Theorem 4.21, so no extra assumptions about $\mathcal{P}$ are necessary.

For $s = r + 1$, part (i) follows from Theorem 4.20 because the single-layer potential $u = \mathrm{SL}\,\psi$ satisfies $\mathcal{P}u = -\mathcal{K}_1\gamma^*\psi$ on $\Omega^\pm$ with $u|_{\Omega_2^\pm} \in H^1(\Omega_2^\pm)^m$, $[u]_\Gamma = 0$, $[\mathcal{B}_\nu u]_{\Gamma_2} = -\psi \in H^{r+1/2}(\Gamma)^m$ and $\mathcal{K}_1\gamma^*\psi \in H^r(\Omega_2^\pm)^m$. We then obtain (i) for the full range of values of s by interpolation, viewing SL as a linear operator from $\widetilde{H}^{s-1/2}(\Gamma_{3/2})^m$ into $H^{s+1}(\Omega_1)^m$.

Similarly, part (ii) follows because if $s = r + 1$, then the double-layer potential $u = \mathrm{DL}\,\psi$ satifies $\mathcal{P}u = -\mathcal{K}_1\widetilde{\mathcal{B}}_\nu^*\psi$ on $\Omega^\pm$ with $u|_{\Omega_2^\pm} \in H^1(\Omega_2^\pm)^m$, $[u]_{\Gamma_1} = \psi \in H^{r+3/2}(\Gamma_2)^m$, $[\mathcal{B}_\nu u]_\Gamma = 0$ and $\mathcal{K}_1\widetilde{\mathcal{B}}_\nu^*\psi \in H^r(\Omega_2^\pm)^m$. $\square$

Corollary 6.14 *Fix a cutoff function $\chi \in C^\infty_{\mathrm{comp}}(\mathbb{R}^n)$. If the whole of Γ is $C^{r+1,1}$ for some integer $r \geq 0$, and if $-\frac{1}{2} < s \leq r + 1$, then*

$$\chi\,\mathrm{SL} : H^{s-1/2}(\Gamma)^m \to H^{s+1}(\Omega^\pm)^m \quad and \quad \chi\,\mathrm{DL} : H^{s+1/2}(\Gamma)^m \to H^{s+1}(\Omega^\pm)^m.$$

Duality Relations

Recall from (4.30) that the first Green identity for $\Omega^{\pm}$ takes the form

$$\Phi^{\pm}(u, v) = (\mathcal{P}u, v)_{\Omega^{\pm}} \mp (\mathcal{B}_{\nu}^{\pm}u, \gamma^{\pm}v)_{\Gamma},$$

and from (4.31) that we have also the dual version

$$\Phi^{\pm}(u, v) = (u, \mathcal{P}^*v)_{\Omega^{\pm}} \mp (\gamma^{\pm}u, \widetilde{\mathcal{B}}_{\nu}^{\pm}v)_{\Gamma}. \tag{6.26}$$

Let $f^{\pm} \in \widetilde{H}^{-1}(\Omega^{\pm})^m$, $v \in L_2(\mathbb{R}^n)^m$ and $v^{\pm} = v|_{\Omega^{\pm}} \in H^1(\Omega^{\pm})^m$, and suppose that

$$\mathcal{P}^*v^{\pm} = f^{\pm} \quad \text{on } \Omega^{\pm}.$$

Putting $f = f^+ + f^- \in H^{-1}(\mathbb{R}^n)^m$, and arguing as in Lemma 4.19, we find that

$$(\mathcal{P}^*v, \phi) = (f, \phi) + ([v]_{\Gamma}, \mathcal{B}_{\nu}\phi)_{\Gamma} - ([\widetilde{\mathcal{B}}_{\nu}v]_{\Gamma}, \gamma\phi)_{\Gamma} \quad \text{for } \phi \in \mathcal{D}(\mathbb{R}^n)^m,$$

or in other words,

$$\mathcal{P}^*v = f + \mathcal{B}_{\nu}^*[v]_{\Gamma} - \gamma^*[\widetilde{\mathcal{B}}_{\nu}v]_{\Gamma} \quad \text{on } \mathbb{R}^n, \tag{6.27}$$

which is the dual version of (6.15).

Recall from Theorem 6.7 that $\mathcal{G}^*$ is a parametrix for $\mathcal{P}^*$. Indeed, if $\mathcal{K}_1$ and $\mathcal{K}_2$ are as in (6.4), then

$$\mathcal{P}^*\mathcal{G}^*u = u - \mathcal{K}_2^*u \quad \text{and} \quad \mathcal{G}^*\mathcal{P}^*u = u - \mathcal{K}_1^*u. \tag{6.28}$$

Accordingly, we define $\widetilde{\mathrm{SL}}$ and $\widetilde{\mathrm{DL}}$, the single- and double-layer potentials associated with $\mathcal{P}^*$, by

$$\widetilde{\mathrm{SL}} = \mathcal{G}^*\gamma^* \quad \text{and} \quad \widetilde{\mathrm{DL}} = \mathcal{G}^*\mathcal{B}_{\nu}^*.$$

Assuming that v has compact support, we may apply $\mathcal{G}^*$ to both sides of (6.27), and obtain another version of the third Green identity,

$$v = \mathcal{G}^*f + \widetilde{\mathrm{DL}}[v]_{\Gamma} - \widetilde{\mathrm{SL}}[\widetilde{\mathcal{B}}_{\nu}v]_{\Gamma} + \mathcal{K}_1^*v \quad \text{on } \mathbb{R}^n; \tag{6.29}$$

cf. Theorem 6.10. The definitions of $\widetilde{\mathrm{SL}}$ and $\widetilde{\mathrm{DL}}$ mean that

$$(\widetilde{\mathrm{SL}}\psi, \phi) = (\psi, \gamma\mathcal{G}\phi)_{\Gamma} \quad \text{and} \quad (\widetilde{\mathrm{DL}}\psi, \phi) = (\psi, \mathcal{B}_{\nu}\mathcal{G}\phi)_{\Gamma} \quad \text{for } \phi \in \mathcal{D}(\mathbb{R}^n)^m,$$

and we see that for $x \in \mathbb{R}^n \setminus \Gamma$,

$$\widetilde{\mathrm{SL}}\,\psi(x) = \int_\Gamma G(y,x)^* \psi(y)\,d\sigma_y \qquad (6.30)$$

and

$$\widetilde{\mathrm{DL}}\,\psi(x) = \int_\Gamma \left[\mathcal{B}_{\nu,y}G(y,x)\right]^* \psi(y)\,d\sigma_y; \qquad (6.31)$$

cf. (6.16) and (6.17). Theorems 6.11–6.13 on the jump relations and mapping properties of SL and DL carry over in the obvious way to $\widetilde{\mathrm{SL}}$ and $\widetilde{\mathrm{DL}}$, with $\widetilde{\mathcal{B}}_\nu$ taking the place of $\mathcal{B}_\nu$; thus,

$$[\widetilde{\mathrm{SL}}\,\psi]_\Gamma = 0 \ \text{ and } \ [\widetilde{\mathcal{B}}_\nu \widetilde{\mathrm{SL}}\,\psi]_\Gamma = -\psi \qquad \text{for } \psi \in H^{-1/2}(\Gamma)^m,$$

and

$$[\widetilde{\mathrm{DL}}\,\psi]_\Gamma = \psi \ \text{ and } \ [\widetilde{\mathcal{B}}_\nu\widetilde{\mathrm{DL}}\psi]_\Gamma = 0 \qquad \text{for } \psi \in H^{1/2}(\Gamma)^m.$$

We remark that if $\mathcal{P}$ is formally self-adjoint, then $\frac{1}{2}(\mathcal{G} + \mathcal{G}^*)$ is a self-adjoint parametrix for $\mathcal{P}$, which means that $\mathcal{G}$ can be chosen to satisfy $\mathcal{G}^* = \mathcal{G}$. Obviously, in this case $\widetilde{\mathrm{SL}} = \mathrm{SL}$ and $\widetilde{\mathrm{DL}} = \mathrm{DL}$.

The traces of the single-layer potentials SL and $\widetilde{\mathrm{SL}}$ satisfy the following duality relation.

Theorem 6.15 *If $\psi, \phi \in H^{-1/2}(\Gamma)^m$, then*

$$(\gamma\,\mathrm{SL}\,\psi, \phi)_\Gamma = (\psi, \gamma\,\widetilde{\mathrm{SL}}\,\phi)_\Gamma.$$

Proof. Fix a cutoff function $\chi \in \mathcal{D}(\mathbb{R}^n)$ with $\chi = 1$ on a neighbourhood of Γ. The operator $\chi\mathcal{G}^*\chi : H^{-1}(\mathbb{R}^n)^m \to H^1(\mathbb{R}^n)^m$ is the adjoint of $\chi\mathcal{G}\chi : H^{-1}(\mathbb{R}^n)^m \to H^1(\mathbb{R}^n)^m$, so, noting that

$$\gamma : H^1(\mathbb{R}^n)^m \to H^{1/2}(\Gamma)^m \quad \text{and} \quad \gamma^* : H^{-1/2}(\Gamma)^m \to H^{-1}(\mathbb{R}^n)^m,$$

we have

$$(\gamma\,\mathrm{SL}\,\psi, \phi)_\Gamma = (\chi\mathcal{G}\chi\gamma^*\psi, \gamma^*\phi) = (\gamma^*\psi, \chi\mathcal{G}^*\chi\gamma^*\phi) = (\psi, \gamma\,\widetilde{\mathrm{SL}}\,\phi)_\Gamma.$$

$\square$

The next lemma will help us to relate the trace of the double-layer potential and the conormal derivative of the single-layer potential. The functions K_1 and K_2 are the kernels of the smoothing operators $\mathcal{K}_1$ and $\mathcal{K}_2$ in (6.4).

Lemma 6.16 Suppose $u \in \mathcal{D}(\mathbb{R}^n)^m$, and put $\psi = \gamma u$. If $x \in \Omega^{\pm}$, then

$$\pm \mathrm{DL}\,\psi(x) = \Phi^{\mp}\big(G(x, \cdot)^*, u\big) + \int_{\Omega^{\mp}} K_2(x, y)u(y)\,dy$$

$$= \pm \mathrm{SL}\,\mathcal{B}_\nu u(x) + \int_{\Omega^{\mp}} \big[G(x, y)\mathcal{P}u(y) + K_2(x, y)u(y)\big]\,dy$$

and

$$\pm \widetilde{\mathrm{DL}}\,\psi(x) = \Phi^{\mp}\big(G(\cdot, x), u\big) + \int_{\Omega^{\mp}} K_1(y, x)^* u(y)\,dy$$

$$= \pm \widetilde{\mathrm{SL}}\,\widetilde{\mathcal{B}}_\nu u(x) + \int_{\Omega^{\mp}} \big[G(y, x)^* \mathcal{P}^* u(y) + K_1(y, x)^* u(y)\big]\,dy$$

Proof. Taking $u = \delta_x$ in (6.28), we see that $\mathcal{P}^* G(x, \cdot)^* = \delta_x - K_2(x, \cdot)^*$ on $\mathbb{R}^n$. Thus, for $x \in \Omega^{\pm}$, the first Green identity gives

$$-\big(K_2(x, \cdot)^*, u\big)_{\Omega^{\mp}} \pm \big(\widetilde{\mathcal{B}}_\nu G(x, \cdot)^*, \psi\big)_\Gamma = \Phi^{\mp}\big(G(x, \cdot)^*, u\big)$$

$$= \big(G(x, \cdot)^*, \mathcal{P}u\big)_{\Omega^{\mp}} \pm \big(G(x, \cdot)^*, \mathcal{B}_\nu u\big)_\Gamma.$$

We see from (6.16) and (6.17) that

$$\big(G(x, \cdot)^*, \mathcal{B}_\nu u\big)_\Gamma = \mathrm{SL}\,\mathcal{B}_\nu u(x) \quad \text{and} \quad \big(\widetilde{\mathcal{B}}_\nu G(x, \cdot)^*, \psi\big)_\Gamma = \mathrm{DL}\,\psi(x),$$

proving the formulae for $\pm \mathrm{DL}\,\psi(x)$. The formulae for $\pm \widetilde{\mathrm{DL}}\,\psi(x)$ follow in the same way, because $\mathcal{P}G(\cdot, x) = \delta_x - K_1(\cdot, x)$ on $\mathbb{R}^n$. $\qquad\square$

Theorem 6.17 Suppose $u \in H^1(\mathbb{R}^n)^m$, and put $\psi = \gamma u \in H^{1/2}(\Gamma)^m$.

(i) For $\phi \in H^{-1/2}(\Gamma)^m$,

$$\pm(\phi, \gamma^{\pm}\,\mathrm{DL}\,\psi)_\Gamma = \Phi^{\mp}(\widetilde{\mathrm{SL}}\,\phi, u) + (\mathcal{K}_2^* \gamma^* \phi, u)_{\Omega^{\mp}} = \pm(\widetilde{\mathcal{B}}_\nu^{\mp}\,\widetilde{\mathrm{SL}}\,\phi, \psi)_\Gamma$$

and

$$\pm(\phi, \gamma^{\pm}\,\widetilde{\mathrm{DL}}\psi)_\Gamma = \Phi^{\mp}(\mathrm{SL}\,\phi, u) + (\mathcal{K}_1 \gamma^* \phi, u)_{\Omega^{\mp}} = \pm(\mathcal{B}_\nu^{\mp}\,\mathrm{SL}\,\phi, \psi)_\Gamma.$$

(ii) For $\phi \in H^{1/2}(\Gamma)^m$,

$$\pm(\phi, \mathcal{B}_\nu\,\mathrm{DL}\psi)_\Gamma = \Phi^{\mp}(\widetilde{\mathrm{DL}}\,\phi, u) + (\mathcal{K}_2^* \mathcal{B}_\nu^* \phi, u)_{\Omega^{\mp}} = \pm(\widetilde{\mathcal{B}}_\nu\,\widetilde{\mathrm{DL}}\,\phi, \psi)_\Gamma$$

and

$$\pm(\phi, \widetilde{\mathcal{B}}_\nu\,\widetilde{\mathrm{DL}}\psi)_\Gamma = \Phi^{\mp}(\mathrm{DL}\,\phi, u) + (\mathcal{K}_1 \mathcal{B}_\nu^* \phi, u)_{\Omega^{\mp}} = \pm(\mathcal{B}_\nu\,\mathrm{DL}\,\phi, \psi)_\Gamma.$$

Proof. Theorem 6.3 implies that $\partial_{y_k} G(x, y)$ is locally integrable on $\mathbb{R}^n$ for $1 \le k \le n$. Hence, the function $x \mapsto \Phi^{\mp}(G(x, \cdot)^*, u)$ is continuous on $\mathbb{R}^n$, and we can show that

$$\int_\Gamma \phi^*(x)\Phi^{\mp}(G(x, \cdot)^*, u)\, d\sigma_x = \Phi^{\mp}(\widetilde{SL}\,\phi, u).$$

For instance, the first of the three terms arising from the definition of $\Phi^{\mp}$ is

$$\int_\Gamma \phi(x)^* \int_{\Omega^{\mp}} \left(A_{jk}(y)\partial_{y_k} G(x, y)^*\right)^* \partial_j u(y)\, dy\, d\sigma_x$$

$$= \int_{\Omega^{\mp}} \partial_{y_k}\left(\int_\Gamma G(x, y)^*\phi(x)\, d\sigma_x\right)^* A_{jk}^*(y)\partial_j u(y)\, dy$$

$$= \int_{\Omega^{\mp}} \left(A_{jk}\partial_k \widetilde{SL}\,\phi\right)^* \partial_j u\, dy.$$

Thus, by Lemma 6.16,

$$\pm(\phi, \gamma^{\pm}\,\mathrm{DL}\,\psi)_\Gamma = \Phi^{\mp}(\widetilde{SL}\,\phi, u) + \int_\Gamma \phi(x)^* \int_{\Omega^{\mp}} K_2(x, y)u(y)\, dy\, d\sigma_x$$

$$= \Phi^{\mp}(\widetilde{SL}\,\phi, u) + \int_{\Omega^{\mp}} \left(\int_\Gamma K_2(x, y)^*\phi(x)\, d\sigma_x\right)^* u(y)\, dy$$

$$= \Phi^{\mp}(\widetilde{SL}\,\phi, u) + (\mathcal{K}_2^*\gamma^*\phi, u)_{\Omega^{\mp}},$$

and the first Green identity (6.26) gives

$$\Phi^{\mp}(\widetilde{SL}\,\phi, u) = (\mathcal{P}^*\,\widetilde{SL}\,\phi, u) \pm (\widetilde{\mathcal{B}}_\nu^{\mp}\,\widetilde{SL}\,\phi, \gamma^{\mp}u)_\Gamma$$

$$= -(\mathcal{K}_2^*\gamma^*\phi, u)_{\Omega^{\mp}} \pm (\widetilde{\mathcal{B}}_\nu^{\mp}\,\widetilde{SL}\,\phi, \psi)_\Gamma,$$

proving the first half of part (i); the second half holds by a similar argument.

To prove part (ii), we use the second formula for $\mathrm{DL}\,\psi$ in Lemma 6.16, followed by part (i) and the first Green identity. Indeed,

$$\pm(\phi, \mathcal{B}_\nu^{\pm}\,\mathrm{DL}\,\psi)_\Gamma = \pm(\phi, \mathcal{B}_\nu^{\pm}\,\mathrm{SL}\,\mathcal{B}_\nu u)_\Gamma + \int_\Gamma \phi(x)^* \mathcal{B}_{\nu,x} \int_{\Omega^{\mp}} \left[G(x, y)\mathcal{P}u(y)\right.$$

$$\left. + K_2(x, y)u(y)\right] dy\, d\sigma_x$$

$$= \pm(\gamma^{\mp}\,\widetilde{\mathrm{DL}}\,\phi, \mathcal{B}_\nu u)_\Gamma + (\widetilde{\mathrm{DL}}\,\phi, \mathcal{P}u)_{\Omega^{\mp}} + (\mathcal{K}_2^*\mathcal{B}_\nu^*\phi, u)_{\Omega^{\mp}}$$

$$= \Phi^{\mp}(\widetilde{\mathrm{DL}}\,\phi, u) + (\mathcal{K}_2^*\mathcal{B}_\nu^*\phi, u)_{\Omega^{\mp}},$$

and since $\mathcal{P}^*\,\widetilde{\mathrm{DL}}\,\phi = -\mathcal{K}_2^*\mathcal{B}_\nu^*\phi$ on $\Omega^{\mp}$, another application of the first Green identity gives

$$\Phi^{\mp}(\widetilde{\mathrm{DL}}\,\phi, u) + (\mathcal{K}_2^* \mathcal{B}_\nu^* \phi, u)_{\Omega^{\mp}} = \pm(\widetilde{\mathcal{B}}_\nu^{\mp}\, \widetilde{\mathrm{DL}}\,\phi, \psi)_\Gamma.$$

The second half of part (ii) is proved in the same way. $\qquad\square$

Exercises

6.1 With the notation of Theorem 6.3, show that if $\mathcal{P}^* = \mathcal{P}$, then $\mathsf{G}_1 = 0$.

6.2 By thinking of $G(y, x)^*$ as a parametrix for $\mathcal{P}^*$, we can apply Theorem 6.3 to obtain an expansion

$$G(y, x)^* = \sum_{j=0}^{N} \widetilde{\mathsf{G}}_j(x, x - y) + \widetilde{R}_N(x, y),$$

where the $\widetilde{\mathsf{G}}_j$ have the obvious properties. Show that $\widetilde{\mathsf{G}}_0(x, z) = \mathsf{G}_0(x, -z)^*$.

6.3 Here is another way of deriving the jump relations for the single- and double-layer potentials, assuming that the basic mapping properties of Theorem 6.11 are known.

(i) Show that if $\psi \in H^{-1/2}(\Gamma)^m$ and $u = \mathrm{SL}\,\psi$, then

$$\big([u]_\Gamma, \widetilde{\mathcal{B}}_\nu\phi\big)_\Gamma = \big([\mathcal{B}_\nu u]_\Gamma + \psi, \gamma\phi\big)_\Gamma \quad \text{for } \phi \in \mathcal{D}(\mathbb{R}^n)^m.$$

[Hint: use Lemma 6.9 in combination with (6.18) and (6.19).]

(ii) Show that if $\psi \in H^{1/2}(\Gamma)^m$ and $u = \mathrm{DL}\,\psi$, then

$$\big([u]_\Gamma - \psi, \widetilde{\mathcal{B}}_\nu\phi\big)_\Gamma = \big([\mathcal{B}_\nu u]_\Gamma, \gamma\phi\big)_\Gamma \quad \text{for } \phi \in \mathcal{D}(\mathbb{R}^n)^m.$$

(iii) Assume that Ω^- is C^2. Show that if $f \in H^{1/2}(\Gamma)^m$ and $g \in H^{-1/2}(\Gamma)^m$ satisfy

$$\big(f, \widetilde{\mathcal{B}}_\nu\phi\big)_\Gamma = (g, \gamma\phi)_\Gamma \quad \text{for } \phi \in \mathcal{D}(\mathbb{R}^n)^m,$$

then $f = 0$ and $g = 0$. [Hint: Since $\mathcal{D}(\mathbb{R}^n)^m$ is dense in $H^2(\mathbb{R}^n)^m$, we can use a C^2 coordinate transformation to reduce to the case $\Omega^\pm = \mathbb{R}^n_\pm$. For η_1 as in Lemma 3.36, if $\theta \in \mathcal{D}(\mathbb{R}^{n-1})^m$ and $\phi = \eta_1\psi$, then $\gamma\phi = 0$ and $\widetilde{\mathcal{B}}_\nu\phi = A_{nn}\psi$. Costabel [14, Lemma 3.5] gives a proof for Lipschitz domains.]

6.4 Fix a cutoff function $\chi \in C^\infty_{\mathrm{comp}}(\mathbb{R}^n)$, and show that

$$\chi\,\mathrm{DL} : H^{s-1/2}(\Gamma)^m \to H^s(\mathbb{R}^n)^m \quad \text{for } -\tfrac{1}{2} < s < \tfrac{1}{2}.$$

[Hint: adapt the proof of Theorem 6.12 (i).]

6.5 Let the hypotheses of Theorem 6.13 be satisfied, and suppose that $f^\pm \in \widetilde{H}^{-1}(\Omega^\pm)^m$ with f^+ having compact support. As in Lemma 6.9, we put $f = f^+ + f^- \in H^{-1}(\mathbb{R}^n)^m$, and note that $\mathcal{G}f \in H^1_{\mathrm{loc}}(\mathbb{R}^n)^m$ by (6.10), and $[\mathcal{G}f]_\Gamma = 0$ by Exercise 4.5.

(i) Show that $[\mathcal{B}_\nu \mathcal{G}f]_\Gamma = 0$ if $f \in L_2(\mathbb{R}^n)^m$.

(ii) Prove regularity of the volume potential up to the boundary: if $f^\pm|_{\Omega_2^\pm} \in H^r(\Omega_2^\pm)^m$, then $\mathcal{G}f^\pm \in H^{r+2}(\Omega_1^\pm)^m$ and

$$\|\mathcal{G}f^\pm\|_{H^{r+2}(\Omega_1^\pm)^m} \le C\|f^\pm\|_{\widetilde{H}^{-1}(\Omega^\pm)^m} + C\|f\|_{H^r(\Omega_2^\pm)^m}.$$

[Hint: use Theorem 4.20.]

7

Boundary Integral Equations

Using the properties of the surface potentials established in Chapter 6, we can reformulate boundary value problems over the domain Ω^- or Ω^+ as integral equations over the boundary Γ. To describe these reformulations, we begin by defining four boundary operators (three if $\mathcal{P}$ is formally self-adjoint) in terms of traces and conormal derivatives of surface potentials, and by showing how to write them as integral operators, in some cases with non-integrable kernels. Next, the pure Dirichlet and Neumann problems for the interior domain Ω^- are shown to be equivalent to boundary integral equations of the first kind, for which the Fredholm alternative is valid. The case of mixed boundary conditions is more complicated, because one obtains a 2×2 system of integral equations. We establish the Fredholm alternative for this system only when $\mathcal{P}$ is formally self-adjoint. The next section treats exterior problems, i.e., boundary value problems for the unbounded domain Ω^+. In such cases, a suitable radiation condition must be specified, to force appropriate behaviour of the solution at infinity. Finally, we study regularity of the solution to the integral equation when the surface and the data are suitably smooth (at least locally).

Throughout this chapter, G is always a fundamental solution (not just a parametrix) for $\mathcal{P}$, and we implicitly assume whenever $\mathcal{P}$ is formally self-adjoint that $G(y, x)^* = G(x, y)$.

Operators on the Boundary

Recall from (6.16) and (6.17) that the single- and double-layer potentials associated with $\mathcal{P}$ are given by

$$\mathrm{SL}\,\psi(x) = \int_\Gamma G(x, y)\psi(y)\, d\sigma_y,$$

$$\mathrm{DL}\,\psi(x) = \int_\Gamma \big[\widetilde{\mathcal{B}}_{v, y} G(x, y)^*\big]^* \psi(y)\, d\sigma_y,$$

(7.1)

and recall from (6.30) and (6.31) that the ones associated with $\mathcal{P}^*$ are given by

$$\widetilde{\mathrm{SL}}\,\psi(x) = \int_\Gamma G(y,x)^* \psi(y)\,d\sigma_y,$$
$$\widetilde{\mathrm{DL}}\,\psi(x) = \int_\Gamma \left[\mathcal{B}_{v,y} G(y,x)\right]^* \psi(y)\,d\sigma_y,$$

(7.2)

for $x \in \mathbb{R}^n \setminus \Gamma$. We will see that all traces and conormal derivatives of these potentials can be expressed in terms of four boundary operators, namely

$$R = -\mathcal{B}_v\,\mathrm{DL} : H^{1/2}(\Gamma)^m \to H^{-1/2}(\Gamma)^m,$$
$$S = \gamma\,\mathrm{SL} : H^{-1/2}(\Gamma)^m \to H^{1/2}(\Gamma)^m,$$
$$T = \gamma^+\,\mathrm{DL} + \gamma^-\,\mathrm{DL} : H^{1/2}(\Gamma)^m \to H^{1/2}(\Gamma)^m,$$
$$\widetilde{T} = \gamma^+\,\widetilde{\mathrm{DL}} + \gamma^-\,\widetilde{\mathrm{DL}} : H^{1/2}(\Gamma)^m \to H^{1/2}(\Gamma)^m.$$

(7.3)

These mapping properties were proved in Theorem 6.11. (The reader may now wish to turn to the first section of Chapter 8 and look at the explicit forms for R, S and T in the simplest and most familiar case, i.e., when $\mathcal{P} = -\Delta$.) The duality relations in Theorems 6.15 and 6.17 show that the adjoints of the operators in (7.3) are given by

$$R^* = -\widetilde{\mathcal{B}}_v\,\widetilde{\mathrm{DL}} : H^{1/2}(\Gamma)^m \to H^{-1/2}(\Gamma)^m,$$
$$S^* = \gamma\,\widetilde{\mathrm{SL}} : H^{-1/2}(\Gamma)^m \to H^{1/2}(\Gamma)^m,$$
$$T^* = \widetilde{\mathcal{B}}_v^+\,\widetilde{\mathrm{SL}} + \widetilde{\mathcal{B}}_v^-\,\widetilde{\mathrm{SL}} : H^{-1/2}(\Gamma)^m \to H^{-1/2}(\Gamma)^m,$$
$$\widetilde{T}^* = \mathcal{B}_v^+\,\mathrm{SL} + \mathcal{B}_v^-\,\mathrm{SL} : H^{-1/2}(\Gamma)^m \to H^{-1/2}(\Gamma)^m.$$

(7.4)

From the definitions above, and the jump relations in Theorem 6.11, we obtain the following expressions for the traces and conormal derivatives of the single- and double-layer potentials:

$$
\begin{aligned}
\gamma\,\mathrm{SL}\,\psi &= S\psi, & \gamma\,\widetilde{\mathrm{SL}}\,\psi &= S^*\psi, \\
\mathcal{B}_v^\pm\,\mathrm{SL}\,\psi &= \tfrac{1}{2}(\mp\psi + \widetilde{T}^*\psi), & \widetilde{\mathcal{B}}_v^\pm\,\widetilde{\mathrm{SL}}\,\psi &= \tfrac{1}{2}(\mp\psi + T^*\psi), \\
\gamma^\pm\,\mathrm{DL}\,\psi &= \tfrac{1}{2}(\pm\psi + T\psi), & \gamma^\pm\,\widetilde{\mathrm{DL}}\,\psi &= \tfrac{1}{2}(\pm\psi + \widetilde{T}\psi), \\
\mathcal{B}_v\,\mathrm{DL}\,\psi &= -R\psi, & \widetilde{\mathcal{B}}_v\,\widetilde{\mathrm{DL}}\,\psi &= -R^*\psi.
\end{aligned}
$$

(7.5)

If the partial differential operator $\mathcal{P}$ is formally self-adjoint, then

$$\widetilde{\mathrm{SL}} = \mathrm{SL}, \quad \widetilde{\mathrm{DL}} = \mathrm{DL}, \quad \widetilde{\mathcal{B}}_v = \mathcal{B}_v, \quad \widetilde{T} = T, \quad S^* = S, \quad R^* = R,$$

and so the eight relations in (7.5) reduce to four:

$$\gamma \operatorname{SL} \psi = S\psi, \qquad\qquad \gamma^{\pm} \operatorname{DL} \psi = \tfrac{1}{2}(\pm\psi + T\psi),$$
$$\mathcal{B}_{\nu}^{\pm} \operatorname{SL} \psi = \tfrac{1}{2}(\mp\psi + T^*\psi), \qquad \mathcal{B}_{\nu} \operatorname{DL} \psi = -R\psi.$$

Theorem 6.12 implies at once that the mapping properties in (7.3) and (7.4) extend to a range of Sobolev spaces as stated in the theorem below. Note, however, our discussion of the end-point cases $s = \pm\tfrac{1}{2}$ following the proof of Theorem 6.12.

Theorem 7.1 For $-\tfrac{1}{2} < s < \tfrac{1}{2}$,

$$S : H^{s-1/2}(\Gamma)^m \to H^{s+1/2}(\Gamma)^m \quad and \quad S^* : H^{s-1/2}(\Gamma)^m \to H^{s+1/2}(\Gamma)^m,$$
$$(7.6)$$

and if $\mathcal{P}$ satisfies the assumptions of Theorem 4.25, then

$$R : H^{s+1/2}(\Gamma)^m \to H^{s-1/2}(\Gamma)^m, \quad R^* : H^{s+1/2}(\Gamma)^m \to H^{s-1/2}(\Gamma)^m,$$
$$T : H^{s+1/2}(\Gamma)^m \to H^{s+1/2}(\Gamma)^m, \quad T^* : H^{s-1/2}(\Gamma)^m \to H^{s-1/2}(\Gamma)^m, \quad (7.7)$$
$$\widetilde{T} : H^{s+1/2}(\Gamma)^m \to H^{s+1/2}(\Gamma)^m, \quad \widetilde{T}^* : H^{s-1/2}(\Gamma)^m \to H^{s-1/2}(\Gamma)^m.$$

For smooth domains, a larger range of values of s is allowed; cf. Exercise 7.8.

Theorem 7.2 If Γ is $C^{r+1,1}$ for some integer $r \ge 0$, then the mapping properties in (7.6) and (7.7) hold for $-r - 1 \le s \le r + 1$.

Proof. The mapping properties for $0 \le s \le r + 1$ follow from Theorems 6.13 and 3.37. We then get the estimates for $-r - 1 \le s \le 0$ by duality. $\quad\square$

Integral Representations

We shall now derive integral representations for each of the eight boundary operators in (7.3) and (7.4).

For $\rho > 0$ and $x \in \mathbb{R}^n$, let $B_\rho(x)$ denote the open ball with centre x and radius ρ. If $n \ge 3$, then, by Theorem 6.3, the leading term G_0 in the homogeneous expansion of G has degree $2 - n$. If $n = 2$, then G_0 contains a logarithm. Consequently,

$$\int_{\Gamma \cap B_\epsilon(x)} |G(z, y)|\, d\sigma_y \le \begin{cases} C\epsilon & \text{for } z \in \mathbb{R}^n \text{ and } n \ge 3, \\ C\epsilon(1 + |\log \epsilon|) & \text{for } z \in B_1(x) \text{ and } n = 2, \end{cases}$$

and it is easy to see that if, say, $\psi \in L_\infty(\Gamma)^m$, then

$$S\psi(x) = \int_\Gamma G(x, y)\psi(y)\,d\sigma_y \quad \text{and} \quad S^*\psi(x) = \int_\Gamma G(y, x)^*\psi(y)\,d\sigma_y$$

$$(7.8)$$

for $x \in \Gamma$. Hence, S and S^* are integral operators on Γ with weakly singular kernels. To handle the other six boundary operators, we define

$$T_\epsilon\psi(x) = 2\int_{\Gamma\backslash B_\epsilon(x)} \left[\tilde{\mathcal{B}}_{\nu,y}G(x, y)^*\right]^*\psi(y)\,d\sigma_y,$$

$$T_\epsilon^*\psi(x) = 2\int_{\Gamma\backslash B_\epsilon(x)} \tilde{\mathcal{B}}_{\nu,x}G(y, x)^*\psi(y)\,d\sigma_y,$$

$$\tilde{T}_\epsilon\psi(x) = 2\int_{\Gamma\backslash B_\epsilon(x)} \left[\mathcal{B}_{\nu,y}G(y, x)\right]^*\psi(y)\,d\sigma_y,$$

$$\tilde{T}_\epsilon^*\psi(x) = 2\int_{\Gamma\backslash B_\epsilon(x)} \mathcal{B}_{\nu,x}G(x, y)\psi(y)\,d\sigma_y,$$

$$R_\epsilon\psi(x) = -\int_{\Gamma\backslash B_\epsilon(x)} \mathcal{B}_{\nu,x}\left[\tilde{\mathcal{B}}_{\nu,y}G(x, y)^*\right]^*\psi(y)\,d\sigma_y,$$

$$R_\epsilon^*\psi(x) = -\int_{\Gamma\backslash B_\epsilon(x)} \tilde{\mathcal{B}}_{\nu,x}\left[\mathcal{B}_{\nu,y}G(y, x)\right]^*\psi(y)\,d\sigma_y;$$

cf. the integral formulae for the single- and double-layer potentials given in (7.1) and (7.2). Recalling the definition of $\mathcal{B}_\nu$ from (4.3) and (4.4), and the definition of $\tilde{\mathcal{B}}_\nu$ from (4.5), we see that the kernels of the last six integral operators above are given explicitly as follows:

$$2\left[\tilde{\mathcal{B}}_{\nu,y}G(x, y)^*\right]^* = 2\left[\partial_{n+k}G(x, y)A_{kj}(y) + G(x, y)A_j(y)\right]\nu_j(y),$$

$$2\tilde{\mathcal{B}}_{\nu,x}G(y, x)^* = 2\left[\partial_{n+k}G(y, x)A_{kj}(x) + G(y, x)A_j(x)\right]^*\nu_j(x),$$

$$2\left[\mathcal{B}_{\nu,y}G(y, x)\right]^* = 2\nu_j(y)\left[A_{jk}(y)\partial_k G(y, x)\right]^*,$$

$$2\mathcal{B}_{\nu,x}G(x, y) = 2\nu_j(x)A_{jk}(x)\partial_k G(x, y),$$

$$-\mathcal{B}_{\nu,x}\left[\tilde{\mathcal{B}}_{\nu,y}G(x, y)^*\right]^* = -\nu_j(x)\left[A_{jk}(x)\partial_k\partial_{n+m}G(x, y)A_{ml}(y)\right]\nu_l(y)$$
$$- \nu_j(x)\left[A_{jk}(x)\partial_k G(x, y)A_l(y)\right]\nu_l(y),$$

$$-\tilde{\mathcal{B}}_{\nu,x}\left[\mathcal{B}_{\nu,y}G(y, x)\right]^* = -\nu_j(y)\left[A_{jk}(y)\partial_k\partial_{n+m}G(y, x)A_{ml}(x)\right]^*\nu_l(x)$$
$$- \nu_j(y)\left[A_{jk}(y)\partial_k G(y, x)A_l(x)\right]^*\nu_l(x).$$

$$(7.9)$$

Here, we have used the summation convention, and note that

$$\partial_{n+k}G(x, y) = \partial_{y_k}G(x, y).$$

In general, the six kernels in (7.9) are all strongly singular on the $(n-1)$-dimensional surface Γ, because the leading term in the homogeneous expansion of $\partial^\alpha G$ is of degree $2 - n - |\alpha|$.

To investigate what happens as $\epsilon \downarrow 0$, suppose that Ω^- is given locally by $x_n < \zeta(x')$, and define the directional derivative

$$d\zeta(x', h') = \lim_{t \downarrow 0} \frac{\zeta(x' + th') - \zeta(x')}{t}.$$

For $x \in \Gamma$, i.e., for $x_n = \zeta(x')$, we shall say that Γ is *uniformly directionally differentiable* at x if

$$\zeta(x' + h') = \zeta(x') + d\zeta(x', h') + o(|h'|) \quad \text{as } |h'| \to 0. \tag{7.10}$$

Note that $d\zeta(x', h')$ is homogeneous of degree 1, but not necessarily linear, in h'. In order to state our next theorem, we define two subsets of the unit sphere $\mathbb{S}^{n-1} \subseteq \mathbb{R}^n$,

$$\Upsilon^+(x) = \{\omega \in \mathbb{S}^{n-1} : \omega_n > d\zeta(x', \omega')\}$$

and

$$\Upsilon^-(x) = \{\omega \in \mathbb{S}^{n-1} : \omega_n < d\zeta(x', \omega')\}.$$

In the simplest case, when ζ is differentiable at x', we have $d\zeta(x', h') = h' \cdot \operatorname{grad} \zeta(x')$, and so by (3.28),

$$\Upsilon^\pm(x) = \{\omega \in \mathbb{S}^{n-1} : \pm\omega \cdot \nu(x) > 0\}. \tag{7.11}$$

Theorem 7.3 Let $x \in \Gamma$, suppose that Γ is uniformly directionally differentiable at x, and define

$$a^\pm(x) = \int_{\Upsilon^\mp(x)} \partial_{n+k} G_0(x, \omega) A_{kj}(x)\omega_j \, d\omega,$$

$$\tilde{a}^\pm(x) = \int_{\Upsilon^\mp(x)} \partial_{n+k} G_0(x, -\omega)^* A_{jk}^*(x)\omega_j \, d\omega,$$

where $G_0(x, x - y)$ is the leading term in the homogeneous expansion of $G(x, y)$. For $\psi \in \mathcal{D}(\Gamma)^m$,

$$\gamma^\pm \operatorname{DL} \psi(x) = \pm a^\pm(x)\psi(x) + \lim_{\epsilon \downarrow 0} \tfrac{1}{2} T_\epsilon \psi(x),$$

$$\gamma^\pm \widetilde{\operatorname{DL}} \psi(x) = \pm \tilde{a}^\pm(x)\psi(x) + \lim_{\epsilon \downarrow 0} \tfrac{1}{2} \widetilde{T}_\epsilon \psi(x),$$

and so

$$T\psi(x) = [a^+(x) - a^-(x)]\psi(x) + \lim_{\epsilon \downarrow 0} T_\epsilon \psi(x),$$

$$\widetilde{T}\psi(x) = [\tilde{a}^+(x) - \tilde{a}^-(x)]\psi(x) + \lim_{\epsilon \downarrow 0} \widetilde{T}_\epsilon \psi(x).$$

Proof. Suppose $\psi = \gamma u$ where $u \in \mathcal{D}(\mathbb{R}^n)^m$. We know from Lemma 6.16 that

$$\gamma^\pm \mathrm{DL}\, \psi(x) = \pm \lim_{\epsilon \downarrow 0} \Phi_{\Omega^\mp \setminus B_\epsilon(x)}\big(G(x, \cdot)^*, u\big),$$

and since $\mathcal{P}^* G(x, \cdot)^* = 0$ on $\Omega^\mp \setminus B_\epsilon(x)$, the first Green identity gives

$$\pm \Phi_{\Omega^\mp \setminus B_\epsilon(x)}\big(G(x, \cdot), u\big) = \tfrac{1}{2} T_\epsilon \psi(x) + \int_{\Omega^\mp \cap \partial B_\epsilon(x)} \big[\widetilde{\mathcal{B}}_{\nu,y} G(x, y)^*\big]^* u(y)\, d\sigma_y,$$

where ν is the *out*ward unit normal to $\Omega^- \setminus B_\epsilon(x)$ and the *in*ward unit normal to $\Omega^+ \setminus B_\epsilon(x)$. Suppose $y \in \Omega^\mp \cap \partial B_\epsilon(x)$ and put $y = x + \epsilon \omega$, where $\omega \in \mathbb{S}^{n-1}$. Observe that $\nu(y) = \mp\omega$, so by (7.9) and Theorem 6.3,

$$\big[\widetilde{\mathcal{B}}_{\nu,y} G(x, y)^*\big]^* = \big[\partial_{n+k} G(x, x + \epsilon\omega) A_{kj}(x + \epsilon\omega)$$

$$+ G(x, x + \epsilon\omega) A_j(x + \epsilon\omega)\big](\mp\omega_j)$$

$$= \mp \partial_{n+k} G_0(x, -\epsilon\omega) A_{kj}(x)\omega_j + \begin{cases} O(1 + |\log\epsilon|) & \text{if } n = 2, \\ O(\epsilon^{2-n}) & \text{if } n \geq 3. \end{cases}$$

Now put

$$\Upsilon_\epsilon^\pm(x) = \{\omega \in \mathbb{S}^n : x + \epsilon\omega \in \Omega^\pm\},$$

so that, noting $\partial_{n+k} G_0(x, -z) = -\partial_{n+k} G_0(x, z)$,

$$\int_{\Omega^\mp \cap \partial B_\epsilon(x)} \big[\widetilde{\mathcal{B}}_{\nu,y} G(x, y)^*\big]^* u(y)\, d\sigma_y$$

$$= \pm \int_{\Upsilon_\epsilon^\mp(x)} \partial_{n+k} G_0(x, \epsilon\omega) A_{kj}(x)\omega_j u(x + \epsilon\omega)\epsilon^{n-1}\, d\omega$$

$$+ \begin{cases} O\big(\epsilon(1 + |\log\epsilon|)\big) & \text{if } n = 2, \\ O(\epsilon) & \text{if } n \geq 3. \end{cases} \tag{7.12}$$

Since $\partial_{n+k} G_0(x, z)$ is homogeneous in z of degree $1 - n$, and since (7.10) implies that

$$\lim_{\epsilon \downarrow 0} \sigma\Big([\Upsilon^\pm(x) \setminus \Upsilon_\epsilon^+(x)] \cup [\Upsilon_\epsilon^\pm(x) \setminus \Upsilon^\perp(x)]\Big) = 0,$$

we see that

$$\lim_{\epsilon \downarrow 0} \int_{\Omega^{\mp} \cap \partial B_{\epsilon}(x)} [\widetilde{\mathcal{B}}_{\nu,y} G(x, y)^*]^* u(y) \, d\sigma_y = \pm a^{\pm}(x) \psi(x),$$

giving the formula for $\gamma^{\pm} \mathrm{DL} \, \psi(x)$. The expression for $T \psi(x)$ then follows immediately from the definition of T in (7.3).

The formulae for $\gamma^{\pm} \widetilde{\mathrm{DL}} \, \psi(x)$ and $\widetilde{T} \psi(x)$ follow by a similar argument, with the help of Exercise 6.2. $\qquad \square$

When Γ is sufficiently smooth, the preceding results for T and $\widetilde{T}$ simplify, and we can deal with the other four boundary operators; cf. Theorem 5.23.

Theorem 7.4 Let $x \in \Gamma$ and $\psi \in \mathcal{D}(\Gamma)^m$.

(i) If Γ has a tangent plane at x, then

$$T \psi(x) = \lim_{\epsilon \downarrow 0} T_{\epsilon} \psi(x) \quad and \quad \widetilde{T} \psi(x) = \lim_{\epsilon \downarrow 0} \widetilde{T}_{\epsilon} \psi(x).$$

(ii) If Γ is $C^{1,\mu}$ (with $0 < \mu < 1$) on a neighbourhood of x, then

$$T^* \psi(x) = \lim_{\epsilon \downarrow 0} T_{\epsilon}^* \psi(x) \quad and \quad \widetilde{T}^* \psi(x) = \lim_{\epsilon \downarrow 0} \widetilde{T}_{\epsilon}^* \psi(x).$$

(iii) If Γ is C^2 on a neighbourhood of x, then

$$R \psi(x) = \underset{\epsilon \downarrow 0}{\mathrm{f.p.}} \, R_{\epsilon} \psi(x) \quad and \quad R^* \psi(x) = \underset{\epsilon \downarrow 0}{\mathrm{f.p.}} \, R_{\epsilon}^* \psi(x).$$

Proof. Since $\partial_{n+k} G_0(x, -\omega) = -\partial_{n+k} G_0(x, \omega)$, and since $\Upsilon^{\pm}(x)$ is given by (7.11), we see that $a^-(x) = a^+(x)$ and so $T_{\epsilon} \psi(x) \to T \psi(x)$. In the same way, $\tilde{a}^+(x) = \tilde{a}^-(x)$, so $\widetilde{T}_{\epsilon} \psi(x) \to \widetilde{T} \psi(x)$.

Part (ii) follows from part (i) because, cf. (7.12), the combination

$$\begin{aligned} \widetilde{\mathcal{B}}_{\nu,x} G(y, x)^* + \widetilde{\mathcal{B}}_{\nu,y} G(x, y)^* &= \left[\nu_j(x) A_{kj}(x)^* - \nu_j(y) A_{kj}(y)^* \right] \\ &\quad \times \partial_{n+k} G_0(x, x - y)^* \\ &\quad + \begin{cases} O\big(1 + |\log|x - y|\,|\big) & \text{if } n = 2, \\ O\big(|x - y|^{2-n}\big) & \text{if } n \geq 3, \end{cases} \\ &= O\big(|x - y|^{\mu+1-n}\big), \end{aligned}$$

is only a weakly singular kernel on Γ, and $[\widetilde{\mathcal{B}}_{\nu,y} G(x, y)^*]^*$ is the kernel of T.

We now deal with the hypersingular operators. By Lemma 6.16,

$$R\psi(x) = -\mathcal{B}_v^{\pm} \, \mathrm{DL} \, \psi(x) = -\mathcal{B}_v^{\pm} \, \mathrm{SL} \, \mathcal{B}_v u(x) \mp \lim_{\epsilon \downarrow 0} \left(\left[\mathcal{B}_{v,x} G(x, \cdot) \right]^*, \mathcal{P}u \right)_{\Omega_\epsilon^{\mp}},$$

where $\Omega_\epsilon^{\pm} = \Omega^{\pm} \setminus B_\epsilon(x)$. Since $\mathcal{P}^* [\mathcal{B}_{v,x} G(x, \cdot)]^* = 0$ on $\Omega_\epsilon^{\pm}$, the second Green identity (Theorem 4.4) gives

$$\mp \left(\left[\mathcal{B}_{v,x} G(x, \cdot) \right]^*, \mathcal{P}u \right)_{\Omega_\epsilon^{\mp}} = -\left(\widetilde{\mathcal{B}}_v \left[\mathcal{B}_{v,x} G(x, \cdot) \right]^*, \gamma u \right)_{\partial \Omega_\epsilon^{\mp}}$$
$$+ \left(\left[\mathcal{B}_{v,x} G(x, \cdot) \right]^*, \mathcal{B}_v u \right)_{\partial \Omega_\epsilon^{\mp}}.$$

From (7.9), we see that

$$-\left(\widetilde{\mathcal{B}}_v \left[\mathcal{B}_{v,x} G(x, \cdot) \right]^*, \gamma u \right)_{\partial \Omega_\epsilon^{\mp}} = R_\epsilon \psi(x)$$
$$\pm \int_{\Omega^{\pm} \cap \partial B_\epsilon(x)} \mathcal{B}_{v,x} \left[\widetilde{\mathcal{B}}_{v,y} G(x, y)^* \right]^* u(y) \, d\sigma_y$$

and

$$\left(\left[\mathcal{B}_{v,x} G(x, \cdot) \right]^*, \mathcal{B}_v u \right)_{\partial \Omega_\epsilon^{\mp}} = \tfrac{1}{2} \widetilde{T}_\epsilon^* \mathcal{B}_v u(x) \mp \int_{\Omega^{\mp} \cap \partial B_\epsilon(x)} \mathcal{B}_{v,x} G(x, y) \mathcal{B}_v u(y) \, d\sigma_y,$$

where $v(y)$ is the *out*ward unit normal to $B_\epsilon(x)$. Thus,

$$R\psi(x) = -\tfrac{1}{2} [\mp \mathcal{B}_v u(x) + \widetilde{T}^* \mathcal{B}_v u(x)] + \lim_{\epsilon \downarrow 0} \Big(R_\epsilon \psi(x) + \tfrac{1}{2} \widetilde{T}_\epsilon^* \mathcal{B}_v u(x)$$
$$\pm \int_{\Omega^{\mp} \cap \partial B_\epsilon(x)} \big\{ \mathcal{B}_{v,x} \left[\widetilde{\mathcal{B}}_{v,y} G(x, y)^* \right]^* u(y)$$
$$- \mathcal{B}_{v,x} G(x, y) \mathcal{B}_v u(y) \big\} \, d\sigma_y \Big), \qquad (7.13)$$

and by arguing as in the proof of Theorem 7.3 and noting that $\partial_{n+k} G_0(x, -\omega) = -\partial_{n+k} G_0(x, \omega)$, we find that

$$\int_{\Omega^{\mp} \cap \partial B_\epsilon(x)} \mathcal{B}_{v,x} G(x, y) \mathcal{B}_v u(y) \, d\sigma_y$$
$$= \int_{\Upsilon_\epsilon^{\mp}(x)} v_j(x) A_{jk}(x) \partial_k G(x, x + \epsilon \omega) \omega_l \partial_l u(x + \epsilon \omega) \epsilon^{n-1} \, d\omega.$$

Differentiating the expansion in Theorem 6.3 with respect to x, one obtains as the leading term $\partial_{n+k} G_0(x, x - y)$, and since $\partial_{n+k} G_0(x, z)$ is odd and homogeneous of degree $1 - n$ as a function of z, we have

$$\int_{\Omega^{\mp} \cap \partial B_\epsilon(x)} \mathcal{B}_{v,x} G(x, y) \mathcal{B}_v u(y) \, d\sigma_y$$

$$= -\frac{1}{2} \int_{|\omega|=1} v_j(x) A_{jk}(x) \partial_{n+k} G_0(x, \omega) \partial_l u(x) \omega_l \, d\omega + O(\epsilon).$$

Hence, taking the average of the $+$ and $-$ expressions for $R\psi(x)$ in (7.13), we are left with

$$R\psi(x) = \lim_{\epsilon \downarrow 0} \left(R_\epsilon \psi(x) + \frac{1}{2} \int_{\Omega^- \cap \partial B_\epsilon(x)} \mathcal{B}_{v,x} \big[\widetilde{\mathcal{B}}_{v,y} G(x, y)^* \big]^* u(y) \, d\sigma_y \right.$$

$$\left. - \frac{1}{2} \int_{\Omega^+ \cap \partial B_\epsilon(x)} \mathcal{B}_{v,x} \big[\widetilde{\mathcal{B}}_{v,y} G(x, y)^* \big]^* u(y) \, d\sigma_y \right).$$

In view of Theorem 6.3 and (7.9), if we let $u_{ml}(y) = A_{ml}(y)u(y)$ and $u_l(y) = A_l(y)u(y)$, then

$$-\mathcal{B}_{v,x} \big[\widetilde{\mathcal{B}}_{v,y} G(x, y)^* \big]^* u(y) = v_j(x) A_{jk}(x) \big[\partial_{n+k} \partial_{n+m} G_0(x, x - y) u_{ml}(x)$$

$$+ \partial_k \partial_{n+m} G_0(x, x - y) u_{ml}(x)$$

$$+ \partial_{n+k} \partial_{n+m} G_1(x, x - y) u_{ml}(x)$$

$$+ \partial_{n+k} \partial_{n+m} G_0(x, x - y) \partial_p u_{ml}(x)(y_p - x_p)$$

$$- \partial_{n+k} G_0(x, x - y) u_l(x) \big] v_l(y)$$

$$+ O(|x - y|^{2-n}).$$

For the leading term, we apply Exercise 7.2 with $f(\omega) = \partial_{n+k} \partial_{n+m} G_0(x, -\omega) A_{ml}(x) \omega_l$ to obtain

$$\int_{\Omega^{\mp} \cap \partial B_\epsilon(x)} \partial_{n+k} \partial_{n+m} G_0(x, x - y) v_l(y) \, d\sigma_y$$

$$= \epsilon^{-1} \int_{\Upsilon^{\mp}(x)} \partial_{n+k} \partial_{m+k} G_0(x, \omega) \omega_l \, d\omega + O(\epsilon).$$

Each of the remaining strongly singular terms in the integrand has the form $f(x, x - y)$, where $f(x, z)$ is even and homogeneous of degree $1 - n$ as a function of z. Since

$$\lim_{\epsilon \downarrow 0} \int_{\Omega^{\mp} \cap \partial B_\epsilon(x)} f(x, x - y) \, d\sigma_y = \int_{\Upsilon^{\mp}(x)} f(x, \omega) \, d\omega,$$

the contributions from $\Upsilon^+(x)$ and $\Upsilon^-(x)$ cancel, and part (iii) follows. $\square$

The Dirichlet Problem

We now show how the single- and double-layer potentials allow a pure Dirichlet problem to be reformulated as a boundary integral equation of the first kind with a weakly singular kernel.

Theorem 7.5 *Let $f \in \tilde{H}^{-1}(\Omega^-)^m$ and $g \in H^{1/2}(\Gamma)^m$.*

(i) If $u \in H^1(\Omega^-)^m$ is a solution of the interior Dirichlet problem

$$\begin{aligned} \mathcal{P}u &= f \quad \text{on } \Omega^-, \\ \gamma^- u &= g \quad \text{on } \Gamma, \end{aligned} \tag{7.14}$$

then the conormal derivative $\psi = \mathcal{B}_\nu^- u \in H^{-1/2}(\Gamma)^m$ is a solution of the boundary integral equation

$$S\psi = \tfrac{1}{2}(g + Tg) - \gamma \mathcal{G} f \quad \text{on } \Gamma, \tag{7.15}$$

and u has the integral representation

$$u = \mathcal{G}f - \mathrm{DL}\, g + \mathrm{SL}\, \psi \quad \text{on } \Omega^-. \tag{7.16}$$

(ii) Conversely, if $\psi \in H^{-1/2}(\Gamma)^m$ is a solution of the boundary integral equation (7.15), then the formula (7.16) defines a solution $u \in H^1(\Omega^-)^m$ of the interior Dirichlet problem (7.14).

Proof. As in Theorem 6.10, we view f as a distribution in $H^{-1}(\mathbb{R}^n)^m$ with supp $f \subseteq \overline{\Omega^-}$; cf. Theorem 3.29 (ii).

Suppose that $u \in H^1(\Omega^-)$ satisfies (7.14), and define $u = 0$ on the exterior domain Ω^+. Applying Theorem 6.10, we obtain the representation formula,

$$u = \mathcal{G}f - \mathrm{DL}\, \gamma^- u + \mathrm{SL}\, \mathcal{B}_\nu^- u \quad \text{on } \Omega^-, \tag{7.17}$$

and then by (7.5),

$$\gamma^- u = \gamma \mathcal{G} f - \tfrac{1}{2}(-\gamma^- u + T\gamma^- u) + S\mathcal{B}_\nu^- u \quad \text{on } \Gamma.$$

Part (i) now follows from the boundary condition $\gamma^- u = g$.

To prove (ii), suppose that $\psi \in H^{-1/2}(\Gamma)^m$ satisfies (7.15), and define u by (7.16). The mapping property (6.10) of the volume potential, together with those of the surface potentials given in Theorem 6.11, imply that $\mathcal{G}f$, DL g and SL ψ all belong to $H^1(\Omega^-)^m$, so $u \in H^1(\Omega^-)^m$. By (6.2), we have $\mathcal{P}\mathcal{G}f = f$ on $\mathbb{R}^n$, and by (6.19), we have $\mathcal{P}\,\mathrm{DL}\,g = \mathcal{P}\,\mathrm{SL}\,\psi = 0$ on Ω^-, so $\mathcal{P}u = f$ on Ω^-. Finally, $\gamma^- u = g$ by (7.5). $\square$

The next theorem shows that the boundary integral equation (7.15) satisfies the Fredholm alternative; cf. Theorem 2.33. The method of proof was first used by Nedelec and Planchard [76], [74], Le Roux [56], [57], and Hsiao and Wendland [42], for the case when $\mathcal{P}$ is the Laplacian. These authors were all concerned with error estimates for Galerkin boundary element methods, in which context positivity up to a compact perturbation is of fundamental importance for establishing stability.

Theorem 7.6 *The boundary operator* $S = \gamma$ SL *admits a decomposition*

$$S = S_0 + L,$$

in which $S_0 : H^{-1/2}(\Gamma)^m \to H^{1/2}(\Gamma)^m$ *is positive and bounded below, i.e.,*

$$\mathrm{Re}(S_0\psi, \psi)_\Gamma \geq c\|\psi\|^2_{H^{-1/2}(\Gamma)^m} \quad for \ \psi \in H^{-1/2}(\Gamma)^m,$$

and in which $L : H^{-1/2}(\Gamma)^m \to H^{1/2}(\Gamma)^m$ *is a compact linear operator. Hence,*

$$S : H^{-1/2}(\Gamma)^m \to H^{1/2}(\Gamma)^m$$

is a Fredholm operator with index zero.

Proof. Put $u = \chi \, \mathrm{SL} \, \psi$ and $v = \chi \, \mathrm{SL} \, \phi$, where $\psi, \phi \in H^{-1/2}(\Gamma)^m$ and $\chi \in C^\infty_{\mathrm{comp}}(\mathbb{R}^n)$ is a cutoff function satisfying $\chi = 1$ on a neighbourhood of $\overline{\Omega^-}$. Since $S\psi = \gamma u$ and $\phi = -[\mathcal{B}_\nu v]_\Gamma$, and since $\mathcal{P}v = 0$ on Ω^-, the first Green identity implies that

$$(S\psi, \phi)_\Gamma = (\gamma u, \mathcal{B}_\nu^- v - \mathcal{B}_\nu^+ v)_\Gamma = \Phi_{\Omega^-}(u, v) + \Phi_{\Omega^+}(u, v) + (L_1\psi, \phi)_\Gamma,$$

where

$$(L_1\psi, \phi)_\Gamma = -(u, \mathcal{P}v)_{\Omega^+}. \tag{7.18}$$

We have $[u]_\Gamma = 0$ so $u \in H^1(\mathbb{R}^n)^m$, and hence the strong ellipticity of $\mathcal{P}$ implies that

$$\mathrm{Re}\,\Phi_{\Omega^-}(u, u) + \mathrm{Re}\,\Phi_{\Omega^+}(u, u) = \mathrm{Re}\,\Phi_{\mathbb{R}^n}(u, u) \geq c\|u\|^2_{H^1(\mathbb{R}^n)^m} + (L_2\psi, \psi)_\Gamma, \tag{7.19}$$

where

$$(L_2\psi, \phi)_\Gamma = -C(u, v)_{\mathbb{R}^n}. \tag{7.20}$$

Furthermore,

$$\|\psi\|^2_{H^{-1/2}(\Gamma)^m} = \|\mathcal{B}_\nu^- u - \mathcal{B}_\nu^+ u\|^2_{H^{-1/2}(\Gamma)^m} \leq C\|u\|^2_{H^1(\mathbb{R}^n)^m},$$

so if $L = L_1 + L_2$ and $S_0 = S - L$, then S_0 is positive and bounded below, as required. By Theorem 3.27, to show that $L : H^{-1/2}(\Gamma)^m \to H^{1/2}(\Gamma)^m$ is compact, it suffices to show that

$$L : H^{\epsilon-1}(\Gamma)^m \to H^{1-\epsilon}(\Gamma)^m$$

is bounded for $0 < \epsilon < 1$. In fact, L_1 has the form

$$L_1 \psi(x) = \int_\Gamma K_1(x, y) \psi(y) \, d\sigma_y,$$

where K_1 is C^∞ on a neighbourhood of $\Gamma \times \Gamma$, because $G(x, y)$ is C^∞ for $x \neq y$, and $\mathcal{P}v$ has compact support in Ω^+. To deal with L_2, we apply the Cauchy–Schwarz inequality and obtain

$$|(L_2 \psi, \phi)_\Gamma| \leq C \|u\|_{L_2(\mathbb{R}^n)^m} \|v\|_{L_2(\mathbb{R}^n)^m}.$$

By the mapping property of the single-layer potential in Theorem 6.12,

$$\|u\|_{L_2(\mathbb{R}^n)^m} \leq C \|u\|_{H^{\epsilon+1/2}(\mathbb{R}^n)^m} \leq C \|\psi\|_{H^{\epsilon-1}(\Gamma)^m},$$

so

$$|(L_2 \psi, \phi)_\Gamma| \leq C \|\psi\|_{H^{\epsilon-1}(\Gamma)^m} \|\phi\|_{H^{\epsilon-1}(\Gamma)^m},$$

and hence $\|L_2 \psi\|_{H^{1-\epsilon}(\Gamma)^m} \leq C \|\psi\|_{H^{\epsilon-1}(\Gamma)^m}$. $\square$

Applying the Fredholm alternative to the boundary integral equation (7.15), we see that a solution exists if and only if

$$\left(\tfrac{1}{2}(g + Tg) - \gamma \mathcal{G}f, \phi\right)_\Gamma = 0 \tag{7.21}$$

for every solution $\phi \in H^{-1/2}(\Gamma)^m$ of the homogeneous adjoint equation $S^* \phi = 0$. Every such ϕ has the form $\phi = \widetilde{B}_\nu v$ where $v \in H^1(\Omega^-)^m$ satisfies

$$\mathcal{P}^* v = 0 \quad \text{on } \Omega^-,$$

$$\gamma^- v = 0 \quad \text{on } \Gamma,$$

and since $\tfrac{1}{2}(g + Tg) = g + \gamma^- \mathrm{DL}\, g$, the condition (7.21) is equivalent to

$$
\begin{aligned}
(g, \widetilde{B}_\nu v)_\Gamma &= \left(\gamma^-[\mathcal{G}f - \mathrm{DL}\, g], \widetilde{B}_\nu^- v\right)_\Gamma \\
&= \Phi(\mathcal{G}f - \mathrm{DL}\, g, v) &&\text{by (6.26), since } \mathcal{P}^* v = 0, \\
&= \left(\mathcal{P}[\mathcal{G}f - \mathrm{DL}\, g], v\right)_{\Omega^-} &&\text{since } \gamma^- v = 0, \\
&= (f, v)_{\Omega^-},
\end{aligned}
$$

which is the same as the condition obtained in Theorem 4.10 for solvability of the (pure) Dirichlet problem.

The Neumann Problem

The pure Neumann problem can also be reformulated as a boundary integral equation of the first kind, but this time the kernel is hypersingular.

Theorem 7.7 Let $f \in \widetilde{H}^{-1}(\Omega^-)^m$ and $g \in H^{-1/2}(\Gamma)^m$.

(i) If $u \in H^1(\Omega^-)^m$ is a solution of the interior Neumann problem

$$\begin{aligned}
\mathcal{P}u &= f \quad on\ \Omega^-, \\
\mathcal{B}_\nu^- u &= g \quad on\ \Gamma,
\end{aligned} \tag{7.22}$$

then the trace $\psi = \gamma^- u \in H^{1/2}(\Gamma)^m$ is a solution of the boundary integral equation

$$R\psi = \tfrac{1}{2}(g - \widetilde{T}^* g) - \mathcal{B}_\nu^- \mathcal{G} f \quad on\ \Gamma, \tag{7.23}$$

and u has the integral representation

$$u = \mathcal{G} f - \mathrm{DL}\,\psi + \mathrm{SL}\,g \quad on\ \Omega^-. \tag{7.24}$$

(ii) Conversely, if $\psi \in H^{1/2}(\Gamma)^m$ is a solution of the boundary integral equation (7.23), then the formula (7.24) defines a solution $u \in H^1(\Omega^-)^m$ of the interior Neumann problem (7.22).

Proof. Essentially, one repeats the proof of Theorem 7.5, interchanging g and ψ, and taking the conormal derivative of the Green representation formula (7.17), instead of its trace. In fact, by (7.5),

$$\mathcal{B}_\nu^- u = \mathcal{B}_\nu^- \mathcal{G} f + R\gamma^- u + \tfrac{1}{2}\big(\mathcal{B}_\nu^- u + \widetilde{T}^* \mathcal{B}_\nu^- u\big) \quad on\ \Gamma,$$

giving (7.23). □

Next we show that the Fredholm alternative is valid for the boundary integral equation (7.23). As in the proof of Theorem 6.12, we use the notation

$$\Omega_\rho^+ = \{x \in \Omega^+ : |x| < \rho\} = \Omega^+ \cap B_\rho, \tag{7.25}$$

where the number ρ is large enough so that $|x| < \rho/2$ for all $x \in \Omega^-$. The argument below follows a similar pattern to the one for the Dirichlet problem (Theorem 7.6).

Theorem 7.8 If $\mathcal{P}$ is coercive on $H^1(\Omega^-)^m$ and on $H^1(\Omega_\rho^+)^m$, then the boundary operator $R = -\mathcal{B}_\nu \, \mathrm{DL}$ is coercive on $H^{1/2}(\Gamma)^m$, i.e.,

$$\mathrm{Re}(R\psi, \psi)_\Gamma \ge c\|\psi\|_{H^{1/2}(\Gamma)^m}^2 - C\|\psi\|_{L_2(\Gamma)^m}^2 \quad \text{for } \psi \in H^{1/2}(\Gamma)^m.$$

Hence,

$$R : H^{1/2}(\Gamma)^m \to H^{-1/2}(\Gamma)^m$$

is a Fredholm operator with index zero.

Proof. Let ψ and ϕ belong to $H^{1/2}(\Gamma)^m$ and put $u = \chi \, \mathrm{DL} \, \psi$ and $v = \chi \, \mathrm{DL} \, \phi$, where $\chi \in C_{\mathrm{comp}}^\infty(\mathbb{R}^n)$ is a cutoff function satisfying $\chi = 1$ on a neighbourhood of $\overline{\Omega^-}$, with $\mathrm{supp}\,\chi \Subset B_\rho$. Since $R\psi = -\mathcal{B}_\nu u$ and $\phi = [\gamma v]_\Gamma$, and since $\mathcal{P}u = 0$ on Ω^-, the first Green identity implies that

$$(R\psi, \phi)_\Gamma = (-\mathcal{B}_\nu u, \gamma^+ v - \gamma^- v)_\Gamma = \Phi_{\Omega^-}(u, v) + \Phi_{\Omega_\rho^+}(u, v) + (L_1\psi, \phi)_\Gamma,$$

where $(L_1\psi, \phi)_\Gamma = -(\mathcal{P}u, v)_{\Omega^+}$. By hypothesis,

$$\mathrm{Re}\,\Phi_{\Omega^-}(u, u) + \mathrm{Re}\,\Phi_{\Omega_\rho^+}(u, u) \ge c\big(\|u\|_{H^1(\Omega^-)^m}^2 + \|u\|_{H^1(\Omega_\rho^+)^m}^2\big) + (L_2\psi, \psi)_\Gamma,$$

where $(L_2\psi, \phi)_\Gamma = -C\big[(u, v)_{\Omega^-} + (u, v)_{\Omega_\rho^+}\big]$, and we have

$$\|\psi\|_{H^{1/2}(\Gamma)^m}^2 = \|\gamma^+ u - \gamma^- u\|_{H^{1/2}(\Gamma)^m}^2 \le C\big(\|u\|_{H^1(\Omega^-)^m}^2 + \|u\|_{H^1(\Omega_\rho^+)^m}^2\big).$$

Hence, if we put $L = L_1 + L_2$ and $R_0 = R - L$, then

$$R = R_0 + L \quad \text{and} \quad \mathrm{Re}(R_0\psi, \psi)_\Gamma \ge c\|\psi\|_{H^{1/2}(\Gamma)^m}^2 \qquad \text{for } \psi \in H^{1/2}(\Gamma)^m.$$

To complete the proof, we will show that

$$L : H^{-1/2}(\Gamma)^m \to H^{1/2}(\Gamma)^m$$

is bounded, and so $|(L\psi, \psi)_\Gamma| \le C\|\psi\|_{H^{-1/2}(\Gamma)^m}^2 \le C\|\psi\|_{L_2(\Gamma)^m}^2$. In fact, since L_1 is an integral operator on Γ whose kernel is C^∞ on a neighbourhood of $\Gamma \times \Gamma$, it suffices to consider L_2. By the Cauchy–Schwarz inequality,

$$|(L_2\psi, \phi)_\Gamma| \le C\big(\|u\|_{L_2(\Omega^-)^m}\|v\|_{L_2(\Omega^-)^m} + \|u\|_{L_2(\Omega_\rho^+)^m}\|v\|_{L_2(\Omega_\rho^+)^m}\big),$$

and by the mapping property of the double-layer potential in Exercise 6.4,

$$\|u\|_{L_2(\Omega^-)^m} \le C\|\psi\|_{H^{-1/2}(\Gamma)^m} \quad \text{and} \quad \|u\|_{L_2(\Omega_\rho^+)^m} \le C\|\psi\|_{H^{-1/2}(\Gamma)^m},$$

so

$$|(L_2\psi, \phi)_\Gamma| \le C\|\psi\|_{H^{-1/2}(\Gamma)^m}\|\phi\|_{H^{-1/2}(\Gamma)^m},$$

and hence $\|L_2\psi\|_{H^{1/2}(\Gamma)^m} \le C\|\psi\|_{H^{-1/2}(\Gamma)^m}.$ □

Mixed Boundary Conditions

Elliptic problems with mixed Dirichlet and Neumann boundary conditions can be reformulated as 2×2 sytems of boundary integral equations. As in Chapter 4, let

$$\Gamma = \Gamma_D \cup \Pi \cup \Gamma_N$$

be a Lipschitz dissection of the boundary. We write

$$S_{DD}\psi = (S\psi)|_{\Gamma_D} \text{ and } \widetilde{T}^*_{ND}\psi = (\widetilde{T}^*\psi)|_{\Gamma_N} \qquad \text{when supp } \psi \subseteq \Gamma_D \cup \Pi,$$

and

$$R_{NN}\psi = (R\psi)|_{\Gamma_N} \text{ and } T_{DN}\psi = (T\psi)|_{\Gamma_D} \qquad \text{when supp } \psi \subseteq \Gamma_N \cup \Pi.$$

It follows from (7.3) that, at least for $s = 0$,

$$S_{DD} : \widetilde{H}^{s-1/2}(\Gamma_D)^m \to H^{s+1/2}(\Gamma_D)^m, \quad \widetilde{T}^*_{ND} : \widetilde{H}^{s-1/2}(\Gamma_D)^m \to H^{s-1/2}(\Gamma_N)^m,$$
$$R_{NN} : \widetilde{H}^{s+1/2}(\Gamma_N)^m \to H^{s-1/2}(\Gamma_N)^m, \quad T_{DN} : \widetilde{H}^{s+1/2}(\Gamma_N)^m \to H^{s+1/2}(\Gamma_D)^m;$$

cf. Theorems 7.1 and 7.2.

Theorem 7.9 Let $f \in \widetilde{H}^{-1}(\Omega^-)^m$, $g_D \in H^{1/2}(\Gamma_D)^m$ and $g_N \in H^{-1/2}(\Gamma_N)^m$, and consider the mixed boundary value problem

$$\begin{aligned}
\mathcal{P}u &= f \quad on \ \Omega^-, \\
\gamma^-u &= g_D \quad on \ \Gamma_D, \\
\mathcal{B}^-_\nu u &= g_N \quad on \ \Gamma_N.
\end{aligned} \tag{7.26}$$

Extend the Dirichlet and Neumann data to the whole of Γ, in such a way that $g_D \in H^{1/2}(\Gamma)^m$ and $g_N \in H^{-1/2}(\Gamma)$, and define $h_D \in H^{1/2}(\Gamma_D)^m$ and $h_N \in H^{-1/2}(\Gamma_N)^m$ by

$$\begin{aligned}
h_D &= \tfrac{1}{2}(g_D + Tg_D) - Sg_N - \gamma\mathcal{G}f \quad on \ \Gamma_D, \\
h_N &= \tfrac{1}{2}(g_N - \widetilde{T}^*g_N) - Rg_D - \mathcal{B}^-_\nu\mathcal{G}f \quad on \ \Gamma_N.
\end{aligned}$$

(i) *If* $u \in H^1(\Omega^-)^m$ *is a solution of (7.26), then the differences*

$$\psi_D = \mathcal{B}_\nu^- u - g_N \in \tilde{H}^{-1/2}(\Gamma_D)^m \quad and \quad \psi_N = \gamma^- u - g_D \in \tilde{H}^{1/2}(\Gamma_N)^m$$

satisfy

$$\begin{bmatrix} S_{DD} & -\frac{1}{2}T_{DN} \\ \frac{1}{2}\tilde{T}_{ND}^* & R_{NN} \end{bmatrix} \begin{bmatrix} \psi_D \\ \psi_N \end{bmatrix} = \begin{bmatrix} h_D \\ h_N \end{bmatrix}, \tag{7.27}$$

and u *has the integral representation*

$$u = \mathcal{G}f - \mathrm{DL}(\psi_N + g_D) + \mathrm{SL}(\psi_D + g_N) \quad on \ \Omega^-. \tag{7.28}$$

(ii) *Conversely, if* $\psi_D \in \tilde{H}^{-1/2}(\Gamma_D)^m$ *and* $\psi_N \in \tilde{H}^{1/2}(\Gamma_N)^m$ *satisfy the system of boundary integral equations (7.27), then the formula (7.28) defines a solution* $u \in H^1(\Omega^-)^m$ *of the mixed boundary value problem (7.26).*

Proof. Let $u \in H^1(\Omega^-)^m$ be a solution of (7.26). Since $\mathcal{B}_\nu^- u = \psi_D + g_N$ and $\gamma^- u = \psi_N + g_D$, we see from (7.17) that

$$u = \mathcal{G}f - \mathrm{DL}(\psi_N + g_D) + \mathrm{SL}(\psi_D + g_N) \quad on \ \Omega^-,$$

and since u satisfies the boundary conditions, $\psi_D = 0$ on Γ_N, and $\psi_N = 0$ on Γ_D. Hence, by (7.5),

$$g_D = \gamma^- u = \gamma \mathcal{G}f - \frac{1}{2}(-g_D + T_{DN}\psi_N + Tg_D) + S_{DD}\psi_D + Sg_N \quad on \ \Gamma_D,$$

and

$$g_N = \mathcal{B}_\nu^- u = \mathcal{B}_\nu^- \mathcal{G}f + R_{NN}\psi_N + Rg_D + \frac{1}{2}(g_N + \tilde{T}_{ND}^*\psi_D + \tilde{T}^* g_N) \quad on \ \Gamma_N,$$

giving

$$S_{DD}\psi_D - \frac{1}{2}T_{DN}\psi_N = \frac{1}{2}(g_D + Tg_D) - Sg_N - \gamma \mathcal{G}f \quad on \ \Gamma_D,$$

$$\frac{1}{2}\tilde{T}_{ND}^*\psi_D + R_{NN}\psi_N = \frac{1}{2}(g_N - \tilde{T}^* g_N) - Rg_D - \mathcal{B}_\nu^- \mathcal{G}f \quad on \ \Gamma_N,$$

which is just the 2×2 system (7.27). This argument proves part (i).

Conversely, suppose that $\psi_D \in \tilde{H}^{1/2}(\Gamma_D)^m$ and $\psi_N \in \tilde{H}^{-1/2}(\Gamma_N)^m$ satisfy (7.27). By (6.10) and Theorem 6.11, the equation (7.28) defines a function $u \in H^1(\Omega^-)^m$, and obviously $\mathcal{P}u = f$ on Ω^-. Finally, by working backwards through the calculations above, we see that $\gamma^- u = g_D$ on Γ_D, and $\mathcal{B}_\nu^- u = g_N$ on Γ_N, proving part (ii). □

By putting

$$A = \begin{bmatrix} S_{DD} & -\frac{1}{2}T_{DN} \\ \frac{1}{2}\widetilde{T}^*_{ND} & R_{NN} \end{bmatrix}, \qquad \psi = \begin{bmatrix} \psi_D \\ \psi_N \end{bmatrix}, \qquad h = \begin{bmatrix} h_D \\ h_N \end{bmatrix},$$

we can write the system (7.27) as

$$A\psi = h,$$

and by putting

$$(\psi, \phi)_{\Gamma_D \times \Gamma_N} = (\psi_D, \phi_D)_{\Gamma_D} + (\psi_N, \phi_N)_{\Gamma_N},$$

we have

$$\begin{aligned} (A\psi, \phi)_{\Gamma_D \times \Gamma_N} &= (S_{DD}\psi_D, \phi_D)_{\Gamma_D} - \tfrac{1}{2}(T_{DN}\psi_N, \phi_D)_{\Gamma_D} \\ &\quad + \tfrac{1}{2}(\widetilde{T}^*_{ND}\psi_D, \phi_N)_{\Gamma_N} + (R_{NN}\psi_N, \phi_N)_{\Gamma_D}. \end{aligned}$$

When $\mathcal{P}^* = \mathcal{P}$, a simple argument shows that the Fredholm alternative is valid for (7.27).

Theorem 7.10 Let $H = \widetilde{H}^{-1/2}(\Gamma_D)^m \times \widetilde{H}^{1/2}(\Gamma_N)^m$. If $\mathcal{P}$ is formally self-adjoint, and if $\mathcal{P}$ is coercive on $H^1(\Omega^-)^m$ and on $H^1(\Omega_\rho^+)^m$, then

$$A = A_0 + L,$$

where $A_0 : H \to H^*$ is positive and bounded below, i.e.,

$$\mathrm{Re}(A_0\psi, \psi)_{\Gamma_D \times \Gamma_N} \geq c\|\psi\|_H^2 \quad \text{for } \psi \in H,$$

and where $L : H \to H^*$ is a compact linear operator. Hence,

$$A : H \to H^*$$

is a Fredholm operator with index zero.

Proof. Let S_0 be as in Theorem 7.6, and let R_0 be as in the proof of Theorem 7.8. Thus, $S = S_0 + L_S$ and $R = R_0 + L_R$, where

$$L_S : H^{-1/2}(\Gamma)^m \to H^{1/2}(\Gamma)^m \quad \text{and} \quad L_R : H^{1/2}(\Gamma)^m \to H^{-1/2}(\Gamma)^m$$

are compact linear operators. Noting that $T = \tilde{T}$ because $\mathcal{P}$ is formally self-adjoint, we define

$$A_0\psi = \begin{bmatrix} (S_0\psi_D)|_{\Gamma_D} - \frac{1}{2}T_{DN}\psi_N \\ \frac{1}{2}T_{ND}^*\psi_D + (R_0\psi_N)|_{\Gamma_N} \end{bmatrix} \quad \text{and} \quad L\psi = \begin{bmatrix} (L_S\psi_D)|_{\Gamma_D} \\ (L_R\psi_N)|_{\Gamma_N} \end{bmatrix}$$

In this way, $A = A_0 + L$, the operator $L : H \to H^*$ is compact, and

$$(A_0\psi, \psi)_{\Gamma_D \times \Gamma_N} = \big((S_0\psi_D)|_{\Gamma_D}, \psi_D\big)_{\Gamma_D} - \frac{1}{2}(T_{DN}\psi_N, \psi_D)_{\Gamma_D}$$
$$+ \frac{1}{2}(T_{ND}^*\psi_D, \psi_N)_{\Gamma_N} + \big((R_0\psi_N)|_{\Gamma_N}, \psi_N\big)_{\Gamma_N}.$$

Since

$$(T_{ND}^*\psi_D, \psi_N)_{\Gamma_N} = (\psi_D, T_{DN}\psi_N)_{\Gamma_N} = \overline{(T_{DN}\psi_N, \psi_D)_{\Gamma_D}},$$

and since supp $\psi_D \subseteq \Gamma_D \cup \Pi$ and supp $\psi_N \subseteq \Gamma_N \cup \Pi$, it follows that

$$\mathrm{Re}(A_0\psi, \psi)_{\Gamma_D \times \Gamma_N} = (S_0\psi_D, \psi_D)_\Gamma + (R_0\psi_N, \psi_N)_\Gamma$$
$$\geq c\|\psi_D\|_{\tilde{H}^{-1/2}(\Gamma)^m}^2 + c\|\psi_N\|_{\tilde{H}^{1/2}(\Gamma)^m}^2 = c\|\psi\|_H^2,$$

as required. $\square$

Exterior Problems

Integral equation methods are particularly suited to boundary value problems posed on the exterior domain Ω^+. It turns out that, in general, the solution will not belong to $H^1(\Omega^+)^m$, but only to $H^1(\Omega_\rho^+)^m$ for each finite ρ, where Ω_ρ^+ is defined as in (7.25). Furthermore, to make use of the third Green identity on Ω^+ we require a somewhat stronger result than Theorem 6.10, that incorporates a suitable *radiation condition*. In other words, some assumption about the behaviour of the solution at infinity is needed, and here we shall adopt the approach of Costabel and Dauge [16].

Lemma 7.11 Let $u \in \mathcal{D}^*(\Omega^+)^m$. If $\mathcal{P}u$ has compact support in Ω^+, then there exists a unique function $\mathcal{M}u \in C^\infty(\mathbb{R}^n)^m$ such that

$$\mathcal{M}u(x) = \int_{\Gamma_1} G(x, y)\mathcal{B}_\nu u(y)\, d\sigma_y - \int_{\Gamma_1} \big[\tilde{\mathcal{B}}_{\nu,y} G(x, y)^*\big]^* u(y)\, d\sigma_y \quad (7.29)$$

for x in any bounded Lipschitz domain Ω_1^- such that $\overline{\Omega^-} \cup \mathrm{supp}\,\mathcal{P}u \Subset \Omega_1^-$, and where $\Gamma_1 = \partial\Omega_1^-$.

Proof. First note that, by Theorem 6.4, the distribution u is C^∞ on $\Omega^+ \setminus$ supp $\mathcal{P}u$. Given $x \in \mathbb{R}^n$, we define $\mathcal{M}u(x)$ to equal the right-hand side of (7.29) with Γ_1 the boundary of any ball $\Omega^- = B_\rho$ centred at the origin with radius ρ large enough to ensure that $\overline{\Omega^-} \cup \operatorname{supp} \mathcal{P}u \Subset B_\rho$ and $x \in B_\rho$. By applying the second Green identity over an annular region of the form $B_{\rho_2} \setminus \overline{B_{\rho_1}}$, one sees that the definition of $\mathcal{M}u(x)$ is independent of the choice of ρ, because $\mathcal{P}u = 0 = \mathcal{P}^*G(x, \cdot)^*$ on $B_{\rho_2} \setminus \overline{B_{\rho_1}}$. Similarly, to see that (7.29) holds for x in any bounded Lipschitz domain Ω_1^- with $\overline{\Omega^-} \cup \operatorname{supp} \mathcal{P}u \Subset \Omega_1^-$, we apply the second Green identity over $B_\rho \setminus \overline{\Omega_1^-}$ for any ρ such that $\overline{\Omega_1^-} \subseteq B_\rho$. $\qquad\square$

Notice that $\mathcal{P}\mathcal{M}u = 0$ on $\mathbb{R}^n$. We now give the desired version of the third Green identity for Ω^+.

Theorem 7.12 *Suppose that $f \in \widetilde{H}^{-1}(\Omega^+)^m$ has compact support, and choose ρ_0 large enough so that*

$$\Omega^- \cup \Gamma \Subset B_{\rho_0} \quad and \quad \operatorname{supp} f \Subset \Omega_{\rho_0}^+.$$

If $u \in \mathcal{D}^*(\Omega^+)^m$ satisfies

$$\mathcal{P}u = f \quad on\ \Omega^+,$$

and if the restriction of u to $\Omega_{\rho_0}^+$ belongs to $H^1(\Omega_{\rho_0}^+)^m$, then

$$u = \mathcal{G}f + \operatorname{DL}\gamma^+u - \operatorname{SL}\mathcal{B}_\nu^+u + \mathcal{M}u \quad on\ \Omega^+. \tag{7.30}$$

Proof. By Theorem 6.10 with u^- identically zero, the representation formula (7.30) holds on the bounded Lipschitz domain $\Omega^+ \cap \Omega_1^-$, with $\mathcal{M}u(x)$ given by (7.29) for $x \in \Omega_1^-$. (Keep in mind that ν is the inward unit normal to $\Omega^+ \cap \Omega_1^-$ on Γ, but the *out*ward unit normal on Γ_1.) $\qquad\square$

Before considering boundary value problems on Ω^+, we need to know how $\mathcal{M}$ acts on volume and surface potentials.

Lemma 7.13 *Fix $z \in \mathbb{R}^n$. If $u(y) = G(y, z)$, then $\mathcal{M}u = 0$ on $\mathbb{R}^n$.*

Proof. Let $z, x \in \mathbb{R}^n$ with $x \neq z$. We choose a bounded Lipschitz domain Ω^- such that $z \in \Omega^-$ and $x \in \Omega^+$, and define

$$v(y) = \begin{cases} G(x, y)^* & \text{for } y \in \Omega^- \\ 0 & \text{for } y \in \Omega^+ \end{cases}.$$

Since $\mathcal{P}^*v = 0$ on $\Omega^\pm$, the third Green identity (6.29) gives

$$v(z) = \widetilde{\operatorname{SL}}\,\widetilde{\mathcal{B}}_\nu^- v(z) - \widetilde{\operatorname{DL}}\,\gamma^- v(z),$$

or equivalently,

$$G(x, z)^* = \int_\Gamma G(y, z)^* \tilde{B}_{\nu,y} G(x, y)^* \, d\sigma_y - \int_\Gamma [B_{\nu,y} G(y, z)]^* G(x, y)^* \, d\sigma_y.$$

Thus,

$$G(x, z) = \int_\Gamma [\tilde{B}_{\nu,y} G(x, y)^*]^* G(y, z) \, d\sigma_y - \int_\Gamma G(x, y) B_{\nu,y} G(y, z) \, d\sigma_y,$$

or equivalently,

$$u(x) = \mathrm{DL}\, \gamma u(x) - \mathrm{SL}\, \mathcal{B}_\nu u(x),$$

and therefore $\mathcal{M}u(x) = 0$ by Theorem 7.12, because $\mathcal{P}u = 0$ on Ω^+.   □

Lemma 7.14 Let $f \in \mathcal{E}^*(\mathbb{R}^n)^m$. If $u = \mathcal{G}f$, then $\mathcal{M}u = 0$ on $\mathbb{R}^n$.

Proof. Obviously, $\mathcal{P}u = 0$ on $\mathbb{R}^n \setminus \mathrm{supp}\, f$, and since $G(x, y)$ is C^∞ for $x \neq y$, one sees from Lemma 7.13 that

$$\mathcal{M}u(x) = \langle \mathcal{M}_x G(x, y), f(y) \rangle = 0 \quad \text{for } x \in \mathbb{R}^n \setminus \mathrm{supp}\, f.$$

Since $\mathcal{P}\mathcal{M}u = 0$ on $\mathbb{R}^n$, it follows from the third Green identity that $\mathcal{M}u = 0$ on $\mathbb{R}^n$.   □

To state the main result for this section, it is convenient to introduce the notation

$$H_{\mathrm{loc}}^1(\Omega^+)^m = \{ u \in \mathcal{D}^*(\Omega^+)^m :$$

$$u|_{\Omega_\rho^+} \in H^1(\Omega_\rho^+)^m \text{ for each finite } \rho > 0 \text{ such that } \overline{\Omega^-} \subseteq B_\rho \}.$$

We point out that Exercise 7.4 gives some simple sufficient conditions on u for ensuring that $\mathcal{M}u = 0$, in the case when $\mathcal{P} = \mathcal{P}_0$.

Theorem 7.15 Suppose that $f \in \tilde{H}^{-1}(\Omega^+)^m$ has compact support.

(i) *Let $g \in H^{1/2}(\Gamma)^m$. If $u \in H_{\mathrm{loc}}^1(\Omega^+)^m$ is a solution of the exterior Dirichlet problem*

$$\mathcal{P}u = f \quad \text{on } \Omega^+,$$
$$\gamma^+ u = g \quad \text{on } \Gamma, \tag{7.31}$$
$$\mathcal{M}u = 0 \quad \text{on } \mathbb{R}^n,$$

then the conormal derivative $\psi = \mathcal{B}_\nu^+ u \in H^{-1/2}(\Gamma)^m$ *is a solution of the boundary integral equation*

$$S\psi = \gamma \mathcal{G} f - \tfrac{1}{2}(g - Tg) \quad \text{on } \Gamma, \tag{7.32}$$

and u has the integral representation

$$u = \mathcal{G} f + \mathrm{DL}\, g - \mathrm{SL}\, \psi \quad \text{on } \Omega^+. \tag{7.33}$$

Conversely, if $\psi \in H^{-1/2}(\Gamma)^m$ *is a solution of the boundary integral equation (7.32), then the formula (7.33) defines a solution* $u \in H^1_{\mathrm{loc}}(\Omega^+)^m$ *of the exterior Dirichlet problem (7.31).*

(ii) *Let* $g \in H^{-1/2}(\Gamma)^m$. *If* $u \in H^1_{\mathrm{loc}}(\Omega^+)^m$ *is a solution of the exterior Neumann problem*

$$\begin{aligned} \mathcal{P}u &= f \quad \text{on } \Omega^+, \\ \mathcal{B}_\nu^+ u &= g \quad \text{on } \Gamma, \\ \mathcal{M}u &= 0 \quad \text{on } \mathbb{R}^n, \end{aligned} \tag{7.34}$$

then the trace $\psi = \gamma^+ u \in H^{1/2}(\Gamma)^m$ *is a solution of the boundary integral equation*

$$R\psi = \mathcal{B}_\nu^+ \mathcal{G} f - \tfrac{1}{2}(g + \widetilde{T}^* g) \quad \text{on } \Gamma, \tag{7.35}$$

and u has the integral representation

$$u = \mathcal{G} f + \mathrm{DL}\, \psi - \mathrm{SL}\, g \quad \text{on } \Omega^+. \tag{7.36}$$

Conversely, if $\psi \in H^{1/2}(\Gamma)^m$ *is a solution of the boundary integral equation (7.35), then the formula (7.36) defines a solution* $u \in H^1_{\mathrm{loc}}(\Omega^+)^m$ *of the exterior Neumann problem (7.34).*

(iii) *Let* $g_{\mathrm{D}} \in H^{1/2}(\Gamma)^m$ *and* $g_{\mathrm{N}} \in H^{-1/2}(\Gamma)^m$, *and define* $h_{\mathrm{D}} \in H^{1/2}(\Gamma_{\mathrm{D}})^m$ *and* $h_{\mathrm{N}} \in H^{-1/2}(\Gamma_{\mathrm{N}})^m$ *by*

$$\begin{aligned} h_{\mathrm{D}} &= \gamma \mathcal{G} f - S g_{\mathrm{N}} - \tfrac{1}{2}(g_{\mathrm{D}} - T g_{\mathrm{D}}) \quad \text{on } \Gamma_{\mathrm{D}}, \\ h_{\mathrm{N}} &= \mathcal{B}_\nu^+ \mathcal{G} f - R g_{\mathrm{D}} - \tfrac{1}{2}(g_{\mathrm{N}} + \widetilde{T}^* g_{\mathrm{N}}) \quad \text{on } \Gamma_{\mathrm{N}}. \end{aligned}$$

If $u \in H^1_{\text{loc}}(\Omega^+)^m$ is a solution of the exterior mixed problem

$$\begin{aligned}
\mathcal{P}u &= f & \text{on } \Omega^+, \\
\gamma^+ u &= g_D & \text{on } \Gamma_D, \\
\mathcal{B}_\nu^+ u &= g_N & \text{on } \Gamma_N, \\
\mathcal{M}u &= 0 & \text{on } \mathbb{R}^n,
\end{aligned} \tag{7.37}$$

then the differences

$$\psi_D = \mathcal{B}_\nu^+ u - g_N \in \widetilde{H}^{-1/2}(\Gamma_D)^m \quad \text{and} \quad \psi_N = \gamma^+ u - g_D \in \widetilde{H}^{1/2}(\Gamma_N)^m$$

satisfy

$$\begin{bmatrix} S_{DD} & -\tfrac{1}{2} T_{DN} \\ \tfrac{1}{2} \widetilde{T}^*_{ND} & R_{NN} \end{bmatrix} \begin{bmatrix} \psi_D \\ \psi_N \end{bmatrix} = \begin{bmatrix} h_D \\ h_N \end{bmatrix}, \tag{7.38}$$

and u has the integral representation

$$u = \mathcal{G}f + \text{DL}(\psi_N + g_D) - \text{SL}(\psi_D + g_N) \quad \text{on } \Omega^+. \tag{7.39}$$

Conversely, if $\psi_D \in \widetilde{H}^{-1/2}(\Gamma_D)^m$ and $\psi_N \in \widetilde{H}^{1/2}(\Gamma_N)^m$ satisfy the system of boundary integral equations (7.38), then the formula (7.39) defines a solution $u \in H^1_{\text{loc}}(\Omega^+)^m$ of the exterior mixed problem (7.37).

Proof. Suppose that $u \in H^1_{\text{loc}}(\Omega^+)$ is a solution to the exterior Dirichlet problem (7.31). By Theorem 7.12,

$$u = \mathcal{G}f + \text{DL}\,\gamma^+ u - \text{SL}\,\mathcal{B}_\nu^+ u \quad \text{on } \Omega^+, \tag{7.40}$$

and then by (7.5),

$$\gamma^+ u = \gamma \mathcal{G}f + \tfrac{1}{2}(\gamma^+ u + T\gamma^+ u) - S\mathcal{B}_\nu^+ u \quad \text{on } \Gamma.$$

Using the boundary condition $\gamma^+ u = g$, and putting $\psi = \mathcal{B}_\nu^+ u$, we arrive at the boundary integral equation (7.32) and the integral representation (7.33).

To complete the proof of (i), suppose conversely that $\psi \in H^{-1/2}(\Gamma)^m$ satisfies (7.32), and define u by (7.33). Together, (6.10) and Theorem 6.11 imply that $u \in H^1_{\text{loc}}(\Omega^+)^m$, and it follows from (6.2) and (6.19) that $\mathcal{P}u = f$ on Ω^+. Also, $\gamma^+ u = g$ by (7.5). Finally, Lemma 7.14 shows that $\mathcal{M}\mathcal{G}f = 0$, $\mathcal{M}\,\text{DL}\,g = \mathcal{M}\mathcal{G}\widetilde{\mathcal{B}}_\nu^* g = 0$ and $\mathcal{M}\,\text{SL}\,\psi = \mathcal{M}\mathcal{G}\gamma^*\psi = 0$, so $\mathcal{M}u = 0$, and hence u is a solution of (7.31).

The proof of part (ii) proceeds in the same way, except that one takes the conormal derivative of both sides of (7.40), instead of the trace, to obtain

$$\mathcal{B}_\nu^+ u = \mathcal{B}_\nu^+ \mathcal{G} f - R\gamma^+ u - \tfrac{1}{2}(-\mathcal{B}_\nu^+ u + \widetilde{T}^* \mathcal{B}_\nu^+) \quad \text{on } \Gamma.$$

The proof of part (iii) is similar to that of Theorem 7.9. □

Regularity Theory

In Theorem 4.18, we proved local regularity up to the boundary for solutions to elliptic partial differential equations. A simple argument based on this result yields the following local regularity estimates for the boundary integral equations. Recall the definition (7.25) of the set Ω_ρ^+.

Theorem 7.16 *Let G_1 and G_2 be bounded open subsets of $\mathbb{R}^n$ such that $\overline{G}_1 \Subset G_2$ and G_1 intersects Γ. Put*

$$\Omega_j^\pm = G_j \cap \Omega^\pm \quad \text{and} \quad \Gamma_j = G_j \cap \Omega^\pm \quad \text{for } j = 1, 2,$$

and suppose, for some integer $r \geq 0$, that Γ_2 is $C^{r+1,1}$.

(i) If $\psi \in H^{-1/2}(\Gamma)^m$ and $f \in H^{r+3/2}(\Gamma_2)^m$ satisfy

$$S\psi = f \quad on \ \Gamma_2,$$

then $\psi \in H^{r+1/2}(\Gamma_1)^m$ and

$$\|\psi\|_{H^{r+1/2}(\Gamma_1)^m} \leq C \|\psi\|_{H^{-1/2}(\Gamma)^m} + C \|f\|_{H^{r+3/2}(\Gamma_2)^m}.$$

(ii) If $\mathcal{P}$ is coercive on $H^1(\Omega^-)^m$ and on $H^1(\Omega_\rho^+)^m$, and if $\psi \in H^{1/2}(\Gamma)^m$ and $f \in H^{r+1/2}(\Gamma_2)^m$ satisfy

$$R\psi = f \quad on \ \Gamma_2,$$

then $\psi \in H^{r+3/2}(\Gamma_1)^m$ and

$$\|\psi\|_{H^{r+3/2}(\Gamma_1)^m} \leq C \|\psi\|_{H^{1/2}(\Gamma)^m} + C \|f\|_{H^{r+1/2}(\Gamma_2)^m}.$$

Proof. Let $G_{3/2}$ be a bounded open subset of $\mathbb{R}^n$ such that $\overline{G}_1 \Subset G_{3/2} \subseteq \overline{G}_{3/2} \Subset G_2$, and put $\Omega_{3/2}^\pm = G_{3/2} \cap \Omega^\pm$.

In case (i), the single-layer potential $u = \text{SL}\,\psi \in H^1(\Omega_2^{\pm})^m$ satisfies

$$\mathcal{P}u = 0 \quad \text{on } \Omega_2^{\pm},$$

$$\gamma u = f \quad \text{on } \Gamma_2,$$

so using the jump relation for $\mathcal{B}_\nu^{\pm} \text{SL}\,\psi$ (Theorem 6.11), together with the trace estimates (Theorem 3.37) and elliptic regularity (Theorem 4.18), we find that

$$\|\psi\|_{H^{r+1/2}(\Gamma_1)^m} = \|\mathcal{B}_\nu^{-} u - \mathcal{B}_\nu^{+} u\|_{H^{r+1/2}(\Gamma_1)^m}$$

$$\leq C\|u\|_{H^{r+2}(\Omega_{3/2}^{-})^m} + C\|u\|_{H^{r+2}(\Omega_{3/2}^{+})^m}$$

$$\leq C\|u\|_{H^1(\Omega_2^{-})^m} + C\|u\|_{H^1(\Omega_2^{+})^m} + C\|f\|_{H^{r+3/2}(\Gamma_2)^m}.$$

The estimate follows because $\|u\|_{H^1(\Omega_2^{\pm})^m} \leq C\|\psi\|_{H^{-1/2}(\Gamma)^m}$.

In case (ii), the double-layer potential $u = \text{DL}\,\psi \in H^1(\Omega_2^{\pm})^m$ satisfies

$$\mathcal{P}u = 0 \quad \text{on } \Omega_2^{\pm},$$

$$\mathcal{B}_\nu u = f \quad \text{on } \Gamma_2,$$

so

$$\|\psi\|_{H^{r+3/2}(\Gamma_1)^m} = \|\gamma^{+} u - \gamma^{-} u\|_{H^{r+3/2}(\Gamma_1)^m}$$

$$\leq C\|u\|_{H^{r+2}(\Omega_{3/2}^{+})^m} + C\|u\|_{H^{r+2}(\Omega_{3/2}^{-})^m}$$

$$\leq C\|u\|_{H^1(\Omega_2^{+})^m} + C\|u\|_{H^1(\Omega_2^{-})^m} + C\|f\|_{H^{r+1/2}(\Gamma_2)^m},$$

and finally $\|u\|_{H^1(\Omega_2^{\pm})^m} \leq C\|\psi\|_{H^{1/2}(\Gamma)^m}$. $\qquad\qquad\square$

We saw in Theorem 7.2 that the mapping properties of the boundary integral operators hold for an extended range of Sobolev spaces when Γ is smoother than just Lipschitz. The regularity result just proved allows us also to extend the Fredholm property for R and S.

Theorem 7.17 If Γ is $C^{r+1,1}$ for some integer $r \geq 0$, then

(i) $S : H^{s-1/2}(\Gamma)^m \to H^{s+1/2}(\Gamma)^m$ *is Fredholm with index zero for* $-r \leq s \leq r$, *and* $\ker S$ *does not depend on s in this range;*

(ii) $R : H^{s+1/2}(\Gamma)^m \to H^{s-1/2}(\Gamma)^m$ *is Fredholm with index zero for* $-r - 1 \leq s \leq r + 1$, *and* $\ker R$ *does not depend on s in this range.*

Proof. We know from Theorem 7.6 that S is Fredholm with index zero when $s = 0$. Thus, let $\phi_1, \ldots, \phi_p$ be a basis for $\ker S$ in $H^{-1/2}(\Gamma)^m$. In fact, by Theorem 7.16 these basis functions belong to $H^{r+1/2}(\Gamma)^m$, and we can choose them to be orthonormal in $L_2(\Gamma)^m$. The same reasoning applied to S^* yields an orthonormal basis $\theta_1, \ldots, \theta_p$ for ker S^*, with each $\theta_j \in H^{r+1/2}(\Gamma)^m$. We can therefore define a compact linear operator

$$K : H^{s-1/2}(\Gamma)^m \to H^{r+1/2}(\Gamma)^m \quad \text{for } -r \le s \le r,$$

by

$$K\psi = \sum_{j=1}^{p} (\phi_j, \psi)_\Gamma \theta_j,$$

and a bounded linear operator

$$A = S + K : H^{s-1/2}(\Gamma)^m \to H^{s+1/2}(\Gamma)^m \quad \text{for } -r \le s \le r.$$

The operator A is certainly Fredholm with index zero when $s = 0$, and since

$$(\theta_j, S\psi)_\Gamma = 0 \quad \text{and} \quad (\theta_j, K\psi) = (\phi_j, \psi)_\Gamma \quad \text{for all } \psi \in H^{-1/2}(\Gamma)^m,$$

it is easy to see that the homogeneous equation $A\psi = 0$ has only the trivial solution in $H^{-1/2}(\Gamma)^m$. Thus, the inhomogeneous equation $A\psi = f$ has a unique solution $\psi \in H^{-1/2}(\Gamma)^m$ for every $f \in H^{1/2}(\Gamma)^m$. Furthermore, if $f \in H^{r+1/2}(\Gamma)^m$, then $S\psi \in H^{r+1/2}(\Gamma)^m$ because $K\psi \in H^{r+1/2}(\Gamma)^m$, and so $\psi \in H^{r-1/2}(\Gamma)^m$ by Theorem 7.16. It follows that A has a bounded inverse for $s = r$, and the same is true of A^*. Hence, by interpolation and duality, A is invertible for $-r \le s \le r$. Therefore, since K is compact, the operator S must be Fredholm with index zero for $-r \le s \le r$. Also, ker S does not depend on s, because if $\psi \in H^{-r-1/2}(\Gamma)^m$ satisfies $S\psi = 0$, then $A\psi = K\psi \in H^{r+1/2}(\Gamma)^m$ and thus $\psi = A^{-1}K\psi \in H^{r-1/2}(\Gamma)^m$. The proof of (i) is now complete.

Part (ii) may be proved using the same approach. The allowed range of s is larger because the basis functions for ker R belong to $H^{r+3/2}(\Gamma)^m$. $\qquad\square$

Exercises

7.1 Suppose that Ω is a C^∞ hypograph $x_n < \zeta(x')$, and let $K(x, y)$ be any one of the six kernels in (7.9). Show that

$$\operatorname*{f.p.}_{\epsilon \downarrow 0} \int_{\Gamma \backslash B_\epsilon(x)} K(x, y)\psi(y)\, d\sigma_y$$

$$= \operatorname*{f.p.}_{\epsilon \downarrow 0} \int_{|x'-y'|>\epsilon} K\big(x, y', \zeta(y')\big)\psi\big(y', \zeta(y')\big)\sqrt{1 + |\operatorname{grad}\zeta(y')|^2}\, dy'$$

for $x_n = \zeta(x')$ and $\psi \in \mathcal{D}(\Gamma)^m$. [Hint: apply Theorems 5.15, 5.19 and 6.3.]

7.2 Suppose that Ω^- is a C^2 hypograph given by $y_n < \zeta(y')$, and assume (without loss of generality) that

$$\zeta(0) = 0 \quad \text{and} \quad \text{grad}\, \zeta(0) = 0.$$

Put

$$\Upsilon_\epsilon^+ = \{\omega \in \mathbb{S}^{n-1} : \epsilon\omega \in \Omega^+\},$$

and show the following.

(i) There exists $\alpha = \alpha(\epsilon, \eta)$ such that

$$\Upsilon_\epsilon^+ = \{\omega \in \mathbb{S}^{n-1} :$$
$$\omega = (\eta\cos\theta, \sin\theta),\ \eta \in \mathbb{S}^{n-2},\ \alpha(\epsilon,\eta) < \theta \le \pi/2\}.$$

(If $n = 2$, then $\eta \in \mathbb{S}^0 = \{+1, -1\}$.)

(ii) The function α satisfies

$$\lim_{\epsilon\downarrow 0}\alpha(\epsilon,\eta) = 0 \quad \text{and} \quad \lim_{\epsilon\downarrow 0}\frac{\partial\alpha}{\partial\epsilon} = \frac{1}{2}\sum_{p=1}^{n-1}\sum_{q=1}^{n-1}\partial_p\partial_q\zeta(0)\eta_p\eta_q.$$

[Hint: $\epsilon\sin\alpha = \zeta(\epsilon\eta\cos\alpha)$.]

(iii) With $\mathbb{S}_+^{n-1} = \{\omega \in \mathbb{S}^{n-1} : \omega_n > 0\}$,

$$\int_{\Upsilon_\epsilon^+} f(\omega)\,d\omega = \int_{\mathbb{S}^{n-2}}\int_{\alpha(\epsilon,\eta)}^{\pi/2} f(\eta\cos\theta, \sin\theta)\,d\theta\,d\eta$$
$$= \int_{\mathbb{S}_+^{n-1}} f(\omega)\,d\omega - \frac{\epsilon}{2}\sum_{p=1}^{n-1}\sum_{q=1}^{n-1}\partial_p\partial_q\zeta(0)$$
$$\times \int_{\mathbb{S}^{n-2}} f(\eta, 0)\eta_p\eta_q\,d\eta + O(\epsilon^2).$$

(iv) In part (iii), the term in ϵ vanishes if $f(-\omega) = -f(\omega)$ for all $\omega \in \mathbb{S}^{n-1}$.

7.3 Show that the right-hand side of the boundary integral equation (7.23) satisfies

$$\left(\tfrac{1}{2}(g - \widetilde{T}^*g) - \mathcal{B}_\nu^- \mathcal{G}f, \phi\right)_\Gamma = 0$$

for every solution $\phi \in H^{1/2}(\Gamma)^m$ of the homogeneous adjoint problem

$R^*\phi = 0$ if and only if f and g satisfy

$$(g, \gamma^- v)_\Gamma + (f, v)_{\Omega^-} = 0$$

for every solution $v \in H^1(\Omega^-)^m$ of

$$\mathcal{P}^* v = 0 \quad \text{on } \Omega^-,$$
$$\widetilde{\mathcal{B}}_v^- v = 0 \quad \text{on } \Gamma.$$

[Hint: see the discussion following Theorem 7.6.]

7.4 Assume that $\mathcal{P}$ has no lower-order terms (i.e., $\mathcal{P} = \mathcal{P}_0$), and that $G(x, y) = G(x - y)$ is as in Theorem 6.8. Show that if u satisfies the hypotheses of Lemma 7.11 and if, as $|x| \to \infty$,

$$u(x) = o(1) \quad \text{and} \quad \mathcal{B}_v u(x) = \begin{cases} o\big([|x| \log |x|]^{-1}\big) & \text{when } n = 2, \\ o(|x|^{-1}) & \text{when } n \geq 3, \end{cases}$$

then $\mathcal{M}u = 0$ on $\mathbb{R}^n$. Here, v is the outward unit normal to the ball B_ρ.

7.5 Show that if $\mathcal{P}u = 0$ on $\mathbb{R}^n$, then $\mathcal{M}u = u$ on $\mathbb{R}^n$. [Hint: apply the third Green identity over the ball B_ρ.]

7.6 The *Calderón projection* is the linear operator P_C defined by

$$P_C \psi = \begin{bmatrix} \gamma^- (\mathrm{SL}\, \psi_2 - \mathrm{DL}\, \psi_1) \\ \mathcal{B}_v^- (\mathrm{SL}\, \psi_2 - \mathrm{DL}\, \psi_1) \end{bmatrix}, \quad \text{where } \psi = \begin{bmatrix} \psi_1 \\ \psi_2 \end{bmatrix}.$$

Theorem 6.11 implies the mapping property

$$P_C : H^{1/2}(\Gamma)^m \times H^{-1/2}(\Gamma)^m \to H^{1/2}(\Gamma)^m \times H^{-1/2}(\Gamma)^m$$

and if $\phi = P_C \psi$, then the function $u = \mathrm{SL}\, \psi_2 - \mathrm{DL}\, \psi_1$ satisfies

$$\mathcal{P}u = 0 \quad \text{on } \Omega^-,$$
$$\gamma^- u = \phi_1 \quad \text{on } \Gamma,$$
$$\mathcal{B}_v^- u = \phi_2 \quad \text{on } \Gamma,$$

so $u = \mathrm{SL}\, \phi_2 - \mathrm{DL}\, \phi_1$ by Theorem 6.10. Hence, $P_C \phi = \phi$, or in other words $P_C^2 \psi = P_C \psi$, demonstrating that P_C really is a projection.

(i) Show that

$$P_C = \begin{bmatrix} \frac{1}{2}(I - T) & S \\ R & \frac{1}{2}(I + \widetilde{T}^*) \end{bmatrix}.$$

(ii) Deduce that

$$SR = \tfrac{1}{4}(I - T^2), \quad S\widetilde{T}^* = TS, \quad RT = \widetilde{T}^*R, \quad RS = \tfrac{1}{4}\big[I - (\widetilde{T}^*)^2\big].$$

7.7 Recall the definition of the Steklov–Poincaré operators $\mathcal{B}_\nu\mathcal{U}$ and $\widetilde{\mathcal{B}}_\nu\mathcal{V}$ from (4.38). Show that if $S : H^{-1/2}(\Gamma)^m \to H^{1/2}(\Gamma)^m$ is invertible, then we have the representations

$$\mathcal{B}_\nu\mathcal{U} = R + \tfrac{1}{4}(I + \widetilde{T}^*)S^{-1}(I + T) = \tfrac{1}{2}(I + \widetilde{T}^*)S^{-1}$$

and

$$\widetilde{\mathcal{B}}_\nu\mathcal{V} = R^* + \tfrac{1}{4}(I + T^*)(S^*)^{-1}(I + \widetilde{T}) = \tfrac{1}{2}(I + T^*)(S^*)^{-1}.$$

7.8 Let $\Omega^\pm = \mathbb{R}^n_\pm$, and think of Γ as $\mathbb{R}^{n-1}$. Also, assume $\mathcal{P} = \mathcal{P}_0$ with constant coefficients, so that $P(\xi)$ is a homogeneous quadratic polynomial, and let $G(x, y) = \mathsf{G}(x - y)$ be the fundamental solution given by Theorem 6.8.

(i) By applying Lemma 5.21, show that if $n \geq 3$, then

$$S\psi(x') = \int_{\mathbb{R}^{n-1}} m_S(\xi')\hat{\psi}(\xi')e^{\mathrm{i}2\pi\xi\cdot x}\, d\xi' \quad \text{for } x' \in \mathbb{R}^{n-1},$$

where, with the notation (5.28),

$$m_S(\xi') = \int^\pm P(\xi', \xi_n)^{-1}\, d\xi_n.$$

The function m_S is called the *symbol* of S.

(ii) In the same fashion, show that for $n \geq 2$,

$$R\psi(x') = \int_{\mathbb{R}^{n-1}} m_R(\xi')\hat{\psi}(\xi')e^{\mathrm{i}2\pi\xi\cdot x}\, d\xi',$$

where

$$m_R(\xi') = \int^\pm \sum_{j=1}^n \sum_{k=1}^n A_{nj}(\mathrm{i}2\pi\xi_j)P(\xi)^{-1}(\mathrm{i}2\pi\xi_k)A_{kn}\, d\xi_n$$

$$= -(2\pi)^2 \sum_{j=1}^n \sum_{k=1}^n A_{nj}\left(\int^\pm \xi_j\xi_k P(\xi', \xi_n)^{-1}\, d\xi_n\right)A_{kn}.$$

(iii) From the homogeneity properties

$$m_S(t\xi') = t^{-1}m_S(\xi') \quad \text{and} \quad m_R(t\xi') = t m_S(\xi') \qquad \text{for } t > 0,$$

deduce that

$$S : H^{s-1/2}(\mathbb{R}^{n-1})^m \rightarrow H^{s+1/2}(\mathbb{R}^{n-1})^m$$

and

$$R : H^{s+1/2}(\mathbb{R}^{n-1})^m \rightarrow H^{s-1/2}(\mathbb{R}^{n-1})^m$$

for all $s \in \mathbb{R}$.

(iv) Let $\chi \in C_{\text{comp}}^{\infty}(\mathbb{R}^n)$ satisfy $\chi = 1$ on a neighbourhood of zero, and consider

$$u(x) = \mathcal{F}_{\xi \rightarrow x}^* \{[1 - \chi(\xi)]v(\xi)\},$$

where v is rational and homogeneous of degree $j - 1$, as in Assumption 5.20, but with $j \leq -1 + n$. (Thus, v is not locally integrable on $\mathbb{R}^n$.) Modify Lemma 5.21 accordingly, and hence show that the mapping property of S in part (iii) holds also when $n = 2$.

8

The Laplace Equation

Our development of the general theory of elliptic systems and boundary integral equations is now complete. In this and the remaining two chapters, we concentrate on three specific examples of elliptic operators that are important in applications. This chapter deals with the Laplace operator in $\mathbb{R}^n$, denoted by

$$\Delta = \sum_{j=1}^{n} \partial_j^2.$$

The Laplacian constitutes the simplest example of an elliptic partial differential operator, and its historical role was discussed already in Chapter 1. After deriving the fundamental solution for the Laplacian, we shall introduce a classical tool from potential theory: spherical harmonics. These functions turn out to be eigenfunctions of the boundary integral operators associated with the Laplace equation on the unit ball. They are also useful for studying the behaviour of harmonic functions at infinity, leading to simple radiation conditions. The final section of the chapter investigates ker S and ker R, and the sense in which S and R are positive-definite.

The operator $\mathcal{P} = -\Delta$ has the form (4.1) with constant, scalar coefficients

$$A_{jk} = \delta_{jk}, \quad A_j = 0, \quad A = 0.$$

Associated with $-\Delta$ is the *Dirichlet form*

$$\Phi_\Omega(u, v) = \int_\Omega \operatorname{grad} \bar{u} \cdot \operatorname{grad} v \, dx,$$

and the conormal derivative (4.4) is simply the normal derivative,

$$\mathcal{B}_\nu u = \partial_\nu u = \frac{\partial u}{\partial \nu}.$$

Obviously, $-\Delta$ is formally self-adjoint and strongly elliptic; see (4.6) and (4.7).

Fundamental Solutions

The Fourier transform of $-\Delta u$ is $P(\xi)\hat{u}(\xi)$ where $P(\xi) = (2\pi)^2|\xi|^2$, so by Theorem 6.8 a fundamental solution for $-\Delta$ is

$$G(x, y) = G(x - y), \qquad \text{where } G(x) = \frac{1}{4\pi|x|} \text{ when } n = 3.$$

Exercise 8.1 is the corresponding calculation for $n = 2$. To give a fundamental solution for a general n, we denote the surface area of $\mathbb{S}^{n-1}$, the unit sphere in $\mathbb{R}^n$, by

$$\Upsilon_n = \int_{\mathbb{S}^{n-1}} d\sigma = \int_{|\omega|=1} d\omega = 2\frac{\pi^{n/2}}{\Gamma(n/2)}; \qquad (8.1)$$

see Exercise 8.2.

Theorem 8.1 A fundamental solution for the operator $-\Delta$ is given by

$$G(x) = \frac{1}{(n-2)\Upsilon_n}\frac{1}{|x|^{n-2}} \quad \text{when } n \geq 3,$$

and, for any constant $r > 0$, by

$$G(x) = \frac{1}{2\pi}\log\frac{r}{|x|} \quad \text{when } n = 2.$$

Proof. The Laplacian is radially symmetric (see Exercise 8.3), so it is natural to seek G in the form $G(x) = w(\rho)$ where $\rho = |x|$. Since $\Delta G = 0$ on $\mathbb{R}^n \setminus \{0\}$, Exercise 8.4 shows that w must satisfy

$$\frac{1}{\rho^{n-1}}\frac{d}{d\rho}\left(\rho^{n-1}\frac{dw}{d\rho}\right) = 0 \quad \text{for } \rho > 0,$$

so

$$w(\rho) = \frac{a_n}{n-2}\frac{1}{\rho^{n-2}} + b_n \quad \text{when } n \geq 3,$$

or

$$w(\rho) = a_2\log\frac{1}{\rho} + b_2 \quad \text{when } n = 2,$$

for some constants a_n and b_n. The choice of b_n is arbitrary, but a_n is fixed by the requirement that G satisfy (6.12), i.e., by the requirement that $-\Delta G = \delta$

on $\mathbb{R}^n$, or in other words

$$-\langle G, \Delta\phi \rangle = \phi(0) \quad \text{for } \phi \in \mathcal{D}(\mathbb{R}^n). \tag{8.2}$$

Any test function $\phi \in \mathcal{D}(\mathbb{R}^n)$ has compact support, so we can apply the second Green identity (1.9) over the unbounded domain $\{x : |x| > \epsilon\}$, and arrive at the formula

$$-\int_{|x|>\epsilon} G(x)\Delta\phi(x)\,dx = \int_{|x|=\epsilon} \phi(x)\partial_\nu G(x)\,d\sigma_x - \int_{|x|=\epsilon} G(x)\partial_\nu\phi(x)\,d\sigma_x, \tag{8.3}$$

where $\nu_x = -x/\epsilon$. Since

$$\text{grad}\, G(x) = \frac{dw}{d\rho}\frac{x}{|x|} = -\frac{a_n x}{|x|^n} \quad \text{for } n \geq 2, \tag{8.4}$$

we have $\partial_\nu G(x) = -(x/\epsilon) \cdot \text{grad}\, G(x) = a_n \epsilon^{1-n}$ for $|x| = \epsilon$. Thus, by the mean–value theorem for integrals,

$$\int_{|x|=\epsilon} \phi(x)\partial_\nu G(x)\,d\sigma_x = \frac{a_n}{\epsilon^{n-1}}\int_{|x|=\epsilon} \phi(x)\,d\sigma_x = a_n \Upsilon_n \phi(x_\epsilon)$$

for some x_ϵ satisfying $|x_\epsilon| = \epsilon$, whereas

$$\int_{|x|=\epsilon} G(x)\partial_\nu\phi(x)\,d\sigma_x = \begin{cases} O(\epsilon) & \text{if } n \geq 3, \\ O(\epsilon|\log \epsilon|) & \text{if } n = 2. \end{cases}$$

Thus, if $a_n = 1/\Upsilon_n$, then (8.2) follows from (8.3) after sending $\epsilon \to 0$.

An alternative method of determining a_n is to apply the third Green identity to the constant function $u = 1$ over the unit ball. One obtains the formula $DL\,1(0) = -1$, from which it again follows that $a_n = 1/\Upsilon_n$. $\qquad\square$

Throughout the remainder of this chapter, $G(x, y) = G(x - y)$ will always denote the fundamental solution from Theorem 8.1. Recalling the definitions of the boundary integral operators S, R and T given in (7.3), we see first that by (7.8),

$$S\psi(x) = \frac{1}{(n-2)\Upsilon_n}\int_\Gamma \frac{\psi(y)}{|x-y|^{n-2}}\,d\sigma_y \quad \text{when } n \geq 3,$$

and

$$S\psi(x) = \frac{1}{2\pi}\int_\Gamma \psi(y)\log\frac{r}{|x-y|}\,d\sigma_y \quad \text{when } n = 2,$$

for $x \in \Gamma$. By (7.9) and (8.4), the kernel of T is

$$2\partial_{v,y} G(x,y) = \frac{2}{\Upsilon_n} \frac{v_y \cdot (x-y)}{|x-y|^n},$$

and the kernel of R is

$$-\partial_{v,x}\partial_{v,y} G(x,y) = \frac{-1}{\Upsilon_n} \left(\frac{v_x \cdot v_y}{|x-y|^n} + n \frac{v_x \cdot (y-x) v_y \cdot (x-y)}{|x-y|^{n+2}} \right),$$

for $n \geq 2$. Notice that if Γ is $C^{1,\mu}$, then

$$|v_y \cdot (x-y)| \leq C|x-y|^{1+\mu} \quad \text{for } x, y \in \Gamma,$$

so T has an integrable kernel, allowing us to write

$$T\psi(x) = \frac{2}{\Upsilon_n} \int_\Gamma \frac{v_y \cdot (x-y)}{|x-y|^n} \psi(y) \, d\sigma_y \quad \text{for } x \in \Gamma;$$

see Theorem 7.4, and cf. Exercise 8.6. If Γ is C^2, then

$$R\psi(x) = \frac{-1}{\Upsilon_n} \text{f.p.} \int_{\Gamma \setminus B_\epsilon(x)} \frac{v_x \cdot v_y}{|x-y|^n} \psi(y) \, d\sigma_y$$
$$- \frac{n}{\Upsilon_n} \int_\Gamma \frac{v_x \cdot (y-x) v_y \cdot (x-y)}{|x-y|^{n+2}} \psi(y) \, d\sigma_y$$

for $x \in \Gamma$; see (7.9).

For later use, we conclude this section by considering the eigenvalues of $-\Delta$.

Theorem 8.2 *Let Ω be a bounded Lipschitz domain and let $\Gamma = \Gamma_D \cup \Pi \cup \Gamma_N$ be a Lipschitz dissection of $\Gamma = \partial\Omega$. If $u \in H^1(\Omega)$ is a non-trivial solution of*

$$-\Delta u = \lambda u \quad \text{on } \Omega,$$
$$\gamma u = 0 \quad \text{on } \Gamma_D,$$
$$\partial_v u = 0 \quad \text{on } \Gamma_N,$$

then $\lambda \geq 0$. If, in addition, Γ_D intersects every component of Γ, then $\lambda > 0$.

Proof. Applying the first Green identity, we have

$$\Phi_\Omega(u,u) = (\lambda u, u)_\Omega + (\partial_v u, \gamma u)_\Gamma = \lambda \|u\|_{L_2(\Omega)}^2,$$

and obviously $\Phi_\Omega(u,u) = \|\text{grad } u\|_{L_2(\Omega)}^2 \geq 0$ and $\|u\|_{L_2(\Omega)} > 0$, so $\lambda \geq 0$. Moreover, if $\lambda = 0$ then $\Phi_\Omega(u,u) = 0$, implying that grad $u = 0$ on Ω, and hence u is constant on each component of Ω. $\square$

The same argument yields a uniqueness theorem.

Corollary 8.3 Let Ω be a bounded Lipschitz domain. If $u \in H^1(\Omega)$ is a solution of the homogeneous mixed problem

$$-\Delta u = 0 \quad on\ \Omega,$$

$$\gamma u = 0 \quad on\ \Gamma_D,$$

$$\partial_\nu u = 0 \quad on\ \Gamma_N,$$

then u is constant on each component of Ω. If, in addition, Γ_D intersects every component of Γ, then u is identically zero on Ω.

We remark that, for the first part of each of the preceding two results, the homogeneous boundary conditions could be replaced by the weaker assumption that $(\partial_\nu u, \gamma u)_\Gamma \leq 0$.

Spherical Harmonics

For each integer $m \geq 0$, let $\mathcal{P}_m(\mathbb{R}^n)$ denote the set of homogeneous polynomials of degree m in n variables, i.e., the set of functions u of the form

$$u(x) = \sum_{|\alpha|=m} a_\alpha x^\alpha \quad \text{for } x \in \mathbb{R}^n, \tag{8.5}$$

with coefficients $a_\alpha \in \mathbb{C}$. A *solid spherical harmonic of degree m* is an element of the subspace

$$\mathcal{H}_m(\mathbb{R}^n) = \{ u \in \mathcal{P}_m(\mathbb{R}^n) : \Delta u = 0 \text{ on } \mathbb{R}^n \}.$$

Apart from the results involving the boundary integral operators, our general approach to the study of spherical harmonics is essentially that of Müller [70]. Let

$$M(n, m) = \dim \mathcal{P}_m(\mathbb{R}^n) \quad \text{and} \quad N(n, m) = \dim \mathcal{H}_m(\mathbb{R}^n)$$

$$\text{for } n \geq 1 \text{ and } m \geq 0. \quad (8.6)$$

By a standard combinatorial argument, the number of non-negative integer solutions $\alpha_1, \ldots, \alpha_n$ to the equation $\alpha_1 + \cdots + \alpha_n = m$ is

$$M(n, m) = \binom{m+n-1}{n-1}, \tag{8.7}$$

and since each $u \in \mathcal{P}_m(\mathbb{R}^n)$ has a unique representation

$$u(x) = \sum_{k=0}^{m} v_k(x') x_n^{m-k} \quad \text{with } v_k \in \mathcal{P}_k(\mathbb{R}^{n-1}), \tag{8.8}$$

we see at once that

$$M(n, m) = \sum_{k=0}^{m} M(n-1, k) \quad \text{for } n \geq 2 \text{ and } m \geq 0. \tag{8.9}$$

Also, $\mathcal{P}_0 = \mathcal{H}_0$ is just the space of constant functions, and $\mathcal{P}_1 = \mathcal{H}_1$ is just the space of homogeneous linear functions, so

$$M(n, 0) = N(n, 0) = 1 \text{ and } M(n, 1) = N(n, 1) = n \quad \text{for } n \geq 1.$$

Taking the Laplacian of (8.8), we find after some simple manipulations that

$$\Delta u(x) = \sum_{k=2}^{m} \left[\Delta' v_k(x') + (m - k + 2)(m - k + 1) v_{k-2}(x') \right] x_n^{m-k}$$

$$\text{for } n \geq 2 \text{ and } m \geq 2,$$

where Δ' is the Laplacian on $\mathbb{R}^{n-1}$. Thus, $u \in \mathcal{H}_m(\mathbb{R}^n)$ if and only if

$$v_{k-2}(x') = \frac{-\Delta' v_k(x')}{(m - k + 2)(m - k + 1)} \quad \text{for } 2 \leq k \leq m, \tag{8.10}$$

with the choice of $v_{m-1} \in \mathcal{P}_{m-1}(\mathbb{R}^{n-1})$ and $v_m \in \mathcal{P}_m(\mathbb{R}^{n-1})$ being arbitrary. Hence,

$$N(n, m) = M(n-1, m-1) + M(n-1, m) \quad \text{for } n \geq 2 \text{ and } m \geq 1, \tag{8.11}$$

and so it follows from (8.9) that

$$N(n, m) = \sum_{k=0}^{m} N(n-1, k) \quad \text{for } n \geq 1 \text{ and } m \geq 1. \tag{8.12}$$

Furthermore, since

$$N(1, m) = \begin{cases} 1 & \text{if } m = 0 \text{ or } 1, \\ 0 & \text{if } m \geq 2, \end{cases}$$

we have

$$N(2, m) = \begin{cases} 1 & \text{if } m = 0, \\ 2 & \text{if } m \geq 1, \end{cases} \tag{8.13}$$

and in view of (8.7) and (8.11),

$$N(n, m) = \frac{2m + n - 2}{n - 2} \binom{m + n - 3}{n - 3} \qquad \text{for } n \geq 3 \text{ and } m \geq 0. \quad (8.14)$$

In particular, $N(3, m) = 2m + 1$.

Put

$$\mathcal{H}_m(\mathbb{S}^{n-1}) = \{\psi : \psi = u|_{\mathbb{S}^{n-1}} \text{ for some } u \in \mathcal{H}_m(\mathbb{R}^n)\}.$$

Corollary 8.3 implies that the restriction map $u \mapsto u|_{\mathbb{S}^{n-1}}$ is one–one, and hence is an isomorphism from $\mathcal{H}_m(\mathbb{R}^n)$ onto $\mathcal{H}_m(\mathbb{S}^{n-1})$, so

$$\dim \mathcal{H}_m(\mathbb{S}^{n-1}) = N(n, m) \quad \text{for } n \geq 2 \text{ and } m \geq 0.$$

An element of $\mathcal{H}_m(\mathbb{S}^{n-1})$ is called a *surface spherical harmonic of degree m.*

We now show that $\mathcal{H}_m(\mathbb{S}^{n-1})$ is an eigenspace of each of the four boundary integral operators on $\mathbb{S}^{n-1}$. Later, in Theorem 8.17, we shall see that the spherical harmonics account for *all* of the eigenfunctions of these operators.

Theorem 8.4 *If* $\Gamma = \mathbb{S}^{n-1}$ *for* $n \geq 3$, *then* $T = T^* = -(n - 2)S$,

$$R\psi = \frac{m(m + n - 2)}{2m + n - 2}\psi \quad \text{and} \quad S\psi = \frac{1}{2m + n - 2}\psi \qquad \text{for } \psi \in \mathcal{H}_m(\mathbb{S}^{n-1}).$$
$$(8.15)$$

Proof. Observe that

$$\nu_y = y \quad \text{and} \quad |x - y|^2 = |x|^2 - 2x \cdot y + |y|^2 = 2(1 - x \cdot y) \qquad \text{for } x, y \in \Gamma,$$
$$(8.16)$$

so the kernel of T is

$$\frac{2}{\Upsilon_n} \frac{\nu_y \cdot (x - y)}{|x - y|^n} = -(n - 2)G(x, y),$$

and therefore $T = T^* = -(n - 2)S$. Now suppose that $\psi = \gamma u \in \mathcal{H}_m(\mathbb{S}^{n-1})$ with $u \in \mathcal{H}_m(\mathbb{R}^n)$. Euler's relation for homogeneous functions, $\sum_{j=1}^{n} y_j \partial_j u(y) = mu(y)$, implies that

$$\frac{\partial u}{\partial \nu} = m\psi \quad \text{on } \Gamma, \qquad (8.17)$$

and therefore, applying Theorems 7.5 and 7.7,

$$S(m\psi) = \tfrac{1}{2}(\psi + T\psi) \quad \text{and} \quad R\psi = \tfrac{1}{2}[(m\psi) - T^*(m\psi)]. \quad (8.18)$$

Since $T = T^* = -(n-2)S$, it follows that

$$mS\psi = \tfrac{1}{2}[\psi - (n-2)S\psi] \quad \text{and} \quad R\psi = \tfrac{1}{2}m[\psi + (n-2)S\psi],$$

giving (8.15). □

The same method of proof yields the following result in two dimensions.

Theorem 8.5 If $\Gamma = \mathbb{S}^1$, then

$$T\psi = T^*\psi = \frac{-1}{2\pi} \int_\Gamma \psi(y)\, d\sigma_y \quad \text{for } \psi \in L_2(\Gamma), \tag{8.19}$$

and when $\psi \in \mathcal{H}_m(\mathbb{S}^1)$,

$$R\psi = 0, \quad S\psi = (\log r)\psi, \quad T\psi = T^*\psi = -\psi \quad \text{for } m = 0; \tag{8.20}$$

$$R\psi = \frac{m}{2}\psi, \quad S\psi = \frac{1}{2m}\psi, \quad T\psi = T^*\psi = 0 \quad \text{for } m \geq 1. \tag{8.21}$$

Proof. The relations (8.16) remain valid for $x, y \in \mathbb{S}^{n-1}$ when $n = 2$, but the kernel of T is now

$$\frac{1}{\pi} \frac{\nu_y \cdot (x-y)}{|x-y|^2} = -\frac{1}{2\pi} \quad \text{for } x, y \in \Gamma,$$

implying (8.19). To prove (8.20), suppose that $m = 0$. We see at once from (8.18) that $R\psi = 0$. Furthermore, ψ is constant, so by symmetry $S\psi$ is also constant, and then uniqueness for the solution of the interior Dirichlet problem shows that $\mathrm{SL}\,\psi$ is constant on the disc Ω^-. At the origin, we have

$$\mathrm{SL}\,\psi(0) = \frac{1}{2\pi} \int_{|y|=1} \log\frac{r}{|y|} \psi(y)\, d\sigma_y = (\log r)\psi,$$

so $\mathrm{SL}\,\psi = (\log r)\psi$ on Ω^-, and hence $S\psi = \gamma^- \mathrm{SL}\,\psi = (\log r)\psi$ on Γ. If $m \geq 1$, then using (8.17) and the divergence theorem,

$$T\psi = T^*\psi = \frac{-1}{2\pi m} \int_\Gamma \frac{\partial u}{\partial \nu}\, d\sigma = \frac{-1}{2\pi m} \int_{\Omega^-} \Delta u\, dx = 0.$$

Thus, (8.21) follows at once from (8.18). □

We now consider spherical harmonics that are invariant under rotation about the nth coordinate axis.

Lemma 8.6 *Given* $m \geq 0$, *there exists a unique function u satisfying*

(i) $u \in \mathcal{H}_m(\mathbb{R}^n)$;
(ii) *if* $A \in \mathbb{R}^{n \times n}$ *is an orthogonal matrix satisfying* $Ae_n = e_n$, *then*

$$u(Ax) = u(x) \quad \text{for } x \in \mathbb{R}^n;$$

(iii) $u(e_n) = 1$.

In fact,

$$u(x) = \frac{1}{\Upsilon_{n-1}} \int_{\mathbb{S}^{n-2}} (x_n + ix' \cdot \eta)^m \, d\eta. \qquad (8.22)$$

Proof. One easily verifies that (8.22) defines a function u satisfying (i), (ii) and (iii). To show uniqueness, suppose that $A \in \mathbb{R}^{n \times n}$ satisfies the assumptions of (ii). It follows that A has the block structure

$$A = \begin{bmatrix} A' & 0 \\ 0 & 1 \end{bmatrix},$$

where $A' \in \mathbb{R}^{(n-1) \times (n-1)}$ is orthogonal. Thus, with v_k as in (8.8),

$$u(Ax) = \sum_{k=0}^{m} v_k(A'x') x_n^{m-k},$$

and so the conclusion of (ii) means that $v_k(A'x') = v_k(x')$ for all $x' \in \mathbb{R}^{n-1}$, which in turn means that $v_k(x')$ depends only on $|x'|$. Hence, every $u \in \mathcal{P}_m(\mathbb{R}^n)$ satisfying (ii) has the form (8.8) with $v_k(x') = d_k|x'|^k$, where $d_k = 0$ if k is odd. Exercise 8.4 shows that

$$\Delta'|x'|^r = r(n + r - 3)|x'|^{r-2},$$

so the condition (8.10) for $u \in \mathcal{H}_m(\mathbb{R}^n)$ holds if and only if the coefficients satisfy

$$d_{k-2} = \frac{-k(n + k - 3)}{(m - k + 2)(m - k + 1)} d_k \quad \text{for } 2 \leq k \leq m. \qquad (8.23)$$

If m is even, then d_m is arbitrary and $d_{m-1} = 0$, whereas if m is odd, then d_{m-1} is arbitrary and $d_m = 0$. Since $u(e_n) = d_0$, the normalisation condition (iii) fixes u with the requirement that $d_0 = 1$. $\qquad \square$

With the notation of Lemma 8.6, we define $P_m(t) = P_m(n, t)$, the *Legendre polynomial of degree m for the dimension $n \geq 2$*, by

$$P_m(t) = u(\omega), \quad \text{where } \omega = \sqrt{1 - t^2}\, \eta + t e_n,$$

$$\text{for } -1 \leq t \leq 1 \text{ and } \eta \in \mathbb{S}^{n-2},$$

noting that $u(\omega)$ is independent of η. Observe that since $u(e_n) = 1$ and $u(-x) = (-1)^m u(x)$, the Legendre polynomials satisfy

$$P_m(1) = 1 \quad \text{and} \quad P_m(-t) = (-1)^m P_m(t) \qquad \text{for } m \geq 0 \text{ and } n \geq 2.$$

Also, we have the explicit representation

$$P_m(n, t) = \sum_{k=0}^{m} d_k(n, m)(1 - t^2)^{k/2} t^{m-k},$$

where the coefficients $d_k = d_k(n, m)$ are determined by the recurrence relation (8.23) with the starting values $d_0 = 1$ and $d_1 = 0$.

When $n = 2$, the integral (8.22) becomes just the sum over $\eta \in \mathbb{S}^0 = \{-1, +1\}$, with $\Upsilon_0 = 2$, so we find that

$$P_m(2, t) = \frac{1}{2}\left[\left(t + i\sqrt{1 - t^2}\right)^m + \left(t - i\sqrt{1 - t^2}\right)^m \right]. \tag{8.24}$$

If $t = \cos\phi$, then $t \pm i\sqrt{1 - t^2} = e^{\pm i\phi}$, so

$$P_m(2, \cos\phi) = \cos m\phi,$$

and therefore $P_m(2, t)$ is the mth Chebyshev polynomial of the first kind. Exercise 8.8 shows that $P_m(3, t)$ is the usual Legendre polynomial of degree m. The fundamental solution for the Laplacian can be expanded in terms of these polynomials, as follows.

Theorem 8.7 *Let $b_m = b_m(n)$ be the coefficients in the Taylor expansion*

$$\frac{1}{(1 - z)^{n-2}} = \sum_{m=0}^{\infty} b_m z^m \quad \text{for } |z| < 1. \tag{8.25}$$

If $0 < |x| < |y|$ and $x \cdot y = |x||y| \cos\theta$, then

$$\frac{1}{|x - y|^{n-2}} = \sum_{m=0}^{\infty} b_m P_m(n, \cos\theta) \frac{|x|^m}{|y|^{n-2+m}} \quad \text{for } n \geq 3, \tag{8.26}$$

and

$$\log \frac{1}{|x - y|} = \log \frac{1}{|y|} + \sum_{m=1}^{\infty} \frac{1}{m} P_m(2, \cos\theta) \frac{|x|^m}{|y|^m}.$$

Proof. By Taylor expansion about $x = 0$, we have

$$\frac{1}{|x - y|^{n-2}} = \sum_{m=0}^{\infty} F_m(x, y) \quad \text{for } |x| < |y|, \tag{8.27}$$

where

$$F_m(x, y) = \sum_{|\alpha|=m} a_\alpha(y) x^\alpha \quad \text{and} \quad a_\alpha(y) = \frac{(-1)^{|\alpha|}}{\alpha!} \partial_y^\alpha \frac{1}{|y|^{n-2}}.$$

Note that $F_m(x, y)$ is homogeneous of degree m in x, and of degree $-(n-2)-m$ in y. Taking the Laplacian of (8.27) with respect to x, we see that

$$\sum_{m=0}^{\infty} \Delta_x F_m(x, y) = 0 \quad \text{for } |x| < |y|,$$

so by uniqueness of the coefficients in a Taylor expansion, $F_m(\cdot, y) \in \mathcal{H}_m(\mathbb{R}^n)$ for each integer $m \geq 0$. Moreover, if $A \in \mathbb{R}^{n \times n}$ is an orthogonal matrix, then $|Ax - Ay| = |x - y|$, and so

$$F_m(Ax, Ay) = F_m(x, y).$$

In particular, $F_m(Ax, y) = F_m(x, y)$ if $Ay = y$, and therefore by Lemma 8.6,

$$F_m(\omega, \zeta) = b_m P_m(\omega \cdot \zeta) \quad \text{for } \omega, \zeta \in \mathbb{S}^{n-1},$$

and for some constant b_m. Thus,

$$F_m(x, y) = |x|^m |y|^{-(n-2)-m} F_m\left(\frac{x}{|x|}, \frac{y}{|y|}\right) = b_m P_m(\cos\theta) \frac{|x|^m}{|y|^{n-2+m}}, \tag{8.28}$$

and by choosing x and y so that $|y| = 1$ and $\cos\theta = 1$, we have $|x - y| = (|x|^2 - 2|x| + 1)^{1/2} = 1 - |x|$, so the b_m are as in (8.25). The proof of (8.26) is now complete.

When $n = 2$, we proceed in the same way, except that now

$$a_\alpha(y) = \frac{(-1)^{|\alpha|}}{\alpha!} \partial_y^\alpha \log \frac{1}{|y|},$$

so if $m \geq 1$, then $F_m(x, y)$ is homogeneous of degree m in x, and of degree $-m$ in y. It follows that for some constants b_m,

$$F_0(x, y) = \log \frac{1}{|y|} \quad \text{and} \quad F_m(x, y) = b_m P_m(2, \cos\theta) \frac{|x|^m}{|y|^m} \qquad \text{for } m \geq 1,$$

and by choosing $|y| = 1$ and $\cos\theta = 1$, we see that $b_m = 1/m$ because

$$\log \frac{1}{1 - |x|} = -\log(1 - |x|) = \sum_{m=1}^{\infty} \frac{|x|^m}{m} \qquad \text{for } |x| < 1. \qquad \square$$

The next theorem gives some expansions of general harmonic functions in terms of spherical harmonics. Recall the definition of $\mathcal{M}u$ given in Lemma 7.11.

Theorem 8.8 *Write $x = \rho\omega$, where $\rho = |x|$ and $\omega = x/\rho$.*

(i) If $\Delta u(x) = 0$ for $\rho < \rho_0$, then there exist $\psi_m \in \mathcal{H}_m(\mathbb{S}^{n-1})$ such that

$$u(x) = \sum_{m=0}^{\infty} \rho^m \psi_m(\omega) \quad \text{for } \rho < \rho_0.$$

(ii) If $\Delta u(x) = 0$ for $\rho > \rho_0$, and if $\mathcal{M}u = 0$ on $\mathbb{R}^n$, then there exist $\psi_m \in \mathcal{H}_m(\mathbb{S}^{n-1})$ such that

$$u(x) = \sum_{m=0}^{\infty} \rho^{2-n-m} \psi_m(\omega) \quad \text{for } \rho > \rho_0, \qquad \text{when } n \geq 3,$$

and

$$u(x) = (\log \rho)\psi_0(\omega) + \sum_{m=1}^{\infty} \rho^{-m} \psi_m(\omega) \quad \text{for } \rho > \rho_0, \qquad \text{when } n = 2.$$

Proof. Let $\Omega^- = B_{\rho_0}$, the open ball of radius ρ_0 and centre 0, and consider the special case when $u = \text{SL}\,\phi$ for some $\phi \in L_1(\Gamma)$. By (8.27), we get the desired expansion for $\rho < \rho_0$, with

$$\psi_m(\omega) = \frac{1}{(n-2)\Upsilon_n} \int_\Gamma F_m(\omega, y)\phi(y)\, d\sigma_y.$$

If $n \geq 3$, then by interchanging x and y in (8.27) we see that

$$\frac{1}{|x - y|^{n-2}} = \sum_{m=0}^{\infty} F_m(y, x) \quad \text{for } |x| > |y|,$$

and by (8.28),

$$F_m(y, x) = \left(\frac{|y|}{|x|}\right)^{n-2+2m} F_m(x, y),$$

so u has the desired expansion for $\rho > \rho_0$, with

$$\psi_m(\omega) = \frac{1}{(n-2)\Upsilon_n} \int_\Gamma |y|^{n-2+2m} F_m(\omega, y)\phi(y)\, d\sigma_y.$$

When $n = 2$, the only essential change is to the terms with $m = 0$, which are

$$\int_\Gamma F_0(x, y)\phi(y)\, d\sigma_y = -\int_\Gamma (\log|y|)\phi(y)\, d\sigma_y \quad \text{if } \rho < \rho_0,$$

and

$$\int_\Gamma F_0(y, x)\phi(y)\, d\sigma_y = -(\log\rho)\int_\Gamma \phi(y)\, d\sigma_y \quad \text{if } \rho > \rho_0.$$

Likewise, any double-layer potential $u = \mathrm{DL}\,\phi$ has expansions of the desired form, with $\psi_0 = 0$ in part (ii), because if $n \geq 3$, then

$$\frac{\partial}{\partial v_y} \frac{1}{|x - y|^{n-2}} = \sum_{m=0}^\infty \frac{\partial}{\partial v_y} F_m(x, y) \quad \text{for } |x| < |y|,$$

and

$$\frac{\partial}{\partial v_y} \frac{1}{|x - y|^{n-2}} = \sum_{m=1}^\infty \frac{1}{|x|^{n-2+2m}} \frac{\partial}{\partial v_y}\left[|y|^{n-2+2m} F_m(x, y)\right] \quad \text{for } |x| > |y|.$$

In the general case, if $\Delta u(x) = 0$ for $|x| < \rho_0$, then by applying the third Green identity over $\Omega^- = B_{\rho_1}$, where $\rho_1 < \rho_0$, we see that an expansion of the desired form holds for $\rho < \rho_1$. In fact, this expansion is valid for $\rho < \rho_0$, because the ψ_m cannot depend on the choice of ρ_1. This completes the proof of part (i), and part (ii) follows in a similar fashion with the help of Theorem 7.12.

$\square$

Behaviour at Infinity

Fix $\rho_0 > 0$, and define

$$x^\sharp = \left(\frac{\rho_0}{|x|}\right)^2 x \quad \text{for } x \neq 0. \tag{8.29}$$

We call $x^\sharp$ the *inverse point* of x with respect to the sphere ∂B_{ρ_0}. Notice

$$(x^\sharp)^\sharp = x \quad \text{and} \quad |x^\sharp||x| = \rho_0^2,$$

so $|x^\sharp| > \rho_0$ if and only if $|x| < \rho_0$, $x^\sharp = x$ if and only if $|x| = \rho_0$, and $|x^\sharp| < \rho_0$ if and only if $|x| > \rho_0$. For any subset $E \subseteq \mathbb{R}^n \setminus \{0\}$, we write $E^\sharp = \{x^\sharp : x \in E\}$, and for any function u defined on E, we define $u^\sharp$, the *Kelvin transform* of u, by

$$u^\sharp(x^\sharp) = \left(\frac{|x|}{\rho_0} \right)^{n-2} u(x). \tag{8.30}$$

Exercise 8.9 shows that

$$\Delta u^\sharp(x^\sharp) = \left(\frac{|x|}{\rho_0} \right)^{n+2} \Delta u(x),$$

a fact that yields a simple characterisation for the radiation condition $\mathcal{M}u = 0$; see (7.29) for the definition of $\mathcal{M}u$.

Theorem 8.9 *Suppose that $\Delta u(x) = 0$ for $|x| > \rho_0$.*

(i) *When $n \geq 3$, the function u satisfies $\mathcal{M}u = 0$ on $\mathbb{R}^n$ if and only if*

$$u(x) = O(|x|^{2-n}) \quad \text{as } |x| \to \infty.$$

(ii) *When $n = 2$, the function u satisfies $\mathcal{M}u = 0$ on $\mathbb{R}^n$ if and only if there is a constant b such that*

$$u(x) = b \log |x| + O(|x|^{-1}) \quad \text{as } |x| \to \infty.$$

(iii) *When $n = 2$, the function u satisfies $u(x) = O(1)$ as $|x| \to \infty$ if and only if there is a constant b such that*

$$u(x) = b + O(|x|^{-1}) \quad \text{as } |x| \to \infty.$$

In this case, $\mathcal{M}u = b$.

Proof. The Kelvin transform $u^\sharp$ is harmonic on the punctured ball $B_{\rho_0} \setminus \{0\}$, so by applying the third Green identity over an annular domain $B_{\rho_2} \setminus \overline{B_{\rho_1}}$, with $0 < \rho_1 < \rho_2 < \rho_0$, we see that $u^\sharp = v^0 + v^\infty$, where $v^0(x^\sharp)$ is harmonic for $|x^\sharp| < \rho_0$, and $v^\infty(x^\sharp)$ is harmonic for $|x^\sharp| > 0$, with $\mathcal{M}v^\infty = 0$ on $\mathbb{R}^n$.

The Laplace Equation

To prove part (i), let $n \geq 3$ and write $\omega = x/|x| = x^{\sharp}/|x^{\sharp}|$. By Theorem 8.8, there are surface spherical harmonics ψ_m^0 and ψ_m^{∞} of degree m, such that

$$v^0(x^{\sharp}) = \sum_{m=0}^{\infty} |x^{\sharp}|^m \psi_m^0(\omega) \quad \text{for } |x^{\sharp}| < \rho_0,$$

and

$$v^{\infty}(x^{\sharp}) = \sum_{m=0}^{\infty} |x^{\sharp}|^{2-n-m} \psi_m^{\infty}(\omega) \quad \text{for } |x^{\sharp}| > 0.$$

Suppose that $u(x) = O(|x|^{2-n})$ as $|x| \to \infty$. This assumption means that $u^{\sharp}$ is bounded at zero, and thus ψ_m^{∞} must be identically zero for all $m \geq 0$. Hence, $u^{\sharp} = v^0$ and

$$u(x) = \left(\frac{\rho_0}{|x|} \right)^{n-2} u^{\sharp}(x^{\sharp}) = \sum_{m=0}^{\infty} \rho_0^{n-2+2m} |x|^{2-n-m} \psi_m^0(\omega) \quad \text{for } |x| > \rho_0.$$

We conclude that $\partial^{\alpha} u(x) = O(|x|^{2-n-|\alpha|})$ for all α, and so $\mathcal{M}u = 0$ on $\mathbb{R}^n$ by Exercise 7.4. Conversely, if $\mathcal{M}u = 0$ on $\mathbb{R}^n$ then $u(x) = O(|x|^{2-n})$ by Theorem 7.12.

Now suppose that $n = 2$. By arguing as above, it is easy to see that if $u(x) = b_1 \log |x| + b_2 + O(|x|^{-1})$ as $|x| \to \infty$, then the function $\tilde{u}(x) = u(x) - b_1 \log |x| - b_2$ satisfies $\mathcal{M}\tilde{u} = 0$, and therefore $\mathcal{M}u = b_2 \mathcal{M}1 = b_2$ by Lemma 7.13 and Exercise 7.5. The converse again follows by Theorem 7.12. $\qquad\square$

Solvability for the Dirichlet Problem

We know from Theorems 7.6 and 7.8 that, for any bounded Lipschitz domain Ω^-, the boundary integral operators

$$S : H^{-1/2}(\Gamma) \to H^{1/2}(\Gamma) \quad \text{and} \quad R : H^{1/2}(\Gamma) \to H^{-1/2}(\Gamma)$$

are Fredholm with index zero. The following uniqueness theorem for the exterior Dirichlet problem will help us to investigate $\ker S$. We shall see that complications arise when $n = 2$.

Theorem 8.10 *A function $u \in H^1_{\text{loc}}(\Omega^+)$ satisfies*

$$\begin{aligned} \Delta u &= 0 && \text{on } \Omega^+, \\ \gamma^+ u &= 0 && \text{on } \Gamma, \\ u(x) &= O(|x|^{2-n}) && \text{as } |x| \to \infty \end{aligned} \tag{8.31}$$

if and only if $u = 0$ on Ω^+.

Proof. Suppose that (8.31) holds. Applying the first Green identity over $\Omega_\rho^+ = \Omega^+ \cap B_\rho$ for ρ sufficiently large, we have

$$\Phi_{\Omega_\rho^+}(u, u) = -(\partial_\nu^+ u, \gamma^+ u)_\Gamma - \int_{\partial B_\rho} (\partial_\nu \bar{u})u\, d\sigma$$

$$= \int_{|\omega|=1} \left(\frac{d}{d\rho}\bar{u}(\rho\omega) \right) u(\rho\omega)\rho^{n-1}\, d\omega.$$

Theorems 8.8 and 8.9 imply that

$$u(\rho\omega) = O(\rho^{2-n}) \quad \text{and} \quad \frac{d}{d\rho}u(\rho\omega) = \begin{cases} O(\rho^{1-n}) & \text{if } n \geq 3, \\ O(\rho^{-2}) & \text{if } n = 2, \end{cases}$$

so

$$\int_{\Omega_\rho^+} |\operatorname{grad} u|^2\, dx = \Phi_{\Omega_\rho^+}(u, u) = \begin{cases} O(\rho^{2-n}) & \text{if } n \geq 3, \\ O(\rho^{-1}) & \text{if } n = 2. \end{cases}$$

Sending $\rho \to \infty$, we deduce that $\operatorname{grad} u = 0$ on Ω^+, and thus u is constant on each component of Ω^+. Since $\gamma^+ u = 0$, it follows that $u = 0$ on Ω^+. The converse is obvious. □

Corollary 8.11 Let $\psi \in H^{-1/2}(\Gamma)$ satisfy $S\psi = 0$ on Γ.

(i) If $n \geq 3$, then $\psi = 0$.
(ii) If $n = 2$ and $(1, \psi)_\Gamma = 0$, then $\psi = 0$.

Proof. The single-layer potential $u = \operatorname{SL}\psi$ satisfies

$$\Delta u = 0 \quad \text{on } \Omega^\pm,$$
$$\gamma^\pm u = 0 \quad \text{on } \Gamma,$$

and as $|x| \to \infty$, we have $u(x) = O(|x|^{2-n})$ when $n \geq 3$, but

$$u(x) = \frac{1}{2\pi}(1, \psi)_\Gamma \log\frac{r}{|x|} + O(|x|^{-1}) \quad \text{when } n = 2.$$

Thus, provided we assume that $(1, \psi)_\Gamma = 0$ when $n = 2$, it follows from Theorem 8.10 that u is identically zero, and hence $\psi = -[\partial_\nu u]_\Gamma = 0$. □

For the Laplacian, we can prove a stronger version of Theorem 7.6.

Theorem 8.12 Let

$$V = H^{-1/2}(\Gamma) \quad \text{if } n \geq 3,$$

and

$$V = \{ \psi \in H^{-1/2}(\Gamma) : (1, \psi)_\Gamma = 0 \} \quad if\, n = 2.$$

The boundary operator S satisfies

$$(S\psi, \phi)_\Gamma = \int_{\mathbb{R}^n} \operatorname{grad} \operatorname{SL} \bar{\psi} \cdot \operatorname{grad} \operatorname{SL} \phi \, dx \quad for\, \psi \in V\, and\, \phi \in H^{-1/2}(\Gamma),$$

(8.32)

and is strictly positive-definite on V, i.e.,

$$(S\psi, \psi)_\Gamma > 0 \quad for\, all\, \psi \in V \setminus \{0\}.$$

Proof. Let $\psi \in V$ and $\phi \in H^{-1/2}(\Gamma)$. If ρ is sufficiently large, and if we put $u = \operatorname{SL} \psi$ and $v = \operatorname{SL} \phi$, then (as in the proof of Theorem 7.6) the jump relations and the first Green identity imply that

$$(S\psi, \phi)_\Gamma = (\gamma u, \partial_\nu^- v - \partial_\nu^+ v)_\Gamma = \Phi_{\Omega^-}(u, v) + \Phi_{\Omega_\rho^+}(u, v) + \int_{\partial B_\rho} \bar{u} \partial_\nu v \, d\sigma.$$

The integral over ∂B_ρ is $O(\rho^{2-n})$ if $n \geq 3$, and is $O(\rho^{-1})$ if $n = 2$. Thus, in either case, we obtain (8.32) after sending $\rho \to \infty$.

It is obvious from (8.32) that $(S\psi, \psi)_\Gamma \geq 0$. Moreover, if $(S\psi, \psi)_\Gamma = 0$, then $\operatorname{grad} u = 0$ on $\mathbb{R}^n \setminus \Gamma$, so $\psi = -[\partial_\nu u]_\Gamma = 0$ on Γ. $\square$

Corollary 8.13 *If $n \geq 3$, then S is positive and bounded below on $H^{-1/2}(\Gamma)$, i.e.,*

$$(S\psi, \psi)_\Gamma \geq c\|\psi\|^2_{H^{-1/2}(\Gamma)} \quad for\, all\, \psi \in H^{-1/2}(\Gamma).$$

(8.33)

Proof. By Theorem 7.6, $S : H^{-1/2}(\Gamma) \to H^{1/2}(\Gamma)$ is Fredholm with zero index, and since S is strictly positive-definite, $\ker S = \{0\}$. Hence, S has a bounded inverse. Since S^{-1} is self-adjoint, and since the inclusion $H^{1/2}(\Gamma) \subseteq L_2(\Gamma)$ is compact by Theorem 3.27, the result follows from Corollary 2.38 with $A = S^{-1}$. $\square$

By modifying the boundary integral equation $S\psi = f$ and adding a side condition, we obtain a system that is always uniquely solvable, even when $n = 2$.

Lemma 8.14 *Given any $f \in H^{1/2}(\Gamma)$ and $b \in \mathbb{C}$, the system of equations*

$$S\psi + a = f \quad and \quad (1, \psi)_\Gamma = b,$$

has a unique solution $\psi \in H^{-1/2}(\Gamma)$ and $a \in \mathbb{C}$.

Proof. Introduce the Hilbert space $H = H^{-1/2}(\Gamma) \times \mathbb{C}$, identify the dual space H^* with $H^{1/2}(\Gamma) \times \mathbb{C}$ by writing

$$\big((\psi, a), (\phi, b)\big) = (\psi, \phi)_\Gamma + \bar{a}b,$$

and define a bounded linear operator $A : H \to H^*$ by

$$A(\psi, a) = \big(S\psi + a, (1, \psi)_\Gamma\big).$$

In this way, A is self-adjoint, and we now show that A has a bounded inverse.

Let S_0 and L be as in Theorem 7.6, so that $S = S_0 + L$ with S_0 invertible and L compact as operators from $H^{-1/2}(\Gamma)$ to $H^{1/2}(\Gamma)$. We define

$$A_0(\psi, a) = (S_0\psi, a) \quad \text{and} \quad K(\psi, a) = \big(a + L\psi, (1, \psi)_\Gamma - a\big),$$

so that $A = A_0 + K$, with $A_0 : H \to H^*$ invertible, and $K : H \to H^*$ compact. By Theorem 2.26, A is invertible if the homogeneous system $A(\psi, a) = (0, 0)$ has only the trivial solution. In fact, if

$$S\psi + a = 0 \quad \text{and} \quad (1, \psi)_\Gamma = 0,$$

then $(S\psi, \psi)_\Gamma = (-a, \psi)_\Gamma = -\bar{a}(1, \psi)_\Gamma = 0$, so $\psi = 0$ by Theorem 8.12, and in turn $a = -S\psi = 0$. $\qquad\square$

Theorem 8.15 *There exists a unique distribution $\psi_{eq} \in H^{-1/2}(\Gamma)$ such that $S\psi_{eq}$ is constant on Γ, and $(1, \psi_{eq})_\Gamma = 1$. If $n \geq 3$, then $S\psi_{eq} > 0$.*

Proof. Let ψ_{eq} be the solution of the system in Lemma 8.14 when $f = 0$ and $b = 1$. Thus, $S\psi_{eq} = -a$ is constant on Γ, and by Theorem 8.12, if $n \geq 3$, then $-a = -a(1, \psi_{eq})_\Gamma = (S\psi_{eq}, \psi_{eq})_\Gamma > 0$. $\qquad\square$

The distribution ψ_{eq} is real-valued, and is called the *equilibrium density* for Γ. If $n \geq 3$, then the reciprocal of the positive constant $S\psi_{eq}$ is called the *capacity* of Γ, a quantity we denote by Cap_Γ, so that

$$\frac{1}{\text{Cap}_\Gamma} = S\psi_{eq} \quad \text{when } n \geq 3.$$

This terminology has its origins in electrostatics: if an isolated conductor carries a charge Q in equilibrium, so that the potential V is constant throughout the conductor, then the ratio Q/V does not depend on Q, and is called the capacitance. Mutual repulsion causes all of the charge to lie on the boundary of the conductor, so (with appropriately normalised units) the electrostatic potential is

SL ψ, where ψ is the surface charge density. Thus, $Q = \int_\Gamma \psi \, d\sigma$ and $V = S\psi$, and in the case of a unit charge $Q = 1$, we have $\psi = \psi_{eq}$, so the capacitance is the reciprocal of $S\psi_{eq}$.

Now consider the case $n = 2$, and write $S = S_r$ to indicate the dependence on the choice of the parameter r in the fundamental solution from Theorem 8.1. The equilibrium density ψ_{eq} is the same for all r, but not so the constant $S_r \psi_{eq}$. Since $S_r \psi_{eq}$ is not always positive, one introduces the *logarithmic capacity*, $\mathrm{Cap}_\Gamma = e^{-2\pi S_1 \psi_{eq}}$, so that

$$\frac{1}{2\pi} \log \frac{1}{\mathrm{Cap}_\Gamma} = S_1 \psi_{eq} \quad \text{when } n = 2.$$

Notice that

$$S_r \psi = S_1 \psi + \frac{(1, \psi)_\Gamma}{2\pi} \log r,$$

and hence

$$S_r \psi_{eq} = \frac{1}{2\pi} \log \frac{r}{\mathrm{Cap}_\Gamma}.$$

In particular, $S_r \psi_{eq} = 0$ if and only if $r = \mathrm{Cap}_\Gamma$.

Theorem 8.16 *Consider $S_r : H^{-1/2}(\Gamma) \to H^{1/2}(\Gamma)$ when $n = 2$.*

(i) *The operator S_r is positive and bounded below on the whole of $H^{-1/2}(\Gamma)$ if and only if $r > \mathrm{Cap}_\Gamma$.*

(ii) *The operator S_r has a bounded inverse if and only if $r \neq \mathrm{Cap}_\Gamma$.*

Proof. For brevity, put $a_r = S_r \psi_{eq} = (2\pi)^{-1} \log(r/\mathrm{Cap}_\Gamma)$. Let $\psi \in H^{-1/2}(\Gamma)$, define $\psi_0 = \psi - (1, \psi)_\Gamma \psi_{eq}$, and observe that

$$\psi = \psi_0 + (1, \psi)_\Gamma \psi_{eq}, \quad (1, \psi_0)_\Gamma = 0 \quad \text{and} \quad S_r \psi = S_r \psi_0 + a_r (1, \psi)_\Gamma.$$

Also, since $(S_r \psi_0, \psi_{eq}) = (\psi_0, S_r \psi_{eq})_\Gamma = 0$, we have

$$(S_r \psi, \psi)_\Gamma = (S_r \psi_0, \psi_0)_\Gamma + a_r (\psi, 1)_\Gamma (1, \psi)_\Gamma. \tag{8.34}$$

If $r \leq \mathrm{Cap}_\Gamma$, then $(S_r \psi_{eq}, \psi_{eq})_\Gamma = a_r \leq 0$. To complete the proof of (i), suppose that $r > \mathrm{Cap}_\Gamma$, or equivalently, $a_r > 0$. By Theorem 8.12, both terms on the right-hand side of (8.34) are non-negative, and the first is zero if and only if $\psi_0 = 0$. Thus, $(S_r \psi, \psi)_\Gamma \geq 0$, with equality if and only if $\psi_0 = 0$ and $(1, \psi)_\Gamma = 0$, i.e., if and only if $\psi = 0$. Hence, S_r is strictly positive-definite on the whole of $H^{-1/2}(\Gamma)$. Arguing as in the proof of Corollary 8.13, we conclude that S_r is positive and bounded below on $H^{-1/2}(\Gamma)$.

Turning to part (ii), we note that if $r = \mathrm{Cap}_\Gamma$, then S_r cannot be invertible because $S_r \psi_{\mathrm{eq}} = 0$. Thus, suppose that $r \neq \mathrm{Cap}_\Gamma$ and $S_r \psi = 0$. We have $S_r \psi_0 = -a_r(1, \psi)_\Gamma$, hence $(S_r \psi_0, \psi_0)_\Gamma = 0$, and therefore $\psi_0 = 0$ by Theorem 8.12. In turn, $(1, \psi)_\Gamma = 0$ because $a_r \neq 0$, giving $\psi = 0$. Thus, the homogeneous equation has only the trivial solution, and S_r is invertible. $\square$

In the case of the unit sphere $\Gamma = \mathbb{S}^{n-1}$, it is clear from symmetry that ψ_{eq} takes the constant value $1/\Upsilon_n$, and in view of Theorems 8.4 and 8.5,

$$
\mathrm{Cap}_{\mathbb{S}^{n-1}} = \begin{cases} 1 & \text{if } n = 2, \\ (n-2)\Upsilon_n & \text{if } n \geq 3. \end{cases}
$$

Further properties of Cap_Γ are given in Exercises 8.10 and 8.11, and in the books of Hille [40, pp. 280–289] and Landkof [52].

We conclude this section with an interesting application of Theorem 2.36.

Theorem 8.17 *If* $m \neq l$, *then* $\mathcal{H}_m(\mathbb{S}^{n-1})$ *and* $\mathcal{H}_l(\mathbb{S}^{n-1})$ *are orthogonal to each other as subspaces of* $L_2(\mathbb{S}^{n-1})$. *Furthermore, the orthogonal direct sum* $\bigoplus_{m=0}^\infty \mathcal{H}_m(\mathbb{S}^{n-1})$ *is dense in* $L_2(\mathbb{S}^{n-1})$.

Proof. Recall from Theorems 8.4 and 8.5 that $\mathcal{H}_m(\mathbb{S}^{n-1})$ is an eigenspace of S. Hence, the orthogonality of $\mathcal{H}_m(\mathbb{S}^{n-1})$ and $\mathcal{H}_l(\mathbb{S}^{n-1})$ follows at once from the fact that S is self-adjoint. We may assume, by choosing $r > 1$ if $n = 2$, that S is strictly positive-definite on $H^{-1/2}(\mathbb{S}^{n-1})$. Since the inclusion $L_2(\mathbb{S}^{n-1}) \subseteq H^{-1/2}(\mathbb{S}^{n-1})$ is compact, the operator $S : L_2(\mathbb{S}^{n-1}) \to L_2(\mathbb{S}^{n-1})$ is compact with $\ker S = \{0\}$. Hence, the eigenfunctions of S span a dense subspace of $L_2(\mathbb{S}^{n-1})$, and to complete the proof it suffices to show that *every* eigenfunction of S is a spherical harmonic.

Suppose for a contradiction that $\psi \in L_2(\Gamma)$ is a non-trivial solution of $S\psi = \mu\psi$ on $\mathbb{S}^{n-1}$ for some (necessarily positive) μ, and that $\psi \perp \mathcal{H}_m(\mathbb{S}^{n-1})$ for every $m \geq 0$. It follows from Theorem 7.2 that $\psi \in C^\infty(\mathbb{S}^{n-1})$, so by Theorem 6.13 the single-layer potential $u = \mathrm{SL}\,\psi$ is C^∞ up to the boundary of the unit ball. If $\psi_m \in \mathcal{H}_m(\mathbb{S}^{n-1})$ is as in part (i) of Theorem 8.8, then for $0 < \rho < 1$,

$$
\int_{|\omega|=1} \overline{\mathrm{SL}\,\psi(\rho\omega)}\psi(\omega)\,d\omega = \sum_{m=1}^\infty \rho^m \int_{|\omega|=1} \overline{\psi_m(\omega)}\psi(\omega)\,d\omega = 0,
$$

and so, sending $\rho \uparrow 1$, we see that $(S\psi, \psi)_{\mathbb{S}^{n-1}} = 0$, implying, $\psi = 0$, a contradiction. $\square$

Solvability for the Neumann Problem

Solutions of the Neumann problem for the Laplacian are unique only modulo constants; more precisely, the following holds.

Theorem 8.18 A function $u \in H^1(\Omega^-)$ satisfies

$$\Delta u = 0 \quad on \ \Omega^-,$$
$$\partial_\nu^- u = 0 \quad on \ \Gamma \tag{8.35}$$

if and only if u is constant on each component of Ω^-. Likewise, $u \in H^1_{\text{loc}}(\Omega^+)$ satisfies

$$\Delta u = 0 \qquad on \ \Omega^+,$$
$$\partial_\nu^+ u = 0 \qquad on \ \Gamma,$$
$$u(x) = O(|x|^{2-n}) \quad as \ |x| \to \infty$$

if and only if u is constant on each component of Ω^+ and, when $n \geq 3$, is zero on the unbounded component of Ω^+.

Proof. The first part of the theorem is a special case of Corollary 8.3. The exterior problem is handled in a similar fashion, by applying the first Green identity over Ω_ρ^+, and arguing as in the proof of Theorem 8.10. $\quad\square$

Let $\Omega_1^-, \ldots, \Omega_p^-$ be the components (i.e., the maximal connected subsets) of Ω^-, and define

$$v_j = \begin{cases} 1 & on \ \Omega_j^-, \\ 0 & on \ \Omega^- \setminus \Omega_j^-, \end{cases}$$

for $1 \leq j \leq p$. Thus, the functions $v_1, \ldots, v_p$ form a basis for the solution space of the homogeneous interior Neumann problem (8.35).

Theorem 8.19 Let $f \in \widetilde{H}^{-1}(\Omega^-)$ and $g \in H^{-1/2}(\Gamma)$. The interior Neumann problem

$$-\Delta u = f \quad on \ \Omega^-,$$
$$\partial_\nu^- u = g \quad on \ \Gamma, \tag{8.36}$$

has a solution $u \in H^1(\Omega^-)$ if and only if the data f and g satisfy

$$\int_{\Omega_j^-} f \, dx + \int_{\partial\Omega_j^-} g \, d\sigma = 0 \quad for \ 1 \leq j \leq p, \tag{8.37}$$

in which case u is unique modulo the subspace $\text{span}\{v_1, \ldots, v_p\}$.

Proof. In the present case, the conditions in part (ii) of Theorem 4.10 reduce to

$$(v_j, f)_{\Omega^-} + (\gamma^- v_j, g)_\Gamma = 0 \quad \text{for } 1 \le j \le p. \qquad \square$$

Let $\Gamma_1, \ldots, \Gamma_q$ denote the components of Γ, and define the function χ_j on Γ by

$$\chi_j = \begin{cases} 1 & \text{on } \Gamma_j, \\ 0 & \text{on } \Gamma \setminus \Gamma_j, \end{cases}$$

for $1 \le j \le q$. The lack of uniqueness for the Neumann problem gives rise to a lack of uniqueness for the corresponding boundary integral equation; cf. Exercise 8.17.

Theorem 8.20 *A function $\psi \in H^{1/2}(\Gamma)$ satisfies $R\psi = 0$ on Γ if and only if ψ is constant on each component of Γ. Thus, the functions $\chi_1, \ldots, \chi_q$ form a basis for* ker R.

Proof. If $R\psi = 0$, then the double-layer potential $u = \mathrm{DL}\,\psi$ satisfies

$$\begin{aligned} \Delta u &= 0 && \text{on } \Omega^\pm, \\ \partial_\nu^\pm u &= 0 && \text{on } \Gamma, \\ u(x) &= O(|x|^{2-n}) && \text{as } |x| \to \infty, \end{aligned}$$

so u is constant on each component of $\mathbb{R}^n \setminus \Gamma$, and thus $\psi = [u]_\Gamma$ is constant on each component of Γ. To prove the converse, it suffices to observe that, by the third Green identity,

$$\mathrm{DL}\,\chi_j = \begin{cases} -1 & \text{inside } \Gamma_j, \\ 0 & \text{outside } \Gamma_j, \end{cases} \tag{8.38}$$

so $R\chi_j = \partial_\nu \mathrm{DL}\,\chi_j = 0$ on Γ. $\qquad \square$

Corresponding to Theorem 8.12 for the operator S, we have the following result on the positivity of R; see also Exercise 8.17.

Theorem 8.21 *The operator $R : H^{1/2}(\Gamma) \to H^{-1/2}(\Gamma)$ satisfies*

$$(R\psi, \phi)_\Gamma = \int_{\mathbb{R}^n \setminus \Gamma} \mathrm{grad}\,\mathrm{DL}\,\bar{\psi} \cdot \mathrm{grad}\,\mathrm{DL}\,\phi\, dx \quad \text{for } \psi, \phi \in H^{1/2}(\Gamma) \tag{8.39}$$

and

$$(R\psi, \psi)_\Gamma \geq c\|\psi\|^2_{H^{1/2}(\Gamma)}$$

for $\psi \in H^{1/2}(\Gamma)$ *such that* $(\chi_j, \psi)_\Gamma = 0$ *for* $1 \leq j \leq q$.

Proof. Let $\psi, \phi \in H^{1/2}(\Gamma)$, and put $u = DL\,\psi$ and $v = DL\,\phi$. As in the proof of Theorem 7.8, we find that for ρ sufficiently large,

$$(R\psi, \phi)_\Gamma = (-\partial_\nu u, \gamma^+ v - \gamma^- v)_\Gamma$$

$$= \Phi_{\Omega_\rho^+}(u, v) + \Phi_{\Omega^-}(u, v) + \int_{\partial B_\rho} (\partial_\nu \bar{u})v\, d\sigma.$$

Since $\operatorname{grad} u(x) = O(|x|^{-n})$ and $v(x) = O(|x|^{1-n})$ as $|x| \to \infty$, the integral over ∂B_ρ is $O(\rho^{-n})$, and we obtain (8.39) by sending $\rho \to \infty$.

It follows at once from (8.39) that $(R\psi, \psi)_\Gamma \geq 0$. Moreover, if $(R\psi, \psi)_\Gamma = 0$, then $\operatorname{grad} u = 0$ on $\mathbb{R}^n \backslash \Gamma$, so u is constant on each component of $\mathbb{R}^n \backslash \Gamma$, implying that $\psi = [u]_\Gamma$ is constant on each component of Γ. Hence, by Theorem 8.20, $(R\psi, \psi)_\Gamma = 0$ implies $R\psi = 0$. Therefore, the coercive, self-adjoint operator R is strictly positive-definite on the orthogonal complement of $\ker R$, and we may appeal to part (iii) of Exercise 2.17.   □

Exercises

8.1 Let $G(x, y) = G(x - y)$ be the fundamental solution given by Theorem 6.8 when $n = 2$ and $P(\xi) = (2\pi)^2 |\xi|^2$.
   (i) Show that

$$G(x) = \frac{1}{2\pi} \log \frac{1}{|x|} + \frac{\Gamma'(1) - \log 2\pi}{2\pi} - \frac{1}{(2\pi)^2} \int_0^{2\pi} \log|\sin\theta|\, d\theta.$$

   (ii) Use the substitution $\theta = 2\phi$ to show that

$$\int_0^{2\pi} \log|\sin\theta|\, d\theta = -2\pi \log 2.$$

8.2 Deduce (8.1) from

$$\left(\int_{-\infty}^\infty e^{-t^2}\, dt \right)^n = \Upsilon_n \int_0^\infty e^{-\rho^2} \rho^{n-1}\, d\rho = \tfrac{1}{2} \Upsilon_n \Gamma(n/2).$$

8.3 Show that if $A \in \mathbb{R}^{n \times n}$ is an orthogonal matrix, and if $v(x) = u(Ax)$, then $\triangle v(x) = (\triangle u)(Ax)$. In particular, when u is harmonic, so is v.

8.4 Show that if $u(x) = w(\rho)$, where $\rho = |x|$, then

$$\Delta u(x) = \frac{d^2w}{d\rho^2} + \frac{n-1}{\rho}\frac{dw}{d\rho} = \frac{1}{\rho^{n-1}}\frac{d}{d\rho}\left(\rho^{n-1}\frac{dw}{d\rho}\right).$$

8.5 Show that $T1 = -1$ and $T^*\psi_{eq} = -\psi_{eq}$.

8.6 Assume the hypotheses of Theorem 7.3, take $\mathcal{P} = -\Delta$, and let $x \in \Gamma$.
 (i) Show that $a^\pm(x) = -\sigma[\Upsilon^\mp(x)]/\Upsilon_n$.
 (ii) Show that $\lim_{\epsilon\downarrow 0} T_\epsilon 1(x) = -1 - [a^+(x) - a^-(x)] = -2a^+(x)$.
 (iii) Deduce that

$$T\psi(x) = -\psi(x) + \frac{2}{\Upsilon_n}\int_\Gamma \frac{\nu_y\cdot(x-y)}{|x-y|^n}[\psi(y)-\psi(x)]\,d\sigma_y.$$

8.7 Deduce from (8.12) that the numbers $N(n,m)$ have the generating function

$$\sum_{m=0}^\infty N(n,m)z^m = \frac{1+z}{(1-z)^{n-1}} \quad \text{for } |z| < 1.$$

8.8 Let $b_m = b_m(n)$ be as in Theorem 8.7. Show that when $n \geq 3$, the Legendre polynomials have the generating function

$$\sum_{m=0}^\infty b_m(n)P_m(n,t)z^m = \frac{1}{(1-2tz+z^2)^{(n-2)/2}} \quad \text{for } |z| < 1,$$

and by differentiating both sides with respect to z, derive the three-term recurrence

$$P_0(t) = 1, \quad b_1 P_1(t) = (n-2)t,$$
$$(m+1)b_{m+1}P_{m+1}(t) - (2m+n-2)tb_m P_m(t)$$
$$+(m+n-3)b_{m-1}P_{m-1}(t)=0 \quad \text{for } m \geq 1.$$

Show likewise that when $n = 2$,

$$\sum_{m=1}^\infty \frac{1}{m}P_m(2,t)z^m = \log\frac{1}{\sqrt{1-2tz+z^2}} \quad \text{for } |z| < 1$$

and

$$P_0(2,t) = 1, \quad P_1(2,t) = t,$$
$$P_{m+1}(2,t) - 2tP_m(2,t) + P_{m-1}(2,t) = 0 \quad \text{for } m \geq 1.$$

8.9 Recall the definition (8.29) of the inverse point $x^\sharp$ with respect to the sphere ∂B_{ρ_0}.

(i) Show that

$$\frac{\partial x_j^\sharp}{\partial x_k} = \left(\frac{\rho_0}{|x|}\right)^2 \left(\delta_{jk} - 2\frac{x_j x_k}{|x|^2}\right)$$

and

$$\sum_{l=1}^{n} \frac{\partial x_l^\sharp}{\partial x_j} \frac{\partial x_l^\sharp}{\partial x_k} = \left(\frac{\rho_0}{|x|}\right)^4 \delta_{jk}.$$

(ii) Consider two curves $x = x(t)$ and $y = y(t)$ that intersect at a point $a \in \mathbb{R}^n$ when $t = 0$. Show that

$$\frac{dx^\sharp}{dt} \cdot \frac{dy^\sharp}{dt} = \left(\frac{\rho_0}{|a|}\right)^4 \frac{dx}{dt} \cdot \frac{dy}{dt} \quad \text{when } t = 0,$$

and deduce that the mapping $x \mapsto x^\sharp$ is *conformal*, i.e., angle-preserving.

(iii) Show that if E is a plane or sphere, then so is $E^\sharp$. [Hint: consider the equation $a|x|^2 + b \cdot x + c = 0$, where the coefficients a and c are scalars, but b is a vector.]

(iv) Think of $\xi = x^\sharp$ as a system of orthogonal curvilinear coordinates for $x = \xi^\sharp$, and deduce that

$$\Delta u = \left(\frac{\rho_0}{|\xi|}\right)^{-2n} \sum_{j=1}^{n} \frac{\partial}{\partial \xi_j}\left[\left(\frac{\rho_0}{|\xi|}\right)^{2n-4} \frac{\partial u}{\partial \xi_j}\right].$$

Next, establish the identity

$$\left(\frac{\rho_0}{|\xi|}\right)^{2-n} \frac{\partial}{\partial \xi_j}\left[\left(\frac{\rho_0}{|\xi|}\right)^{2n-4} \frac{\partial u}{\partial \xi_j}\right]$$

$$= 2\frac{\partial u}{\partial \xi_j}\frac{\partial}{\partial \xi_j}\left(\frac{\rho_0}{|\xi|}\right)^{n-2} + \frac{\partial^2 u}{\partial \xi_j^2}\left(\frac{\rho_0}{|\xi|}\right)^{n-2}$$

$$= \frac{\partial^2}{\partial \xi_j^2}\left[\left(\frac{\rho_0}{|\xi|}\right)^{n-2} u\right] - u\frac{\partial^2}{\partial \xi_j^2}\left(\frac{\rho_0}{|\xi|}\right)^{n-2},$$

and finally conclude that

$$\Delta u = \left(\frac{|\xi|}{\rho_0}\right)^{n+2} \sum_{j=1}^{n} \frac{\partial^2}{\partial \xi_j^2}\left[\left(\frac{\rho_0}{|\xi|}\right)^{n-2} u\right] = \left(\frac{\rho_0}{|x|}\right)^{n+2} \sum_{j=1}^{n} \frac{\partial^2 u^\sharp}{\partial \xi_j^2},$$

where $u^\sharp$ is the Kelvin transform (8.30).

8.10 Let $n \geq 3$, and consider the exterior Dirichlet problem

$$\Delta u = 0 \qquad \text{on } \Omega^{+},$$
$$\gamma^{+} u = 1 \qquad \text{on } \Gamma,$$
$$u(x) = O(|x|^{2-n}) \quad \text{as } |x| \to \infty.$$

Show that the unique solution is $u = \text{Cap}_{\Gamma} \text{ SL } \psi_{\text{eq}}$, and deduce that

$$\text{Cap}_{\Gamma} = -\int_{\Gamma} \partial_{\nu}^{+} u \, d\sigma.$$

This result can sometimes be used to compute Cap_{Γ}; see Landkof [52, p. 165].

8.11 Suppose that Γ is a simple, closed curve in the complex z-plane, and let $w = f(z)$ define a conformal mapping of $\Omega^{+} \cup \Gamma$ onto $|w| \geq 1$. Since f is one–one on Ω^{+}, it must have a simple pole at ∞; see Markushevich [63, pp. 90–91]. Thus, there is a constant ρ_{Γ} such that

$$f(z) = \frac{z}{\rho_{\Gamma}} + O(1) \quad \text{as } z \to \infty.$$

Moreover, we can assume that ρ_{Γ} is real, because the domain $|w| \geq 1$ is invariant under rotation, i.e., under multiplication by $e^{i\theta}$ for any angle θ. The constant ρ_{Γ} is then known as the *external conformal radius*.

(i) Show that the real-valued function u defined by $f = e^{u+iv}$ satisfies

$$\Delta u = 0 \qquad \qquad \text{on } \Omega^{+},$$
$$\gamma^{+} u = 0 \qquad \qquad \text{on } \Gamma,$$
$$u(z) = \log \frac{|z|}{\rho_{\Gamma}} + O(1) \quad \text{as } z \to \infty.$$

(ii) Hence show that

$$u = \log \frac{r}{\text{Cap}_{\Gamma}} - 2\pi \text{ SL } \psi_{\text{eq}},$$

and deduce that $\rho_{\Gamma} = \text{Cap}_{\Gamma}$.

(iii) Suppose that $a > b > 0$, and let Γ be the ellipse

$$|z - c| + |z + c| = 2a, \quad \text{where } c = \sqrt{a^2 - b^2}.$$

Thus, a and b are the semimajor and semiminor axes, respectively, and c/a is the eccentricity. Verify that the formula

$$z = \frac{a+b}{2}w + \frac{a-b}{2w}$$

defines a conformal mapping $w \mapsto z$ of the region $|w| > 1$ onto Ω^+, and deduce that $\text{Cap}_\Gamma = \frac{1}{2}(a+b)$. For further examples, see Landkof [52, p. 172].

8.12 Prove the (crude) bound

$$\text{Cap}_\Gamma \leq \begin{cases} \text{diam}(\Gamma) & \text{if } n = 2, \\ (n-2)\Upsilon_n \, \text{diam}(\Gamma)^{n-2} & \text{if } n \geq 3. \end{cases}$$

8.13 For any $a > 0$, we write $a\Gamma = \{ax : x \in \Gamma\}$. By expressing the equilibrium density for $a\Gamma$ in terms of the equilibrium density for Γ, show that

$$\text{Cap}_{a\Gamma} = \begin{cases} a\,\text{Cap}_\Gamma & \text{if } n = 2, \\ a^{n-2}\,\text{Cap}_\Gamma & \text{if } n \geq 3. \end{cases}$$

8.14 Derive the following variational characterisation of the capacity:

$$\min_{\psi \in H^{-1/2}(\Gamma), \ (1,\psi)_\Gamma = 1} (S\psi, \psi)_\Gamma = (S\psi_{eq}, \psi_{eq})_\Gamma$$

$$= \begin{cases} \dfrac{1}{2\pi} \log \dfrac{r}{\text{Cap}_\Gamma} & \text{if } n = 2, \\ \dfrac{1}{\text{Cap}_\Gamma} & \text{if } n \geq 3. \end{cases}$$

[Hint: if $(1, \psi)_\Gamma = 1$, then $\psi = \psi_0 + \psi_{eq}$ with $(1, \psi_0)_\Gamma = 0$.]

8.15 Let $\Omega^+ = \mathbb{C} \setminus [-1, 1]$. The formula

$$z = \frac{1}{2}\left(w + \frac{1}{w}\right)$$

defines a conformal mapping $w \mapsto z$ of the region $|w| > 1$ onto Ω^+. [This mapping is a degenerate case of the one in Exercise 8.11(iii) above.] Define

$$f_0(z) = \log \frac{2r}{w} \quad \text{and} \quad f_m(z) = \frac{1}{2mw^m} \quad \text{for } m \geq 1,$$

and let

$$u_m(x, y) = \text{Re}\, f_m(z), \qquad \text{where } z = x + iy.$$

(i) Show that $w = z + \sqrt{z^2 - 1}$, and that

$$\lim_{y \to 0^\pm} \sqrt{z^2 - 1} = \pm i\sqrt{1 - x^2} \quad \text{for } -1 \le x \le 1.$$

(ii) Hence show that if $-1 < x < 1$, then

$$\partial_y^\pm u_m(x, 0) = \lim_{y \to 0^\pm} i f_m'(z)$$

$$= \frac{\mp 1}{\sqrt{1 - x^2}} \times \begin{cases} 1 & \text{when } m = 0, \\ \frac{1}{2}\left(x \pm i\sqrt{1 - x^2}\right)^{-m} & \text{when } m \ge 1. \end{cases}$$

(iii) Let $\Gamma = (-1, 1)$, and show that $u_m = \text{SL } \psi_m$, where $\psi_m = -[\partial_y u_m]_\Gamma$ is given by

$$\psi_0(x) = \frac{2}{\sqrt{1 - x^2}} \quad \text{and} \quad \psi_m(x) = \frac{P_m(2, x)}{\sqrt{1 - x^2}} \quad \text{for } m \ge 1.$$

Here, $P_m(2, x)$ is the Chebyshev polynomial (8.24).

(iv) Deduce that for $x \in \Gamma$,

$$\frac{1}{2\pi} \int_{-1}^{1} \log\left(\frac{r}{|x - y|}\right) \frac{P_m(2, y)}{\sqrt{1 - y^2}} \, dy = \begin{cases} \dfrac{1}{2} \log 2r & \text{if } m = 0, \\[2mm] \dfrac{1}{2m} P_m(2, x) & \text{if } m \ge 1. \end{cases}$$

(v) Assume $r \ne \frac{1}{2}$, and express the solution of $S\psi = f$ as a series involving the Fourier–Chebyshev coefficients of f.

8.16 Let Ω^+, Γ and $w \mapsto z$ be as in Exercise 8.15, but now put

$$u_m(x, y) = \text{Re}\left(\frac{-1}{2iw^{m+1}}\right) \quad \text{for } m \ge 1.$$

(i) Show that $u_m = \text{DL } \psi_m$, where $\psi_m = [u_m]_\Gamma$ is given by

$$\psi_m(x) = Q_m(x)\sqrt{1 - x^2} \quad \text{and} \quad Q_m(\cos\theta) = \frac{\sin(m + 1)\theta}{\sin\theta}.$$

(The function Q_m is the mth Chebyshev polynomial of the *second* kind.)

(ii) Deduce that

$$\frac{-1}{2\pi} \operatorname*{f.p.}_{\epsilon \downarrow 0} \int_{(-1,1)\setminus B_\epsilon(x)} \frac{1}{(x - y)^2} Q_m(y)\sqrt{1 - y^2} \, dy = \frac{m + 1}{2} Q_m(x)$$

$$\text{for } -1 < x < 1.$$

(iii) Hence obtain an explicit series solution for the equation $R\psi = f$.

8.17 Let $f \in \widetilde{H}^{-1}(\Omega^-)$ and $g \in H^{-1/2}(\Gamma)$, and put

$$h = \tfrac{1}{2}(g - T^*g) - \partial_\nu^- \mathcal{G} f,$$

so that $R\psi = h$ is the boundary integral equation corresponding to the interior Neumann problem (8.36) for the Laplacian; see Theorem 7.7. Define an operator

$$R_1 : H^{1/2}(\Gamma) \to H^{-1/2}(\Gamma)$$

by

$$R_1\psi = R\psi + \sum_{j=1}^{q}(\chi_j, \psi)_\Gamma \chi_j,$$

where $\chi_1, \ldots, \chi_q$ are as in Theorem 8.20.

(i) Show that R_1 is self-adjoint, and that

$$(R_1\psi, \psi)_\Gamma \geq c\|\psi\|^2_{H^{1/2}(\Gamma)} \quad \text{for all } \psi \in H^{1/2}(\Gamma).$$

(ii) Show that $T\chi_j = -\chi_j$, and deduce that

$$(\chi_j, h)_\Gamma = \pm \int_{\Omega_j \cap \Omega^-} f \, dx + \int_{\Gamma_j} g \, d\sigma \quad \text{for } 1 \leq j \leq q,$$

where Ω_j is the bounded, simply connected domain enclosed by Γ_j, with the $+$ $(-)$ case occurring when ν points out of (into) Ω_j.

(iii) Explain why Ω^+ has $q - p$ bounded components (and one unbounded component).

(iv) Suppose $q = p = 1$. Show that the unique solution ψ of $R_1\psi = h$ satisfies $R\psi = h$ if and only if the data f and g satisfy

$$\int_{\Omega^-} f \, dx + \int_\Gamma g \, d\sigma = 0,$$

the necessary and sufficient condition for the existence of a solution to (8.36).

(v) Now suppose $q = 2$ and $p = 1$, with Γ_1 and Γ_2 labelled so that they are the boundaries of the unbounded and bounded componenents of Ω^+, respectively. Show that the result in (iv) is still valid provided either $f = 0$ and $g|_{\Gamma_1} = 0$, or $g|_{\Gamma_2} = 0$. How can we proceed for general f and g?

8.18 Let Ω^- be the lower half space $\{x \in \mathbb{R}^n : x_n < 0\}$ so that $\Gamma = \mathbb{R}^{n-1}$, and let $\psi, \phi \in \mathcal{D}(\mathbb{R}^{n-1})$.

(i) Show that the double-layer potential for the Laplacian may be written as

$$\mathrm{DL}\,\psi = -(\partial_n G) * (\psi \otimes \delta).$$

(ii) Deduce that

$$\partial_n \,\mathrm{DL}\,\psi = \psi \otimes \delta + G * (\Delta'\psi \otimes \delta),$$

where $\Delta' = \sum_{j=1}^{n-1} \partial_j^2$ is the Laplacian in $\mathbb{R}^{n-1}$.

(iii) Hence derive the identities

$$R\psi = -S\Delta'\psi \quad \text{and} \quad (R\psi, \phi)_\Gamma = \sum_{j=1}^{n-1} (S\partial_j\psi, \partial_j\phi)_\Gamma.$$

8.19 Show that in the case of the Laplacian, the functions m_S and m_R of Exercise 7.8 are given by

$$m_S(\xi') = \frac{1}{4\pi|\xi'|} \quad \text{and} \quad m_R(\xi') = \pi|\xi'|.$$

8.20 Let $\Gamma = \Gamma_0 \cup \Pi \cup \Gamma_1$ be a Lipschitz dissection of Γ, so that Γ_0 is an (oriented) open surface, and define $S_0\psi = (S\psi)|_{\Gamma_0}$ and $R_0\psi = (R\psi)|_{\Gamma_0}$ if $\mathrm{supp}\,\psi \subseteq \Gamma_0$.

(i) Show $S_0 : \widetilde{H}^{-1/2}(\Gamma_0) \to H^{1/2}(\Gamma_0)$ and $R_0 : \widetilde{H}^{1/2}(\Gamma_0) \to H^{-1/2}(\Gamma_0)$ are Fredholm operators with index zero.

(ii) Show that there exists a unique distribution $\psi_{eq}^0 \in \widetilde{H}^{-1/2}(\Gamma_0)$ such that $S_0\psi_{eq}^0$ is constant on Γ_0, and $(1, \psi_{eq}^0)_{\Gamma_0} = 1$. We use ψ_{eq}^0 to define Cap_{Γ_0}, the capacity of Γ_0, in the same way as for a closed surface.

(iii) Prove for $n \ge 3$ that S_0 is positive and bounded below on $\widetilde{H}^{-1/2}(\Gamma_0)$.

(iv) Let $n = 2$. Prove that S_0 is positive and bounded below on $\widetilde{H}^{-1/2}(\Gamma)$ if and only if $r > \mathrm{Cap}_{\Gamma_0}$, and that $S_0 : \widetilde{H}^{-1/2}(\Gamma_0) \to H^{1/2}(\Gamma_0)$ is invertible if and only if $r \ne \mathrm{Cap}_{\Gamma_0}$.

(v) Prove that R_0 is positive and bounded below on $\widetilde{H}^{1/2}(\Gamma_0)$.

8.21 Show that the logarithmic capacity of a line segment equals one-quarter of its length.

[Hint: use Exercises 8.13 and 8.15.]

8.22 As in Theorem 7.10, the interior, mixed problem for the Laplacian leads to a 2×2 system of boundary integral equations of the form $A\psi = h$, where

$$A = \begin{bmatrix} S_{\mathrm{DD}} & -\frac{1}{2}T_{\mathrm{DN}} \\ \frac{1}{2}T_{\mathrm{ND}}^* & R_{\mathrm{NN}} \end{bmatrix}.$$

Show that if Ω^- is connected, and if $r > \mathrm{Cap}_{\Gamma_\mathrm{D}}$ in case $n = 2$, then $A : H \to H^*$ is positive and bounded below on the space

$$H = \widetilde{H}^{-1/2}(\Gamma_\mathrm{D}) \times \widetilde{H}^{1/2}(\Gamma_\mathrm{N}).$$

9

The Helmholtz Equation

Consider the scalar wave equation,

$$\frac{\partial^2 U}{\partial t^2} - c^2 \Delta U = 0. \tag{9.1}$$

We obtain a time-harmonic solution $U(x, t) = \mathrm{Re}[e^{-i\lambda t} u(x)]$ if the space-dependent part u satisfies the *Helmholtz equation*,

$$-\Delta u - k^2 u = 0, \tag{9.2}$$

with $k = \lambda/c$. In this setting, the *wave number k* is real, but in the theory that follows we shall allow the coefficient k^2 to be any non-zero complex number. It is convenient to assume, without loss of generality, that

$$0 \leq \arg k < \pi. \tag{9.3}$$

This chapter begins by showing how separation of variables leads to Bessel's equation, and by deriving a fundamental solution $G(x, y) = \mathrm{G}(x - y)$. (The presence of the lower-order term in the Helmholtz equation means that we cannot apply Theorem 6.8 to obtain G.) Next, we discuss the well-known Sommerfeld radiation condition, and proceed to establish an existence and uniqueness theorem for the exterior problem. The final section of the chapter derives an integral identity that connects the sesquilinear forms associated with the boundary operators S and R. The books of Colton and Kress [12], [13] give more detailed treatments of the Helmholtz equation, emphasising integral equations of the *second* kind.

The sesquilinear form associated with the Helmholtz operator $\mathcal{P} = -\Delta - k^2$ is

$$\Phi_\Omega(u, v) = \int_\Omega \mathrm{grad}\, \bar{u} \cdot \mathrm{grad}\, v \, dx - \int_\Omega \overline{k^2 u} \cdot v \, dx,$$

276

and since $\mathcal{P}^* = -\Delta - \bar{k}^2$,

$$\mathcal{B}_\nu u = \tilde{\mathcal{B}}_\nu u = \partial_\nu u = \frac{\partial u}{\partial \nu}.$$

Obviously, $\mathcal{P}$ is strongly elliptic. Also, $\mathcal{P}$ is self-adjoint if and only if k is real.

Separation of Variables

We begin our investigation of the Helmholtz equation (9.2) by seeking solutions of the form

$$u(x) = f(k\rho)\psi(\omega) \quad \text{where } x = \rho\omega \text{ and } \rho = |x|. \tag{9.4}$$

It will be convenient to introduce the *Beltrami operator*, $\Delta_{\mathbb{S}^{n-1}}$, a differential operator on the unit sphere defined by

$$\Delta_{\mathbb{S}^{n-1}}\psi = (\Delta\tilde{\psi})|_{\mathbb{S}^{n-1}} \quad \text{where} \quad \tilde{\psi}(x) = \psi(\omega) \text{ for } x \neq 0.$$

Thus, $\tilde{\psi}$ is the extension of ψ to a homogeneous function of degree 0.

Lemma 9.1 If u has the form (9.4), and if $z = k\rho$, then

$$\Delta u(x) = k^2\left(f''(z)\psi(\omega) + \frac{n-1}{z}f'(z)\psi(\omega) + \frac{1}{z^2}f(z)\Delta_{\mathbb{S}^{n-1}}\psi(\omega) \right).$$

Proof. Since $\partial\rho/\partial x_j = x_j/\rho$, we find that

$$\partial_j^2 u(x) = \left[\left(\frac{kx_j}{\rho}\right)^2 f''(z) + k\frac{\rho^2 - x_j^2}{\rho^3}f'(z) \right]\tilde{\psi}(x)$$

$$+ 2\frac{kx_j}{\rho}f'(z)\partial_j\tilde{\psi}(x) + f(z)\partial_j^2\tilde{\psi}(x).$$

Being homogeneous of degree 0, the function $\tilde{\psi}$ satisfies $\sum_{j=1}^n x_j\partial_j\tilde{\psi}(x) = 0$, so

$$\Delta u(x) = \left(k^2 f''(z) + k\frac{n-1}{\rho}f'(z) \right)\tilde{\psi}(x) + f(z)\Delta\tilde{\psi}(x).$$

Finally, because $\Delta\tilde{\psi}$ is homogeneous of degree -2,

$$\Delta\tilde{\psi}(x) = \rho^{-2}\Delta\tilde{\psi}(\omega) = \frac{k^2}{z^2}\Delta_{\mathbb{S}^{n-1}}\psi(\omega),$$

giving the desired formula for $\Delta u(x)$. $\qquad\square$

We now restrict our attention to the case when ψ is a surface spherical harmonic, because ψ is then an eigenfunction of the Beltrami operator.

Lemma 9.2 If $\psi \in \mathcal{H}_m(\mathbb{S}^{n-1})$, then $-\Delta_{\mathbb{S}^{n-1}} \psi = m(m + n - 2)\psi$.

Proof. Let $\psi = u|_{\mathbb{S}^n}$ for a solid spherical harmonic $u \in \mathcal{H}_m(\mathbb{R}^n)$. Since $u(x) = \rho^m \psi(\omega)$, we see that u has the form (9.4) with $k = 1$ and $f(z) = z^m$. Applying Lemma 9.1, it follows that $\Delta u(x) = \rho^{m-2}\big[m(m + n - 2)\psi(\omega) + \Delta_{\mathbb{S}^{n-1}} \psi(\omega)\big]$, which is identically zero. $\square$

Lemmas 9.1 and 9.2 show that when $\psi \in \mathcal{H}_m(\mathbb{S}^{n-1})$, the function (9.4) is a solution of the Helmholtz equation (9.2) if and only if f is a solution of

$$f''(z) + \frac{n-1}{z} f'(z) + \left(1 - \frac{m(m + n - 2)}{z^2}\right) f(z) = 0. \qquad (9.5)$$

This ordinary differential equation can be transformed, by putting

$$g(z) = z^{(n/2)-1} f(z),$$

into Bessel's equation of order μ,

$$g''(z) + \frac{1}{z} g'(z) + \left(1 - \frac{\mu^2}{z^2}\right) g(z) = 0, \qquad (9.6)$$

with

$$\mu = m + \frac{n}{2} - 1.$$

Let J_μ denote the usual *Bessel function of the first kind* of order μ, which has the series representation

$$J_\mu(z) = \sum_{p=0}^{\infty} \frac{(-1)^p (z/2)^{\mu+2p}}{p!\,\Gamma(\mu + p + 1)} \quad \text{for } |\arg z| < \pi. \qquad (9.7)$$

The *Bessel function of the second kind*, Y_μ, is defined by

$$Y_\mu(z) = \frac{J_\mu(z) \cos \pi\mu - J_{-\mu}(z)}{\sin \pi\mu} \quad \text{for } |\arg z| < \pi, \qquad \text{if } \mu \notin \mathbb{Z},$$

and by

$$Y_m(z) = \lim_{\mu \to m} Y_\mu(z) \quad \text{for } |\arg z| < \pi, \qquad \text{if } m \in \mathbb{Z}.$$

The functions J_μ and Y_μ form a basis for the solution space of Bessel's equation (9.6), so the functions

$$j_m(z) = j_m(n, z) = \sqrt{\frac{\pi}{2}} \frac{J_{m+(n/2)-1}(z)}{z^{(n/2)-1}}$$

and

$$y_m(z) = y_m(n, z) = \sqrt{\frac{\pi}{2}} \frac{Y_{m+(n/2)-1}(z)}{z^{(n/2)-1}}$$

form a basis for the solution space of the original differential equation (9.5). If $n = 3$, then j_m and y_m are known as *spherical Bessel functions*.

Lemma 9.3 Let u have the form (9.4), where $\psi \in \mathcal{H}_m(\mathbb{S}^{n-1})$.

(i) If $f(z) = j_m(n, z)$, then $-\Delta u - k^2 u = 0$ on $\mathbb{R}^n$.
(ii) If $f(z) = y_m(n, z)$, then $-\Delta u - k^2 u = 0$ on $\mathbb{R}^n \setminus \{0\}$.

Proof. The preceding argument shows that in each case, u is a solution of the Helmholtz equation on $\mathbb{R}^n \setminus \{0\}$.

The series (9.7) shows that $J_\mu(z)/z^\mu$ is an entire function of z^2, and therefore the same is true for $j_m(n, z)/z^m$. Since $\rho^m \psi(\omega)$ is a homogeneous polynomial of degree m in x, it follows that if $f = j_m$, then u is C^∞ on $\mathbb{R}^n$, proving (i). □

As $z \to 0$, the singular behaviour of the non-negative integer- and half-integer-order Bessel functions of the second kind is given by

$$Y_0(z) = \frac{2}{\pi}[-\Gamma'(1) + \log(z/2)][1 + O(z^2)] + O(z^2),$$

and, for any integer $m \geq 1$,

$$Y_\mu(z) = \frac{-\Gamma(\mu)}{\pi(z/2)^\mu}[1 + O(z^2)]$$

$$+ \begin{cases} \dfrac{2}{\pi} \dfrac{(z/2)^m}{m!} \log(z/2) [1 + O(z^2)] & \text{if } \mu = m, \\ 0 & \text{if } \mu = m - \frac{1}{2}, \end{cases}$$

where each $O(z^2)$ term is an entire function of z^2; see [1, p. 360]. We are therefore able to show the following.

Theorem 9.4 For any constant $a \in \mathbb{C}$, a fundamental solution $G(x, y) = G(x - y)$ for the Helmholtz operator $-\Delta - k^2$ on $\mathbb{R}^n$ is given by

$$G(x) = \frac{k^{n-2}}{2(2\pi)^{(n-1)/2}}\big[-y_0(n, k|x|) + a j_0(n, k|x|)\big]. \tag{9.8}$$

If $n = 3$, then this expression reduces to

$$G(x) = \frac{\cos(k|x|) + a\sin(k|x|)}{4\pi|x|}. \tag{9.9}$$

Proof. By (8.1), if $n \geq 3$ and $\mu = (n/2) - 1$, then

$$\frac{\Gamma(\mu)}{\pi 2^{-\mu}} = \frac{\Gamma(1+\mu)/\mu}{\pi 2^{-\mu}} = \frac{\Gamma(n/2)}{(n-2)\pi 2^{-n/2}}$$

$$= \frac{2\pi^{n/2}/\Upsilon_n}{(n-2)\pi 2^{-n/2}} = \frac{2(2\pi)^{(n-1)/2}}{(n-2)\Upsilon_n}\sqrt{\frac{2}{\pi}}.$$

Thus, if $G_\Delta(x - y)$ denotes the fundamental solution for the Laplacian given in Theorem 8.1, then for $n \geq 2$,

$$-\frac{k^{n-2}}{2(2\pi)^{(n-1)/2}}y_0(n, k|x|) = G_\Delta(x)[1 + f(x)] + g(x) \quad \text{as } |x| \to 0,$$

where $f(x) = O(|x|^2)$ and

$$g(x) = \begin{cases} \text{const} + O(|x|^2) & \text{for } n = 2, \\ 0 & \text{for odd } n > 2, \\ \text{const} \times \log(k|x|/2)\,[1 + O(|x|^2)] & \text{for even } n > 2; \end{cases}$$

again, the $O(|x|^2)$ terms are analytic functions of $|x|^2$. To prove that G is a fundamental solution, we let $w = \delta + (\Delta + k^2)G$ and show that $w = 0$ on $\mathbb{R}^n$. In fact, since $\mathcal{H}_0(\mathbb{S}^{n-1})$ consists of the constant functions, Theorem 9.3 implies that w is independent of the coefficient a, and that $w = 0$ on $\mathbb{R}^n \setminus \{0\}$. Thus, it suffices to show that w is locally integrable on $\mathbb{R}^n$. We know already that $-\Delta G_\Delta = \delta$, so $w = k^2 G_\Delta + (\Delta + k^2)(G_\Delta f + g)$, therefore, as $|x| \to 0$,

$$w(x) = \begin{cases} O(\log|x|) & \text{if } n = 2, \\ O(|x|^{2-n}) & \text{if } n \geq 3, \end{cases}$$

and the result for a general n follows. In the particular case $n = 3$, we can use Exercise 9.1. $\qquad\qquad\square$

The Sommerfeld Radiation Condition

When dealing with exterior problems for the Helmholtz equation, it is convenient to introduce the *Hankel functions* of the first and second kind,

$$H_\mu^{(1)}(z) = J_\mu(z) + iY_\mu(z) \quad \text{and} \quad H_\mu^{(2)}(z) = J_\mu(z) - iY_\mu(z),$$

which form an alternative basis for the solution space of Bessel's equation. We also put

$$h_m^{(1)}(z) = j_m(z) + iy_m(z) \quad \text{and} \quad h_m^{(2)}(z) = j_m(z) - iy_m(z),$$

writing $h^{(1)}(n, z)$ and $h^{(2)}(n, z)$ when it is desirable to indicate the dimension n explicitly. If $n = 3$, then and $h_m^{(1)}$ and $h_m^{(2)}$ are known as *spherical Hankel functions*.

From the standard asymptotic expansions for the Bessel functions – see Abramowitz and Stegun [1, p. 364] or Gradshteyn and Ryzhik [29, pp. 961–962] – we find that

$$
\begin{aligned}
h_m^{(1)}(z) &= \frac{1}{z^{(n-1)/2}} \left[e^{i[z-(2m+n-1)\pi/4]} + O\left(\frac{1}{z}\right) \right], \\
h_m^{(2)}(z) &= \frac{1}{z^{(n-1)/2}} \left[e^{-i[z-(2m+n-1)\pi/4]} + O\left(\frac{1}{z}\right) \right],
\end{aligned}
\tag{9.10}
$$

as $z \to \infty$ with $-\pi < \arg z < \pi$. By Lemma 9.3, the function

$$u(x) = h_m^{(1)}(k\rho)\psi(\omega) \quad \text{for } x = \rho\omega, \ \psi \in \mathcal{H}_m(\mathbb{S}^{n-1}) \tag{9.11}$$

is a solution of the Helmholtz equation, and the corresponding time-harmonic solution of the wave equation (9.1) satisfies

$$
\begin{aligned}
&\mathrm{Re}\left[e^{-i\lambda t} h_m^{(1)}(k\rho)\psi(\omega) \right] \\
&= \rho^{-(n-1)/2} \cos[k\rho - \lambda t - (2m + n - 1)\pi/4] + O\left(\rho^{-(n+1)/2}\right)
\end{aligned}
$$

as $\rho \to \infty$. Physically, this solution is an *outgoing* or *radiating* wave; if $h_m^{(2)}$ is used in place of $h_m^{(1)}$, then we obtain an *incoming* wave.

Definition 9.5 *Write* $x = \rho\omega$ *with* $\rho = |x|$ *and* $\omega \in \mathbb{S}^{n-1}$, *and let* $u(x)$ *be a solution of the Helmholtz equation for* ρ *sufficiently large. We call* u *a radiating solution if it satisfies the Sommerfeld radiation condition*

$$\lim_{\rho \to \infty} \rho^{(n-1)/2} \left(\frac{\partial u}{\partial \rho} - iku \right) = 0,$$

uniformly in ω.

The derivatives of $h_m^{(1)}$ and $h_m^{(2)}$ have the asymptotic behaviour

$$\frac{dh_m^{(1)}}{dz} = \frac{i}{z^{(n-1)/2}}\left[e^{i[z-(2m+n-1)\pi/4)]} + O\left(\frac{1}{z}\right)\right],$$

$$\frac{dh_m^{(2)}}{dz} = \frac{-i}{z^{(n-1)/2}}\left[e^{-i[z-(2m+n-1)\pi/4)]} + O\left(\frac{1}{z}\right)\right], \tag{9.12}$$

as $z \to \infty$ with $-\pi < \arg z < \pi$. Hence, u in (9.11) satisfies

$$\rho^{(n-1)/2}\left(\frac{\partial u}{\partial \rho} - iku\right) = O\left(\frac{1}{\rho}\right), \tag{9.13}$$

and is therefore a radiating solution. Also, by taking $a = i$ in (9.8), we obtain a radiating fundamental solution,

$$G(x) = \frac{k^{n-2}}{2(2\pi)^{(n-1)/2}} ih_0^{(1)}(n, k|x|), \tag{9.14}$$

and, as a special case,

$$G(x) = \frac{e^{ik|x|}}{4\pi|x|} \quad \text{when } n = 3. \tag{9.15}$$

Recalling our assumption (9.3), we note that if $\operatorname{Im} k > 0$ then $G(x)$ has exponential decay at infinity.

The following theorem reveals the connection between Definition 9.5 and our earlier treatment of radiation conditions for general elliptic equations; recall the definition of the operator $\mathcal{M}$ given in Lemma 7.11.

Theorem 9.6 *Let u be a solution of the exterior Helmholtz equation*

$$-\Delta u - k^2 u = 0 \quad \text{on } \Omega^+,$$

and suppose that $\mathcal{M}$ is defined using the radiating fundamental solution (9.14).

(i) If $\mathcal{M}u = 0$, then u satisfies (9.13).

(ii) If u satisfies

$$\lim_{\rho \to \infty} \int_{\partial B_\rho} \left|\frac{\partial u}{\partial \rho} - iku\right|^2 d\sigma = 0, \tag{9.16}$$

then $\mathcal{M}u = 0$.

Hence, the requirement $\mathcal{M}u = 0$ is equivalent to the Sommerfeld radiation condition.

Proof. By enlarging Ω^- if necessary, we can assume that u is C^∞ on $\overline{\Omega^+}$. By Theorem 7.12, if $\mathcal{M}u = 0$, then $u = \mathrm{DL}\, \gamma^+u - \mathrm{SL}\, \partial_\nu^+u$ on Ω^+, so part (i) follows from Exercise 9.4.

Assume now that u satisfies (9.16). We claim that

$$\int_{\partial B_\rho} |u|^2\, d\sigma = O(1) \quad \text{as } \rho \to \infty. \tag{9.17}$$

In fact,

$$\left| \frac{\partial u}{\partial \rho} - iku \right|^2 = \left| \frac{\partial u}{\partial \rho} \right|^2 + |k|^2|u|^2 + 2\,\mathrm{Im}\left(k\frac{\partial \bar{u}}{\partial \nu}u \right),$$

and by applying the first Green identity over the bounded domain $\Omega_\rho^+ = \Omega^+ \cap B_\rho$, we see that

$$\int_{\Omega_\rho^+} \left(|\mathrm{grad}\,u|^2 - \bar{k}^2|u|^2 \right) dx = \int_{\partial B_\rho} \frac{\partial \bar{u}}{\partial \nu}u\, d\sigma - \int_\Gamma \frac{\partial \bar{u}}{\partial \nu}u\, d\sigma. \tag{9.18}$$

(In the first integral on the right, ν is the *out*ward unit normal to Ω_ρ^+, but in the second integral ν is the *in*ward unit normal.) Multiplying (9.18) by k, and taking the imaginary part, we obtain

$$-\int_\Gamma \mathrm{Im}\left(k\frac{\partial \bar{u}}{\partial \nu}u \right) d\sigma = \mathrm{Im}(k) \int_{\Omega_\rho^+} \left(|\mathrm{grad}\,u|^2 + |k|^2|u|^2 \right) dx$$

$$- \int_{\partial B_\rho} \mathrm{Im}\left(k\frac{\partial \bar{u}}{\partial \nu}u \right) d\sigma$$

and so

$$\mathrm{Im}(k) \int_{\Omega_\rho^+} \left(|\mathrm{grad}\,u|^2 + |k|^2|u|^2 \right) dx + \frac{1}{2} \int_{\partial B_\rho} \left(\left| \frac{\partial u}{\partial \rho} \right|^2 + |k|^2|u|^2 \right) d\sigma$$

$$\to -\int_\Gamma \mathrm{Im}\left(k\frac{\partial \bar{u}}{\partial \nu}u \right) d\sigma \quad \text{as } \rho \to \infty. \tag{9.19}$$

The claim (9.17) now follows from our assumptions that $k \neq 0$ and $\mathrm{Im}\,k \geq 0$. To complete the proof of (ii), we simply write

$$\mathcal{M}u(x) = \int_{\partial B_\rho} G(x, y)\left[\partial_\nu u(y) - iku(y) \right] d\sigma_y$$

$$- \int_{\partial B_\rho} \left[\partial_{\nu,y} G(x, y) - ikG(x, y) \right]u(y)\, d\sigma_y$$

for ρ sufficiently large, and then apply the Cauchy–Schwarz inequality, noting that the radiating fundamental solution $G(x, y) = \mathsf{G}(x - y)$ satisfies

$$\int_{\partial B_\rho} |G(x, y)|^2 \, d\sigma_y \le C \int_{\partial B_\rho} |x - y|^{-(n-1)} \, d\sigma_y \le C$$

and

$$\int_{\partial B_\rho} |\partial_{\nu,y} G(x, y) - ikG(x, y)|^2 \, d\sigma_y \le C \int_{\partial B_\rho} |x - y|^{-(n+1)} \, d\sigma_y \le C\rho^{-2}.$$

$$\square$$

We are now in a position to give an expansion of the fundamental solution in spherical harmonics or Legendre polynomials.

Theorem 9.7 *For $m \ge 0$, let $\{\psi_{mp} : 1 \le p \le N(n, m)\}$ be an orthonormal basis for $\mathcal{H}_m(\mathbb{S}^{n-1})$. The radiating fundamental solution $G(x, y) = \mathsf{G}(x - y)$ given by (9.14) has the expansion*

$$G(x, y) = ik^{n-2} \sum_{m=0}^{\infty} \sum_{p=1}^{N(n,m)} \mathsf{h}_m^{(1)}(k|x|) \psi_{mp}(x/|x|) \mathsf{j}_m(k|y|) \overline{\psi_{mp}(y/|y|)}$$

$$= \frac{ik^{n-2}}{\Upsilon_n} \sum_{m=0}^{\infty} N(n, m) \mathsf{h}_m^{(1)}(k|x|) \mathsf{j}_m(k|y|) P_m(n, \cos\theta)$$

$$for \ |x| > |y| > 0,$$

where

$$\cos\theta = \frac{x \cdot y}{|x||y|}.$$

Proof. Let $\psi \in \mathcal{H}_m(\mathbb{S}^{n-1})$, and put

$$u(y) = \mathsf{j}_m(k\rho)\psi(\omega) \quad \text{and} \quad v(y) = \mathsf{h}_m^{(1)}(k\rho)\psi(\omega) \qquad \text{for } y = \rho\omega.$$

By Lemma 9.3, we have $-\Delta u - k^2 u = 0$ on $\mathbb{R}^n$, and $-\Delta v - k^2 v = 0$ on $\mathbb{R}^n \setminus \{0\}$. Hence, $\mathcal{M}u = u$ by Exercise 7.5, and therefore by Theorem 7.12,

$$\int_{\partial B_\rho} \left[\partial_{\nu,y} G(x, y) u(y) - G(x, y) \partial_\nu u(y) \right] d\sigma_y = 0 \quad \text{for } |x| > \rho, \qquad (9.20)$$

whereas $\mathcal{M}v = 0$ by Theorem 9.6, so

$$\int_{\partial B_\rho} \left[\partial_{\nu,y} G(x, y) v(y) - G(x, y) \partial_\nu v(y) \right] d\sigma_y = v(x) \quad \text{for } |x| > \rho. \qquad (9.21)$$

It follows from Exercise 9.3 that

$$j_m(z)\frac{dh_m^{(1)}}{dz} - \frac{dj_m}{dz}h_m^{(1)} = \frac{i}{z^{n-1}},$$

so the combination $h_m^{(1)}(k\rho) \times (9.20) - j_m(k\rho) \times (9.21)$ gives the equation

$$\int_{\partial B_\rho} G(x,y)\frac{ik}{(k\rho)^{n-1}}\psi(y/\rho)\,d\sigma_y = -j_m(k\rho)h_m^{(1)}(k|x|)\psi(x/|x|) \quad \text{for } |x| > \rho,$$

or equivalently,

$$\int_{|\omega|=1} G(x,\rho\omega)\psi(\omega)\,d\omega = ik^{n-2}j_m(k\rho)h_m^{(1)}(k|x|)\psi(x/|x|) \quad \text{for } |x| > \rho.$$

By Theorem 8.17, we have

$$G(x,y) = G(x,\rho\omega) = \sum_{m=0}^{\infty}\sum_{p=1}^{N(n,m)}\int_{\mathbb{S}^{n-1}} G(x,\rho\eta)\psi_{mp}(\eta)\,d\eta\,\overline{\psi_{mp}(\omega)},$$

implying the first expansion, and the second then follows by Corollary C.2.

However, so far we have proved only convergence in the sense of $L_2(\mathbb{S}^{n-1})$. It is easy to see from the definition of the Legendre polynomial $P_m(n,t)$ immediately following Lemma 8.6 that

$$|P_m(n,t)| \le 1 \quad \text{for } -1 \le t \le 1,$$

so the mth term in the second expansion is bounded by $N(n,m)|h_m^{(1)}(k|x|)j_m(k|y|)|$. Using the standard large-argument approximations [1, p. 365],

$$J_\mu(z) = \frac{1}{\sqrt{2\pi\mu}}\left(\frac{ez}{2\mu}\right)^\mu[1+o(1)] \quad \text{and}$$

$$Y_\mu(z) = -\sqrt{\frac{2}{\pi\mu}}\left(\frac{2\mu}{ez}\right)^\mu[1+o(1)] \quad \text{as } \mu \to \infty$$

(with fixed z), we find after some calculation that

$$h_m^{(1)}(k|x|)j_m(k|y|) = iy_m(k|x|)j_m(k|y|)[1+o(1)]$$
$$= \frac{-i}{2m+n-2}\frac{(k|y|)^m}{(k|x|)^{m+n-2}}[1+o(1)] \quad \text{as } m \to \infty.$$

Since (8.14) shows that $N(n,m) = O(m^{n-2})$ as $m \to \infty$ with n fixed, the expansions converge pointwise for $|x| > |y|$. $\qquad\square$

Uniqueness and Existence of Solutions

Theorems 4.12 and 8.2 imply that for each Lipschitz dissection $\Gamma = \Gamma_D \cup \Pi \cup \Gamma_N$ there exist *interior eigenvalues* $0 \leq \lambda_1 \leq \lambda_2 \leq \cdots$, and corresponding *interior eigenfunctions* $\phi_1, \phi_2, \ldots$ in $H^1(\Omega^-)$, satisfying

$$-\Delta\phi_j = \lambda_j\phi_j \quad \text{on } \Omega^-,$$

$$\gamma^-\phi_j = 0 \quad \text{on } \Gamma_D,$$

$$\partial_\nu^-\phi_j = 0 \quad \text{on } \Gamma_N,$$

with ϕ_j not identically zero. Therefore, by Theorem 4.10, the interior problem

$$-\Delta u - k^2 u = f \quad \text{on } \Omega^-,$$

$$\gamma^- u = g_D \quad \text{on } \Gamma_D, \qquad\qquad (9.22)$$

$$\partial_\nu^- u = g_N \quad \text{on } \Gamma_N,$$

has a unique solution $u \in H^1(\Omega^-)$ for each $f \in \widetilde{H}^{-1}(\Omega^-)$, $g_D \in H^{1/2}(\Gamma_D)$ and $g_N \in H^{-1/2}(\Gamma_N)$ if and only if k^2 is not an interior eigenvalue. Otherwise, a solution exists if and only if the data satisfy

$$(\phi_j, f)_{\Omega^-} + (\gamma^-\phi_j, g_N)_{\Gamma_N} = (\partial_\nu^-\phi_j, g_D)_{\Gamma_D} \quad \text{for all } j \text{ such that } \lambda_j = k^2.$$

Notice that since $\lambda_j \geq 0$, if $\operatorname{Im} k > 0$ then k^2 cannot be an interior eigenvalue, and so (9.22) is uniquely solvable.

The following result of Rellich [86] will help us to prove uniqueness for exterior problems.

Lemma 9.8 *For any real wave number $k > 0$, if*

$$-\Delta u - k^2 u = 0 \quad \text{on } \mathbb{R}^n \setminus \overline{B_{\rho_0}}$$

and if

$$\lim_{\rho\to\infty} \int_{|x|=\rho} |u(x)|^2 \, d\sigma_x = 0, \qquad\qquad (9.23)$$

then $u = 0$ on $\mathbb{R}^n \setminus \overline{B_{\rho_0}}$.

Proof. Let $\{\psi_{mp} : 1 \leq p \leq N(n, m)\}$ be an orthonormal basis for $\mathcal{H}_m(\mathbb{S}^{n-1})$. Theorem 6.4 shows that $u(x)$ is C^∞ for $|x| > \rho_0$, so

$$u(x) = \sum_{m=0}^{\infty} \sum_{p=1}^{N(n,m)} f_{mp}(k\rho)\psi_{mp}(\omega) \quad \text{for } x = \rho\omega, \rho > \rho_0 \text{ and } \omega \in \mathbb{S}^{n-1},$$

where

$$f_{mp}(t) = \int_{\mathbb{S}^{n-1}} u(k^{-1}t\omega)\overline{\psi_{mp}(\omega)}\, d\omega \quad \text{for } t > k\rho_0.$$

The sum converges in $L_2(\mathbb{S}^{n-1})$, and

$$\int_{|x|=\rho} |u(x)|^2\, d\sigma_x = \int_{\mathbb{S}^{n-1}} |u(\rho\omega)|^2 \rho^{n-1}\, d\omega = \sum_{m=0}^{\infty} \sum_{p=1}^{N(n,m)} \rho^{n-1} |f_{mp}(k\rho)|^2.$$

Since u satisfies the Helmholtz equation, the function f_{mp} is a solution of (9.5), and hence

$$f_{mp}(t) = a_{mp}^{(1)} h_m^{(1)}(t) + a_{mp}^{(2)} h_m^{(2)}(t)$$

for some constants $a_{mp}^{(1)}$ and $a_{mp}^{(2)}$. By (9.10),

$$\rho^{n-1} |f_{mp}(k\rho)|^2 = \left| a_{mp}^{(1)} e^{2i[k\rho - (2m+n-1)\pi/4]} + a_{mp}^{(2)} \right|^2 + O(\rho^{-1}),$$

so the assumption (9.23) implies that $a_{mp}^{(1)} e^{i\theta} + a_{mp}^{(2)} = 0$ for all real θ. Therefore, $a_{mp}^{(1)} = a_{mp}^{(2)} = 0$, which means that f_{mp} is identically zero. $\qquad\square$

Lemma 9.9 Suppose that $u \in H^1_{\text{loc}}(\Omega^+)$ is a radiating solution of the Helmholtz equation, i.e., suppose

$$-\Delta u - k^2 u = 0 \quad \text{on } \Omega^+$$

and

$$\lim_{\rho \to \infty} \rho^{(n-1)/2} \left(\frac{\partial u}{\partial \rho} - iku \right) = 0. \tag{9.24}$$

If

$$\text{Im}\left(k \int_\Gamma (\partial_\nu^+ \bar{u}) u\, d\sigma \right) \geq 0,$$

then $u = 0$ on Ω^+.

Proof. If $\text{Im}\, k > 0$, then we see from (9.19) that $\int_{\Omega_\rho^+} |u(x)|^2\, dx \to 0$ as $\rho \to \infty$, and thus u must be identically zero on Ω^+. If $\text{Im}\, k = 0$, then we can apply Lemma 9.8 because (9.19) shows that $\int_{|x|=\rho} |u(x)|^2\, d\sigma_x \to 0$ as $\rho \to \infty$, and $k > 0$ by our assumption (9.3). $\qquad\square$

The desired uniqueness theorem follows at once.

Theorem 9.10 If $u \in H^1_{loc}(\Omega^+)$ is a solution of the homogeneous exterior mixed problem

$$-\Delta u - k^2 u = 0 \quad on \ \Omega^+,$$

$$\gamma^+ u = 0 \quad on \ \Gamma_D,$$

$$\partial_\nu^+ u = 0 \quad on \ \Gamma_N,$$

and if u satisfies the Sommerfeld radiation condition (9.24), then $u = 0$ on Ω^+.

It is now possible to deduce *existence* results for exterior problems using boundary integral equations. For brevity, we treat only the pure Dirichlet problem; but see also Exercise 9.5.

Theorem 9.11 If $f \in \widetilde{H}^{-1}(\Omega^+)$ has compact support, and if $g \in H^{1/2}(\Gamma)$, then the exterior Dirichlet problem for the Helmholtz equation,

$$-\Delta u - k^2 u = f \quad on \ \Omega^+,$$

$$\gamma^+ u = g \quad on \ \Gamma,$$

has a unique radiating solution $u \in H^1_{loc}(\Omega^+)$.

Proof. We have already proved uniqueness. By Theorems 7.15 and 9.6, a solution exists if and only if there exists $\psi \in H^{-1/2}(\Gamma)$ satisfying

$$S\psi = \gamma \mathcal{G} f - \tfrac{1}{2}(g - Tg) \quad on \ \Gamma. \tag{9.25}$$

(In the usual way, to define $\mathcal{G} f$ we view f as a distribution on $\mathbb{R}^n$ with supp $f \subseteq \overline{\Omega^+}$.) Theorem 7.6 shows that the Fredholm alternative is valid for this boundary integral equation, and by Theorem 7.5 the set ker S^* consists of all functions of the form $\partial_\nu^- v$ where $v \in H^1(\Omega^-)$ is a solution of the interior homogeneous adjoint problem

$$-\Delta v - \bar{k}^2 v = 0 \quad on \ \Omega^-,$$

$$\gamma^- v = 0 \quad on \ \Gamma.$$

Using (7.5) and the second Green identity, we find that for all such v,

$$\begin{aligned}
\left(\partial_\nu^- v, \gamma \mathcal{G} f - \tfrac{1}{2}(g - Tg)\right)_\Gamma &= \left(\partial_\nu^- v, \gamma^-(\mathcal{G} f + \mathrm{DL}\, g)\right)_\Gamma \\
&= \left(\gamma^- v, \partial_\nu^-(\mathcal{G} f + \mathrm{DL}\, g)\right)_\Gamma \\
&\quad - \left((-\Delta - \bar{k}^2)v, \mathcal{G} f + \mathrm{DL}\, g\right)_{\Omega^-} \\
&\quad + \left(v, (-\Delta - k^2)(\mathcal{G} f + \mathrm{DL}\, g)\right)_{\Omega^-}.
\end{aligned}$$

Each of the three terms on the right vanishes, because $\gamma^- v = 0$ on Γ, $(-\Delta - \bar{k}^2)v = 0$ on Ω^-, and $(-\Delta - k^2)(\mathcal{G} f + \mathrm{DL}\, g) = f = 0$ on Ω^-. Thus, ψ exists, as required. $\qquad \Box$

We remark that the solution ψ of the boundary integral equation (9.25) is unique if and only if k^2 is *not* an interior Dirichlet eigenvalue for $-\Delta$.

A Boundary Integral Identity

In this section, we derive a remarkable identity connecting the hypersingular boundary integral operator R with the weakly singular operator S, associated with the Helmholtz equation in $\mathbb{R}^3$. This identity, together with analogous ones for other elliptic equations, was introduced by Nedelec [75] as a way of avoiding the evaluation of hypersingular integrals in Galerkin boundary element methods involving R. It will be convenient in what follows to work with the bilinear form $\langle \cdot, \cdot \rangle$ instead of the sesquilinear form $(\cdot, \cdot)$.

For any scalar test function $w \in \mathcal{D}(\mathbb{R}^3)$ and any vector-valued test function $W \in \mathcal{D}(\mathbb{R}^3)^3$, we have the identities

$$\operatorname{div}(u W) = (\operatorname{grad} u) \cdot W + u \operatorname{div} W,$$

$$\operatorname{div}(w F) = F \cdot \operatorname{grad} w + (\operatorname{div} F)w,$$

$$\operatorname{div}(F \times W) = (\operatorname{curl} F) \cdot W - F \cdot \operatorname{curl} W,$$

if, say, $u : \mathbb{R}^3 \to \mathbb{C}$ and $F : \mathbb{R}^3 \to \mathbb{C}^3$ are C^1. Consequently, the divergence theorem implies that

$$\langle \operatorname{grad} u, W \rangle = - \int_{\mathbb{R}^3} u \operatorname{div} W \, dx,$$

$$\langle \operatorname{div} F, w \rangle = - \int_{\mathbb{R}^3} F \cdot \operatorname{grad} w \, dx,$$

$$\langle \operatorname{curl} F, W \rangle = \int_{\mathbb{R}^3} F \cdot \operatorname{curl} W \, dx,$$

so for distributions $u \in \mathcal{D}^*(\mathbb{R}^3)$ and $F \in \mathcal{D}^*(\mathbb{R}^3)^3$,

$$\langle \operatorname{grad} u, W \rangle = -\langle u, \operatorname{div} W \rangle,$$

$$\langle \operatorname{div} F, w \rangle = -\langle F, \operatorname{grad} w \rangle,$$

$$\langle \operatorname{curl} F, W \rangle = \langle F, \operatorname{curl} W \rangle.$$

Given a scalar test function $\phi \in \mathcal{D}(\mathbb{R}^3)$, we shall write $\phi_\Gamma = \phi|_\Gamma$, and define

γ^{t} and ∂_ν^{t} by

$$\langle \gamma^{\text{t}}\phi_\Gamma, \psi \rangle = \langle \phi_\Gamma, \psi_\Gamma \rangle_\Gamma \text{ and } \langle \partial_\nu^{\text{t}}\phi_\Gamma, \psi \rangle = \langle \phi_\Gamma, \partial_\nu \psi \rangle_\Gamma \qquad \text{for } \psi \in \mathcal{D}(\mathbb{R}^3);$$

cf. (6.14). The following identities hold in the sense of distributions.

Lemma 9.12 If ϕ, $\psi \in \mathcal{D}(\mathbb{R}^3)$, *then*

$$\operatorname{div} \gamma^{\text{t}}(\phi_\Gamma \nu) = -\partial_\nu^{\text{t}}\phi_\Gamma \text{ and } \operatorname{curl} \gamma^{\text{t}}(\phi_\Gamma \nu) = -\gamma^{\text{t}}(\nu \times \gamma \operatorname{grad}\phi) \qquad \text{on } \mathbb{R}^3.$$

Proof. We have

$$\langle \operatorname{div} \gamma^{\text{t}}(\phi_\Gamma \nu), w \rangle = -\langle \gamma^{\text{t}}(\phi_\Gamma \nu), \operatorname{grad} w \rangle = -\langle \phi_\Gamma \nu, \gamma \operatorname{grad} w \rangle_\Gamma$$
$$= -\langle \phi_\Gamma, \partial_\nu w \rangle_\Gamma = \langle -\partial_\nu^{\text{t}}\phi_\Gamma, w \rangle,$$

which proves the first part of the lemma. Next,

$$\langle \operatorname{curl} \gamma^{\text{t}}(\phi_\Gamma \nu), W \rangle = \langle \gamma^{\text{t}}(\phi_\Gamma \nu), \operatorname{curl} W \rangle = \langle \phi_\Gamma \nu, \gamma \operatorname{curl} W \rangle_\Gamma$$
$$= \int_\Gamma \nu \cdot \gamma(\phi \operatorname{curl} W) \, d\sigma,$$

so by the divergence theorem,

$$\langle \operatorname{curl} \gamma^{\text{t}}(\phi_\Gamma \nu), W \rangle = \mp \int_{\Omega^\pm} \operatorname{div}(\phi \operatorname{curl} W) \, dx.$$

From the identities

$$\operatorname{div}(\phi \operatorname{curl} W) = \operatorname{grad}\phi \cdot \operatorname{curl} W + \phi \operatorname{div} \operatorname{curl} W = \operatorname{grad}\phi \cdot \operatorname{curl} W + 0$$

and

$$\operatorname{div}(\operatorname{grad}\phi \times W) = (\operatorname{curl} \operatorname{grad}\phi) \cdot W - \operatorname{grad}\phi \cdot \operatorname{curl} W = 0 - \operatorname{grad}\phi \cdot \operatorname{curl} W,$$

we have $\operatorname{div}(\phi \operatorname{curl} W) = -\operatorname{div}(\operatorname{grad}\phi \times W)$. Thus,

$$\langle \operatorname{curl} \gamma^{\text{t}}(\phi_\Gamma \nu), W \rangle = \pm \int_{\Omega^\pm} \operatorname{div}(\operatorname{grad}\phi \times W) \, dx = -\int_\Gamma \nu \cdot (\operatorname{grad}\phi \times W) \, d\sigma$$
$$= -\int_\Gamma (\nu \times \operatorname{grad}\phi) \cdot W \, d\sigma = \langle -\nu \times \gamma \operatorname{grad}\phi, \gamma W \rangle_\Gamma,$$

which proves the second part of the lemma. $\qquad\square$

Now fix a $\phi \in \mathcal{D}(\mathbb{R}^3)$, and define

$$u = \mathrm{DL}\,\phi_\Gamma \quad \text{on } \mathbb{R}^3 \quad \text{and} \quad F^\pm = \mathrm{grad}\,u^\pm \quad \text{on } \Omega^\pm,$$

using, in the double-layer potential, any fundamental solution of the form (9.9). We also construct a locally integrable vector field $F : \mathbb{R}^3 \to \mathbb{C}^3$ by putting

$$F = \begin{cases} F^+ & \text{on } \Omega^+, \\ F^- & \text{on } \Omega^-. \end{cases}$$

In this way, the support of the distribution $F - \mathrm{grad}\,u$ is a subset of Γ. Two further technical lemmas are required.

Lemma 9.13 As distributions on $\mathbb{R}^3$,

$$u = -\mathrm{div}\,\mathrm{SL}(\phi_\Gamma \nu), \quad \mathrm{grad}\,u = F + \gamma^{\mathrm{t}}(\phi_\Gamma \nu),$$

$$\mathrm{div}\,F = -k^2 u, \quad\quad\quad \mathrm{curl}\,F = \gamma^{\mathrm{t}}(\nu \times \gamma\,\mathrm{grad}\,\phi).$$

Proof. Using Lemma 9.12, we find that since ∂_j commutes with the convolution operator $\mathcal{G}$,

$$\begin{aligned} u = \mathrm{DL}\,\phi_\Gamma &= \mathcal{G}\big(\partial_\nu^{\mathrm{t}}\phi_\Gamma\big) = \mathcal{G}\big(-\mathrm{div}\,\gamma^{\mathrm{t}}(\phi_\Gamma \nu)\big) \\ &= -\mathrm{div}\,\mathcal{G}\big(\gamma^{\mathrm{t}}(\phi_\Gamma \nu)\big) = -\mathrm{div}\,\mathrm{SL}(\phi_\Gamma \nu). \end{aligned}$$

Next, since

$$\langle F - \mathrm{grad}\,u, W \rangle = \int_{\mathbb{R}^3} (F \cdot W + u\,\mathrm{div}\,W)\,dx$$

and

$$\begin{aligned} \int_{\Omega^\pm} (F \cdot W + u\,\mathrm{div}\,W)\,dx &= \int_{\Omega^\pm} \big[(\mathrm{grad}\,u^\pm) \cdot W + u^\pm\,\mathrm{div}\,W\big]\,dx \\ &= \int_{\Omega^\pm} \mathrm{div}(u^\pm W)\,dx = \mp \int_\Gamma \nu \cdot \gamma^\pm(uW)\,d\sigma, \end{aligned}$$

it follows by the jump relation for the double-layer potential, $[u]_\Gamma = [\mathrm{DL}\,\phi_\Gamma]_\Gamma = \phi_\Gamma$, that

$$\langle F - \mathrm{grad}\,u, W \rangle = -\int_\Gamma \nu \cdot (\phi_\Gamma \gamma W)\,d\sigma = -\langle \phi_\Gamma \nu, \gamma W \rangle_\Gamma = \langle -\gamma^{\mathrm{t}}(\phi_\Gamma \nu), W \rangle,$$

so $\mathrm{grad}\,u = F + \gamma^{\mathrm{t}}(\phi_\Gamma \nu)$. Another application of Lemma 9.12 gives

$$\mathrm{div}\,F = \mathrm{div}\big[\mathrm{grad}\,u - \gamma^{\mathrm{t}}(\phi_\Gamma \nu)\big] = \Delta u + \partial_\nu^{\mathrm{t}}\phi_\Gamma,$$

and since $-\Delta u - k^2 u = (-\Delta - k^2)\mathcal{G}\partial_\nu^t \phi_\Gamma = \partial_\nu^t \phi_\Gamma$, we see that div $F = -k^2 u$. Finally, since curl grad $u = 0$, the second part of Lemma 9.12 implies that

$$\mathrm{curl}\, F = \mathrm{curl}\big[\mathrm{grad}\, u - \gamma^t(\phi_\Gamma \nu)\big] = \gamma^t(\nu \times \gamma\, \mathrm{grad}\,\phi).$$

$\square$

Lemma 9.14 The vector potential

$$A = \mathcal{G}(\mathrm{curl}\, F) = \mathrm{SL}(\nu \times \gamma\, \mathrm{grad}\,\phi)$$

satisfies

$$\mathrm{div}\, A = 0 \quad \textit{and} \quad \mathrm{curl}\, A = F - k^2\, \mathrm{SL}(\phi_\Gamma \nu) \qquad \textit{on } \mathbb{R}^3.$$

Proof. Since div curl $F = 0$, we have div $A = \mathrm{div}\,\mathcal{G}(\mathrm{curl}\, F) = \mathcal{G}(\mathrm{div}\,\mathrm{curl}\, F) = 0$, whereas since

$$\begin{aligned}
\mathrm{curl}\,\mathrm{curl}\, F &= \mathrm{grad}\,\mathrm{div}\, F - \Delta F = \mathrm{grad}(-k^2 u) - \Delta F \\
&= -k^2\big[F + \gamma^t(\phi_\Gamma \nu)\big] - \Delta F = (-\Delta F - k^2 F) - k^2 \gamma^t(\phi_\Gamma \nu),
\end{aligned}$$

we have

$$\begin{aligned}
\langle \mathrm{curl}\, A, W\rangle &= \langle \mathrm{curl}\,\mathcal{G}(\mathrm{curl}\, F), W\rangle = \langle \mathcal{G}(\mathrm{curl}\,\mathrm{curl}\, F), W\rangle \\
&= \langle (-\Delta - k^2)F - k^2 \gamma^t(\phi_\Gamma \nu), \mathcal{G}W\rangle \\
&= \langle F, (-\Delta - k^2)\mathcal{G}W\rangle - \langle k^2 \mathcal{G}(\gamma^t \phi_\Gamma \nu), W\rangle \\
&= \langle F - k^2\, \mathrm{SL}(\phi_\Gamma \nu), W\rangle,
\end{aligned}$$

as claimed.

$\square$

We can now prove the main result for this section; see also Exercises 8.18 and 9.6.

Theorem 9.15 Let $G(x, y) = \mathsf{G}(x - y)$ where G is given by (9.9). If $\phi, \psi \in \mathcal{D}(\mathbb{R}^3)$, then

$$\langle R\phi_\Gamma, \psi_\Gamma\rangle_\Gamma = \langle S(\nu \times \gamma\, \mathrm{grad}\,\phi), \nu \times \gamma\, \mathrm{grad}\,\psi\rangle_\Gamma - k^2 \langle S(\phi_\Gamma \nu), \psi_\Gamma \nu\rangle_\Gamma.$$

Proof. Using the definition of R and the first Green identity, we see that

$$\langle R\phi_\Gamma, \psi_\Gamma\rangle_\Gamma = \langle -\partial_\nu^\pm u, \psi_\Gamma\rangle_\Gamma = \pm \int_{\Omega^\pm} \big(\mathrm{grad}\, u \cdot \mathrm{grad}\,\psi - k^2 u\psi\big)\, dx,$$

and on $\Omega^{\pm}$ we have

$$
\begin{aligned}
\operatorname{grad} u \cdot \operatorname{grad} \psi - k^2 u \psi &= \left[\operatorname{curl} A + k^2 \operatorname{SL}(\phi_\Gamma v) \right] \cdot \operatorname{grad} \psi + k^2 \operatorname{div} \operatorname{SL}(\phi_\Gamma v) \psi \\
&= \operatorname{curl} A \cdot \operatorname{grad} \psi + k^2 \left\{ \operatorname{SL}(\phi_\Gamma v) \cdot \operatorname{grad} \psi \right. \\
&\quad \left. + [\operatorname{div} \operatorname{SL}(\phi_\Gamma v)] \psi \right\} \\
&= \operatorname{div} \left[A \times \operatorname{grad} \psi + k^2 \operatorname{SL}(\phi_\Gamma v) \psi \right].
\end{aligned}
$$

Hence, the divergence theorem gives

$$
\begin{aligned}
\langle R\phi_\Gamma, \psi_\Gamma \rangle_\Gamma &= - \int_\Gamma v \cdot \gamma \left[A \times \operatorname{grad} \psi + k^2 \operatorname{SL}(\phi_\Gamma v) \psi \right] d\sigma \\
&= \langle \gamma A, v \times \gamma \operatorname{grad} \psi \rangle_\Gamma - k^2 \langle S(\phi_\Gamma v), \psi_\Gamma v \rangle_\Gamma,
\end{aligned}
$$

and finally $\gamma A = S(v \times \gamma \operatorname{grad} \phi)$. $\qquad\qquad \square$

Exercises

9.1 Show from the series definitions of $J_{1/2}(z)$ and $Y_{1/2}(z)$ that the zero-order spherical Bessel functions may be written as

$$
j_0(3, z) = \frac{\sin z}{z} \quad \text{and} \quad y_0(3, z) = -\frac{\cos z}{z}.
$$

9.2 Prove Theorem 9.4 by showing that as $|x| \to 0$,

$$
\partial_j G(x) = \partial_j G_\Delta(x) + \begin{cases} O(|x| \log |x|) & \text{if } n = 2, \\ O(|x|^{3-n}) & \text{if } n \geq 3, \end{cases}
$$

and then arguing as in the proof of Theorem 8.1.

9.3 Show that if f_1 and f_2 are solutions of the differential equation (9.5), then their Wronskian

$$
W = W(f_1, f_2) = \begin{vmatrix} f_1 & f_2 \\ f_1' & f_2' \end{vmatrix}
$$

is a solution of

$$
\frac{dW}{dz} + \frac{n-1}{z} W = 0.
$$

Deduce that $W = \text{const}/z^{n-1}$, and in particular

$$
W\left(h_m^{(1)}, h_m^{(2)}\right) = \frac{-2i}{z^{n-1}} \quad \text{and} \quad W(j_m, y_m) = \frac{1}{z^{n-1}}.
$$

[Hint: use (9.10) and (9.12).]

9.4 Let $G(x, y) = G(x - y)$ be the radiating fundamental solution given by (9.14), and write $x = \rho\omega$ with $\rho = |x|$ and $\omega \in \mathbb{S}^{n-1}$.

   (i) Show that $|x - y| = |x| - \omega \cdot y + O(|x|^{-1})$ as $|x| \to \infty$, uniformly for $y \in \Gamma$.

   (ii) Use (9.10) and (9.12) to show that

$$\text{SL}\,\psi(x) = \frac{k^{(n-3)/2}e^{-i(n-3)\pi/4}}{2(2\pi)^{(n-1)/2}} \frac{e^{ik\rho}}{\rho^{(n-1)/2}}$$
$$\times \left(\int_\Gamma e^{-ik\omega \cdot y} \psi(y)\, d\sigma_y + O(\rho^{-1}) \right)$$

and

$$\text{DL}\,\psi(x) = \frac{k^{(n-3)/2}e^{-i(n-3)\pi/4}}{2(2\pi)^{(n-1)/2}} \frac{e^{ik\rho}}{\rho^{(n-1)/2}}$$
$$\times \left(\int_\Gamma (\partial_{\nu,y} e^{-ik\omega \cdot y}) \psi(y)\, d\sigma_y + O(\rho^{-1}) \right)$$

as $|x| \to \infty$, uniformly in ω.

  (iii) Show that SL ψ and DL ψ satisfy the Sommerfeld radiation condition.

  (iv) Deduce that if $u \in H^1_{\text{loc}}(\Omega^+)$ is a radiating solution of the $-\Delta u - k^2 u = 0$ on Ω^+, then there exists a unique function $u_\infty \in C^\infty(\mathbb{S}^{n-1})$ such that

$$u(x) = \frac{e^{ik\rho}}{\rho^{(n-1)/2}} [u_\infty(\omega) + O(\rho^{-1})] \quad \text{as } \rho \to \infty,$$

uniformly in ω. The function u_∞ is called the *far-field pattern* of u. [Hint: use Theorem 7.12.]

   (v) Show that if $u_\infty = 0$ on $\mathbb{S}^{n-1}$, then $u = 0$ on Ω^+. [Hint: use Lemma 9.8.]

9.5 Show that if $f \in \widetilde{H}^{-1}(\Omega^+)$ has compact support, and if $g \in H^{-1/2}(\Gamma)$, then the exterior Neumann problem for the Helmholtz equation,

$$-\Delta u - k^2 u = f \quad \text{on } \Omega^+,$$
$$\partial_\nu^+ u = g \quad \text{on } \Gamma,$$

has a unique radiating solution $u \in H^1_{\text{loc}}(\Omega^+)$. [Hint: reformulate the

problem as a boundary integral equation involving the hypersingular operator R, and apply the Fredholm alternative as in the proof of Theorem 9.11.]

9.6 Show that, in two dimensions, the identity of Theorem 9.15 takes the form

$$\langle R\phi_\Gamma, \psi_\Gamma\rangle_\Gamma = \langle S\partial_\tau\phi, \partial_\tau\psi\rangle_\Gamma - k^2\langle S(\phi_\Gamma\nu), \psi_\Gamma\nu\rangle_\Gamma \quad \text{for } \phi, \psi \in \mathcal{D}(\mathbb{R}^2),$$

where $\tau = (-\nu_2, \nu_1)$ is the tangent vector to Γ satisfying $\nu \times \tau = e_3$, and
$\partial_\tau\phi = \tau \cdot \text{grad}\,\phi$ denotes the tangential derivative of ϕ.

10

Linear Elasticity

In the preceding two chapters, we considered the simplest and most important examples of scalar elliptic equations. Now we turn to the best-known example of an elliptic *system*, namely, the equilibrium equations of linear elasticity. For the history of these equations, we refer to the article by Cross in [30, pp. 1023–1033], the introduction of the textbook by Love [60], and the collection of essays by Truesdell [101]. Our aim in what follows is simply to show how the elasticity equations fit into the general theory developed in earlier chapters. Nečas and Hlaváček [73] give a much more extensive but still accessible treatment of these equations, without, however, discussing boundary integral formulations.

Let u denote the displacement field of an elastic medium. Mathematically, $u : \Omega \to \mathbb{C}^n$, so $m = n$ in our usual notation, and physically u is $\mathbb{R}^n$-valued and the dimension n equals 3. In Cartesian coordinates, the components of the (infinitesimal) strain tensor are given by

$$E_{jk}(u) = \tfrac{1}{2}(\partial_j u_k + \partial_k u_j) \quad \text{for } j, k \in \{1, 2, \ldots, n\},$$

and we denote the components of the stress tensor by Σ_{jk}. Thus, using the summation convention, the kth component of the traction over Γ is $\nu_j \Sigma_{jk}$. (The traction is the force per unit area acting on Ω through the surface Γ.) If f is the body force density, then in equilibrium we have

$$\partial_j \Sigma_{jk} + f_k = 0. \tag{10.1}$$

For a linear homogeneous and isotropic elastic medium, the stress–strain relation is

$$\Sigma_{jk}(u) = 2\mu E_{jk}(u) + \lambda(\text{div } u)\delta_{jk},$$

where the *Lamé coefficients* μ and λ are real constants. We can write the

296

equilibrium equations (10.1) in our standard form $\mathcal{P}u = f$ by putting

$$\mathcal{P}u = -\partial_j \mathcal{B}_j u \quad \text{and} \quad (\mathcal{B}_j u)_k = \Sigma_{jk}(u).$$

Notice that the conormal derivative has a direct physical meaning:

$$\mathcal{B}_\nu u = \text{traction on } \Gamma,$$

and since

$$\partial_j \Sigma_{jk} = \mu \partial_j (\partial_j u_k + \partial_k u_j) + \lambda \partial_k (\text{div } u) = \mu \partial_j \partial_j u_k + (\mu + \lambda) \partial_k (\text{div } u),$$

the second-order partial differential operator $\mathcal{P}$ can be written in the form

$$\mathcal{P}u = -\mu \Delta u - (\mu + \lambda) \,\text{grad}(\text{div } u). \tag{10.2}$$

This chapter begins with a proof of Korn's inequality, thereby establishing that $\mathcal{P}$ is coercive on $H^1(\Omega)^n$. After that, we derive the standard two- and three-dimensional fundamental solutions. The third and final section discusses uniqueness theorems and the positivity of the boundary integral operators, but only for the three-dimensional case.

Korn's Inequality

We see from (10.2) that the Fourier transform of $\mathcal{P}u$ is $P(\xi)\hat{u}(\xi)$, where $P(\xi)$ is the homogeneous, $\mathbb{R}^{n \times n}$-valued quadratic polynomial with jk-entry

$$P_{jk}(\xi) = (2\pi)^2 [\mu |\xi|^2 \delta_{jk} + (\mu + \lambda) \xi_k \xi_j],$$

or, letting I_n denote the $n \times n$ identity matrix,

$$P(\xi) = (2\pi)^2 [\mu |\xi|^2 I_n + (\mu + \lambda) \xi \xi^T]. \tag{10.3}$$

Thus,

$$\eta^* P(\xi) \eta = (2\pi)^2 [\mu |\xi|^2 |\eta|^2 + (\mu + \lambda) |\eta \cdot \xi|^2] \quad \text{for } \xi \in \mathbb{R}^n \text{ and } \eta \in \mathbb{C}^n,$$

and therefore $\mathcal{P}$ is strongly elliptic if and only if

$$\mu > 0 \quad \text{and} \quad 2\mu + \lambda > 0; \tag{10.4}$$

cf. (6.5). Landau and Lifshitz [54, p. 11] explain the physical significance of (10.4). Since

$$(\mathcal{B}_j u)^* \partial_j v = \Sigma_{jk}(\bar{u}) \partial_j v_k = 2\mu E_{jk}(\bar{u}) \partial_j v_k + \lambda (\text{div } \bar{u}) \partial_k v_k$$
$$= 2\mu E_{jk}(\bar{u}) E_{jk}(v) + \lambda (\text{div } \bar{u})(\text{div } v),$$

the sesquilinear form associated with $\mathcal{P}$ is

$$\Phi_\Omega(u, v) = \int_\Omega \left[2\mu E_{jk}(\bar{u}) E_{jk}(v) + \lambda (\text{div } \bar{u})(\text{div } v) \right] dx.$$

For a physical displacement field $u : \Omega \to \mathbb{R}^3$, the quadratic form

$$\frac{1}{2} \Phi_\Omega(u, u) = \frac{1}{2} \int_\Omega \Sigma_{jk}(u) E_{jk}(u) \, dx$$

is the free energy of the elastic medium within Ω; see Landau and Lifshitz [53, p. 12]. It will be convenient to let grad u denote the $n \times n$ matrix whose jk-entry entry is $\partial_j u_k$, and to write

$$\| E(u) \|^2_{L_2(\Omega)^{n \times n}} = \int_\Omega E_{jk}(\bar{u}) E_{jk}(u) \, dx \quad \text{and}$$

$$\| \text{grad } u \|^2_{L_2(\Omega)^{n \times n}} = \int_\Omega \partial_j \bar{u}_k \partial_j u_k \, dx.$$

Notice that

$$\Phi_\Omega(u, u) = 2\mu \| E(u) \|^2_{L_2(\Omega)^{n \times n}} + \lambda \| \text{div } u \|^2_{L_2(\Omega)}. \tag{10.5}$$

If the Lamé coefficients satisfy (10.4), so that $\mathcal{P}$ is strongly elliptic, then we know from Theorem 4.6 that Φ is coercive on $H^1_0(\Omega)^n$. In fact, a stronger result known as *Korn's first inequality* holds. Recall that, by Theorem 3.33, $\widetilde{H}^1(\Omega)^n = H^1_0(\Omega)^n$ if Ω is Lipschitz.

Theorem 10.1 If Ω is an open subset of $\mathbb{R}^n$, then

$$\| E(u) \|^2_{L_2(\Omega)^{n \times n}} \geq \tfrac{1}{2} \| \text{grad } u \|^2_{L_2(\Omega)^{n \times n}} \quad \text{for } u \in \widetilde{H}^1(\Omega)^n.$$

Proof. It suffices to consider $u \in \mathcal{D}(\Omega)^n$. As in the first part of the proof of Theorem 4.6, we apply Plancherel's theorem to obtain

$$\int_{\mathbb{R}^n} E_{jk}(\bar{u}) E_{jk}(u) \, dx = \int_{\mathbb{R}^n} \overline{(i\pi)(\xi_j \hat{u}_k + \xi_k \hat{u}_j)} (i\pi)(\xi_j \hat{u}_k + \xi_k \hat{u}_j) \, d\xi$$

$$= \int_{\mathbb{R}^n} 2\pi^2 \left(|\xi|^2 |\hat{u}|^2 + |\xi \cdot \hat{u}|^2 \right) d\xi$$

$$\geq \frac{1}{2} \int_{\mathbb{R}^n} |2\pi \xi|^2 |\hat{u}|^2 \, d\xi = \frac{1}{2} \sum_{j=1}^n \int_{\mathbb{R}^n} |\partial_j u|^2 \, dx.$$

The result follows, because we can replace $\mathbb{R}^n$ by Ω in both of the integrals with respect to x. $\square$

A much deeper result is *Korn's second inequality* (or just Korn's inequality).

Theorem 10.2 *If Ω is a Lipschitz domain, then*

$$\|E(u)\|^2_{L_2(\Omega)^{n\times n}} \geq c\|\operatorname{grad} u\|^2_{L_2(\Omega)^{n\times n}} - C\|u\|^2_{L_2(\Omega)^n} \quad \text{for } u \in H^1(\Omega)^n.$$

Proof. The left-hand side has the form (4.10) with $L = n^2$, and for convenience we change the index set, writing $\mathcal{N}_{jk}u = E_{jk}(u)$ instead of $\mathcal{N}_l u$. Since

$$\mathcal{N}_{jk}u = \tfrac{1}{2}(\partial_j u_k + \partial_k u_j) = \tfrac{1}{2}\big(e_k^T \partial_j + e_j^T \partial_k\big)u,$$

we have $\mathcal{F}_{x\to\xi}\{\mathcal{N}_{jk}u(x)\} = N_{jk}(\mathrm{i}2\pi\xi)\hat{u}(\xi)$, where

$$N_{jk}(\xi) = \tfrac{1}{2}\big(e_k^T \xi_j + e_j^T \xi_k\big).$$

The desired inequality holds because the hypotheses of Theorem 4.9 are satisfied, with $q_r = 1$ for $1 \leq r \leq m = n$. For instance, when $n = 3$ and $r = 1$,

$$\xi_1 N_{11}(\xi)^T = \xi_1^2 e_1, \qquad -\xi_1 N_{22}(\xi)^T + \xi_2 N_{12}(\xi)^T + \xi_2 N_{21}(\xi)^T = \xi_2^2 e_1,$$
$$\xi_2 N_{11}(\xi)^T = \xi_1\xi_2 e_1, \qquad -\xi_1 N_{23}(\xi)^T + \xi_3 N_{12}(\xi)^T + \xi_2 N_{13}(\xi)^T = \xi_2\xi_3 e_1,$$
$$\xi_3 N_{11}(\xi)^T = \xi_1\xi_3 e_1, \qquad -\xi_1 N_{33}(\xi)^T + \xi_3 N_{13}(\xi)^T + \xi_3 N_{31}(\xi)^T = \xi_3^2 e_1,$$

and the other cases can be handled in the same way. $\qquad\square$

In view of (10.5), Korn's second inequality implies the following result, which allows us to apply the general theory of elliptic systems. Nečas and Hlaváček [73, pp. 46–49] give simple physical arguments showing that both Lamé coefficients should be positive for typical elastic materials.

Theorem 10.3 *For any Lipschitz domain Ω, the elasticity operator (10.2) is coercive on $H^1(\Omega)^n$ if $\mu > 0$ and $\lambda \geq 0$.*

Fundamental Solutions

Since the polynomial (10.3) is homogeneous, we may obtain a fundamental solution for the elasticity operator using Theorem 6.8; see Landau and Lifshitz [54, p. 30] for an alternative approach.

Theorem 10.4 If $\mu \neq 0$ and $2\mu + \lambda \neq 0$, then a fundamental solution $G(x, y) = G(x - y)$ for the elasticity operator (10.2) is given by

$$G(x) = \frac{1}{8\pi\mu(2\mu + \lambda)}\left((3\mu + \lambda)\frac{1}{|x|}I_3 + (\mu + \lambda)\frac{xx^T}{|x|^3}\right) \quad \text{when } n = 3,$$

$$(10.6)$$

and by

$$G(x) = \frac{1}{4\pi\mu(2\mu + \lambda)}\left((3\mu + \lambda)\log\frac{1}{|x|}I_2 + (\mu + \lambda)\frac{xx^T}{|x|^2}\right) \quad \text{when } n = 2.$$

Proof. By (10.3), if $\omega \in \mathbb{S}^{n-1}$, then

$$P(\omega) = (2\pi)^2[\mu I_n + (\mu + \lambda)\omega\omega^T]$$

and so by Exercise 10.1,

$$P(\omega)^{-1} = \frac{(2\pi)^{-2}}{\mu(2\mu + \lambda)}[(2\mu + \lambda)I_n - (\mu + \lambda)\omega\omega^T].$$

Assume that $n = 3$, and choose an orthonormal basis $\eta_1, \eta_2 \in \mathbb{R}^3$ for the plane normal to x, so that the unit circle in this plane has the parametric representation

$$\mathbb{S}_x^{\perp} : \omega_{\perp} = (\cos\theta)\eta_1 + (\sin\theta)\eta_2 \quad \text{for } -\pi \leq \theta \leq \pi.$$

By Theorem 6.8(iii),

$$G(x) = \frac{1}{2|x|}\frac{(2\pi)^{-2}}{\mu(2\mu + \lambda)}\int_{\mathbb{S}_x^{\perp}}[(2\mu + \lambda)I_3 - (\mu + \lambda)\omega_{\perp}\omega_{\perp}^T]\,d\omega_{\perp},$$

and since

$$\int_{\mathbb{S}_x^{\perp}}\omega_{\perp}\omega_{\perp}^T\,d\omega_{\perp} = \int_{-\pi}^{\pi}[(\cos^2\theta)\eta_1\eta_1^T + (\cos\theta\sin\theta)(\eta_1\eta_2^T + \eta_2\eta_1^T)$$

$$+ (\sin^2\theta)\eta_2\eta_2^T]\,d\theta = \pi(\eta_1\eta_1^T + \eta_2\eta_2^T) = \pi\left(I_3 - \frac{xx^T}{|x|^2}\right),$$

the formula for $G(x)$ follows.

Suppose now that $n = 2$; by (6.13), the formula

$$G(x) = \frac{-(2\pi)^{-2}}{\mu(2\mu + \lambda)}\int_{\mathbb{S}^1}(\log|\omega \cdot x|)[(2\mu + \lambda)I_2 - (\mu + \lambda)\omega\omega^T]\,d\omega$$

defines a fundamental solution. The vectors

$$\eta_1 = \frac{x}{|x|} = \frac{1}{|x|} \begin{bmatrix} x_1 \\ x_2 \end{bmatrix} \quad \text{and} \quad \eta_2 = \frac{1}{|x|} \begin{bmatrix} -x_2 \\ x_1 \end{bmatrix}$$

form an orthonormal basis for $\mathbb{R}^2$, and by putting $\omega = (\cos\theta)\eta_1 + (\sin\theta)\eta_2$
we see that

$$\int_{\mathbb{S}^1} \log|\omega \cdot x|\, d\omega = \int_{-\pi}^{\pi} \log(|x\cos\theta|)\, d\theta = 2\pi\log|x| + \text{const}$$

and

$$\int_{\mathbb{S}^1} (\log|\omega \cdot x|)\omega\omega^T\, d\omega$$

$$= \int_{-\pi}^{\pi} \log(|x\cos\theta|)\left[(\cos^2\theta)\eta_1\eta_1^T + (\cos\theta\sin\theta)(\eta_1\eta_2^T + \eta_2\eta_1^T)\right.$$

$$\left. + (\sin^2\theta)\eta_2\eta_2^T\right] d\theta$$

$$= \int_{-\pi}^{\pi} \log(|x\cos\theta|)\left(\frac{1+\cos2\theta}{2}\eta_1\eta_1^T + \frac{1-\cos2\theta}{2}\eta_2\eta_2^T\right) d\theta.$$

An integration by parts gives

$$\int_{-\pi}^{\pi} \log(|x\cos\theta|)\cos2\theta\, d\theta = \int_{-\pi}^{\pi} \frac{\sin\theta}{\cos\theta}\frac{\sin2\theta}{2}\, d\theta = \int_{-\pi}^{\pi} \sin^2\theta\, d\theta = \pi,$$

and one easily verifies that

$$\eta_1\eta_1^T + \eta_2\eta_2^T = I_2 \quad \text{and} \quad \eta_1\eta_1^T - \eta_2\eta_2^T = 2\frac{xx^T}{|x|^2} - I_2,$$

so

$$\int_{\mathbb{S}^1} (\log|\omega \cdot x|)\omega\omega^T\, d\omega = \frac{1}{2}\int_{-\pi}^{\pi} \log(|x\cos\theta|)\, d\theta\, I_2 + \frac{1}{2}\pi\left(\frac{2xx^T}{|x|^2} - I_2\right)$$

$$= \pi\log|x|I_2 + \pi\frac{xx^T}{|x|^2} + \text{const},$$

leading to the stated formula for $G(x)$. $\square$

Uniqueness Results

Throughout this section, we assume that the components of u are real, and treat
only three-dimensional problems.

To apply the Fredholm alternative (Theorem 4.10) to the mixed boundary value problem in linear elasticity, we must determine all solutions of the homogeneous problem. As a first step, we show that the only strain-free displacement fields are the infinitesimal rigid motions, and that such displacement fields are also stress-free.

Lemma 10.5 Let Ω be a connected open subset of $\mathbb{R}^3$. A distribution $u \in \mathcal{D}^*(\Omega; \mathbb{R}^3)$ satisfies $E(u) = 0$ on Ω if and only if there exist constant vectors $a, b \in \mathbb{R}^3$ such that

$$u(x) = a + b \times x \quad for \ x \in \Omega. \tag{10.7}$$

Moreover, in this case $\Sigma(u) = 0$ on Ω.

Proof. Let $B = [b_{ij}] \in \mathbb{R}^{3 \times 3}$ denote the skew-symmetric matrix defined by $Bx = b \times x$, i.e.,

$$B = \begin{bmatrix} 0 & -b_3 & b_2 \\ b_3 & 0 & -b_1 \\ -b_2 & b_1 & 0 \end{bmatrix}.$$

If $u(x) = a + b \times x$, then $\partial_j u_k = b_{kj}$, so $E_{jk}(u) = \frac{1}{2}(b_{kj} + b_{jk}) = 0$.

To prove the converse, assume that $E(u) = 0$ on Ω. Since, with the notation of (3.9) and (3.10),

$$E(\psi_\epsilon * u) = \psi_\epsilon * E(u) \quad \text{on } \{x \in \Omega : \text{dist}(x, \Gamma) > \epsilon\},$$

we can assume that $u \in C^\infty(\Omega)^3$. The diagonal entries of the strain tensor are just $E_{jj}(u) = \partial_j u_j$ (no sum over j), so

$$\partial_1 u_1 = \partial_2 u_2 = \partial_3 u_3 = 0 \quad \text{on } \Omega.$$

Since the off-diagonal entries of the strain tensor also vanish, we can show that

$$\partial_2^2 u_1 = \partial_3^2 u_1 = 0, \quad \partial_1^2 u_2 = \partial_3^2 u_2 = 0, \quad \partial_1^2 u_3 = \partial_2^2 u_3 = 0 \qquad \text{on } \Omega;$$

for instance, $\partial_1^2 u_k = \partial_1(2E_{1k}(u) - \partial_k u_1) = -\partial_k(\partial_1 u_1) = 0$. Therefore, $\partial^\alpha u = 0$ on Ω if $|\alpha| \geq 3$, implying that u is a quadratic polynomial in x. In fact, the vanishing of the partial derivatives listed above shows that u must have the form

$$u_1(x) = a_1 + b_{12}x_2 + b_{13}x_3 + c_1 x_2 x_3,$$

$$u_2(x) = a_2 + b_{21}x_1 + b_{23}x_3 + c_2 x_1 x_3,$$

$$u_3(x) = a_3 + b_{31}x_1 + b_{32}x_2 + c_3 x_1 x_2,$$

for some constants a_j, b_{jk} and c_j. Since $E(u)$ equals

$$\frac{1}{2} \begin{bmatrix} 0 & b_{12} + b_{21} & b_{13} + b_{31} \\ b_{12} + b_{21} & 0 & b_{23} + b_{32} \\ b_{13} + b_{31} & b_{23} + b_{32} & 0 \end{bmatrix}$$

$$+ \frac{1}{2} \begin{bmatrix} 0 & (c_1 + c_2)x_3 & (c_1 + c_3)x_2 \\ (c_1 + c_2)x_3 & 0 & (c_2 + c_3)x_1 \\ (c_1 + c_3)x_2 & (c_2 + c_3)x_1 & 0 \end{bmatrix},$$

we conclude that $b_{jk} = -b_{kj}$ and $c_j = 0$ for all j, k.

Finally, if u has the form (10.7), then $\Sigma(u) = 2\mu E(u) + \lambda(\operatorname{div} u)I_3 = 0$ because $E(u) = 0$ and $\operatorname{div} u = E_{jj}(u) = 0$. $\qquad\square$

Theorem 10.6 *Assume that Ω is a bounded, connected Lipschitz domain in $\mathbb{R}^3$, and that the Lamé coefficients satisfy*

$$\mu > 0 \quad and \quad \lambda \geq 0. \tag{10.8}$$

Let W denote the set of solutions in $H^1(\Omega; \mathbb{R}^3)$ to the homogeneous, mixed boundary value problem

$$-\mu \Delta u - (\mu + \lambda) \operatorname{grad}(\operatorname{div} u) = 0 \quad on \ \Omega,$$
$$\gamma u = 0 \quad on \ \Gamma_D, \tag{10.9}$$
$$\mathcal{B}_\nu u = 0 \quad on \ \Gamma_N.$$

(i) *If $\Gamma_D \neq \emptyset$, then $W = \{0\}$, i.e., (10.9) has only the trivial solution.*

(ii) *If $\Gamma_D = \emptyset$, so that (10.9) is a pure Neumann problem, then W consists of all functions of the form (10.7) for $a, b \in \mathbb{R}^3$.*

Proof. If $u \in H^1(\Omega)^3$ is a solution of (10.9), then $\Phi(u, v) = 0$ for all $v \in H^1(\Omega)^3$ such that $\gamma v = 0$ on Γ_D. In particular, by taking $v = u$ and recalling (10.5), we see that

$$2\mu \|E(u)\|_{L_2(\Omega)^{3\times 3}}^2 + \lambda \|\operatorname{div} u\|_{L_2(\Omega)}^2 = 0.$$

Our assumptions on μ and λ then imply that $E(u) = 0$ on Ω, and so u has the form (10.7).

If $\Gamma_D \neq \emptyset$, then, because Γ_D is relatively open in Γ, we can find $x, y, z \in \Gamma_D$ such that $x - y$ and $z - y$ are linearly independent. From the three equations

$$a + b \times x = 0, \quad a + b \times y = 0, \quad a + b \times z = 0,$$

it follows that $b \times (x - y) = 0$ and $b \times (z - y) = 0$, and we conclude that $b = 0$. (Otherwise, $x - y$ and $z - y$ would both be scalar multiples of b.) In turn, $a = 0$, and part (i) is proved.

If $\Gamma_D = \emptyset$ and u has the form (10.7), then $\Sigma(u) = 0$ on Ω, by Lemma 10.5, and so u is a solution of the homogeneous Neumann problem, i.e., $u \in W$. □

To conclude this section, we consider the boundary integral operators

$$S : H^{-1/2}(\Gamma; \mathbb{R}^3) \to H^{1/2}(\Gamma; \mathbb{R}^3) \quad \text{and} \quad R : H^{1/2}(\Gamma; \mathbb{R}^3) \to H^{-1/2}(\Gamma; \mathbb{R}^3)$$

associated with the three-dimensional elasticity operator (10.2), and defined using the standard fundamental solution (10.6).

Theorem 10.7 *Assume that the Lamé coefficients satisfy (10.8).*

(i) *The weakly singular boundary integral operator is positive and bounded below on the whole of its domain, i.e.,*

$$(S\psi, \psi)_\Gamma \geq c\|\psi\|^2_{H^{-1/2}(\Gamma)^3} \quad \textit{for all } \psi \in H^{-1/2}(\Gamma; \mathbb{R}^3).$$

In particular, $\ker S = \{0\}$.

(ii) *The hypersingular boundary integral operator is positive and bounded below on the orthogonal complement of its null space. Indeed, if Ω^- is connected, then the six functions $\chi_j : \Gamma \to \mathbb{R}^3$ $(1 \leq j \leq 6)$ defined by*

$$\chi_l(x) = e_l \quad \textit{and} \quad \chi_{l+3}(x) = e_l \times x \qquad \textit{for } x \in \Gamma \textit{ and } 1 \leq l \leq 3$$

form a basis for $\ker R$, and we have

$$(R\psi, \psi)_\Gamma \geq c\|\psi\|^2_{H^{1/2}(\Gamma)^3}$$

$$\textit{for } \psi \in H^{1/2}(\Gamma; \mathbb{R}^3) \textit{ such that } (\chi_j, \psi)_\Gamma = 0 \quad \textit{for } 1 \leq j \leq 6.$$

Proof. By Theorem 10.3, the assumptions on μ and λ ensure that the elasticity operator (10.2) is coercive on $H^1(\Omega^\pm; \mathbb{R}^3)$. Since, in three dimensions, the elastic single-layer potential satisfies

$$\mathrm{SL}\, \psi(x) = O(|x|^{-1}) \quad \text{and} \quad \mathrm{grad}\, \mathrm{SL}\, \psi(x) = O(|x|^{-2}) \qquad \text{as } |x| \to \infty,$$

we can argue as in the case of the Laplacian (Theorem 8.12) to show that

$$(S\psi, \phi)_\Gamma = \Phi^+(\mathrm{SL}\, \psi, \mathrm{SL}\, \phi) + \Phi^-(\mathrm{SL}\, \psi, \mathrm{SL}\, \phi)$$

$$= \int_{\mathbb{R}^3} \left[2\mu E_{jk}(\mathrm{SL}\, \psi) E_{jk}(\mathrm{SL}\, \phi) + \lambda (\mathrm{div}\, \mathrm{SL}\, \psi)(\mathrm{div}\, \mathrm{SL}\, \phi) \right] dx$$

for all $\psi, \phi \in H^{-1/2}(\Gamma)^3$. In particular,

$$(S\psi, \psi)_\Gamma = 2\mu \|E(SL\,\psi)\|^2_{L_2(\mathbb{R}^3)^{3\times3}} + \lambda \|\text{div SL}\,\psi\|^2_{L_2(\mathbb{R}^3)} \geq 0.$$

Moreover, if $(S\psi, \psi)_\Gamma = 0$ then $E(SL\,\psi) = 0$ on $\mathbb{R}^3$, so $SL\,\psi(x) = a + b \times x$ for some $a, b \in \mathbb{R}^3$. In fact, $a = b = 0$ because $SL\,\psi(x) \to 0$ as $|x| \to \infty$, and hence $\psi = -[\mathcal{B}_\nu\, SL\,\psi]_\Gamma = 0$. Part (i) now follows by arguing as in the case of the Laplacian (Corollary 8.13).

Turning to the hypersingular operator, we can argue as in Theorem 8.21 that

$$(R\psi, \phi)_\Gamma = \Phi^+(DL\,\psi, DL\,\phi) + \Phi^-(DL\,\psi, DL\,\phi) \quad \text{for all } \psi, \phi \in H^{1/2}(\Gamma)^3,$$

and in particular $(R\psi, \psi)_\Gamma \geq 0$. Moreover, if $(R\psi, \psi)_\Gamma = 0$, then $E(DL\,\psi) = 0$ on $\Omega^\pm$, implying that $\Sigma(DL\,\psi) = 0$ on $\Omega^\pm$ and hence $R\psi = -\mathcal{B}_\nu\, DL\,\psi = 0$ on Γ. Exercise 2.17 then shows that R is positive and bounded below on the orthogonal complement of $\ker R$.

To complete the proof of part (ii), we assume that Ω^- is connected. It suffices to show that $\psi \in \ker R$ if and only if $\psi = u|_\Gamma$ for some u of the form (10.7).

Let $R\psi = 0$ on Γ. Since $(R\psi, \psi)_\Gamma = 0$ we have $E(DL\,\psi) = 0$ on $\Omega^\pm$ as above, and so by Theorem 10.6, there are vectors $a_\pm, b_\pm \in \mathbb{R}^3$ such that $DL\,\psi(x) = a_\pm + b_\pm \times x$ for $x \in \Omega^\pm$. In fact, $a_+ = b_+ = 0$ because $DL\,\psi(x) \to 0$ as $|x| \to \infty$, and hence $\psi = [DL\,\psi]_\Gamma = u|_\Gamma$ where u is given by (10.7) with $a = -a_-$ and $b = -b_-$.

Conversely, suppose that $\psi = u|_\Gamma$ where $u(x) = a + b \times x$ for $x \in \mathbb{R}^3$. Since $\mathcal{P}u = 0$ on Ω, and $\mathcal{B}_\nu u = 0$ on Γ, the third Green identity for u reduces to $u = -DL\,\psi$ on Ω^-, implying that $0 = -\mathcal{B}_\nu\, DL\,\psi = R\psi$ on Γ, i.e., $\psi \in \ker R$. $\qquad\square$

Exercises

10.1 Show that if $a \neq 0$, $a + b \neq 0$ and $\omega \in \mathbb{S}^{n-1}$, then

$$\left(aI + b\omega\omega^T\right)^{-1} = \frac{1}{a(a+b)}\left[(a+b)I - b\omega\omega^T\right].$$

10.2 Nitsche [79] gives the following elementary proof of Korn's second inequality for C^1 domains. (The same paper also extends the method of proof to cover Lipschitz domains, with the help of a regularised distance.) Suppose first that Ω^- is the hypograph $x_n < \zeta(x')$, and let $u \in H^1(\Omega^-)^n$. Given $s > 0$, we define $u^s \in H^1(\Omega^+)^n$ by reflection in the surface $x_n = \zeta(x')$, i.e.,

$$u^s(x) = u\big(x', \zeta(x') - s[x_n - \zeta(x')]\big) \quad \text{for } x_n > \zeta(x'),$$

and then consider $v \in H^1(\Omega^+)^n$ of the form

$$v_j = \begin{cases} au_j^s + bu_j^t & \text{if } 1 \le j \le n-1, \\ pu_n^s + qu_n^t & \text{if } j = n, \end{cases}$$

for unspecified constants a, b, p, q, s and t such that $s > 0$ and $t > 0$.

(i) Show that $\gamma^+ v = \gamma^- u$ if and only if

$$a + b = 1 \quad \text{and} \quad p + q = 1.$$

(ii) Show that if $x \in \Omega^+$, then

$$(\partial_j u^s)(x) = \begin{cases} (\partial_j u)^s(x) \\ \quad + (1+s)(\partial_n u)^s(x)\partial_j \zeta(x') & \text{for } 1 \le j \le n-1, \\ -s(\partial_n u)^s(x) & \text{for } j = n. \end{cases}$$

(iii) Deduce that for $1 \le j \le n-1$ and $1 \le k \le n-1$,

$$E_{jk}(v) = aE_{jk}(u)^s + bE_{jk}(u)^t + \frac{a}{2}(1+s)[(\partial_n u_k)^s \partial_j \zeta$$

$$+ (\partial_n u_j)^s \partial_k \zeta] + \frac{b}{2}(1+t)[(\partial_n u_k)^t \partial_j \zeta + (\partial_n u_j)^t \partial_k \zeta],$$

whereas

$$E_{nn}(v) = -spE_{nn}(u)^s - tqE_{nn}(u)^t.$$

(iv) Show that if

$$p = -sa \quad \text{and} \quad q = -tb,$$

then for $1 \le j \le n-1$,

$$E_{jn}(v) = pE_{jn}(u)^s + qE_{jn}(u)^t + \frac{p}{2}(1+s)E_{nn}(u)^s \partial_j \zeta$$

$$+ \frac{q}{2}(1+t)E_{nn}(u)^t \partial_j \zeta.$$

(v) Show that

$$\|u^s\|^2_{L_2(\Omega^+)^n} = s^{-1}\|u\|^2_{L_2(\Omega^-)^n},$$

and deduce that, with $M_\zeta = \max_{1 \le j \le n-1} \|\partial_j \zeta\|_{L_\infty(\mathbb{R}^{n-1})}$,

$$\|E(v)\|^2_{L_2(\Omega^+)^{n \times n}} \le C_1 \|E(u)\|^2_{L_2(\Omega^-)^{n \times n}} + C_2 M_\zeta^2 \|\text{grad}\, u\|^2_{L_2(\Omega^-)^{n \times n}},$$

where C_1 and C_2 depend only on a, b, p, q, s, t and n.

(vi) Choose (say) $a = -2, b = 3, p = 4, q = -3, s = 2$ and $t = 1$, so that these constants satisfy the conditions in parts (i) and (iv), and define $U \in H^1(\mathbb{R}^n)^n$ by

$$U = \begin{cases} u & \text{on } \Omega^-, \\ v & \text{on } \Omega^+. \end{cases}$$

Show, by applying Korn's first inequality (Theorem 10.1) to U, that

$$\|\text{grad}\, u\|^2_{L_2(\Omega^-)^{n \times n}} \le 2(1 + C_1)\|E(u)\|^2_{L_2(\Omega^-)^{n \times n}}$$
$$+ 2C_2 M_\zeta^2 \|\text{grad}\, u\|^2_{L_2(\Omega^-)^{n \times n}},$$

and deduce that if $M_\zeta < 1/\sqrt{2C_2}$ then $\|\text{grad}\, u\|_{L_2(\Omega^-)^{n \times n}} \le C\|E(u)\|_{L_2(\Omega^-)^{n \times n}}$.

(vii) Finally, use a partition of unity to prove Korn's second inequality (Theorem 10.2) for a C^1 domain (with compact boundary).

10.3 For $n = 2$, the elasticity operator (10.2) has the form

$$\mathcal{P}u = -\sum_{j=1}^2 \sum_{k=1}^2 \partial_j(A_{jk}\partial_k u)$$

where

$$A_{11} = \begin{bmatrix} 2\mu + \lambda & 0 \\ 0 & \mu \end{bmatrix}, \qquad A_{12} = A_{21} = \frac{\mu + \lambda}{2}\begin{bmatrix} 0 & 1 \\ 1 & 0 \end{bmatrix},$$

$$A_{22} = \begin{bmatrix} \mu & 0 \\ 0 & 2\mu + \lambda \end{bmatrix}.$$

Thus, for $\xi_1 = [\xi_{11}\ \xi_{12}]^T, \xi_2 = [\xi_{21}\ \xi_{22}]^T \in \mathbb{R}^2$,

$$\sum_{j=1}^2 \sum_{k=1}^2 (A_{jk}\xi_k)^T \xi_j$$

$$= \frac{1}{2}[\xi_{11}\ \xi_{22}\ \xi_{12}\ \xi_{21}]\begin{bmatrix} 4\mu + 2\lambda & \mu + \lambda & 0 & 0 \\ \mu + \lambda & 4\mu + 2\lambda & 0 & 0 \\ 0 & 0 & 2\mu & \mu + \lambda \\ 0 & 0 & \mu + \lambda & 2\mu \end{bmatrix}\begin{bmatrix} \xi_{11} \\ \xi_{22} \\ \xi_{12} \\ \xi_{21} \end{bmatrix}.$$

Assuming $\mathcal{P}$ is strongly elliptic, so that the Lamé coefficients satisfy (10.4), show that the condition (4.9) fails to hold if $\lambda \geq \mu$ or if $-2\mu < \lambda \leq -5\mu/3$.

10.4 Show for $n = 3$ and for $u \in C^1(\overline{\Omega})^3$ that the surface traction can be written as

$$\mathcal{B}_\nu u = 2\mu\partial_\nu u + \lambda(\operatorname{div} u)|_\Gamma \nu - \mu(\nu \times \operatorname{curl} u)|_\Gamma.$$

[Hint: first establish the identity $(\nu \times \operatorname{curl} u)_k = \nu_j\partial_k u_j - \nu_j\partial_j u_k$.]

10.5 Consider the pure Neumann (i.e., pure traction) problem in linear elasticity:

$$-\mu\Delta u - (\mu + \lambda)\operatorname{grad}(\operatorname{div} u) = f \quad \text{on } \Omega,$$
$$\mathcal{B}_\nu u = g \quad \text{on } \Gamma.$$

Show that a solution exists if and only if the resultant force and the resultant torque both vanish, i.e.,

$$\int_\Omega f(x)\,dx + \int_\Gamma g(x)\,d\sigma_x = 0$$

and

$$\int_\Omega x \times f(x)\,dx + \int_\Gamma x \times g(x)\,d\sigma_x = 0.$$

Appendix A

Extension Operators for Sobolev Spaces

We say that $E : W_p^s(\Omega) \to W_p^s(\mathbb{R}^n)$ is an *extension operator* if E is bounded and satisfies

$$Eu|_\Omega = u \quad \text{for } u \in W_p^s(\Omega).$$

For our purposes, the significance of such operators stems from Theorem 3.18.

If Ω is Lipschitz, then a construction due to Calderón [8] yields, for each integer $k \geq 1$, an extension operator $E_k : W_p^k(\Omega) \to W_p^k(\mathbb{R}^n)$ that is bounded for $1 < p < \infty$. Using a different method, Stein [96, p. 181] obtained an extension operator $E : W_p^k(\Omega) \to W_p^k(\mathbb{R}^n)$, not depending on $k \geq 0$, and bounded for $1 \leq p \leq \infty$. In the case when Ω is smooth, there is a simpler construction, due to Seeley [93]; see Exercise A.3. In the main result of this appendix, Theorem A.4, we shall use a modified version of Calderón's extension.

Suppose that Ω is a hypograph,

$$\Omega = \{ x \in \mathbb{R}^n : x_n < \zeta(x') \}, \tag{A.1}$$

where $x' = (x_1, \ldots, x_{n-1})$, and the function $\zeta : \mathbb{R}^{n-1} \to \mathbb{R}$ is Lipschitz:

$$|\zeta(x') - \zeta(y')| \leq M|x' - y'| \quad \text{for } x', y' \in \mathbb{R}^{n-1}. \tag{A.2}$$

A crude extension operator is obtained simply by reflection in the boundary of Ω.

Theorem A.1 *If Ω is the Lipschitz hypograph (A.1), and if*

$$E_0 u(x) = \begin{cases} u(x) & \text{for } x \in \Omega, \\ u(x', 2\zeta(x') - x_n) & \text{for } x \in \mathbb{R}^n \setminus \Omega, \end{cases}$$

then $E_0 : W_p^s(\Omega) \to W_p^s(\mathbb{R}^n)$ is bounded for $0 \leq s \leq 1$ and $1 \leq p \leq \infty$.

Proof. Put $\tilde{x} = \left(x', 2\zeta(x') - x_n\right)$. One easily verifies that $\tilde{x} \in \Omega$ if and only if $x \in \mathbb{R}^n \setminus \Omega$, and vice versa, with $\tilde{\tilde{x}} = x$. Moreover, the Jacobian determinant of the transformation $x \mapsto \tilde{x}$ is identically equal to -1. Thus, $\|E_0 u\|_{L_p(\mathbb{R}^n \setminus \Omega)} = \|u\|_{L_p(\Omega)}$, and $\|E_0 u\|_{L_p(\mathbb{R}^n)} = 2\|u\|_{L_p(\Omega)}$. Also, if $x \in \mathbb{R}^n \setminus \Omega$, then

$$\partial_j E_0 u(x) = \begin{cases} \partial_j u(\tilde{x}) + 2\partial_n u(\tilde{x})\partial_j \zeta(x') & \text{for } 1 \le j \le n-1, \\ -\partial_n u(\tilde{x}) & \text{for } j = n, \end{cases}$$

implying that

$$\|\partial_j E_0 u\|_{L_p(\mathbb{R}^n \setminus \Omega)} \le \begin{cases} \|\partial_j u\|_{L_p(\Omega)} + 2M \|\partial_n u\|_{L_p(\Omega)} & \text{for } 1 \le j \le n-1, \\ \|\partial_n u\|_{L_p(\Omega)} & \text{for } j = n. \end{cases}$$

Hence, $E_0 : W_p^s(\Omega) \to W_p^s(\mathbb{R}^n)$ is bounded if $s = 0$ or 1.

Assume now that $0 < s < 1$ and $1 \le p < \infty$. Recalling the definition (3.18) of the Slobodeckiĭ seminorm, we write

$$|E_0 u|_{s,p,\mathbb{R}^n}^p = |u|_{s,p,\Omega}^p + I_1 + I_2 + I_3,$$

where

$$I_1 = \iint_{x_n > \zeta(x'), \, y_n > \zeta(y')} \frac{|u(\tilde{x}) - u(\tilde{y})|^p}{|x - y|^{n+ps}} \, dx \, dy,$$

$$I_2 = \iint_{x_n > \zeta(x'), \, y_n < \zeta(y')} \frac{|u(\tilde{x}) - u(y)|^p}{|x - y|^{n+ps}} \, dx \, dy,$$

$$I_3 = \iint_{x_n < \zeta(x'), \, y_n > \zeta(y')} \frac{|u(x) - u(\tilde{y})|^p}{|x - y|^{n+ps}} \, dx \, dy.$$

Since $|\tilde{x}_n - \tilde{y}_n| \le 2|\zeta(x') - \zeta(y')| + |x_n - y_n| \le \sqrt{1 + 4M^2}|x - y|$, we see that

$$|\tilde{x} - \tilde{y}| \le 2\sqrt{1 + M^2}|x - y|,$$

and so

$$I_1 = \iint_{\tilde{x}_n < \zeta(x'), \, \tilde{y}_n < \zeta(y')} \frac{|u(\tilde{x}) - u(\tilde{y})|^p}{|x - y|^{n+ps}} \, d\tilde{x} \, d\tilde{y}$$

$$\le \left(2\sqrt{1 + M^2}\right)^{n+ps} \iint_{\tilde{x}_n < \zeta(x'), \, \tilde{y}_n < \zeta(y')} \frac{|u(\tilde{x}) - u(\tilde{y})|^p}{|\tilde{x} - \tilde{y}|^{n+ps}} \, d\tilde{x} \, d\tilde{y} = C|u|_{s,p,\Omega}^p.$$

If $x_n > \zeta(x')$ and $y_n < \zeta(y')$, then

$$0 \le x_n - \zeta(x') \le x_n - y_n + \zeta(y') - \zeta(x') \le |x_n - y_n| + M|x' - y'|,$$

so

$$|\tilde{x}_n - y_n| = |y_n + x_n - 2\zeta(x')| \le |y_n - x_n| + 2|x_n - \zeta(x')| \le C|x - y|,$$

and hence

$$I_2 = \iint_{\tilde{x}_n < \zeta(x'), \; y_n < \zeta(y')} \frac{|u(\tilde{x}) - u(y)|^p}{|x - y|^{n+ps}} \, d\tilde{x} \, dy,$$

$$\le C \iint_{\tilde{x}_n < \zeta(x'), \; y_n < \zeta(y')} \frac{|u(\tilde{x}) - u(y)|^p}{|\tilde{x} - y|^{n+ps}} \, d\tilde{x} \, dy = C|u|^p_{s,p,\Omega}.$$

Similarly, $I_3 \le C|u|^p_{s,p,\Omega}$, giving the desired bound for $E_0 u$.

Finally, consider the remaining case $0 < s < 1$ and $p = \infty$. If $x_n > \zeta(x')$ and $y_n > \zeta(y')$, then

$$|E_0 u(x) - E_0 u(y)| = |u(\tilde{x}) - u(\tilde{y})| \le |u|_{s,\infty,\Omega} |\tilde{x} - \tilde{y}|^s$$

$$\le [2\sqrt{1 + M^2}]^s |u|_{s,\infty,\Omega} |x - y|^s.$$

If $x_n > \zeta(x')$ and $y_n < \zeta(y')$, then

$$|E_0 u(x) - E_0 u(y)| = |u(\tilde{x}) - u(y)| \le |u|_{s,\infty,\Omega} |\tilde{x} - y|^s \le C|u|_{s,\infty,\Omega} |x - y|^s,$$

and similarly if $x_n < \zeta(x')$ and $y_n > \zeta(y')$, then $|E_0 u(x) - E_0 u(y)| \le C|u|_{s,\infty,\Omega} |x - y|^s$. □

To obtain an extension operator for $s > 1$, we shall use the *Sobolev representation formula*; recall the notation of (3.7).

Lemma A.2 *For each integer $k \ge 1$, if $\psi \in C^\infty(\mathbb{S}^{n-1})$ satisfies*

$$\int_{|\omega|=1} \psi(\omega) \, d\omega = \frac{(-1)^k}{(k-1)!}, \qquad (A.3)$$

and if $u \in C^k_{\text{comp}}(\mathbb{R}^n)$, then

$$u(x) = \int_{\mathbb{R}^n} \psi\left(\frac{y}{|y|}\right) \frac{1}{|y|^n} u^{(k)}(x + y; y) \, dy \quad \text{for } x \in \mathbb{R}^n.$$

Proof. By Exercises A.1 and 3.5,

$$u(x) = \frac{(-1)^k}{(k-1)!} \int_0^\infty \rho^{k-1} \left(\frac{d}{d\rho}\right)^k u(x + \rho\omega) \, d\rho$$

$$= \frac{(-1)^k}{(k-1)!} \int_0^\infty \rho^{k-1} u^{(k)}(x + \rho\omega; \omega) \, d\rho.$$

Multiplying both sides by $\psi(\omega)$ and integrating with respect to ω, we see that

$$u(x) = \int_{|\omega|=1} \psi(\omega) \int_0^\infty \rho^{-n} u^{(k)}(x + \rho\omega; \rho\omega) \rho^{n-1} \, d\rho \, d\omega. \qquad \square$$

We use the abbreviation $|u|_\mu$ for the Slobodeckiĭ seminorm on $\mathbb{R}^n$ with $p = 2$.

Lemma A.3 If $K \in C^\infty(\mathbb{R}^n \setminus \{0\})$ *is homogeneous of degree* $1 - n$, *and if* $u \in \mathcal{D}(\mathbb{R}^n)$, *then*

$$\|\partial_j(K * u)\|_{H^s(\mathbb{R}^n)} \le C \|u\|_{H^s(\mathbb{R}^n)} \quad \text{for } -\infty < s < \infty,$$

and

$$|\partial_j(K * u)|_\mu \le C|u|_\mu \quad \text{for } 0 < \mu < 1.$$

Proof. Since K is locally integrable on $\mathbb{R}^n$,

$$\partial_j(K * u)(x) = (K * \partial_j u)(x) = \lim_{\epsilon \downarrow 0} \int_{|y-x|>\epsilon} K(x - y)\partial_j u(y) \, dy,$$

and since

$$\partial_{j,y}[K(x - y)u(y)] = -\partial_j K(x - y)u(y) + K(x - y)\partial_j u(y),$$

the divergence theorem gives

$$\int_{|y-x|=\epsilon} \frac{x_j - y_j}{\epsilon} K(x - y)u(y) \, d\sigma_y = -\int_{|y-x|>\epsilon} \partial_j K(x - y)u(y) \, dy$$
$$+ \int_{|y-x|>\epsilon} K(x - y)\partial_j u(y) \, dy.$$

Making the substitution $y = x - \epsilon\omega$ and using the homogeneity of K, we see that the left-hand side equals

$$\int_{|\omega|=1} \omega_j K(\epsilon\omega)u(x - \epsilon\omega)\epsilon^{n-1} \, d\omega = \int_{|\omega|=1} \omega_j K(\omega)u(x - \epsilon\omega) \, d\omega,$$

so, defining $a_j = \int_{|\omega|=1} \omega_j K(\omega) \, d\omega$, we have

$$\partial_j(K * u)(x) = a_j u(x) + \lim_{\epsilon \downarrow 0} \int_{|y-x|>\epsilon} \partial_j K(x - y)u(y) \, dy.$$

Hence, by Exercise 5.14, the desired estimate for $\partial_j(K * u)$ will follow at once if we show that

$$\int_{|\omega|=1} \partial_j K(\omega)\, d\omega = 0. \tag{A.4}$$

Choose a cutoff function $\chi \in C^\infty_{\text{comp}}(0, \infty)$ satisfying $\int_0^\infty \chi(\rho)\, d\rho/\rho = 1$. Integration by parts gives

$$\int_{\mathbb{R}^n} \partial_j K(x)\chi(|x|)\, dx = -\int_{\mathbb{R}^n} K(x)\chi'(|x|)\frac{x_j}{|x|}\, dx,$$

and by transforming to polar coordinates, we have

$$\int_{\rho>0} \int_{|\omega|=1} \partial_j K(\rho\omega)\chi(\rho)\rho^{n-1}\, d\rho\, d\omega$$

$$= -\int_{\rho>0} \int_{|\omega|=1} K(\rho\omega)\chi'(\rho)\omega_j \rho^{n-1}\, d\rho\, d\omega,$$

which, using the homogeneity of $\partial_j K$ and K, simplifies to

$$\int_0^\infty \chi(\rho)\, \frac{d\rho}{\rho} \int_{|\omega|=1} \partial_j K(\omega)\, d\omega = -\int_0^\infty \chi'(\rho)\, d\rho \int_{|\omega|=1} K(\omega)\omega_j\, d\omega.$$

Since $\int_0^\infty \chi(\rho)\, d\rho/\rho = 1$ and $\int_0^\infty \chi'(\rho)\, d\rho = -\chi(0) = 0$, the function K_j satisfies (A.4), as required. $\qquad\square$

We are now in a position to obtain the desired extension operator.

Theorem A.4 Assume that Ω is a Lipschitz domain. For each integer $k \geq 0$, there exists an extension operator $E_k : W_2^s(\Omega) \to W_2^s(\mathbb{R}^n)$ that is bounded for $k \leq s < k + 1$.

Proof. We may assume that Ω has the form (A.1). Define the semi-infinite cone

$$V = \{y \in \mathbb{R}^n : y_n < -M|y'|\},$$

and observe that by (A.2),

$$x + V \subseteq \Omega \quad \text{for } x \in \Omega \cup \Gamma.$$

Fix an integer $k \geq 1$ and a function $\psi \in C^\infty(\mathbb{S}^{n-1})$ such that (A.3) holds, with

$$\operatorname{supp} \psi \subseteq \mathbb{S}^{n-1} \cap V.$$

Let $\chi \in C_{\text{comp}}^\infty[0, \infty)$ be a cutoff function satisfying $\chi = 1$ on a neighbourhood of 0, and replace $u(x + y)$ by $\chi(|y|)u(x + y)$ in the proof of Lemma A.2 to obtain the identity

$$u(x) = \sum_{l=0}^{k} \int_V \psi_l(y)u^{(l)}(x + y; y) \, dy \quad \text{for } x \in \Omega,$$

where

$$\psi_l(y) = \binom{k}{l} |y|^{k-l-n} \chi^{(k-l)}(|y|)\psi\left(\frac{y}{|y|}\right) \quad \text{for } 0 \le l \le k.$$

Let E_0 be the extension operator from Theorem A.1, assume $u \in \mathcal{D}(\overline{\Omega})$, and put

$$u_\alpha = E_0(\partial^\alpha u).$$

We define

$$E_k u(x) = \sum_{l=0}^{k} \int_V \psi_l(y) \sum_{|\alpha|=l} \frac{l!}{\alpha!} u_\alpha(x + y)y^\alpha \, dy = \sum_{|\alpha| \le k} (\Psi_\alpha * u_\alpha)(x)$$
$$\text{for } x \in \mathbb{R}^n,$$

where

$$\Psi_\alpha(y) = \frac{l!}{\alpha!} \psi_l(-y)(-y)^\alpha \quad \text{for } |\alpha| = l,$$

so that $E_k u|_\Omega = u$. If $|\alpha| < k$, then $\Psi_\alpha \in C_{\text{comp}}^\infty(\mathbb{R}^n)$, because if $l < k$, then $\chi^{(k-l)}$ vanishes on a neighbourhood of 0. However, if $|\alpha| = k$, then Ψ_α coincides on a neighbourhood of 0 with a homogeneous function of degree $k - n$. It follows from Lemma A.3 that

$$\|\partial^\beta(\Psi_\alpha * f)\|_{L_2(\mathbb{R}^n)} \le C\|f\|_{L_2(\mathbb{R}^n)} \quad \text{for } |\alpha| \le k \text{ and } |\beta| \le k,$$

so by Theorem A.1,

$$\|E_k u\|_{W_2^k(\mathbb{R}^n)}^2 = \sum_{|\beta| \le k} \|\partial^\beta E_k u\|_{L_2(\mathbb{R}^n)}^2 \le C \sum_{|\alpha| \le k} \|E_0(\partial^\alpha u)\|_{L_2(\mathbb{R}^n)}^2$$
$$\le C \sum_{|\alpha| \le k} \|\partial^\alpha u\|_{L_2(\Omega)}^2 = C\|u\|_{W_2^k(\Omega)}^2.$$

Also, if $0 < \mu < 1$ then

$$|\partial^\beta(\Psi_\alpha * f)|_{\mu,\mathbb{R}^n} \leq \begin{cases} C\|f\|_{L_2(\mathbb{R}^n)} & \text{if } |\alpha| < k \text{ or } |\beta| < k, \\ C|f|_{\mu,\mathbb{R}^n} & \text{if } |\alpha| = k \text{ and } |\beta| = k, \end{cases}$$

implying that

$$\sum_{|\beta|=k} |\partial^\beta E_k u|^2_{\mu,\mathbb{R}^n} \leq C \sum_{|\alpha|\leq k} \|E_0(\partial^\alpha u)\|^2_{L_2(\mathbb{R}^n)} + C \sum_{|\alpha|=k} |E_0(\partial^\alpha u)|^2_{\mu,\mathbb{R}^n}$$

$$\leq C\|u\|^2_{W_2^k(\Omega)} + C \sum_{|\alpha|=k} |\partial^\alpha u|^2_{\mu,\Omega} = C\|u\|^2_{W_2^{k+\mu}(\Omega)}. \qquad \square$$

Exercises

A.1 Show that, for any integer $k \geq 1$, if $f : [0, \infty) \to \mathbb{C}$ is a C^k function with compact support, then

$$f(0) = \frac{(-1)^k}{(k-1)!} \int_0^\infty t^{k-1} f^{(k)}(t)\, dt.$$

A.2 In Exercise A.3, we shall need a sequence $(\lambda_k)_{k=0}^\infty$ satisfying

$$\sum_{k=0}^\infty 2^{jk}\lambda_k = (-1)^j \quad \text{for all } j \geq 0. \tag{A.5}$$

(i) Let $a_0, \ldots, a_{N-1}$ be distinct complex numbers, let b be any complex number, and consider the $N \times N$ linear system

$$\sum_{k=0}^{N-1} (a_k)^j x_k = b^j \quad \text{for } 0 \leq j \leq N-1.$$

Show that the unique solution is given by

$$x_k = \prod_{\substack{j=0 \\ j \neq k}}^{N-1} \frac{a_j - b}{a_j - a_k} \quad \text{for } 0 \leq k \leq N-1.$$

[Hint: by Cramer's rule, x_k is the ratio of two Vandermonde determinants.]

(ii) Show that

$$1 \leq \prod_{j=0}^\infty (1 + 2^{-j}) \leq C \quad \text{and} \quad c \leq \prod_{j=1}^\infty (1 - 2^{-j}) \leq 1.$$

(iii) Deduce that the sequence

$$\lambda_k = \prod_{\substack{j=0 \\ j \neq k}}^{\infty} \frac{1 + 2^{-j}}{1 - 2^{k-j}}$$

satisfies

$$|\lambda_k| \leq C \prod_{j=1}^{k} \frac{1}{2^j - 1} \leq C \prod_{j=1}^{k} \frac{1}{2^{j-1}} = \frac{C}{2^{k(k-1)/2}}$$

and is a solution of the infinite linear system (A.5).

A.3 Seeley [93] has given a simple construction of an extension operator for the half space $\Omega = \mathbb{R}^n_-$. Let the sequence $(\lambda_k)_{k=1}^{\infty}$ be as in Exercise A.2, and define

$$Eu(x) = \begin{cases} u(x) & \text{if } x_n \leq 0, \\ \displaystyle\sum_{k=0}^{\infty} \lambda_k u(x', -2^k x_n) & \text{if } x_n > 0. \end{cases}$$

(i) Show that $E : \mathcal{D}(\overline{\Omega}) \to \mathcal{D}(\mathbb{R}^n)$.
(ii) Show that $E : W_p^s(\Omega) \to W_p^s(\mathbb{R}^n)$ is bounded for $s \geq 0$ and $1 \leq p \leq \infty$.

A.4 Let $0 < \mu < 1$ and $\epsilon > 0$, and choose a cutoff function $\chi \in C^{\infty}_{\text{comp}}[0, \infty)$ satisfying $\chi(y) = 1$ for $0 \leq y < 1$. Define the $C^{0,\mu}$ epigraph

$$\Omega = \{(x, y) \in \mathbb{R}^2 : y > |x|^{\mu}\},$$

and the function

$$u(x, y) = y^{1-\epsilon} \chi(y) \quad \text{for } (x, y) \in \Omega.$$

(i) Show that $u \in W_2^2(\Omega)$ if $\epsilon < \frac{1}{2}(\mu^{-1} - 1)$.
(ii) Show that $u \notin C^{0,\lambda}(\overline{\Omega})$ if $1 - \epsilon < \lambda \leq 1$.
(iii) Deduce that no extension operator from $W_2^2(\Omega)$ to $W_2^2(\mathbb{R}^2)$ exists. [Hint: use Theorem 3.26.]

Appendix B

Interpolation Spaces

Suppose that X_0 and X_1 are normed spaces, and that both are subspaces of some larger (not necessarily normed) vector space. In this case, X_0 and X_1 are said to form a *compatible pair* $X = (X_0, X_1)$, and we equip the subspaces $X_0 \cap X_1$ and $X_0 + X_1$ with the norms

$$\|u\|_{X_0 \cap X_1} = \left(\|u\|_{X_0}^2 + \|u\|_{X_1}^2 \right)^{1/2}$$

and

$$\|u\|_{X_0 + X_1} = \inf \left\{ \left(\|u_0\|_{X_0}^2 + \|u_1\|_{X_1}^2 \right)^{1/2} : \right.$$
$$\left. u = u_0 + u_1 \text{ where } u_0 \in X_0 \text{ and } u_1 \in X_1 \right\}.$$

Notice that for $j = 0$ and 1,

$$X_0 \cap X_1 \subseteq X_j \subseteq X_0 + X_1,$$

and these inclusions are continuous because

$$\|u\|_{X_0 + X_1} \leq \|u\|_{X_j} \leq \|u\|_{X_0 \cap X_1}.$$

If $X_1 \subseteq X_0$, then $X_0 \cap X_1 = X_1$ and $X_0 + X_1 = X_0$.

In this appendix, we present a general method for constructing, from any given compatible pair X, a family of normed spaces

$$X_{\theta,q} = (X_0, X_1)_{\theta,q} \qquad \text{for } 0 < \theta < 1 \text{ and } 1 \leq q \leq \infty,$$

each of which is intermediate with respect to X_0 and X_1, in the sense that

$$X_0 \cap X_1 \subseteq X_{\theta,q} \subseteq X_0 + X_1. \tag{B.1}$$

Moreover, we shall see that $X_{\theta,q}$ has the following *interpolation* property. Take a second compatible pair of normed spaces $Y = (Y_0, Y_1)$, and two bounded linear operators

$$A_0 : X_0 \to Y_0 \quad \text{and} \quad A_1 : X_1 \to Y_1.$$

If

$$A_0 u = A_1 u \quad \text{for } u \in X_0 \cap X_1,$$

then A_0 and A_1 are said to be *compatible*, and there is a unique bounded linear operator

$$A_\theta : X_{\theta,q} \to Y_{\theta,q}$$

such that

$$A_\theta u = A_0 u = A_1 u \quad \text{for } u \in X_0 \cap X_1. \tag{B.2}$$

We will also show that if X_0 and X_1 are Sobolev spaces based on L_2, then so is $X_{\theta,2}$.

For technical reasons, it is convenient to construct $X_{\theta,q}$ in two different ways. Thus, we shall define two spaces, $K_{\theta,q}(X)$ and $J_{\theta,q}(X)$, and show

$$X_{\theta,q} = K_{\theta,q}(X) = J_{\theta,q}(X),$$

with equivalent norms. The K-method will be used to prove the interpolation properties of $H^s(\Omega)$, after which the interpolation properties of $\widetilde{H}^s(\Omega)$ follow by a duality argument that relies on the J-method. We conclude by considering the interpolation properties of $H^s(\Gamma)$. For more on the theory of interpolation spaces, see Bergh and Löfström [5].

The K-Method

The K-*functional* is defined for $t > 0$ and $u \in X_0 + X_1$ by

$$K(t, u) = \inf \left\{ \left(\|u_0\|_{X_0}^2 + t^2 \|u_1\|_{X_1}^2 \right)^{1/2} : \right.$$
$$\left. u = u_0 + u_1 \text{ where } u_0 \in X_0 \text{ and } u_1 \in X_1 \right\}.$$

When necessary, we write $K(t, u; X)$ to show explicitly the choice of the compatible pair $X = (X_0, X_1)$. For fixed $t > 0$, the K-functional is an equivalent norm on $X_0 + X_1$:

$$K(t, \lambda u) = |\lambda| K(t, u), \qquad K(t, u + v) \le K(t, u) + K(t, v)$$

and

$$\min(1, t)\|u\|_{X_0+X_1} \le K(t, u) \le \max(1, t)\|u\|_{X_0+X_1}.$$

If we fix u, then the K-functional is a non-decreasing function of t, and in fact, for all positive s and t,

$$\min(1, t/s)K(s, u) \le K(t, u) \le \max(1, t/s)K(s, u).$$

Next, we define a weighted L_q-norm,

$$\|f\|_{\theta,q} = \left(\int_0^\infty |t^{-\theta} f(t)|^q \frac{dt}{t} \right)^{1/q} \qquad \text{for } 0 < \theta < 1 \text{ and } 1 \le q < \infty,$$

with the obvious modification when $q = \infty$,

$$\|f\|_{\theta,\infty} = \operatorname*{ess\,sup}_{t>0} |t^{-\theta} f(t)|.$$

This weighted norm has an important dilatation property, namely,

$$\|t \mapsto f(at)\|_{\theta,q} = a^\theta \|f\|_{\theta,q} \quad \text{for } a > 0. \tag{B.3}$$

Now define

$$K_{\theta,q}(X) = \{u \in X_0 + X_1 : \|K(\cdot, u)\|_{\theta,q} < \infty\},$$

and put

$$\|u\|_{K_{\theta,q}(X)} = N_{\theta,q} \|K(\cdot, u)\|_{\theta,q},$$

where the constant $N_{\theta,q} > 0$ may be any desired normalisation factor. As the default value, we take

$$N_{\theta,q} = \frac{1}{\|\min(1, \cdot)\|_{\theta,q}} = \begin{cases} [q\theta(1 - \theta)]^{1/q} & \text{if } 1 \le q < \infty, \\ 1 & \text{if } q = \infty, \end{cases} \tag{B.4}$$

and thereby simplify the statement of the next lemma.

Lemma B.1 Assume the normalisation (B.4).

(i) If $u \in X_0 \cap X_1$, then $u \in K_{\theta,q}(X)$ and

$$\|u\|_{K_{\theta,q}(X)} \le \|u\|_{X_0}^{1-\theta} \|u\|_{X_1}^{\theta} \le \|u\|_{X_0 \cap X_1}.$$

(ii) If $u \in K_{\theta,q}(X)$, then $u \in X_0 + X_1$, with

$$K(t, u) \le t^{\theta} \|u\|_{K_{\theta,q}(X)} \quad and \quad \|u\|_{X_0+X_1} \le \|u\|_{K_{\theta,q}(X)}.$$

Proof. We can assume $u \ne 0$. Put $a = \|u\|_{X_1}/\|u\|_{X_0}$, so that

$$K(t, u) \le \min\big(\|u\|_{X_0}, t\|u\|_{X_1}\big) = \|u\|_{X_0} \min(1, at), \tag{B.5}$$

and hence $\|K(\cdot, u)\|_{\theta,q} \le \|u\|_{X_0} a^{\theta} \|\min(1, \cdot)\|_{\theta,q} = \|u\|_{X_0}^{1-\theta} \|u\|_{X_1}^{\theta}/N_{\theta,q}$, implying the inequality $\|u\|_{K_{\theta,q}(X)} \le \|u\|_{X_0}^{1-\theta} \|u\|_{X_1}^{\theta}$. Next, taking $p = (1-\theta)^{-1}$, we have

$$\|u\|_{X_0}^{1-\theta} \|u\|_{X_1}^{\theta} \le \frac{\big(\|u\|_{X_0}^{1-\theta}\big)^p}{p} + \frac{\big(\|u\|_{X_1}^{\theta}\big)^{p^*}}{p^*} = (1-\theta)\|u\|_{X_0} + \theta\|u\|_{X_1}$$

$$\le \big[(1-\theta)^2 + \theta^2\big]^{1/2} \|u\|_{X_0 \cap X_1},$$

completing the proof of (i). In a similar fashion, the inequality $\min(1, s/t) K(t, u) \le K(s, u)$ implies that

$$t^{-\theta} \|\min(1, \cdot)\|_{\theta,q} K(t, u) \le \|K(\cdot, u)\|_{\theta,q},$$

and so $K(t, u) \le t^{\theta} \|u\|_{X_{\theta,q}}$. Finally, because $\min(1, t)\|u\|_{X_0+X_1} \le K(t, u)$ we have $\|u\|_{X_0+X_1}/N_{\theta,q} \le \|K(\cdot, u)\|_{\theta,q}$, which completes the proof of (ii). □

Lemma B.1 shows that $X_{\theta,q} = K_{\theta,q}(X)$ satisfies (B.1), with continuous inclusions. The next theorem establishes the interpolation property. Here, the choice of normalisation is irrelevant (provided it is the same for X and Y).

Theorem B.2 If the bounded linear operators $A_0 : X_0 \to Y_0$ and $A_1 : X_1 \to Y_1$ are compatible, then there exists a unique bounded linear operator $A_{\theta} : K_{\theta,q}(X) \to K_{\theta,q}(Y)$ satisfying (B.2). In this case, if M_0 and M_1 are positive constants such that

$$\|A_j u\|_{Y_j} \le M_j \|u\|_{X_j} \quad for\ u \in X_j\ and\ j = 0, 1,$$

then

$$\|A_{\theta} u\|_{K_{\theta,q}(Y)} \le M_0^{1-\theta} M_1^{\theta} \|u\|_{K_{\theta,q}(X)} \quad for\ u \in K_{\theta,q}(X).$$

Proof. If A_θ exists, then it must satisfy $A_\theta u = A_0 u_0 + A_1 u_1$ whenever $u = u_0 + u_1$ with $u_j \in X_j$, so uniqueness is clear, and Exercise B.1 shows that $A_\theta u$ is in fact well defined in $Y_0 + Y_1$. Put $a = M_1/M_0$, so that

$$K(t, A_\theta u; Y) \le \left(\|A_0 u_0\|_{Y_0}^2 + t^2 \|A_1 u_1\|_{Y_1}^2 \right)^{1/2} \le \left(M_0^2 \|u_0\|_{X_0}^2 + t^2 M_1^2 \|u_1\|_{X_1}^2 \right)^{1/2}$$

$$\le M_0 \left(\|u_0\|_{X_0}^2 + (at)^2 \|u_1\|_{X_1}^2 \right)^{1/2}.$$

Hence, $K(t, Au : Y) \le M_0 K(at, u; X)$, and so by (B.3),

$$\|A_\theta u\|_{K_{\theta,q}(Y)} \le M_0 a^\theta \|u\|_{K_{\theta,q}(X)} = M_0^{1-\theta} M_1^\theta \|u\|_{K_{\theta,q}(X)}. \qquad \square$$

The *J*-Method

Our second construction of $X_{\theta,q}$ uses the Bochner integral for functions taking values in a normed space, and we digress briefly to review a few pertinent definitions and facts; see Yosida [106, pp. 130–136] for more details.

Let (S, μ) be a measure space and let Z be a normed space. We say that a function $f : S \to Z$ is *finite-valued* if there exist finitely many vectors $u_k \in Z$ and mutually disjoint μ-measurable sets $E_k \subseteq S$ with $\mu(E_k) < \infty$ such that f takes the constant value u_k on E_k, and is identically zero on $S \setminus \bigcup_k E_k$. In this case, the integral of f is a well-defined vector in Z, given by

$$\int_S f(t) \, d\mu_t = \sum_k u_k \mu(E_k).$$

A function $f : S \to Z$ is said to be *strongly Z-measurable* if there exists a sequence $f_m : S \to Z$ of finite-valued functions such that $f_m(t)$ converges to $f(t)$ in the norm of Z for μ-almost all $t \in S$. In this case, the real-valued function $t \mapsto \|f_m(t) - f(t)\|_Z$ is measurable, and if the condition

$$\lim_{m \to \infty} \int_S \|f_m(t) - f(t)\|_Z \, d\mu_t = 0$$

is satisfied, then f is said to be *Z-integrable*. The integral of such an f is well defined, as a vector in the *completion* of Z, by taking the limit of the integrals of the f_m. A strongly Z-measurable function $f : S \to Z$ is Z-integrable if and only if the real-valued function $t \mapsto \|f(t)\|_Z$ is integrable, in which case

$$\left\| \int_S f(t) \, d\mu_t \right\|_Z \le \int_S \|f(t)\|_Z \, d\mu_t.$$

Resuming our consideration of $X_{\theta,q}$, we define the *J-functional*,

$$J(t, u) = \left(\|u\|_{X_0}^2 + t^2 \|u\|_{X_1}^2 \right)^{1/2} \qquad \text{for } t > 0 \text{ and } u \in X_0 \cap X_1.$$

For each fixed $t > 0$, the J-functional is equivalent to the usual norm on $X_0 \cap X_1$, and for each fixed $u \in X_0 \cap X_1$, the function $t \mapsto J(t, u)$ is non-decreasing. Furthermore,

$$\min(1, t/s) J(s, u) \leq J(t, u) \leq \max(1, t/s) J(s, u)$$

and

$$K(t, u) \leq \min(1, t/s) J(s, u). \tag{B.6}$$

We define $J_{\theta,q}(X)$ to be the subspace of $X_0 + X_1$ consisting of those vectors u that possess a representation

$$u = \int_0^\infty f(t) \frac{dt}{t} \tag{B.7}$$

for some function f satisfying

$$\int_0^\infty \|f(t)\|_{X_0 + X_1} \frac{dt}{t} < \infty \quad \text{and} \quad \int_a^b \|f(t)\|_{X_0 \cap X_1} \frac{dt}{t} < \infty$$

$$\text{for } 0 < a < b < \infty.$$

Of course, f is assumed to be $(X_0 \cap X_1)$-measurable, and hence also $(X_0 + X_1)$-measurable. The space $J_{\theta,q}(X)$ is equipped with the norm

$$\|u\|_{J_{\theta,q}(X)} = \inf \left\| t \mapsto J\big(t, f(t)\big) \right\|_{\theta,q},$$

where the infimum is taken over all possible representations (B.7).

Theorem B.3 If $0 < \theta < 1$ and $1 \leq q \leq \infty$, then $J_{\theta,q}(X) = K_{\theta,q}(X)$ with equivalent norms.

Proof. Suppose that $u \in J_{\theta,q}(X)$ has a representation (B.7). Using (B.6), we see that

$$K(t, u) \leq \int_0^\infty K\big(t, f(s)\big) \frac{ds}{s} \leq \int_0^\infty \min(1, t/s) J\big(s, f(s)\big) \frac{ds}{s}$$

$$= \int_0^\infty \min(1, 1/s) J\big(ts, f(ts)\big) \frac{ds}{s},$$

so

$$\|K(\cdot, u)\|_{\theta,q} \leq \int_0^\infty \min(1, 1/s)s^\theta \|J(\cdot, f(\cdot))\|_{\theta,q} \frac{ds}{s}$$

and hence

$$\|K(\cdot, u)\|_{\theta,q} \leq \|u\|_{J_{\theta,q}(X)} \int_0^\infty \min(1, 1/s)s^\theta \frac{ds}{s} = \frac{\|u\|_{J_{\theta,q}(X)}}{\theta(1-\theta)}.$$

Conversely, suppose that $u \in K_{\theta,q}(X)$, and let $\epsilon > 0$. For each $m \in \mathbb{Z}$ there is a decomposition $u = u_{0m} + u_{1m}$ with $u_{jm} \in X_j$ and

$$\|u_{0m}\|_{X_0}^2 + t_m^2 \|u_{1m}\|_{X_1}^2 \leq (1+\epsilon)K(t_m, u)^2, \qquad \text{where } t_m = 2^m.$$

By Lemma B.1, $K(t_m, u) \leq Ct_m^\theta \|u\|_{K_{\theta,q}(X)}$, so $\|u_{0m}\|_{X_0} = O(t_m^\theta)$ and $\|u_{1m}\|_{X_1} = O(t_m^{\theta-1})$. In particular, $\|u_{0m}\|_{X_0} \to 0$ as $m \to -\infty$, and $\|u_{1m}\|_{X_1} \to 0$ as $m \to +\infty$. Define $f : (0, \infty) \to X_0 \cap X_1$ by

$$f(t) = \frac{u_{0m} - u_{0,m-1}}{\log 2} = \frac{u_{1,m-1} - u_{1m}}{\log 2} \quad \text{for } t_{m-1} \leq t < t_m,$$

so that

$$
\begin{aligned}
\left\| u - \int_{t_{-M}}^{t_{M'}} f(t) \frac{dt}{t} \right\|_{X_0+X_1} &= \left\| u - \sum_{-M < m \leq M'} (u_{0m} - u_{0,m-1}) \right\|_{X_0+X_1} \\
&= \|u - u_{0,M'} + u_{0,-M}\|_{X_0+X_1} \\
&= \|u_{0,-M} + u_{1,M'}\|_{X_0+X_1} \\
&\leq \left(\|u_{0,-M}\|_{X_0}^2 + \|u_{1,M'}\|_{X_1}^2 \right)^{1/2},
\end{aligned}
$$

implying the representation formula (B.7). Moreover, if $t_{m-1} \leq t < t_m$, then

$$
\begin{aligned}
(\log 2)^2 J\big(t, f(t)\big)^2 &= \|u_{0m} - u_{0,m-1}\|_{X_0}^2 + t^2 \|u_{1,m-1} - u_{1m}\|_{X_1}^2 \\
&\leq 2\big[\|u_{0m}\|_{X_0}^2 + t_m^2 \|u_{1m}\|_{X_1}^2 + \|u_{0,m-1}\|_{X_0}^2 \\
&\quad + 4t_{m-1}^2 \|u_{1,m-1}\|_{X_1}^2 \big] \\
&\leq 2\big[(1+\epsilon)K(t_m, u)^2 + 4(1+\epsilon)K(t_{m-1}, u)^2 \big] \\
&\leq 2(1+\epsilon)\big[(t_m/t)^2 + 4 \big] K(t, u)^2 \leq 16(1+\epsilon)K(t, u)^2,
\end{aligned}
$$

and so $(\log 2)\|J(\cdot, f(\cdot))\|_{\theta,q} \leq 4\sqrt{1+\epsilon}\|K(\cdot, u)\|_{\theta,q}$. Hence, $u \in J_{\theta,q}(X)$ and $\|u\|_{J_{\theta,q}(X)} \leq (4/\log 2)\|K(\cdot, u)\|_{\theta,q}$. $\qquad \square$

We now show that the K- and J-functionals are dual to each other.

Lemma B.4 If $X_0 \cap X_1$ is dense in X_0 and in X_1, then $X_0 \cap X_1$ is dense in $X_0 + X_1$, and $X^* = (X_0^*, X_1^*)$ is a compatible pair of Banach spaces. Moreover,

$$X_0^* + X_1^* = (X_0 \cap X_1)^* \quad and \quad X_0^* \cap X_1^* = (X_0 + X_1)^*,$$

with equal norms. In fact, for each $t > 0$,

$$K(t, g; X^*) = \sup_{0 \neq u \in X_0 \cap X_1} \frac{|\langle g, u \rangle|}{J(t^{-1}, u; X)} \quad and$$

$$J(t, g; X^*) = \sup_{0 \neq u \in X_0 + X_1} \frac{|\langle g, u \rangle|}{K(t^{-1}, u; X)}.$$

Proof. Given $u = u_0 + u_1 \in X_0 + X_1$, the density assumption means that we can find sequences u_{0m} and u_{1m} in $X_0 \cap X_1$ such that $u_{jm} \to u_j$ in X_j as $m \to \infty$. Define $u_m = u_{0m} + u_{1m} \in X_0 \cap X_1$, and observe that $u_m - u = (u_{0m} - u_0) + (u_{1m} - u_1)$, so $\|u_m - u\|_{X_0 + X_1}^2 \leq \|u_{0m} - u_0\|_{X_0}^2 + \|u_{1m} - u_1\|_{X_1}^2$. Hence $u_m \to u$ in $X_0 + X_1$, showing that $X_0 \cap X_1$ is dense in $X_0 + X_1$. Also, the continuous dense inclusion $X_0 \cap X_2 \subseteq X_j$ gives an imbedding $X_j^* \subseteq (X_0 \cap X_1)^*$, so X_0^* and X_1^* can be viewed as subspaces of $(X_0 \cap X_1)^*$, and hence form a compatible pair.

Let $g \in X_0^* + X_1^*$, and take any decomposition $g = g_0 + g_1$ with $g_j \in X_j^*$. For $u \in X_0 \cap X_1$, the Cauchy–Schwarz inequality implies that

$$|\langle g, u \rangle| = |\langle g_0, u \rangle + \langle g_1, u \rangle| \leq \left(\|g_0\|_{X_0^*}^2 + t^2 \|g_1\|_{X_1^*}^2 \right)^{1/2} \left(\|u\|_{X_0}^2 + t^{-2} \|u\|_{X_1}^2 \right)^{1/2},$$

so $|\langle g, u \rangle| \leq K(t, g; X^*) J(t^{-1}, u; X)$. In particular, by putting $t = 1$ we see that $g \in (X_0 \cap X_1)^*$ and $\|g\|_{(X_0 \cap X_1)^*} \leq \|g\|_{X_0^* + X_1^*}$.

Conversely, let $g \in (X_0 \cap X_1)^*$ and put

$$M_t = \sup_{0 \neq u \in X_0 \cap X_1} \frac{|\langle g, u \rangle|}{J(t^{-1}, u, ; X)}.$$

We equip the product space $X_0 \times X_1$ with the norm

$$\|(u_0, u_1)\|_{X_0 \times X_1} = \left(\|u_0\|_{X_0}^2 + t^{-2} \|u_1\|_{X_1}^2 \right)^{1/2},$$

and denote the diagonal subspace by $W = \{(u, u) : u \in X_0 \cap X_1\}$. Now define a linear functional g_W on W by

$$\langle g_W, (u, u) \rangle = \langle g, u \rangle \quad \text{for } u \in X_0 \cap X_1,$$

and observe that

$$|\langle g_W, (u, u)\rangle| \le M_t J(t^{-1}, u; X) = M_t \|(u, u)\|_{X_0 \times X_1},$$

so the Hahn–Banach theorem gives $(g_0, g_1) \in X_0^* \times X_1^* = (X_0 \times X_1)^*$ such that

$$\langle g_W, (u, u)\rangle = \langle (g_0, g_1), (u, u)\rangle = \langle g_0, u\rangle + \langle g_1, u\rangle \quad \text{for } u \in X_0 \cap X_1,$$

and

$$\|(g_0, g_1)\|_{X_0^* \times X_1^*} = \left(\|g_0\|_{X_0^*}^2 + t^2 \|g_1\|_{X_1^*}^2\right)^{1/2} = M_t.$$

Hence, $g = g_0 + g_1 \in X_0^* + X_1^*$, and $K(t, g; X^*) \le M_t$. In particular, putting $t = 1$ gives $\|g\|_{X_0^* + X_1^*} \le M_1 = \|g\|_{(X_0 \cap X_1)^*}$.

Next, let $g \in X_0^* \cap X_1^*$. If $u = u_0 + u_1 \in X_0 + X_1$, then

$$|\langle g, u\rangle| = |\langle g, u_0\rangle + \langle g, u_1\rangle| \le \left(\|g\|_{X_0^*}^2 + t^2 \|g\|_{X_1^*}^2\right)^{1/2} \left(\|u_0\|_{X_0}^2 + t^{-2} \|u_1\|_{X_1}^2\right)^{1/2},$$

so $|\langle g, u\rangle| \le J(t, g; X^*) K(t^{-1}, u; X)$, and by putting $t = 1$ we see that $g \in (X_0 + X_1)^*$ with $\|g\|_{(X_0 + X_1)^*} \le \|g\|_{X_0^* \cap X_1^*}$.

Conversely, let $g \in (X_0 + X_1)^*$, and put

$$M_t = \sup_{0 \ne u \in X_0 + X_1} \frac{|\langle g, u\rangle|}{K(t^{-1}, u; X)}.$$

If $u \in X_j$, then $|\langle g, u\rangle| \le M_t K(t^{-1}, u; X) \le M_t \min \left(\|u\|_{X_0}, t^{-1} \|u\|_{X_1}\right)$, showing that $g \in X_0^* \cap X_1^*$. Given $\epsilon > 0$, we can find $u_j \in X_j$ such that

$$\langle g, u_j\rangle \le \|g\|_{X_j^*} \le (1 + \epsilon)\langle g, u_j\rangle \quad \text{and} \quad \|u_j\|_{X_j} = 1.$$

(In particular, $\langle g, u_j\rangle$ is *real*.) Put $\lambda_j = \|g\|_{X_j^*}$, so that $\|g\|_{X_j^*}^2 \le (1 + \epsilon)\langle g, \lambda_j u\rangle$ and

$$
\begin{aligned}
J(t, g; X^*)^2 &\le (1 + \epsilon)\langle g, \lambda_0 u_0 + t^2 \lambda_1 u_1\rangle \\
&\le (1 + \epsilon) M_t K(t^{-1}, \lambda_0 u_0 + t^2 \lambda_1 u_1; X) \\
&\le (1 + \epsilon) M_t \left(\|\lambda_0 u_0\|_{X_0}^2 + t^{-2} \|t^2 \lambda_1 u_1\|_{X_1}^2\right)^{1/2} \\
&= (1 + \epsilon) M_t \left(\|g\|_{X_0^*}^2 + t^2 \|g\|_{X_1^*}^2\right)^{1/2} = (1 + \epsilon) M_t J(t, g; X^*).
\end{aligned}
$$

Hence, $J(t, g; X^*) \le M_t$, and in particular $\|g\|_{X_0^* \cap X_1^*} \le \|g\|_{(X_0 + X_1)^*}$. $\qquad\square$

Theorem B.5 Assume that $X_0 \cap X_1$ is dense in X_0 and in X_1. If $0 < \theta < 1$ and $1 \leq q < \infty$, then $X_0 \cap X_1$ is dense in $X_{\theta,q}$ and

$$(X_0, X_1)^*_{\theta,q} = (X_0^*, X_1^*)_{\theta,q^*}, \qquad \text{where } \frac{1}{q^*} + \frac{1}{q} = 1,$$

with equivalent norms.

Proof. Let $u \in J_{\theta,q}(X)$ have a representation (B.7), and define $u_m \in X_0 \cap X_1$ by

$$u_m = \int_{1/m}^{m} f(t) \frac{dt}{t}.$$

Since

$$\|u_m - u\|_{J_{\theta,q}(X)} \leq \left(\int_{(0,1/m) \cup (m,\infty)} \left| t^{-\theta} J\left(t, f(t)\right) \right|^q \frac{dt}{t} \right)^{1/q},$$

we see that $u_m \to u$ in $J_{\theta,q}(X)$, and hence $X_0 \cap X_1$ is dense in $X_{\theta,q}$. Thus, the inclusions $X_0 \cap X_1 \subseteq X_{\theta,q} \subseteq X_0 + X_1$ are continuous and have dense images, so by Lemma B.4,

$$X_0^* \cap X_1^* \subseteq (X_{\theta,q})^* \subseteq X_0^* + X_1^*.$$

It suffices to show that there are continuous inclusions

$$J_{\theta,q^*}(X^*) \subseteq K_{\theta,q}(X)^* \quad \text{and} \quad J_{\theta,q}(X)^* \subseteq K_{\theta,q^*}(X^*).$$

Let $g \in J_{\theta,q^*}(X^*)$, and take any $\phi : (0, \infty) \to X_0^* \cap X_1^*$ such that

$$g = \int_0^{\infty} \phi(t) \frac{dt}{t},$$

with the integral converging in $X_0^* + X_1^* = (X_0 \cap X_1)^*$. For any $u \in K_{\theta,q}(X)$,

$$|\langle g, u \rangle| \leq \int_0^{\infty} |\langle \phi(t), u \rangle| \frac{dt}{t} \leq \int_0^{\infty} J\left(t, \phi(t); X^*\right) K(t^{-1}, u; X) \frac{dt}{t}$$

$$\leq \left\| t \mapsto J\left(t, \phi(t); X^*\right) \right\|_{\theta,q^*} \left\| t \mapsto K(t^{-1}, u; X) \right\|_{-\theta,q},$$

so $|\langle g, u \rangle| \leq \|g\|_{J_{\theta,q^*}(X^*)} \left\| t \mapsto K(t^{-1}, u; X) \right\|_{-\theta,q} = \|g\|_{J_{\theta,q^*}(X^*)} \|K(\cdot, u; X)\|_{\theta,q}$, showing that $g \in K_{\theta,q}(X)^*$ and $\|g\|_{K_{\theta,q}(X)^*} \leq N_{\theta,q}^{-1} \|g\|_{J_{\theta,q^*}(X^*)}$.

Now let $g \in J_{\theta,q}(X)^*$. Given $\epsilon > 0$, we can use Lemma B.4 to find a piecewise-constant function $\psi : (0, \infty) \to X_0 \cap X_1$ such that $\langle g, \psi(t) \rangle$ is real,

and

$$K(t^{-1}, g; X^*) \le (1 + \epsilon) \frac{\langle g, \psi(t) \rangle}{J(t, \psi(t); X)}.$$

For any measurable function $f : (0, \infty) \to (0, \infty)$ such that $\|f\|_{\theta,q} < \infty$, the integral

$$u_f = \int_0^\infty \frac{f(t) \psi(t)}{J(t, \psi(t); X)} \frac{dt}{t}$$

defines a vector $u_f \in J_{\theta,q}(X)$ satisfying $\|u_f\|_{J_{\theta,q}(X)} \le \|f\|_{\theta,q}$. We choose f so that

$$\int_0^\infty K(t^{-1}, g; X^*) f(t) \frac{dt}{t} = \left\| t \mapsto K(t^{-1}, g; X^*) \right\|_{-\theta,q^*} = \|K(\cdot, g; X^*)\|_{\theta,q^*}$$

and $\|f\|_{\theta,q} = 1$. Thus,

$$\|K(\cdot, g; X^*)\|_{\theta,q^*} \le \int_0^\infty (1 + \epsilon) \frac{\langle g, \psi(t) \rangle}{J(t, \psi(t); X)} f(t) \frac{dt}{t} = (1 + \epsilon) \langle g, u_f \rangle$$

$$\le (1 + \epsilon) \|g\|_{J_{\theta,q}(X)^*} \|u_f\|_{J_{\theta,q}(X)} \le (1 + \epsilon) \|g\|_{J_{\theta,q}(X)^*},$$

showing that $g \in K_{\theta,q^*}(X^*)$ and $\|g\|_{K_{\theta,q^*}(X^*)} \le N_{\theta,q^*} \|g\|_{J_{\theta,q}(X)^*}$. $\square$

The next result is known as the reiteration theorem.

Theorem B.6 *Let $1 \le q \le \infty$, and for $j = 0$ and 1 put*

$$Y_j = \begin{cases} X_{\theta_j, q} & \text{if } 0 < \theta_j < 1, \\ X_j & \text{if } \theta_j = j. \end{cases}$$

If $\theta_0 \ne \theta_1$, then

$$(Y_0, Y_1)_{\eta,q} = (X_0, X_1)_{\theta,q} \qquad \text{for } \theta = (1 - \eta)\theta_0 + \eta\theta_1 \text{ and } 0 < \eta < 1.$$

Proof. Let $u \in K_{\eta,q}(Y)$. For any $u_j \in Y_j$ such that $u = u_0 + u_1$, we have

$$K(t, u; X) \le K(t, u_0; X) + K(t, u_1; X) \le Ct^{\theta_0} \|u_0\|_{Y_0} + Ct^{\theta_1} \|u_1\|_{Y_1}$$

$$\le Ct^{\theta_0} \left[\|u_0\|_{Y_0}^2 + t^{2(\theta_1 - \theta_0)} \|u_1\|_{Y_1}^2 \right]^{1/2},$$

where, in the second step, we use Lemma B.1(ii) if $0 < \theta_j < 1$, or (B.5) if $\theta_j \in \{0, 1\}$. Hence,

$$K(t, u; X) \le Ct^{\theta_0} K(t^{\theta_1 - \theta_0}, u; Y),$$

and therefore the substitution $\tau = t^{\theta_1 - \theta_0}$ gives

$$\|u\|^q_{K_{\theta,q}(X)} \le C \int_0^\infty \left| t^{-\theta} t^{\theta_0} K(t^{\theta_1 - \theta_0}, u; Y) \right|^q \frac{dt}{t}$$

$$= C \int_0^\infty \left| \tau^{-\eta} K(\tau, u; Y) \right|^q |\theta_1 - \theta_0| \frac{d\tau}{\tau} \qquad \text{(B.8)}$$

$$= C |\theta_1 - \theta_0| \|u\|^q_{K_{\eta,q}(Y)}$$

for $1 \le q < \infty$, with

$$\|u\|_{K_{\theta,\infty}(X)} \le C \sup_{t>0} t^{-\theta} t^{\theta_0} K(t^{\theta_1 - \theta_0}, u; Y) = C \sup_{\tau>0} \tau^{-\eta} K(\tau, u; Y) = C \|u\|_{K_{\eta,\infty}(Y)}$$

in the limiting case $q = \infty$. Thus, $Y_{\eta,q} \subseteq X_{\theta,q}$.

Conversely, let $u \in X_{\theta,q}$, and take any representation of the type (B.7). By (B.6),

$$t^{\theta_0} K(t^{\theta_1 - \theta_0}, u; Y) \le \int_0^\infty t^{\theta_0} K\left(t^{\theta_1 - \theta_0}, f(s); Y\right) \frac{ds}{s}$$

$$\le \int_0^\infty t^{\theta_0} \min\left(1, (t/s)^{\theta_1 - \theta_0}\right) J\left(s^{\theta_1 - \theta_0}, f(s); Y\right) \frac{ds}{s}.$$

Using Exercise B.2 if $0 < \theta_j < 1$, or just the definition of the J-functional if $\theta_j = j$, we see that $\|f(s)\|_{Y_j} \le C s^{-\theta_j} J(s, f(s); X)$ and so

$$J\left(s^{\theta_1 - \theta_0}, f(s); Y\right) = \left[\|f(s)\|^2_{Y_0} + s^{2(\theta_1 - \theta_0)} \|f(s)\|^2_{Y_1} \right]^{1/2} \le C s^{-2\theta_0} J(s, f(s); X).$$

Thus, with the help of the substitution $s = \sigma t$,

$$t^{\theta_0} K(t^{\theta_1 - \theta_0}, u; Y) \le C \int_0^\infty \min\left((t/s)^{\theta_0}, (t/s)^{\theta_1}\right) J(s, f(s); X) \frac{ds}{s}$$

$$= C \int_0^\infty \min(\sigma^{-\theta_0}, \sigma^{-\theta_1}) J(\sigma t, f(\sigma t); X) \frac{d\sigma}{\sigma},$$

and hence by (B.8),

$$\|u\|_{K_{\eta,q}(Y)} = \frac{1}{|\theta_1 - \theta_0|} \|t \mapsto t^{\theta_0} K(t^{\theta_1 - \theta_0}, u; Y)\|_{\theta,q}$$

$$\le C \int_0^\infty \min(\sigma^{-\theta_0}, \sigma^{-\theta_1}) \|t \mapsto J(\sigma t, f(\sigma t); X)\|_{\theta,q} \frac{d\sigma}{\sigma}$$

$$= C \int_0^\infty \sigma^\theta \min(\sigma^{-\theta_0}, \sigma^{-\theta_1}) \frac{d\sigma}{\sigma} \|t \mapsto J(t, f(t); X)\|_{\theta,q}.$$

Finally, taking the infimum over all f satisfying (B.7) gives $\|u\|_{K_{\eta,q}(Y)} \leq C\|u\|_{J_{\theta,q}(X)}$, proving that $X_{\theta,q} \subseteq Y_{\eta,q}$. $\qquad\square$

Interpolation of Sobolev Spaces

If X_0 and X_1 are Sobolev spaces on $\mathbb{R}^n$, then the K-functional can be computed explicitly via the Fourier transform, allowing us to prove the following.

Theorem B.7 If s_0, $s_1 \in \mathbb{R}$, then

$$\left(H^{s_0}(\mathbb{R}^n), H^{s_1}(\mathbb{R}^n)\right)_{\theta,2} = H^s(\mathbb{R}^n) \quad \text{for } s = (1-\theta)s_0 + \theta s_1 \text{ and } 0 < \theta < 1.$$

Moreover, the Sobolev norm (3.21) equals the $K_{\theta,2}$-norm if, instead of the default normalisation (B.4), we take

$$N_{\theta,q} = \left(\frac{2\sin\pi\theta}{\pi}\right)^{1/2}. \tag{B.9}$$

Proof. Let $u = u_0 + u_1$ with $u_j \in H^{s_j}(\mathbb{R}^n)$ for $j = 0$ and 1, and observe that

$$\|u_0\|_{H^{s_0}(\mathbb{R}^n)}^2 + t^2\|u_1\|_{H^{s_1}(\mathbb{R}^n)}^2 = \int_{\mathbb{R}^n} \Big[(1+|\xi|^2)^{s_0}|\hat{u}_0(\xi)|^2 + t^2(1+|\xi|^2)^{s_1}|\hat{u}_1(\xi)|^2\Big]\,d\xi.$$

Since $\hat{u}(\xi) = \hat{u}_0(\xi) + \hat{u}_1(\xi)$, we see from Exercise B.4 that for each ξ the integrand is minimized when u_0 and u_1 are such that

$$(1+|\xi|^2)^{s_0}\hat{u}_0(\xi) = t^2(1+|\xi|^2)^{s_1}\hat{u}_1(\xi) = \frac{t^2(1+|\xi|^2)^{s_0+s_1}\hat{u}(\xi)}{(1+|\xi|^2)^{s_0} + t^2(1+|\xi|^2)^{s_1}}.$$

It follows that

$$K\left(t,u;H^{s_0}(\mathbb{R}^n),H^{s_1}(\mathbb{R}^n)\right)^2 = \int_{\mathbb{R}^n} \frac{t^2(1+|\xi|^2)^{s_0+s_1}}{(1+|\xi|^2)^{s_0} + t^2(1+|\xi|^2)^{s_1}}|\hat{u}(\xi)|^2\,d\xi$$

$$= \int_{\mathbb{R}^n} (1+|\xi|^2)^{s_0} f[a(\xi)t]^2 |\hat{u}(\xi)|^2\,d\xi,$$

where

$$a(\xi) = (1+|\xi|^2)^{(s_1-s_0)/2} \quad \text{and} \quad f(t) = \frac{t}{\sqrt{1+t^2}}.$$

Thus,

$$\|K(\cdot, u)\|_{\theta,2}^2 = \int_{\mathbb{R}^n} (1 + |\xi|^2)^{s_0} [a(\xi)^\theta \|f\|_{\theta,2}]^2 |\hat{u}(\xi)|^2 \, d\xi = \|f\|_{\theta,2}^2 \|u\|_{H^s(\mathbb{R}^n)}^2,$$

and we have only to apply Exercise B.5. $\qquad\qquad\qquad\qquad\square$

We also have an interpolation theorem for Sobolev spaces on domains.

Theorem B.8 *Let Ω be a non-empty open subset of $\mathbb{R}^n$. If s_0, $s_1 \in \mathbb{R}$, then*

$$\left(H^{s_0}(\Omega), H^{s_1}(\Omega)\right)_{\theta,2} = H^s(\Omega) \qquad \text{for } s = (1 - \theta)s_0 + \theta s_1 \text{ and } 0 < \theta < 1.$$

Moreover, the Sobolev norm (3.23) equals the $K_{\theta,q}$-norm if the latter is normalised by (B.9).

Proof. Put $X_j = H^{s_j}(\Omega)$ and $Y_j = H^{s_j}(\mathbb{R}^n)$.

Let $u \in H^s(\Omega)$, and choose $U \in H^s(\mathbb{R}^n)$ such that $u = U|_\Omega$ and $\|u\|_{H^s(\Omega)} = \|U\|_{H^s(\mathbb{R}^n)}$. Take any decomposition $U = U_0 + U_1$ with $U_j \in Y_j$, and observe that since $u = u_0 + u_1$, where $u_j = U_j|_\Omega \in X_j$, we have

$$K(t, u; X)^2 \le \|u_0\|_{X_0}^2 + t^2 \|u_1\|_{X_1}^2 \le \|U_0\|_{Y_0}^2 + t^2 \|U_1\|_{Y_1}^2,$$

and thus $K(t, u; X) \le K(t, U; Y)$. It follows by Theorem B.7 that $u \in K_{\theta,2}(X)$ with

$$\|u\|_{K_{\theta,2}(X)} \le \|U\|_{K_{\theta,2}(Y)} = \|U\|_{H^s(\mathbb{R}^n)} = \|u\|_{H^s(\Omega)}.$$

Conversely, let $u \in K_{\theta,2}(X)$, and take a decomposition $u = u_0 + u_1$ with $u_j \in X_j$. Choose $U_j \in Y_j$ such that $u_j = U_j|_\Omega$ and $\|u_j\|_{X_j} = \|U_j\|_{Y_j}$, and put $U = U_0 + U_1 \in Y_0 + Y_1$. We have

$$K(t, U; Y)^2 \le \|U_0\|_{Y_0}^2 + t^2 \|U_1\|_{Y_1}^2 = \|u_0\|_{X_0}^2 + t^2 \|u_1\|_{X_1}^2,$$

so $K(t, U; Y) \le K(t, u; X)$ and hence $\|U\|_{K_{\theta,2}(Y)} \le \|u\|_{K_{\theta,2}(X)}$. Since $u = U|_\Omega$, Theorem B.7 implies that $\|u\|_{H^s(\Omega)} \le \|U\|_{H^s(\mathbb{R}^n)} = \|U\|_{K_{\theta,2}(Y)} \le \|u\|_{K_{\theta,2}(X)}$. $\qquad\square$

An interpolation result for $\widetilde{H}^s(\Omega)$ follows at once, by duality; see Corollary 3.30 and Theorem B.5.

Theorem B.9 *Assume that Ω is a Lipschitz domain. If s_0, $s_1 \in \mathbb{R}$, then*

$$\left(\widetilde{H}^{s_0}(\Omega), \widetilde{H}^{s_1}(\Omega)\right)_{\theta,2} = \widetilde{H}^s(\Omega) \qquad \text{for } s = (1 - \theta)s_0 + \theta s_1 \text{ and } 0 < \theta < 1,$$

with equivalent norms.

Our final result deals with Sobolev spaces on the boundary $\Gamma = \partial\Omega$. The proof makes use of the reiteration theorem (Theorem B.6) and a simple interpolation property of pivot spaces.

Lemma B.10 *If H is a pivot space for V, then $H = (V, V^*)_{1/2, 2}$.*

Proof. Let $Y = (V, V^*)_{1/2, 2}$. By Theorem B.5 and Exercise B.7,

$$Y^* = (V^*, V)_{1/2, 2} = (V, V^*)_{1-1/2, 2} = Y,$$

so if $u \in Y$, then $\|u\|_H^2 = (u, u) \leq \|u\|_{Y^*} \|u\|_Y \leq C\|u\|_Y^2$. Hence, we have a continuous, dense inclusion $Y \subseteq H$, and dually, $H = H^* \subseteq Y^* = Y$. $\qquad\square$

Theorem B.11 *Assume that Ω is a $C^{r-1,1}$ domain for some integer $r \geq 1$. If s_0, $s_1 \in \mathbb{R}$ satisfy $|s_0| \leq r$ and $|s_1| \leq r$, then*

$$\left(H^{s_0}(\Gamma), H^{s_1}(\Gamma)\right)_{\theta, 2} = H^s(\Gamma) \qquad \text{for } s = (1 - \theta)s_0 + \theta s_1 \text{ and } 0 < \theta < 1,$$

$$(B.10)$$

with equivalent norms.

Proof. Suppose in the first instance that Γ is a $C^{r-1,1}$ hypograph, and recall the definition of $H^s(\Gamma)$ for the two cases $0 \leq s \leq r$ and $-r \leq s < 0$, given in the discussion after Theorem 3.34. The interpolation property follows at once from Theorem B.7 if both s_0 and s_1 belong to $[0, r]$, or if both belong to $[-r, 0]$. Furthermore, we claim that

$$\left(H^{-r}(\Gamma), H^r(\Gamma)\right)_{\theta, 2} = H^{2\theta-1}(\Gamma) \quad \text{for } 0 < \theta < 1.$$

Indeed, the case $\theta = \frac{1}{2}$ follows from Lemma B.10, after which the cases $0 < \theta < \frac{1}{2}$ and $\frac{1}{2} < \theta < 1$ follow with the help of Theorem B.6. Another application of Theorem B.6 then shows that (B.10) holds when one of the s_j belongs to $[0, r]$ and the other to $[-r, 0)$.

Suppose now that Γ is the boundary of Lipschitz domain in the sense of Definition 3.28. Thus, recalling (3.29),

$$\|u\|_{H^s(\Gamma)}^2 = \sum_l \|\phi_l u\|_{H^s(\Gamma_l)}^2.$$

If we put

$$X = \left(H^{s_0}(\Gamma), H^{s_1}(\Gamma)\right) \quad \text{and} \quad X_l = \left(H^{s_0}(\Gamma_l), H^{s_1}(\Gamma_l)\right),$$

then $K_{\theta,2}(X_l) = H^s(\Gamma_l)$ for each l, and the problem is to show that $K_{\theta,2}(X) = H^s(\Gamma)$.

Let $u \in H^s(\Gamma)$. We choose cutoff functions $\chi_l \in C^\infty_{\mathrm{comp}}(\mathbb{R}^n)$ satisfying $\chi_l = 1$ on a neighbourhood of supp ϕ_l, and supp $\chi_l \subseteq W_l$, so that $\phi_l u = \chi_l \phi_l u$. Take any decomposition $\phi_l u = u_{l0} + u_{l1}$ with $u_{lj} \in H^{s_j}(\Gamma_l)$, and observe that

$$u = \sum_l \phi_l u = \sum_l \chi_l u_{l0} + \sum_l \chi_l u_{l1},$$

so

$$K(t, u; X)^2 \leq \left\| \sum_l \chi_l u_{l0} \right\|^2_{H^{s_0}(\Gamma)} + t^2 \left\| \sum_l \chi_l u_{l1} \right\|^2_{H^{s_1}(\Gamma)}$$

$$\leq C \sum_l \left(\|\chi_l u_{l0}\|^2_{H^{s_0}(\Gamma)} + t^2 \|\chi_l u_{l1}\|^2_{H^{s_1}(\Gamma)} \right)$$

$$\leq C \sum_l \left(\|u_{l0}\|^2_{H^{s_0}(\Gamma_l)} + t^2 \|u_{l1}\|^2_{H^{s_1}(\Gamma_l)} \right).$$

Thus,

$$K(t, u; X)^2 \leq C \sum_l K(t, \phi_l u; X_l)^2,$$

and therefore $u \in K_{\theta,q}(X)$ with

$$\|u\|^2_{K_{\theta,q}(X)} \leq C \sum_l \|\phi_l u\|^2_{K_{\theta,q}(X_l)} \sim \sum_l \|\phi_l u\|^2_{H^s(\Gamma_l)} = \|u\|^2_{H^s(\Gamma)}.$$

Conversely, let $u \in K_{\theta,q}(X) \subseteq H^{\min(s_0,s_1)}(\Gamma)$. Taking any decomposition $u = u_0 + u_1$ with $u_j \in H^{s_j}(\Gamma)$, we have $\phi_l u = \phi_l u_0 + \phi_l u_1$ with $\phi_l u_j \in H^{s_j}(\Gamma_l)$, so

$$\sum_l K(t, \phi_l u; X_l)^2 \leq \sum_l \left[\|\phi_l u_0\|^2_{H^{s_0}(\Gamma_l)} + t^2 \|\phi_l u_1\|^2_{H^{s_1}(\Gamma_l)} \right]$$

$$= \|u_0\|^2_{H^{s_0}(\Gamma)} + t^2 \|u_1\|^2_{H^{s_1}(\Gamma)},$$

and hence $\sum_l K(t, \phi_l u; X_l)^2 \leq K(t, u; X)^2$. It follows that $\phi_l u \in K_{\theta,2}(X_l) = H^s(\Gamma_l)$ for each l, with

$$\|u\|^2_{H^s(\Gamma)} \sim \sum_l \|\phi_l u\|^2_{K_{\theta,2}(X_l)} \leq \|u\|^2_{K_{\theta,2}(X)}. \qquad \square$$

Exercises

B.1 Suppose that $A_0 : X_0 \to Y_0$ and $A_1 : X_1 \to Y_1$ are compatible, and let $u \in X_0 + X_1$. Show that $A_0 u_0 + A_1 u_1$ does not depend on the choice of $u_0 \in X_0$ and $u_1 \in X_1$ satisfying $u = u_0 + u_1$.

B.2 Show that $J(t, u) \geq t^\theta \|u\|_{K_{\theta,q}(X)}$. [Hint: use (B.6).]

B.3 Give a direct proof of the interpolation property for $J_{\theta,q}(X)$, i.e., show that Theorem B.2 holds if $K_{\theta,q}$ is replaced by $J_{\theta,q}$. [Hint: $J\big(t, Af(t); Y\big) \leq M_0 J\big(at, f(t); X\big)$, where $a = M_1/M_0$.]

B.4 Fix real numbers $A_0 > 0$ and $A_1 > 0$, and a complex number z. Show that

$$\min_{z=z_0+z_1} \left(A_0 |z_0|^2 + A_1 |z_1|^2 \right) = \frac{A_0 A_1}{A_0 + A_1} |z|^2,$$

and that the minimum is achieved when $A_0 z_0 = A_1 z_1 = A_0 A_1 z / (A_0 + A_1)$.

B.5 Use contour integration to show that

$$\int_0^\infty \frac{t^{1-2\theta}}{1+t^2} \, dt = \frac{\pi}{2 \sin \pi \theta} \quad \text{for } 0 < \theta < 1.$$

B.6 Show that if X_0 and X_1 are complete, then so are $X_0 \cap X_1$, $X_0 + X_1$ and $X_{\theta,q}$ for $0 < \theta < 1$ and $1 \leq q \leq \infty$. [Hint: use Exercise 2.1.]

B.7 Show that $(X_1, X_0)_{\theta,q} = (X_0, X_1)_{1-\theta,q}$ for $0 < \theta < 1$ and $1 \leq q \leq \infty$.

B.8 Assume that H, V and A satisfy the hypotheses of Corollary 2.38, and equip V with the energy norm $\| \cdot \|_A$. By arguing as in the proof of Theorem B.7, show

$$K(t, u; H, V)^2 = \sum_{j=1}^\infty \frac{\lambda_j t^2}{1 + \lambda_j t^2} |(\phi_j, u)|^2,$$

and deduce that for the normalisation (B.9),

$$\|u\|_{K_{\theta,2}(H,V)}^2 = \sum_{j=1}^\infty \lambda_j^\theta |(\phi_j, u)|^2 = \|A^{\theta/2} u\|^2.$$

Appendix C
Further Properties of Spherical Harmonics

We shall prove a result (Corollary C.2) used in Chapter 9, and also construct the classical spherical harmonics, which form an orthogonal basis for $\mathcal{H}_m(\mathbb{S}^2)$. Recall the definition of the Legendre polynomial $P_m(n, t)$ in the discussion following Lemma 8.6.

Theorem C.1 *The orthogonal projection* $Q_m : L_2(\mathbb{S}^{n-1}) \to \mathcal{H}_m(\mathbb{S}^{n-1})$ *is given by the formula*

$$Q_m \psi(\omega) = \frac{N(n, m)}{\Upsilon_n} \int_{\mathbb{S}^{n-1}} P_m(n, \omega \cdot \eta) \psi(\eta) \, d\eta$$

for $m \geq 0$, $\omega \in \mathbb{S}^{n-1}$ *and* $\psi \in L_2(\mathbb{S}^{n-1})$.

Proof. Let $\{\psi_{mp} : 1 \leq p \leq N(n, m)\}$ be an orthonormal basis for $\mathcal{H}_m(\mathbb{S}^{n-1})$. By part (i) of Exercise C.1,

$$Q_m \psi(\omega) = \int_{\mathbb{S}^{n-1}} K(\omega, \eta) \psi(\eta) \, d\eta, \qquad \text{where}$$

$$K(\omega, \eta) = \sum_{p=1}^{N(n,m)} \overline{\psi_{mp}(\omega)} \psi_{mp}(\eta).$$

If $A \in \mathbb{R}^{n \times n}$ is an orthogonal matrix, then

$$\int_{\mathbb{S}^{n-1}} \psi(A\omega) \, d\omega = \int_{\mathbb{S}^{n-1}} \psi(\omega) \, d\omega \quad \text{for } \psi \in L_1(\mathbb{S}^{n-1}),$$

and, by Exercise 8.3, the function $\omega \mapsto u(A\omega)$ belongs to $\mathcal{H}_m(\mathbb{S}^{n-1})$ whenever $u \in \mathcal{H}_m(\mathbb{S}^{n-1})$. It follows by part (iii) of Exercise C.1 that

$$K(A\omega, A\eta) = K(\omega, \eta) \quad \text{for } \omega, \eta \in \mathbb{S}^{n-1},$$

and in particular, when $Ae_n = e_n$ we have $K(A\omega, e_n) = K(\omega, e_n)$. Since $\omega \mapsto K(\omega, e_n)$ belongs to $\mathcal{H}_m(\mathbb{S}^{n-1})$, if u is as in Lemma 8.6 then

$$K(\omega, e_n) = au(\omega) = aP_m(n, \omega \cdot e_n) \quad \text{for } \omega \in \mathbb{S}^{n-1},$$

where $a = 1/K(e_n, e_n)$. Given $\eta \in \mathbb{S}^{n-1}$, choose an orthogonal matrix $A \in \mathbb{R}^{n \times n}$ such that $A\eta = e_n$; then

$$\begin{aligned} K(\omega, \eta) &= K(A\omega, e_n) = aP_m(n, A\omega \cdot e_n) \\ &= aP_m(n, \omega \cdot A^T e_n) = aP_m(n, \omega \cdot \eta). \end{aligned}$$

Finally, since $K(\omega, \omega) = aP_m(n, 1) = a$, we see that

$$a\Upsilon_n = \int_{\mathbb{S}^{n-1}} K(\omega, \omega)\, d\omega = \sum_{p=1}^{N(n,m)} \|\psi_{mp}\|_{L_2(\mathbb{S}^{n-1})}^2 = N(n, m),$$

giving $a = N(n, m)/\Upsilon_n$. $\qquad\qquad\qquad\qquad\qquad\qquad\qquad\qquad\square$

The proof above also establishes the *addition theorem*.

Corollary C.2 *If $\{\psi_{mp} : 1 \leq p \leq N(n, m)\}$ is an orthonormal basis for $\mathcal{H}_m(\mathbb{S}^{n-1})$, then*

$$\sum_{p=1}^{N(n,m)} \overline{\psi_{mp}(\omega)}\psi_{mp}(\eta) = \frac{N(n, m)}{\Upsilon_n} P_m(n, \omega \cdot \eta) \quad \text{for } \omega, \eta \in \mathbb{S}^{n-1}.$$

The next result is known as the *Funk–Hecke formula*.

Theorem C.3 *Let $f : [-1, 1] \to \mathbb{C}$ be a continuous function. If $m \geq 0$, then*

$$\int_{\mathbb{S}^{n-1}} f(\omega \cdot \eta)\psi(\eta)\, d\eta = \lambda\psi(\omega) \quad \text{for } \psi \in \mathcal{H}_m(\mathbb{S}^{n-1}),$$

where

$$\lambda = \Upsilon_{n-1} \int_{-1}^{1} f(t) P_m(n, t)(1 - t^2)^{(n-3)/2}\, dt. \tag{C.1}$$

Proof. We begin by showing that

$$\int_{\mathbb{S}^{n-1}} f(\xi \cdot \omega) P_m(n, \eta \cdot \omega)\, d\omega = \lambda P_m(n, \xi \cdot \eta) \quad \text{for } \xi, \eta \in \mathbb{S}^{n-1}. \tag{C.2}$$

Define $F(\xi, \eta)$ to be the left-hand side of (C.2), and observe that for a fixed ξ, the function $\eta \mapsto F(\xi, \eta)$ belongs to $\mathcal{H}_m(\mathbb{S}^{n-1})$. Also, if $A \in \mathbb{R}^{n \times n}$ is an orthogonal matrix, then $F(A\xi, A\eta) = F(\xi, \eta)$, so by arguing as in the proof of Theorem C.1 we see that $F(\xi, \eta) = \lambda P_m(n, \xi \cdot \eta)$ for some constant λ. Since $P_m(n, 1) = 1$,

$$\lambda = \lambda P_m(n, 1) = F(e_n, e_n) = \int_{\mathbb{S}^{n-1}} f(e_n \cdot \omega) P_m(n, e_n \cdot \omega)\, d\omega,$$

and the formula (C.1) follows by Exercise C.2. Thus, (C.2) holds, and upon multiplying both sides of this equation by $\psi(\eta)$, where $\psi \in \mathcal{H}_m(\mathbb{S}^{n-1})$, and integrating with respect to η, we have

$$\int_{\mathbb{S}^{n-1}} f(\xi \cdot \omega)\left(\int_{\mathbb{S}^{n-1}} P_m(n, \eta \cdot \omega)\psi(\eta)\, d\eta\right) d\omega = \lambda \int_{\mathbb{S}^{n-1}} P_m(n, \xi \cdot \eta)\psi(\eta)\, d\eta.$$

Applying Theorem C.1, the result follows at once. $\qquad\square$

We now begin our construction of an explicit orthogonal basis for $\mathcal{H}_m(\mathbb{S}^{n-1})$, starting with the case $n = 2$. Remember that $\mathcal{H}_0(\mathbb{S}^{n-1})$ consists of just the constant functions on $\mathbb{S}^{n-1}$, so it suffices to deal with $m \geq 1$.

Lemma C.4 *Let θ be the polar angle in the usual parametric representation of the unit circle $\mathbb{S}^1$,*

$$\omega = (\cos\theta, \sin\theta).$$

If $m \geq 1$, then $N(2, m) = 2$ and the functions

$$\psi_{m1}(\omega) = \frac{1}{\sqrt{\pi}}\cos m\theta \quad and \quad \psi_{m2}(\omega) = \frac{1}{\sqrt{\pi}}\sin m\theta$$

form an orthonormal basis for $\mathcal{H}_m(\mathbb{S}^1)$.

Proof. We have seen already in (8.13) that $N(2, m) = 2$ for $m \geq 1$. Define two real-valued solid spherical harmonics $u_1, u_2 \in \mathcal{H}_m(\mathbb{R}^2)$ by

$$u_1(x) + iu_2(x) = \frac{1}{\sqrt{\pi}}(x_1 + ix_2)^m,$$

and observe that ψ_{m1} and ψ_{m2} are the corresponding surface spherical harmonics in $\mathcal{H}_m(\mathbb{S}^1)$, i.e., $\psi_{mp} = u_p|_{\mathbb{S}^1}$ for $p = 1$ and 2. One readily verifies that ψ_{m1} and ψ_{m2} are orthonormal in $L_2(\mathbb{S}^1)$. $\qquad\square$

Any non-trivial function $A_{mj}(n, t)$ satisfying the conclusion of the next theorem is called an *associated Legendre function of degree m and order j for the dimension n.*

Theorem C.5 Assume $n \geq 3$, let $0 \leq j \leq m$, and define

$$A_{mj}(n, t) = (1 - t^2)^{j/2} P_{m-j}(n + 2j, t).$$ (C.3)

If $\psi \in \mathcal{H}_j(\mathbb{S}^{n-2})$ and

$$\Psi(\omega) = A_{mj}(n, t)\psi(\eta), \qquad where \ \omega = \sqrt{1 - t^2}\,\eta + te_n \ and \ \eta \in \mathbb{S}^{n-2},$$

then $\Psi \in \mathcal{H}_m(\mathbb{S}^{n-1})$.

Proof. For any $\psi \in L_1(\mathbb{S}^{n-2})$, the formula

$$u(x) = \int_{\mathbb{S}^{n-2}} (x_n + ix' \cdot \xi)^m \psi(\xi) \, d\xi$$

defines a solid spherical harmonic of degree m, whose restriction to $\mathbb{S}^{n-1}$ may be written as

$$u(\omega) = \int_{\mathbb{S}^{n-2}} \left(t + i\sqrt{1 - t^2}\,\eta \cdot \xi\right)^m \psi(\xi) \, d\xi.$$

If we now assume that $\psi \in \mathcal{H}_j(\mathbb{S}^{n-2})$, then by Theorem C.3, Exercise C.4(ii) and Exercise C.5,

$$u(\omega) = \Upsilon_{n-2}\psi(\eta) \int_{-1}^{1} \left(t + i\sqrt{1 - t^2}\,s\right)^m P_j(n - 1, s)(1 - s^2)^{(n-4)/2} \, ds$$

$$= c_1\psi(\eta)(1 - t^2)^{j/2} \int_{-1}^{1} \left(t + i\sqrt{1 - t^2}\,s\right)^{m-j} (1 - s^2)^{j+(n-4)/2} \, ds$$

$$= c_2\psi(\eta)(1 - t^2)^{j/2} P_{m-j}(n + 2j, t) = c_2\Psi(\omega),$$

where the constants c_1 and c_2 depend on n, m and j. Thus, $\Psi \in \mathcal{H}_m(\mathbb{S}^{n-1})$, as claimed. $\qquad\square$

Combining Theorem C.5 and Exercise C.6, and recalling (8.12), we see how to construct an orthogonal basis for $\mathcal{H}_m(\mathbb{S}^{n-1})$ by recursion on the dimension n.

Corollary C.6 If $\{\psi_{jp} : 1 \leq p \leq N(n - 1, j)\}$ is an orthogonal basis for $\mathcal{H}_j(\mathbb{S}^{n-2})$, and if

$$\Psi_{mjp}(\omega) = A_{mj}(n, t)\psi_{jp}(\eta), \qquad where \ \omega = \sqrt{1 - t^2}\,\eta + te_n \ and \ \eta \in \mathbb{S}^{n-2},$$

then $\{\Psi_{mjp} : 0 \leq j \leq m$ and $1 \leq p \leq N(n - 1, j)\}$ is an orthogonal basis *for $\mathcal{H}_m(\mathbb{S}^{n-1})$, and*

$$\|\Psi_{mjp}\|^2_{L_2(\mathbb{S}^{n-1})} = \frac{1}{N(n+2j,m-j)} \frac{\Upsilon_{n+2j}}{\Upsilon_{n+2j-1}} \|\psi_{jp}\|^2_{L_2(\mathbb{S}^{n-2})}.$$

In particular, taking $n = 3$ and using a basis for $\mathcal{H}_j(\mathbb{S}^1)$ of the type in Lemma C.4, we arrive at the classical spherical harmonics.

Theorem C.7 *Use the standard parametric representation for the unit sphere $\mathbb{S}^2$,*

$$\omega = (\sin\phi\cos\theta, \sin\phi\sin\theta, \cos\phi) \quad \text{for } 0 < \theta < 2\pi \text{ and } 0 < \phi < \pi.$$

If $m \geq 1$, then $N(3,m) = 2m+1$ and the functions

$$\psi_{m0}(\omega) = P_m(3,\cos\phi),$$
$$\psi_{mj}(\omega) = (\sin\phi)^j P_{m-j}(3+2j,\cos\phi)\cos j\theta \quad \text{for } 1 \leq j \leq m,$$
$$\psi_{m,m+j}(\omega) = (\sin\phi)^j P_{m-j}(3+2j,\cos\phi)\sin j\theta \quad \text{for } 1 \leq j \leq m,$$

form an orthogonal basis for $\mathcal{H}_m(\mathbb{S}^2)$, with $\|\psi_{m0}\|^2_{L_2(\mathbb{S}^2)} = 4\pi$ and

$$\|\psi_{mj}\|^2_{L_2(\mathbb{S}^2)} = \|\psi_{m,m+j}\|^2_{L_2(\mathbb{S}^2)} = \frac{\pi}{N(3+2j,m-j)} \frac{\Upsilon_{3+2j}}{\Upsilon_{2+2j}} \quad \text{for } 1 \leq j \leq m.$$

Exercises

C.1 Suppose that (S,μ) is a measure space, and let Q be the orthogonal projection from $L_2(S,\mu)$ onto a finite-dimensional subspace V.

(i) Show that if $\{\phi_p\}_{p=1}^N$ is an orthonormal basis for V, and if we define

$$K(x,y) = \sum_{p=1}^N \overline{\phi_p(x)}\phi_p(y),$$

then

$$Qu(x) = \int_S K(x,y)u(y)\,d\mu_y \quad \text{for } x \in S \text{ and } u \in L_2(S,\mu).$$

(ii) Show that the kernel K does not depend on the choice of the orthonormal basis for V.

(iii) Suppose that a group G acts on S, and that μ and V are invariant under G, i.e., if $g \in G$, then

$$\int_S u(gx)\,d\mu_x = \int_S u(x)\,d\mu_x \quad \text{for } u \in L_1(S,\mu),$$

and

$$x \mapsto u(gx) \text{ belongs to } V \text{ whenever } u \in V.$$

Show that the kernel K is invariant under G, i.e., if $g \in G$ then $K(gx, gy) = K(x, y)$ for $x, y \in S$.

C.2 Show that if f is, say, continuous on $\mathbb{S}^{n-1}$, then

$$\int_{\mathbb{S}^{n-1}} f(\omega) \, d\omega = \int_{-1}^{1} \int_{\mathbb{S}^{n-2}} f\left(\sqrt{1-t^2}\,\eta + te_n\right) d\eta \, (1-t^2)^{(n-3)/2} \, dt.$$

C.3 Prove the orthogonality property of the Legendre polynomials:

$$\int_{-1}^{1} P_m(n, t) P_l(n, t)(1-t^2)^{(n-3)/2} \, dt = \frac{1}{N(n, m)} \frac{\Upsilon_n}{\Upsilon_{n-1}} \delta_{ml}$$

for $m \geq 0$ and $l \geq 0$.

[Hint: use Theorems C.1 and C.3.]

C.4 For $n \geq 2$ and $m \geq 0$, let

$$p_m(t) = (1-t^2)^{-(n-3)/2} \frac{d^m}{dt^m} (1-t^2)^{m+(n-3)/2}.$$

(i) Show that p_m is a polynomial of degree m.

(ii) Use integration by parts to show that if, say, $f \in C^m[-1, 1]$, then

$$\int_{-1}^{1} f(t) p_m(t)(1-t^2)^{(n-3)/2} \, dt$$

$$= (-1)^m \int_{-1}^{1} f^{(m)}(t)(1-t^2)^{m+(n-3)/2} \, dt.$$

(iii) Deduce the orthogonality property

$$\int_{-1}^{1} p_m(t) p_l(t)(1-t^2)^{(n-3)/2} \, dt = 0 \quad \text{if } m \neq l.$$

(iv) Find $p_m(1)$, and conclude from Exercise C.3 that

$$P_m(n, t) = \frac{(-1)^m}{(n+2m-3)(n+2m-5)\cdots(n-1)} p_m(t),$$

a result known as the *Rodrigues formula*.

C.5 Show that

$$P_m(n, t) = \frac{1}{\Upsilon_{n-1}} \int_{\mathbb{S}^{n-2}} \left(t + i\sqrt{1-t^2}\,\omega \cdot \zeta\right)^m d\zeta \quad \text{for all } \omega \in \mathbb{S}^{n-2},$$

and then derive the *Laplace representation*,

$$P_m(n, t) = \frac{\Upsilon_{n-2}}{\Upsilon_{n-1}} \int_{-1}^{1} \left(t + i\sqrt{1 - t^2}\, s\right)^m (1 - s^2)^{(n-4)/2}\, ds \quad \text{for } n \geq 3.$$

[Hint: use Theorem C.3 with $f(s) = (t + i\sqrt{1 - t^2}\, s)^m$ and $\psi = 1 \in \mathcal{H}_0(\mathbb{S}^{n-2})$.]

C.6 Suppose $n \geq 3$, and let $A_{mj}(n, t)$ be the associated Legendre function (C.3). Show that for $\psi_1, \psi_2 \in C(\mathbb{S}^{n-2})$, if

$$\Psi_1(\omega) = A_{mj}(n, t)\psi_1(\zeta) \quad \text{and} \quad \Psi_2(\omega) = A_{mj}(n, t)\psi_2(\zeta),$$

where $\omega = \sqrt{1 - t^2}\, \zeta + t e_n$, then

$$(\Psi_1, \Psi_2)_{L_2(\mathbb{S}^{n-1})} = \frac{1}{N(n + 2j, m - j)} \frac{\Upsilon_{n+2j}}{\Upsilon_{n+2j-1}} (\psi_1, \psi_2)_{L_2(\mathbb{S}^{n-2})}.$$

References

[1] M. Abramowitz and I. Stegun, *Handbook of Mathematical Functions*, Dover, New York, 1970.

[2] R. A. Adams, *Sobolev Spaces*, Academic Press, 1975.

[3] K. E. Atkinson, *The Numerical Solution of Integral Equations of the Second Kind*, Cambridge University Press, 1997.

[4] A. Beer, Allgemeine Methode zur Bestimmung der elektrischen und magnetischen Induction, *Ann. Phys. Chem.* **98** (1856), 137–142.

[5] J. Bergh and J. Löfström, *Interpolation Spaces – An Introduction*, Springer, 1976.

[6] Yu. D. Burago and V. G. Maz'ya, Potential theory and function theory for irregular regions, *Seminars in Mathematics, Volume 3*, V. A. Steklov Institute, Leningrad. (English translation: Consultants Bureau, New York, 1969.)

[7] G. Birkhoff and U. Merzbach, *A Source Book in Classical Analysis*, Harvard University Press, 1973.

[8] A. P. Calderón, Lebesgue spaces of differentiable functions and distributions, *Partial Differential Equations, Proc. Sympos. Pure Math.* **4** (1961), 33–49.

[9] A. P. Calderón, Cauchy integrals on Lipschitz curves and related operators, *Proc. Nat. Acad. Sci. U.S.A.* **74** (1977), 1324–1327.

[10] J. Chazarain and A. Piriou, *Introduction to the Theory of Linear Partial Differential Equations*, North–Holland, 1982.

[11] R. R. Coifman, A. McIntosh and Y. Meyer, L'intégrale de Cauchy définit un opérateur borné sur L^2 pour les courbes lipschitziennes, *Ann. of Math.* **116** (1982), 361–387.

[12] D. Colton and R. Kress, *Integral Equation Methods in Scattering Theory*, Wiley, 1983.

[13] D. Colton and R. Kress, *Inverse Acoustic and Electromagnetic Scattering Theory*, Springer, 1992.

[14] M. Costabel, Boundary integral operators on Lipschitz domains: elementary results, *SIAM J. Math. Anal.* **19** (1988), 613–626.

[15] M. Costabel and W. L. Wendland, Strong ellipticity of boundary integral operators, *J. Reine Angew. Math.* **372** (1986), 34–63.

[16] M. Costabel and M. Dauge, On representation formulas and radiation conditions, *Math. Methods Appl. Sci.* **20** (1997), 133–150.

[17] B. E. J. Dahlberg, C. E. Kenig and G. C. Verchota, Boundary value problems for the systems of elastostatics in Lipschitz domains, *Duke Math. J.* **57** (1988), 795–818.

[18] J. W. Dauben, *The History of Mathematics from Antiquity to the Present: A Selective Bibliography*, Garland, New York and London, 1985.

[19] J. Dieudonné, *History of Functional Analysis*, North-Holland, 1981.

[20] P. du Bois Reymond, Bemerkungen über $\Delta z = \partial^2 z/\partial x^2 + \partial^2 z/\partial y^2 = 0$, *J. Reine Angew. Math.* **103** (1888), 204–229.

[21] E. Fabes, M. Jodeit and N. Rivière, Potential techniques for boundary value problems on C^1 domains, *Acta Math.* **141** (1978), 165–186.

[22] E. B. Fabes, C. E. Kenig and G. C. Verchota, The Dirichlet problem for the Stokes system on Lipschitz domains, *Duke Math. J.* **57** (1988), 769–793.

[23] I. Fredholm, Sur une nouvelle méthode pour la résolution du problème de Dirichlet, *Öfersigt Kongl. Vetenskaps-Akad. Förhandlingar* **1** (1900), 61–68.

[24] I. Fredholm, Sur une classe d'équations fonctionelles, *Acta Math.* **27** (1903), 365–390.

[25] L. Gårding, The Dirichlet problem, *Math. Intelligencer* **2** (1979), 43–53.

[26] C. F. Gauss, Allgemeine Lehrsätze in Beziehung auf die im verkehrten Verhältnisse des Quadrats der Entfernung wirkenden Anziehungs- und Abstossungs-Kräfte, *Resultate aus den Beobachtungen des magnetischen Vereins im Jahre 1839*; *Werke, Volume 5*, pp. 195–242.

[27] I. M. Gel'fand and G. E. Shilov, *Generalized Functions, Volume 1*, Academic Press, 1964.

[28] I. C. Gohberg and N. Krupnik, *Einführung in die Theorie der eindimensionalen singulären Integraloperatoren*, Birkhäuser, 1979.

[29] I. S. Gradshteyn and I. M. Ryzhik, *Table of Integrals, Series and Products*, Corrected and Enlarged Edition, Academic Press, 1980.

[30] I. Grattan-Guinness (editor), *Companion Encyclopedia of the History and Philosophy of the Mathematical Sciences*, Routledge, London and New York, 1994.

[31] G. Green, *An Essay on the Application of Mathematical Analysis to the Theories of Electricity and Magnetism*, Nottingham, 1828; reprinted in *J. Reine Angew. Math.* **39** (1850), 73–89, **44** (1852), 356–374, **47** (1854), 161–221; facsimile edition reprinted by Wazäta–Melins Aktienbolag, Göteborg, 1958.

[32] G. Green, On the determination of the exterior and interior attractions of ellipsoids of variable densities, *Trans. Cambridge Philos. Soc.*, 1835 (read May 6, 1833).

[33] G. Green, *Mathematical Papers of the Late George Green*, edited by N. M. Ferrers, Macmillan, London, 1877.

[34] P. Grisvard, *Elliptic Problems in Nonsmooth Domains*, Pitman, 1985.

[35] N. M. Günter, *Potentialtheorie und ihre Anwendung auf Grundaufgaben der mathematischen Physik*, Teubner, Leipzig, 1957.

[36] W. Hackbusch, *Integral Equations: Theory and Numerical Treatment*, Birkhäuser, 1995.

[37] J. Hadamard, *Lectures on Cauchy's Problem in Linear Partial Differential Equations*, Dover, New York, 1952; translation of *Le Problème de Cauchy et les équations aux dériveées partielles linéaires hyperboliques*, Paris, 1932.

[38] E. Hellinger and O. Toeplitz, *Integralgleichungen und Gleichungen mit unendlichvielen Unbekannten*, Chelsea, 1953 (article from the *Encyklopädie der mathematischen Wissenschaften*, pp. 1335–1616).

[39] D. Hilbert, *Grundzüge einer allgemeinen Theorie der Integralgleichungen*, Teubner, Leipzig, 1912; reprinted by Chelsea, 1953.

[40] E. Hille, *Analytic Function Theory, Volume II*, Blaisdell, 1962.

[41] L. Hörmander, *The Analysis of Linear Partial Differential Operators I*, Second Edition, Springer, 1990.

[42] G. C. Hsiao and W. L. Wendland, A finite element method for some integral equations of the first kind, *J. Math. Anal. Appl.* **58** (1977), 449–481.

[43] D. S. Jerison and C. Kenig, Boundary value problems on Lipschitz domains, *Studies in Partial Differential Equations*, ed. W. Littman, Mathematical Association of America, 1982.

[44] D. Jerison and C. Kenig, The inhomogeneous Dirichlet problem in Lipschitz domains, *J. Funct. Anal.* **130** (1995), 161–219.

[45] O. D. Kellogg, *Foundations of Potential Theory*, Dover, New York, 1929.

[46] C. Kenig, Elliptic boundary value problems on Lipschitz domains, Beijing Lectures on Harmonic Analysis, *Ann. of Math. Stud.* **112** (1986), 131–183.

[47] C. Kenig, *Harmonic analysis techniques for second order elliptic boundary value problems*, CBMS Regional Conference Series in Mathematics, **83**, American Mathematical Society, Providence, RI, 1994.

[48] R. Kieser, *Über einseitige Sprungrelationen und hypersinguläre Operatoren in der Methode der Randelemente*, Thesis, Universität Stuttgart, 1990.

[49] J. Král, *Integral Operators in Potential Theory*, Lecture Notes in Mathematics **823**, Springer, 1980.

[50] R. Kress, *Linear Integral Equations*, Springer, 1989.

[51] J. L. Lagrange, Sur l'équation séculaire de la lune, *Mém. Acad. Sci. Paris, Savants Étrangers* **7** (1773); *Oeuvres* **6**, 331–399.

[52] N. S. Landkof, *Foundations of Modern Potential Theory*, Springer, 1972.

[53] L. D. Landau and E. M. Lifshitz, *Theory of Elasticity*, Pergamon, 1970.

[54] P. D. Lax and A. N. Milgram, *Parabolic Equations*, Ann. of Math. Stud. **33** (1954), 167–190.

[55] J. Le Roux, Sur les intégrales des équations linéairs aux dérivées partielles du second ordre à deux variables indépendantes, Thèse, Paris, 1894; *Ann. Éc. Norm* **3** (1895), 227–316.

[56] M. N. Le Roux, Équations intégrales pour le problème du potential électrique dans le plan, *C. R. Acad. Sci. Paris Ser. A* **278** (1974), 541–544.

[57] M. N. Le Roux, *Résolution numérique du problème du potential dans le plan par une méthode variationnelle d'éléments finis*, Thèse 3ᵉ cycle, Rennes, 1974.

[58] M. N. Le Roux, Méthode d'éléments finis pour la résolution numérique de problèmes extérieurs en dimension 2, *R.A.I.R.O. Anal. Numér.* **11** (1977), 27–60.

[59] J. L. Lions and E. Magenes, *Non-homogeneous Boundary Value Problems and Applications*, Springer, 1972.

[60] A. E. H. Love, *A Treatise on the Mathematical Theory of Elasticity*, Fourth Edition, Dover, New York, 1944.

[61] J. Lützen, *The Prehistory of the Theory of Distributions*, Springer, 1982.

[62] J. Lützen, *Joseph Liouville 1809–1882: Master of Pure and Applied Mathematics*, Springer, 1990.

[63] A. I. Markushevich, *Theory of Functions of a Complex Variable*, Chelsea, New York, 1977.

[64] N. Meyers and J. Serrin, $H = W$, *Proc. Nat. Acad. Sci. U.S.A.* **51** (1964), 1055–1056.

[65] S. G. Mikhlin, *Mathematical Physics: An Advanced Course*, North-Holland, 1970.

[66] S. G. Mikhlin and S. Prößdorf, *Singular Integral Operators*, Springer, 1986.

[67] C. Miranda, *Partial Differential Equations of Elliptic Type*, Second Revised Edition, Springer, 1970.

[68] D. Mitrea and M. Mitrea, Generalized Dirac operators on nonsmooth manifolds and Maxwell's equations, preprint.

[69] D. Mitrea, M. Mitrea and M. Taylor, Layer potentials, the Hodge Laplacian, and global boundary value problems in nonsmooth Riemannian manifolds, preprint.

[70] C. Müller, *Spherical Harmonics*, Lecture Notes in Mathematics **17**, Springer, 1966.

[71] N. I. Muskhelishvili, *Singular Integral Equations: Boundary Problems of Function Theory and Their Application to Mathematical Physics*, Nordhoff, 1953.

[72] J. Nečas, *Les Méthodes Directes en Théorie des Équations Elliptiques*, Masson, Paris, 1967.

[73] J. Nečas and I. Hlaváček, *Mathematical Theory of Elastic and Elasto-plastic Bodies: An Introduction*, Elsevier, 1981.

[74] J. C. Nedelec, Curved finite element methods for the solution of singular integral equations on surfaces in R^3, *Comp. Methods Appl. Mech. Engrg.* **8** (1976), 61–80.

[75] J. C. Nedelec, Integral equations with non-integrable kernels, *Integral Equations Operator Theory* **4** (1982), 563–572.

[76] J. C. Nedelec and J. Planchard, Une méthode variationelle d'élements finis pour la résolution numérique d'un problème extérieur dans R^3, *R.A.I.R.O.* **7** (1973) R3, 105–129.

[77] C. G. Neumann, Zur Theorie des logarithmischen und des Newtonschen Potentials, *Ber. Verhandlungen Math.-Phys. Classe Königl. Sächs. Akad. Wiss. Leipzig* **22** (1870), 49–56, 264–321; reprinted in *Math. Ann.* **11** (1877), 558–566.

[78] L. Nirenberg, Remarks on strongly elliptic partial differential equations, *Comm. Pure Appl. Math.* **8** (1955), 648–674.

[79] J. A. Nitsche, On Korn's second inequality, *R.A.I.R.O. Modél. Math. Anal. Numér.* **15** (1981), 237–248.

[80] F. Noether, Über eine Klasse singulärer Integralgleichungen, *Math. Ann.* **82** (1921), 42–63.

[81] L. E. Payne and H. F. Weinberger, New bounds for solutions of second order elliptic partial differential equations, *Pacific J. Math.* **8** (1958), 551–573.

[82] H. Poincaré, La méthode de Neumann et le problème de Dirichlet, *Acta Math.* **20** (1896–1897), 59–142; reprinted in *Oeuvres de Henri Poincaré, Volume 9*, pp. 202–272.

[83] H. Rademacher, Über partielle und totale Differenzierbarkeit von Funktionen mehrer Variabeln und über die Transformation der Doppelintegrale, *Math. Ann.* **79** (1919), 340–359.

[84] F. Rellich, Ein Satz über mittlere Konvergenz, *Göttingen Nachr.* (1930), 30–35.

[85] F. Rellich, Darstellung der Eigenwerte von $\Delta u + \lambda u = 0$ durch ein Randintegral, *Math. Z.* **46** (1940), 635–636.

[86] F. Rellich, Über das asymptotische Verhalten der Lösungen von $\Delta \lambda + \lambda u = 0$ in unendlichen Gebieten, *Jber. Deutsch. Math. Verein* **53** (1943), 57–65.

[87] F. Riesz, Über lineare Funktionalgleichungen, *Acta Math.* **41** (1917), 71–98.

[88] F. Riesz and B. Sz.-Nagy, *Functional Analysis*, Frederick Ungar, 1955.

[89] Ja. A. Roĭtberg and Z. G. Šeftel, On equations of elliptic type with discontinuous coefficients, *Dokl. Akad. Nauk SSSR* **146** (1962), 1275–1278; English summary in *Soviet Math. Dokl.* **3** (1962), 1491–1494.

[90] M. Schechter, A generalization of the problem of transmission, *Ann. Scuola Norm. Sup. Pisa* **14** (1960), 207–236.

[91] M. Schechter, *Principles of Functional Analysis*, Academic Press, 1971.

[92] L. Schwartz, *Théorie des Distributions*, Hermann, Paris, 1966.

[93] R. T. Seeley, Extensions of C^∞ functions defined in a half space, *Proc. Amer. Math. Soc.* **15** (1961), 625–626.

[94] G. F. Simmons, *Introduction to Topology and Modern Analysis*, McGraw-Hill, 1963.

[95] V. I. Smirnov, *A Course of Higher Mathematics, Volume IV*, Pergamon and Addison–Wesley, 1964.

[96] E. M. Stein, *Singular Integrals and Differentiability Properties of Functions*, Princeton University Press, 1970.

[97] J. W. Strutt (Lord Rayleigh), On the theory of resonance, *Phil. Trans. Roy. Soc. London* **161** (1870), 77–118.

[98] W. Ritz, Über eine neue Methode zur Lösung gewisser Variationsprobleme der mathematischen Physik, *J. Reine. Angew. Math.* **135** (1908), 1–61.

[99] S. P. Thompson, *The Life of William Thomson, Baron Kelvin of Largs*, Macmillan, London, 1910.

[100] C. Truesdell, Notes on the history of the general equations of hydrodynamics, *Amer. Math. Monthly* **60** (1953), 445–458.

[101] C. Truesdell, *Essays in the History of Mechanics*, Springer, 1968.

[102] G. Verchota, Layer potentials and regularity for the Dirichlet problem for Laplace's equation in Lipschitz domains, *J. Funct. Anal.* **59** (1984), 572–611.

[103] V. Volterra, *Opera Matematiche*, Accad. Lincei, 5 volumes, 1954–1962.

[104] H. Weber, Über die Integration der partiellen Differentialgleichung $\partial^2 u/\partial x^2 + \partial^2 u/\partial y^2 + k^2 u = 0$, *Math. Ann.* **1** (1869), 1–36.

[105] H. Whitney, *Geometric Integration Theory*, Princeton University Press, 1957.

[106] K. Yosida, *Functional Analysis*, Sixth Edition, Springer, 1980.

Index

Index of Notation

Functional Analysis

A^*	adjoint of A, 37
$A_/$	induced map on cosets modulo ker A, 18
A^t	transpose of A, 22
dist(u, W)	distance from point u to set W, 21
im A	image (range) of linear operator A, 18
(g, u)	same as $\langle \bar{g}, u \rangle$, 37
$(\cdot, \cdot)_A$	energy inner product for A, 44
$\| \cdot \|_A$	energy norm for A, 44
ker A	kernel (null space) of linear operator A, 18
$\mathcal{L}(X, Y)$	space of bounded linear operators from X to Y, 18
$\oplus$	direct sum, 20
$\langle g, u \rangle$	value of functional $g \in X^*$ at $u \in X$, 20
$u \perp v$	u is orthogonal to v, 39
$u \perp W$	u is orthogonal to the set W, 39
$\sim$	equivalence of norms, 17
spec(A)	spectrum of A, 45
W^{a}	subspace of X^* that annihilates $W \subseteq X$, 23
$^{\mathrm{a}}V$	subspace of X that annihilates $V \subseteq X^*$, 23
$u_j \rightharpoonup u$	u_j converges weakly to u, 42
$(X_0, X_1)_{\theta, q}$	interpolation space, 318
X^*	dual space of X, 20

Theory of Distributions

$C^\infty_{\mathrm{comp}}(\Omega)$	space of C^∞ functions with compact support in Ω, 61
$C^r_{\mathrm{comp}}(\Omega)$	space of C^r functions with compact support in Ω, 61

$C^\infty(\Omega)$	space of infinitely differentiable functions on Ω, 61
$C^\infty_K(\Omega)$	space of functions in $C^\infty(\Omega)$ having support in K, 61
$C^r(\Omega)$	space of r times continuously differentiable functions on Ω, 61
$C^r_K(\Omega)$	space of functions in $C^r(\Omega)$ having support in K, 61
$\mathcal{D}(\Omega)$	$C^\infty_{\mathrm{comp}}(\Omega)$ with sequential convergence defined, 65
$\mathcal{D}(\overline{\Omega})$	space of restrictions to Ω of functions in $\mathcal{D}(\mathbb{R}^n)$, 77
δ	same as δ_0, 66
δ_x	Dirac delta function(al) at x, 66
$\mathcal{D}_K(\Omega)$	$C^\infty_K(\Omega)$ with sequential convergence defined, 65
$\mathcal{D}^*(\Omega)$	space of Schwartz distributions on Ω, 65
$\mathcal{E}^*(\Omega)$	space of distributions with compact support in Ω, 66
$\mathcal{E}(\Omega)$	$C^\infty(\Omega)$ with sequential convergence defined, 65
f.p. u	finite-part extension of u, 166
p.v. u	principal value of u, 166
$H_a(\phi)$	finite-part integral of $x^a\phi(x)$ over the half line $x > 0$, 160
(u, v)	abbreviation for $(u, v)_\Omega$ when $\Omega = \mathbb{R}^n$, 68
$(u, v)_\Omega$	same as $\langle \bar{u}, v\rangle_\Omega$, 68, 107
$L_{1,\mathrm{loc}}(\Omega)$	space of locally integrable functions on Ω, 64
M_t	dilatation operator, 158
$\otimes$	tensor product of functions or distributions, 104
$\langle u, v\rangle$	abbreviation for $\langle u, v\rangle_\Omega$ when $\Omega = \mathbb{R}^n$, 66
$\langle u, v\rangle_\Omega$	integral (generalised, if necessary) of $u \cdot v$ over Ω, 58, 66, 106
$\Pi^\pm_a(\xi)$	Fourier transform of f.p. $x^a_\pm$, 169
$R_a\phi$	finite-part integral of $\rho^{a+n-1}\phi(x)$ over $\rho > 0$, 166
$\mathcal{S}(\mathbb{R}^n)$	Schwartz class of rapidly decreasing C^∞ functions on $\mathbb{R}^n$, 72
x^a_+	x^a if $x > 0$, but 0 if $x < 0$, 159
f.p. x^a_+	finite-part extension of x^a_+, 160
x^a_-	$|x|^a$ if $x < 0$, but 0 if $x > 0$, 163
f.p. x^a_-	finite-part extension of x^a_-, 163
f.p. x^{-k-1}	finite-part extension of x^{-k-1} for an integer $k \geq 0$, 164
$(x \pm i0)^a$	164

Sobolev Spaces

H^s_F	space of distributions in $H^s(\mathbb{R}^n)$ with support in F, 76
$H^s(\mathbb{R}^n)$	Sobolev space on $\mathbb{R}^n$ (definition via Bessel potential), 76
$H^s(\Gamma)$	Sobolev space on Γ, 98
$H^s(\Omega)$	space of restrictions to Ω of distributions in $H^s(\mathbb{R}^n)$, 77
$H^s(\Omega)^m$	space of H^s functions on Ω with values in $\mathbb{C}^m$, 107

Differential and Integral Operators

SL   single-layer potential, 3, 202
$\widetilde{SL}$ dual version of SL, 211
T boundary integral operator, sum of one-sided traces
      of DL, 11, 218
$\widetilde{T}$ dual version of T, 218
$\mathcal{U}$ solution operator for the Dirichlet problem, 145
$\mathcal{V}$ solution operator for the adjoint Dirichlet problem, 145

Other Symbols

$\alpha!$ factorial of the multi-index α, 61
$|\alpha|$ order of the partial derivative determined by α, 61
$\partial^{\alpha} u$ partial derivative of u determined by the multi-index α, 61
y^{α} monomial determined by the multi-index α, 61
$\mathbb{C}^{+}$ complex upper half plane Im $z > 0$, 183
$\mathbb{C}^{-}$ complex lower half plane Im $z < 0$, 183
Cap_{Γ} capacity of Γ, 263
$u * v$ convolution of u and v, 58
$\Delta_{l,h}$ difference quotient in lth variable with step size h, 62
$d\sigma$ element of surface area on Γ, 1, 97
$\hat{u} = \mathcal{F}u$ Fourier transform of u, 70
$u = \mathcal{F}^{*}\hat{u}$ inverse Fourier transform of $\hat{u}$, 70
Γ (common) boundary of $\Omega = \Omega^{-}$ and of Ω^{+}, 1, 89, 141
γ trace operator for Ω, 100, 102
γ^{*} adjoint of γ, 201
$\gamma^{\pm}$ trace operator for $\Omega^{\pm}$, 1, 141
Γ_{D} portion of Γ with Dirichlet boundary condition, 128
Γ_{N} portion of Γ with Neumann boundary condition, 128
$h_{m}^{(1)}, h_{m}^{(2)}$ spherical Hankel functions, 281
J_{μ} Bessel function of the first kind, 278
j_{m} spherical Bessel function of the first kind, 279
$L_{p}(\Omega)$ Lebesgue space of pth-power-integrable functions on Ω, 58
$M(n, m)$ dimension of $\mathcal{P}_{m}(\mathbb{R}^{n})$, 250
$N(n, m)$ dimension of $\mathcal{H}_{m}(\mathbb{R}^{n})$, 250
$|x|$ Euclidean norm in $\mathbb{R}^{n}$ or unitary norm in $\mathbb{C}^{m}$, 1
ν outward unit normal to $\Omega = \Omega^{-}$, 1, 97, 141
Ω domain in $\mathbb{R}^{n}$, 1
$\Omega^{\pm}$ interior ($-$) and exterior ($+$) domains, 1, 141
p^{*} conjugate exponent to p, 58

$P_m(n, t)$	generalised Legendre polynomial of degree m for the dimension n, 255
$\mathcal{P}_m(\mathbb{R}^n)$	space of homogeneous polynomials of degree m, 250
$\mathbb{R}^n_+$	upper half space $x_n > 0$, 183
$\mathbb{R}^n_-$	lower half space $x_n < 0$, 183
$\mathcal{H}_m(\mathbb{R}^n)$	space of solid spherical harmonics of degree m, 250
$\mathcal{H}_m(\mathbb{S}^{n-1})$	space of surface spherical harmonics of degree m, 252
u^*	transpose of the complex conjugate of the vector u, 107
$u \cdot v$	dot product of vectors u and v, 107
$u^{(k)}(x; y)$	kth Fréchet derivative of u, 61
$u^\natural$	Kelvin transform of u, 259
Υ_n	surface area of $\mathbb{S}^{n-1}$, 247
$\Upsilon^\pm(x)$	221
$x^\natural$	inverse point of x with respect to a sphere, 258
Y_μ	Bessel function of the second kind, 278
y_m	spherical Bessel function of the second kind, 279
$\Gamma(a)$	gamma function, 169
$H_\mu^{(1)}, H_\mu^{(2)}$	Hankel functions, 280
$\int^\pm$	special contour integral, 183